Lignin Conversion Catalysis

Transformation to Aromatic Chemicals

Chaofeng Zhang
Feng Wang

WILEY-VCH

Authors

Dr. Chaofeng Zhang
College of Light Industry and Food Engineering
Nanjing Forestry University
159 Longpan Road
210037 Nanjing
China

Prof. Feng Wang
Dalian Institute of Chemical Physics
Dalian Nat. Lab. for Clean Energy
Chinese Academy of Science
457 Zhongshan Road
116023 Dalian
China

Cover Image: Courtesy of Chaofeng Zhang and Feng Wang

Library of Congress Card No.: applied for

British Library Cataloguing-in-Publication Data:
A catalogue record for this book is available from the British Library.

Bibliographic information published by the Deutsche Nationalbibliothek
The Deutsche Nationalbibliothek lists this publication in the Deutsche Nationalbibliografie; detailed bibliographic data are available on the Internet at http://dnb.d-nb.de.

Print ISBN 978-3-527-34973-9
ePDF ISBN 978-3-527-83501-0
ePub ISBN 978-3-527-83502-7
oBook ISBN 978-3-527-83503-4

Typesetting Straive, Chennai, India
Printing and Binding CPI Group (UK) Ltd, Croydon, CR0 4YY

C117050_010822

To Yiyao Zhang

Lignin Conversion Catalysis

Contents

Preface

The development of civilization and human society can be regarded as the changing history of utilization of various resources and materials. In history, drilling wood for fire lightened the glimmer of the development of human civilization, heating water for steam power by burning coal allowed development to catch the train of the steam age, and petroleum utilization built the cornerstone of modern civilization. Each technical revolution based on the utilization of fossil resources promoted the qualitative development of human civilization. However, just as coins have two sides, a large number of fossil resources have been quickly consumed in a short time and the carbon element that has been stored in Earth's crust is concurrently converted into CO_2 gas that enters the atmosphere of the ecosphere where its presence causes corresponding environmental and climate problems. When human society is at a loss about this dilemma, nature has already prepared a solution for us with biomass as the renewable carbon resource. Photosynthesis on Earth can convert CO_2 and water into organic carbon resources and O_2 with renewable solar energy, which produces about 170 billion tonnes of lignocellulose every year. If the renewable biomass resource can be efficiently transformed to the desired chemical stocks with sustainable energy input, the increasing CO_2 concentration will be not a hot potato, primarily because of the potential CO_2-neutral cycle that transforms solar energy, H_2O, and CO_2 into chemical stocks and fuels via biomass. Different from the utilization of the fossil resource that can be regarded as the conversion and stitching of small molecules into macromolecules, biomass utilization is more like a process of tailoring and converting natural polymers into corresponding small-molecule chemicals, all of which requires the development of more sophisticated strategies and efficient implementation methods.

Lignin is the third abundant organic carbon resource on the earth. As a precious resource given to mankind by nature, especially for its potential as a treasure house of aromatic chemicals, research on lignin depolymerization and the transformation of its downstream products have drawn much attention from scientists and companies worldwide in the recent decade aimed at harvesting aromatic compounds and hydrocarbon stocks/fuels from this abundant and renewable natural polymer. In November 2020, we received an invitation from Wiley to prepare a book on the topic of lignin conversion/transformation to chemicals. Perhaps it is a coincidence, but our literature retrieval process revealed that the electronic literature on the topic of lignin can be traced back to the 1920s with the SciFinder database, 100 years from 2020. Thus, it is meaningful to receive the invitation to record the research process of lignin conversion. Before setting a topic line of this book from the complex content of lignin catalytic conversion, we first reviewed previous records to determine the likely readership of the book. We hope that this book is useful for readers in academia, industry, and education at the introductory, advanced, and specialist levels who are interested in wood chemistry or fields of lignin utilization research.

For this reason, Chapter I of the book briefly introduces the structure of the book. In Part II, the basic knowledge of lignin, including lignin biosynthesis, lignin structure, and lignin isolation, physicochemistry properties, and chemical properties are introduced. Then, lignin depolymerization issues, involving scientific questions, challenges, classification of lignin conversion methods by reaction type, and current progress, are discussed in Part III. For the main topic, the critical book aims to provide an overview of key advances in the field of lignin depolymerization to aromatic monomers and more attention is paid to the generalization and summarization of the mechanism of lignin linkage cleavage. It is challenging to systemically summarize all of the strategies via the traditional classification method by reaction types. In Chapter 5 of Part IV, inspired by the retrosynthetic analysis of the natural product to confirm the synthons proposed by Professor Cory, we herein put forward an inverse disassembly analysis (IDA) method to explore the efficient strategies for lignin depolymerization, which can be also used in the classification and discussion of the various methods for lignin conversion viewed from the cleavage order and type of corresponding chemical bonds over various catalytic systems. With β-O-4 linkage as the main lignin mode, we further introduce the lignin depolymerization strategies with the IDA thoughts from Chapter 6 to Chapter 11. Finally, in Part V, exploring lignin depolymerization to aromatic chemicals and how to make lignin depolymerization profitable, we provide some viewpoints for outlets of the lignin resource that highlights the importance of high-value medicine and synthetic block preparation or the utilization of lignin from natural polymer to artificial polymer. To accelerate the lignin utilization from lab to industry application, we discussed the critical roles of the standardized lignin substrate and standardized products, during which we support the idea that funneling and functionalization of a mixture of lignin-derived monomers into a single high-value chemical is fascinating and promising.

Finally, given the complexity of the catalytic conversion of lignin and the multiformity of the catalytic methods, which involves the research fields of different disciplines, the discussion of some relevant catalytic methods for lignin conversion may be omitted because of the limited knowledge reserve of the authors. In addition, if there are any improper points in this book, we encourage readers and experts to constructively criticize and correct them with us.

November 2021, in Dalian

Chaofeng Zhang
Feng Wang

List of Abbreviations

A	AC	Activated carbon
	ACT	4-Acetamido-TEMPO
	AFEX	Ammonia fiber explosion
	AFGM	Adjacent functional group modification
B	BDE	Bond dissociation enthalpy
	BNL	Brauns native lignin
	BTX	Benzene–toluene–xylenes
C	C3H	Coumarate 3-hydroxylase
	C4H	Cinnamic acid 4-hydroxylase
	CAD	Cinnamyl alcohol dehydrogenase.
	CB	Conduction band
	CCoAOTM	Caffeoyl-CoA O-methyltransferase
	CCR	Cinnamoyl-CoA reductase
	CCTC	Contact charge transfer complex
	CDE	CO_2 explosion
	CEL	Cellulolytic Enzyme Lignins
	4CL	4-coumarate CoA ligase
	CoA	Coenzyme A
	COMT	CoA O-methyltransferase
	CP/MAS	It is a combination of cross-polarization sequence (CP) and magic angle transformation technology (MAS).
D	DAIB	Diacetoxyiodobenzene
	DBAD	Diisopropyl azodiformate
	DDQ	2,3-Dichloro-5,6-dicyano-1,4-benzoquinone
	DEPT	Distortionless enhancement by polarization transfer
	DES	Deep eutectic solvent
	DFRC	Derivatization followed by reductive cleavage
	DFT	Density functional theory
	DIPEA	N,N-diisopropylethylamine
	DL	Dioxane (acidolysis) lignins
	DMSO	Dimethyl sulfoxide
	DSPEC	Dye-sensitized photoelectrochemical cell

E	E-4-P	Erythrose-4-phosphate
	EMAL	Enzymatic mild acidolysis lignin
	ESR	Electron spin resonance absorption
	EXAFS	Extended X-ray absorption fine structure
F	F5H	Ferulate 5-hydroxylase
	FA	Formic acid
	FTIR	Fourier transform infrared spectroscopy infrared absorption spectroscopy
G	GFC	Gel-filtration chromatography
	GH	Glucosylhydrolases
	GPC	Gel-permeation chromatography
	GS	Glutathione
	GT	Glucosyl transferases
	G-unit	Coniferyl alcohol, or lignin phenylpropanoid unit with one –OMe connecting with the aromatic ring
H	HAADF-STEM	High angle annular dark-field scanning transmission electron microscopy
	HAT	Hydrogen atom transfer
	HBS	High-boiling-point solvent
	HCT	Hydroxycinnamoyltransferase
	HMBC	Heteronuclear multiple bond correlation
	HMQC	Heteronuclear multiple-quantum correlation
	HPAs	Heteropoly acids
	HPSEC	High-performance size-exclusion chromatography
	HTC	Hydrotalcite
	H-unit	*p*-Coumaric alcohol, or lignin phenylpropanoid unit without –OMe connecting with the aromatic ring
I	IDA	Inverse disassembly analysis
	ILs	Ionic liquids
	INADEQUATE	Incredible natural abundance double quantum transfer experiment
L	LA	Lewis acid
	LAR	Lignin isolation from lignocellulose with lignin as the residue
	LBD	Lignin isolation from lignocellulose by first lignin dissolution
	LC	Lignocellulose
	LCC	Lignin-carbohydrate complex
	LDH	Layered-double hydroxide catalyst
	LDP	Lignin depolymerization
	LMCT	Ligand-to-metal charge-transfer
	LS	Lignocellulose
M	MALDI-TOF	Matrix-assisted laser desorption/ionization time-of-flight mass spectrometry
	mCPBA	*m*-Chloroperbenzoic acid
	MDO	Methyldioxorheniu
	MOF	Metal-organic framework
	MTO	Methyltrioxorhenium

	MWD	Molecular weight distribution
	MWL	Milled-wood lignin
N	NCS	N-chlorosuccinimide
	NE	Nucleus exchange
	NHC	N-heterocyclic carbene
	NHPI	N-hydroxyphthalimide
	NMR	Nuclear magnetic resonance spectroscopy
	NPs	Nanoparticles (noble metal)
O	OAT	Oxygen atom transfer
	OMS	Open metal sites
	OPCs	Organic (visible light) photocatalysts
	OSL	Organosolv lignin
P	PAL	Phenylalanine ammonia-lyase;
	PAR	Phenylseleninic acid resin
	PB	*p*-Hydroxybenzoate
	PCET	Proton-coupled electron transfer
	PDI	Perylene diimide
	PE	Polyethylene
	PEDOS	Partial electronic density of states
	PEP	Phosphoenol pyruvate
	POFs	Porous organic frameworks
	POM	Polyoxometalate
	ppu	Phenylpropanoid units
	PTH	Phenothiazine
R	RCF	Reductive catalytic fractionation
	RSA	Retrosynthetic analysis
S	scMeOH	Supercritical methanol
	SEC	Size-exclusion chromatography
	SET	Single-electron transfer
	S-unit	Sinapyl alcohol, or lignin phenylpropanoid unit with two -OMe connecting with the aromatic ring
T	TAL	Tyrosine ammonia-lyase
	TBD	1,5,7-Triazabicyclo[4.4.0]dec-5-ene
	^{t}BuONO	tert-Butyl nitrite
	TDTS	TOF-determining transition state
	TEMPO	2,2,6,6-tetramethyl-1-piperidinyloxy
	THF	Tetrahydrofuran
	TMSI	1-(Trimethylsilyl)imidazole
	TPA	Terephthalic acid
U	UV	Ultraviolet absorption spectroscopy
	UV–vis DRS	The ultraviolet-visible diffuse reflectance spectrum
V	VB	Valence band
X	XPS	X-ray photoelectron spectroscopy
	XRD	X-ray diffraction

List of Symbols

δ	NMR chemical shift
M_n	number-average molar mass
M_w	weight-average molar mass
Đ	dispersity index, Đ=M_w/M_n
λ_{max}	The wavelength at which the largest amount of absorption occurs
pH	acidity or basicity of an aqueous solution
ppm	parts per million
Tg	glass transition temperature
Ts	thermal softening temperature
rt	room temperature
ε	molar absorption coefficient (UV-vis)

Part I

Book Introduction

1

Background and Overview

1.1 Introduction

The world currently is relying more and more on non-renewable fossil resources. In 2019, the global primary energy consumption reached 583.9 EJ (1 EJ=10^{18} J) [1] and the 84.3% of this energy consumption was from traditional fossil fueld resources: coal, natural gas, and crude oil. Besides fuel production and energy output from these resources, billions of tons of additional fossil resources are transformed into chemicals and materials to support the explosive development of human civilization. However, this represents the negative side of a double-edged sword because the immoderate utilization of fossil resources has released billions of tons of CO_2 into the atmosphere within the last century. This CO_2 release has considerably exceeded the capacity of the earth's ecological cycle and therefore caused global climate problems such as global warming and problems deriving from global warming. Against this background of ecological problems, carbon peaking and carbon neutrality have become topics of concern to all those concerned with the future of humanity. Meanwhile, increasing awareness of dwindling fossil resources and the ever-increasing need for more chemical feedstock and energy supply has led to extensive research into more efficient and sustainable methods to meet these demands. To make the carbon cycle more efficient and controllable, optimizing existing technologies and reducing dependence on fossil resources will be insufficient and therefore utilizing sustainable carbon resources is an important and promising option to help address the problem.

As one kind of carbon and solar energy storage form, biomass is any organic matter that is renewable over time and it is only renewable organic carbon resource in nature. Biomass is generally plant-based and includes agricultural residues, forestry wastes, and energy crops [2]. Every year, global photosynthesis can produce about 170 billion tons of lignocellulose (LC) by capturing the low concentration CO_2 from the atmosphere, immobilizing the carbon, and releasing O_2 [3]. However, besides the timber industry and papermaking, only a small proportion of biomass carbon resources is utilized to prepare chemicals. The rest of biomass carbon is primarily

Lignin Conversion Catalysis: Transformation to Aromatic Chemicals, First Edition.
Chaofeng Zhang and Feng Wang.

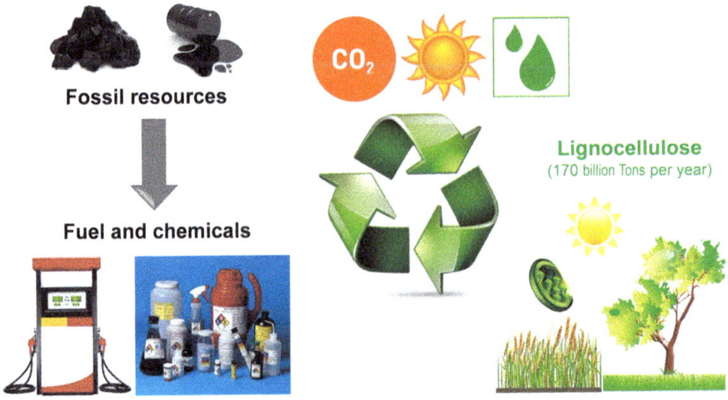

Figure 1.1 The Carbon-Increasing Process Based on the Non-renewable Fossil Resources Compared with the Carbon-neutral Cycle Connecting Solar Energy, H_2O, and CO_2 with Chemicals/Fuels via Biomass. *Source*: Zhang and Wang [38].

used to generate heat and power or abandoned. However, biomass is still the fourth largest source of energy in the world after oil, coal, and methane, supplying 10% of the world's primary energy. Considering another perspective, we can reduce millions of years of fossil resources generation if we can efficiently transform the renewable biomass resource to the desired chemical stocks with an acceptable CO_2 release during their preparation [4–12]. This would reduce the concern regarding increasing CO_2 concentration, primarily because CO_2 released during further utilization of chemicals prepared in this way originates from CO_2 in the modern-day atmosphere. It can be a perfect CO_2-neutral cycle that transforms solar energy, H_2O, and CO_2 into chemical stocks and fuels via biomass (Figure 1.1).

1.2 Lignin: A Natural and Sustainable Aromatic Bank

Aromatic chemicals are important bulk chemicals widely used in the chemical industry for the production of polymers, solvents, medicines, pesticides, and many other materials. With the rapid development of economic and industrial globalization, the demand for aromatic chemicals is increasing sharply every year. At present, aromatic chemicals are primarily obtained through petroleum catalytic reforming or steam cracking followed by a complex functionalization process.

These chemical feedstocks are obtained primarily from fossil resources (coal and oil) formed by the photosynthesis of ancient plants. Ancient plants used solar energy to convert CO_2 into biomass, then the residue of animals and plants in food webs were transformed to fossil resources after millions of years underground.

The technical revolution based on the utilization of these fossil resources to produce power and chemicals promoted the development of human society. However, it is now clear that the inherent non-renewability of fossil resources and tough environmental protection requirements have begun to affect the sustainable production of aromatic chemicals [13]. Therefore, the development of new technologies to replace or supplement fossil resources to prepare aromatic chemicals has become increasingly important [2, 14–16]. In the past decades of research, people have successively developed methane aromatization [17], methanol aromatization [18], CO_2 conversion [19], C_2 compound synthesis [20], and other routes. However, using these low-carbon substrates as a raw material to obtain aromatic hydrocarbon monomers with high selectivity is still a serious challenge [21].

Lignocellulose (LC), with an annual production of around 170 billion tons, is the most abundant biomass form [3]. The potential use of non-edible lignocellulosic biomass for the production of value-added chemicals could provide an attractive alternative for fossil-based processes is therefore of great significance. As shown in Figure 1.2, LC comprises three main components: hemicellulose, cellulose, and lignin. Among them, hemicellulose (20–30 w% of LC) and cellulose (40–50 w% of LC) are the polymers of C_5 and C_6 sugars (Figure 1.2) [22] that can be further converted to the low-carbon alcohols, aldehydes, ketones, or acid. These compounds can be used in some biomass-based reforming processes to obtain aromatic hydrocarbon compounds, as exemplified by the BioFormPX project from the Virient Company. Additionally, the C_5 and C_6 polymers or their monomers can be transformed to furan-based chemicals or other low-carbon chemicals, which can also be used in C_6 aromatization via a typical 4+2 Diels-Alder reaction [23–25]. At the same time, lignin (10-35 w% of LC) is a complex three-dimensional amorphous polymer, consisting of various methoxylated phenylpropanoid units, which can be regarded as a potential resource to replace fossil resources to directly prepare functionalized aromatic chemicals and non-aromatic hydrocarbon stocks/fuels [14, 26–28]. Therefore, the direct conversion of lignin in LC is a more attractive process. However, the current paper industry and, more recently, biorefineries produce large quantities of lignin that is currently considered almost a by-product, not to mention the transformation of the more complex native lignin. Not only from an economic standpoint but also from a sustainability perspective, the misuse or even nonuse of

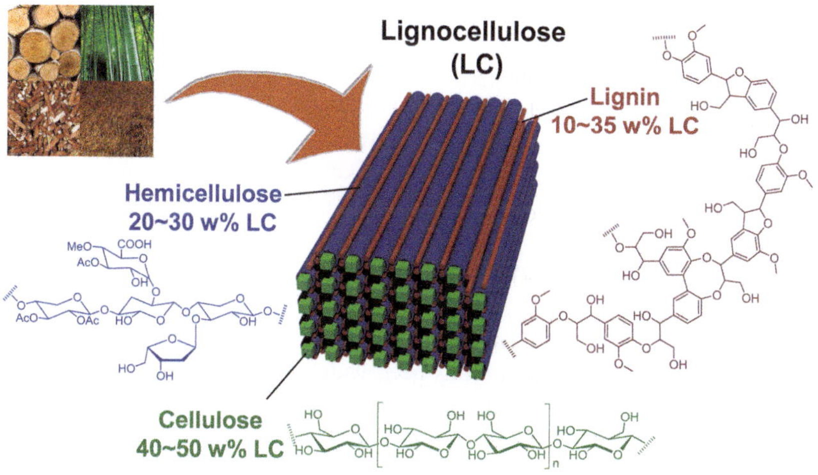

Figure 1.2 Lignocellulose (LC) from Biomass and its Structural Composition.

lignin appears to be a colossal mistake. Lignin, whether as native lignin from plants or partially transformed by industrial separation procedures, is a complex material with great potential by itself or as a source of chemicals. In a world where raw materials are in constant demand, having a renewable source such as lignin should be considered a gift from nature to technology [29].

Unlike other bio-platform molecules with a distinct structure, lignin structures are much more complex. The key issue for the lignin transformation/depolymerization lies in the development of strategies with highly selective and active catalysts to effectively cleave the ubiquitous C–O/C–C bonds [2, 14, 16, 30] while leaving the aromatic benzene rings unconverted. Although lignin research can be traced back to 1819, when the Swiss botanist A. P. Candolle (1778–1841) first used the term lignin [31] from the Latin word lignum meaning wood, and various non-catalytic and catalytic methods have been explored to cleave the various lignin C–C or C–O linkages bonds from the 1920s [29], it remains a big challenge to selectively convert lignin extracts, or even original lignin, to aromatics. For detailed reasons, this can be attributed to the much more complex mechanism caused by disturbing the lignin structure with various targeted linkages, combined with other factors such as transformations of lignin structure and different condensation modes of lignin fragments. With the development of lignin depolymerization in the last decade, many efficient methods and strategies have been reported for lignin utilization to prepare chemicals and materials [2, 14, 16, 32–35]. To better make use of lignin to prepare aromatic chemicals, it is necessary to summarize the existing catalysts and processes. Based on the obtained knowledge about lignin conversion regimens, we can update the catalysts and develop new routes for lignin efficient conversion.

The critical target of this book is to provide an overview of key advances in the field of lignin depolymerization to aromatics. Unlike previous review work, which classified the research by the lignin substrate (native lignin, kraft lignin, organosolv lignin, etc.), cleavage methods (oxidation, hydrogenation, etc.), or main products (arenes, phenol, acid, etc.), we provide a brief introduction to lignin, past technologies, and new catalytic methods or strategies for lignin depolymerization to monomers and then pay more attention to the relationships among the lignin depolymerization strategies/methods, catalysts, and mechanisms viewed from cleavage order and type of corresponding chemical bonds over various catalytic systems [36, 37]. Based on this summary of reported processes and their results, we hope to partly pull off the lid of the black box of lignin depolymerization.

1.3 Structure of This Book

The structure, reactions, and utilizations of the lignins have been studied for about two centuries from 1819 [31]. Lignin has been widely used in the preparation of various functional materials, chemicals, fuels, medicines, polymers, and others. As shown in Figure 1.3, entering the term "lignin into the SciFinder database by the end of December 2020 generated more than 140,000 entries over the last 100 years. Furthermore, the ever-increasing growth curve displays that about half of them

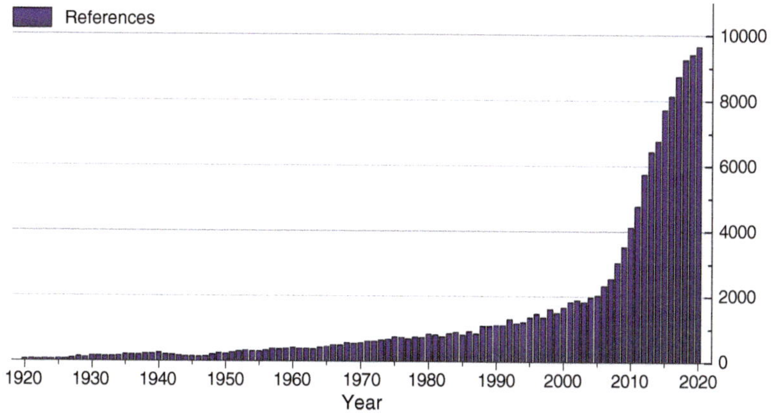

Figure 1.3 Number of Entries Retrieved Using "Lignin" as the Research Topic in SciFinder from 1920 to 2020.

(76554) have been published in the last 10 years and over 8000 references were reported in each of the last four years, reflecting the fact that lignin is a popular and evolving topic. As a star molecule, lignin and its utilization have drawn much attention from academia to industry worldwide.

One important utilization approach is to efficiently depolymerize the natural lignin polymer to easily-handled monomers as this is a precondition of lignin utilization to prepare chemicals that may be potential alternatives to fossil resources. In recent decades, many attempts have been carried out to achieve this target [2, 12, 14, 16, 37]. Given the complexity of lignin structure, numerous strategies and methods taking varying approaches from different research fields have been developed to effectively degrade lignin into monomer chemicals by targeting linkages and chemical bonds cleavages. These include the traditional methods of pyrolysis, gasification, liquid-phase reforming, supercritical solvolysis, chemical oxidation, hydrogenation, reduction, acidolysis, alkaline hydrolysis, and alcoholysis, as well as the newly developed redox-neutral process, biocatalysis, and combinatorial strategies. Therefore, there is a strong demand to summarize this research, providing a research history to show the connections among different research published in different periods and also providing an overview of current research into lignin depolymerization, which may provide useful suggestions for this vigorous research field.

Although it is quite a challenging project to give a systematic summary of so much research from different research fields with different backgrounds, it is possible using a viewpoint that is not just based on the reaction types or what kinds of catalyst are needed to achieve lignin depolymerization. Classifying the related strategies/methods and establishing their connections needs to come from a deeper perspective. Taking the perspective of a detailed discussion of mechanisms, this book examines the science and technology of lignin depolymerization conversion by using a multidisciplinary approach. About 2000 bibliographical references have been compiled to provide the reader with a complete and systematic overview of research into lignin depolymerization to produce chemicals, mainly aromatic chemicals. To handle such a vast amount of information about different lignin conversion strategies/methods, this book is divided into several parts that give a wide vision of the science and technology of lignin.

Considering that not all readers are professionals in the field of lignin research, we added some brief introductions about lignin near the beginning of the book (Part II, Chapter 2, and Chapter 3). In Chapter 2, we briefly introduce the biosynthesis of lignin from monolignols generation to their transport and then polymerization to the lignin molecule. Then, we provide an introduction of the lignin structure, viewing from the lignin linkages, linkages generation mechanism, structure models of different kinds of protolignin, and lignin-carbohydrate complex. In Chapter 3, after a brief introduction about the lignin polymer physical properties and methods to isolate lignin from lignocellulose, we present a basic introduction about the spectroscopy properties for the lignin structure characterization. At the end of Chapter 3, lignin chemical properties related to pre-chemical modification are discussed.

Next, focusing on lignin depolymerization, Part III, Chapter 4 first provides a discussion about scientific questions regarding lignin conversion and new biorefinery approaches. Next, we summarize scientific questions focusing on the topic of lignin depolymerization to aromatic monomers. Then, two different approaches for the foundation of lignin depolymerization strategies are introduced: direct lignin conversion and the bottom-up approach. To provide a macro-level understanding of the research field of lignin conversion, the classification of lignin conversion methods by reaction types is summarized. Then, as an important complementary issue, a brief list of systems for lignin/models conversion based on the catalysts is provided and targeted to provide knowledge about the catalyst/reagent that can be used in the lignin conversion via corresponding reaction types. The chapter concludes with summary of systems recently reported for native/technical lignin conversion.

Part IV, Chapters 5 to 11, comprises the main part of this book. As discussed in Chapter 4, multitudinous methods have been developed to depolymerize lignin to provide aromatic monomers. It is hard work to systemically summarize all of these strategies. Facing this question in Chapter 5, inspired by retrosynthetic analysis of the natural product to confirm the synthons and synthetic equivalents, we put forward an inverse disassembly analysis (IDA) method to explore the efficient strategies for lignin depolymerization. The IDA method, which can be used in the classification and discussion of various methods of lignin conversion, considers the cleavage order and type of corresponding chemical bonds over various catalytic systems. Then, to generalize as many lignin conversion methods with various routes and intermediates as possible, four thinking modes based on the IDA for lignin conversion are emphasized: (1) tailoring mode; (2) wedging mode; (3) protection mode; and (4) cascaded mode. With the β-O-4 linkage as the main lignin model, we further introduce the lignin depolymerization strategies with the IDA method from Chapters 6 to 11.

In Chapter 6, we firstly review the stoichiometric reactions and catalytic systems for the direct cleavage of ether bonds with a strong nucleophile or a strong electrophilic reagent. Then, modifying and updating the above methods, Brönsted/Lewis acid and metal systems, base/organometallic systems, and other heterogeneous catalysts used in the direct hydrogenative cleavage of C–OAr ether bonds are systemically introduced. In addition, for the direct reductive cleavage of ether bond, a brief review of the lignin conversion methods with hydride reagents, and the e^- species from chemical conversion or surface of the cathode or semiconductors is provided.

In Chapter 7, lignin depolymerization strategies beginning with the first activation/transformation of the neighboring lignin phenolic hydroxyl group or aromatic rings to the active radicals are summarized, which includes the following strategies: (1) lignin C_{Ar}–C bond cleavage of after the first phenolic hydroxyl group dehydrogenation to the phenolic radical; (2) lignin C–C bond cleavage *via* the first single-electron transfer from the aromatic ring; (3) lignin C–O bond cleavage of *via* the first semi-hydrogenation/addition of one aromatic ring to a cyclohexenyl ether; and (4) lignin C–O or C–C bond cleavage *via* the first inserting a cleavage reagent (metal catalyst center or extra radical) between the target bond with the assistance of the aromatic rings conjoint groups.

In Chapter 8, strategies for lignin linkages cleavage beginning with C_αO–H/ArO–H heterolysis or C_α–OH bond heterolysis are summarized, which is mainly focusing on the lignin alkaline hydrolysis, acidolysis, and their corresponding tandem processes.

In Chapter 9, strategies for lignin catalytic depolymerization which begin with the activation of C_α–H, C_α–OH, or C_αO–H bonds via a non-ionized route are mainly summarized. The first cleavage of C_αO–H and C_α–H bonds to a β-O-4 ketone intermediate can weaken the C_β–OAr bond and make a slightly stronger C_α–C_β bond, but with an active C_β–H. Therefore, focusing on this molecule intermediate, the lignin depolymerization strategies of catalytic transfer hydrogenation, dehydrogenation/oxidation-hydrogenation (reduction), and the multiple oxidation process are first summarized. Given that the Baeyer-Villiger oxidation and Beckmann rearrangement can be used in the cleavage of C–C bonds in lignin linkages that contain a ketone substrate site, the corresponding strategies and their derivative methods are then summarized. In addition, the first oxidation of C_α–OH can promote lignin hydrolysis, so an embellished β-O-4 linkage hydrolysis method involving the C_α–OH first oxidation and C_γ–OH transformation is discussed. In addition to the first cleavage of C_αO–H and C_α–H bonds to C_α=O at the beginning of the β-O-4 model transformation, some catalytic strategies beginning with the C_α–H or C_αO–H activation to the corresponding active radical intermediates can also lead to the cleavage of the lignin fragments. The first activation of C_α–OH(–H) to a $C_\alpha^\bullet$ radical intermediate can reduce the bond dissociation enthalpy (BDE) of the C_β–OAr bond for the following self-cleavage of β-O-4 linkage. Additionally, the first activation of C_αO–H bond to the $C\alpha O^\bullet$ via a hydrogen atom transfer (HAT), proton-coupled electron transfer (PCET) oxidation, or redox-neutral ligand-to-metal charge transfer (LMCT) process can lead to the selective cleavage of C_α–C_β bonds.

In Chapter 10, lignin cleavage strategies beginning with C_β–H, C_γ–H, or C_γO–H direct activation are summarized. These mainly describe some photocatalytic strategies with direct activation of C_β–H bonds that introduce the cleavage agent O*, leading to the cleavage of the C_α–C_β bonds in β-1 and β-O-4 linkages, and some retro-aldol strategies with first -$C_\gamma H_2$OH activation to -C_γHO/-C_γOOR (R=H or alkyl).

In Chapter 11, the main topic is promoting the lignin linkages cleavage with considering the potential lignin fragments condensation. After a summary of the potential lignin fragments condensation mechanism, we briefly introduce the developed methods for restraining the potential condensation of lignin fragments. The methods are divided into six categories as follows: (1) pre-protection of the active groups to inhibit the condensation; (2) first transformation of the active groups to promote the depolymerization; (3) in-situ converting the active intermediates and scavenging the unwanted species; (4) modifying and designing the catalysts; and (5) intensifying reaction systems; (6) employing "ideal lignins" as the substrates.

Finally, in Part V (Chapter 12), we first briefly introduce the research process of lignin utilization to prepare versatile materials. Then, considering the steps from lignin depolymerization to aromatic chemicals and the issue of how to make lignin depolymerization profitable, we provide our viewpoint for the outlets of lignin resources and emphasize the importance of high-value medicine and synthetic block preparation or the lignin utilization from natural polymer to artificial polymer. To accelerate the lignin utilization from lab to industry application, we discuss the critical roles of standardized lignin substrate and standardized products, during which we hold the idea that funneling and functionalization of a mixture of lignin-derived monomers into a single high-value chemical is fascinating and promising.

References

1 *BP statistical review of world energy* (2020). BP p.l.c. London.

2 Li, C., Zhao, X., Wang, A. et al. (2015). Catalytic transformation of lignin for the production of chemicals and fuels. *Chem. Rev.* 115: 11559–11624.

3 Amidon, T.E. and Liu, S. (2009). Water-based woody biorefinery. *Biotechnol. Adv.* 27: 542–550.

4 Shylesh, S., Gokhale, A.A., Ho, C.R., and Bell, A.T. (2017). Novel strategies for the production of fuels, lubricants, and chemicals from biomass. *Acc. Chem. Res.* 50: 2589–2597.

5 Barta, K. and Ford, P.C. (2014). Catalytic conversion of nonfood woody biomass solids to organic liquids. *Acc. Chem. Res.* 47: 1503–1512.

6 Besson, M., Gallezot, P., and Pinel, C. (2014). Conversion of biomass into chemicals over metal catalysts. *Chem. Rev.* 114: 1827–1870.

7 Huber, G.W., Iborra, S., and Corma, A. (2006). Synthesis of transportation fuels from biomass: Chemistry, catalysts, and engineering. *Chem. Rev.* 106: 4044–4098.

8 Corma, A., Iborra, S., and Velty, A. (2007). Chemical routes for the transformation of biomass into chemicals. *Chem. Rev.* 107: 2411–2502.

9 Zhang, Z.R., Song, J.L., and Han, B.X. (2017). Catalytic transformation of lignocellulose into chemicals and fuel products in ionic liquids. *Chem. Rev.* 117: 6834–6880.

10 Jing, Y.X., Guo, Y., Xia, Q.N. et al. (2019). Catalytic production of value-added chemicals and liquid fuels from lignocellulosic biomass. *Chem* 5: 2520–2546.

11 Luterbacher, J.S., Alonso, D.M., and Dumesic, J.A. (2014). Targeted chemical upgrading of lignocellulosic biomass to ionic platform molecules. *Green Chem.* 16: 4816–4838.

12 Galkin, M.V. and Samec, J.S.M. (2016). Lignin valorization through catalytic lignocellulose fractionation: A fundamental platform for the future biorefinery. *ChemSusChem* 9: 1544–1558.

13 Li, N., Zhang, X., Shi, M., and Zhou, S. (2017). The prospects of China's long-term economic development and CO_2 emissions under fossil fuel supply constraints. *Resour. Conserv. Recycl.* 121: 11–22.

14 Zakzeski, J., Bruijnincx, P.C.A., Jongerius, A.L., and Weckhuysen, B.M. (2010). The catalytic valorization of lignin for the production of renewable chemicals. *Chem. Rev.* 110: 3552–3599.

15 Vispute, T.P., Zhang, H., Sanna, A. et al. (2010). Renewable chemical commodity feedstocks from integrated catalytic processing of pyrolysis oils. *Science* 330: 1222–1227.

16 Sun, Z., Fridrich, B., de Santi, A. et al. (2018). Bright side of lignin depolymerization: toward new platform chemicals. *Chem. Rev.* 118: 614–678.

17 Guo, X., Fang, G., Li, G. et al. (2014). Direct, nonoxidative conversion of methane to ethylene, aromatics, and hydrogen. *Science* 344: 616–619.

18 Gao, P., Xu, J., Qi, G. et al. (2018). A mechanistic study of methanol-to-aromatics reaction over Ga-modified ZSM-5 zeolites: Understanding the dehydrogenation process. *ACS Catal.* 8: 9809–9820.

19 Xu, Y., Wang, T., Shi, C. et al. (2020). Experimental investigation on the two-sided effect of acidic HZSM-5 on the catalytic performance of composite Fe-based Fischer–Tropsch catalysts and HZSM-5 zeolite in the production of aromatics from CO_2/H_2. *Ind. Eng. Chem. Res.* 59: 8581–8591.

20 Lyons, T.W., Guironnet, D., Findlater, M., and Brookhart, M. (2012). Synthesis of p-xylene from ethylene. *J. Am. Chem. Soc.* 134: 15708–15711.

21 Wang, Y., Gao, W., Kazumi, S. et al. (2019). Direct and oriented conversion of CO_2 into value-added aromatics. *Chem. Eur. J.* 25: 5149–5153.

22 Carpenter, D., Westover, T.L., Czernik, S., and Jablonski, W. (2014). Biomass feedstocks for renewable fuel production: a review of the impacts of feedstock and pretreatment on the yield and product distribution of fast pyrolysis bio-oils and vapors. *Green Chem.* 16: 384–406.

23 Lu, R., Lu, F., Chen, J.Z. et al. (2016). Production of diethyl terephthalate from biomass-derived muconic acid. *Angew. Chem. Int. Ed.* 55: 249–253.

24 Dai, T., Li, C.Z., Li, L. et al. (2018). Selective production of renewable *para*-xylene by tungsten carbide catalyzed atom-economic cascade reactions. *Angew. Chem. Int. Ed.* 57: 1808–1812.

25 Qi, X.D. and Fan, W. (2019). Selective production of aromatics by catalytic fast pyrolysis of furan with in situ dehydrogenation of propane. *ACS Catal.* 9: 2626–2632.

26 Xu, C., Arancon, R.A.D., Labidi, J., and Luque, R. (2014). Lignin depolymerisation strategies: Towards valuable chemicals and fuels. *Chem. Soc. Rev.* 43: 7485–7500.

27 Ragauskas, A.J., Beckham, G.T., Biddy, M.J. et al. (2014). Lignin Valorization: Improving lignin processing in the biorefinery. *Science* 344: 1246843.

28 Van den Bosch, S., Schutyser, W., Vanholme, R. et al. (2015). Reductive lignocellulose fractionation into soluble lignin-derived phenolic monomers and dimers and processable carbohydrate pulps. *Energy Environ. Sci.* 8: 1748–1763.

29 Calvo-Flores, F. G., Dobado, J. A., Isac-García, J., and Martín-MartíNez, F. J. (2015) Lignin and lignans as renewable raw materials: Chemistry, technology and applications. Wiley.

30 Rinaldi, R. (2014). Plant biomass fractionation meets catalysis. *Angew. Chem. Int. Ed.* 53: 8559–8560.

31 Jouanin, L. and C., L. (2012). Lignins: Biosynthesis, biodegradation and bioengineering. *Academic Press* 61.

32 Renders, T., Van den Bosch, S., Koelewijn, S.F. et al. (2017). Lignin-first biomass fractionation: the advent of active stabilisation strategies. *Energy Environ. Sci.* 10: 1551–1557.

33 Shen, X., Xin, Y., Liu, H., and Han, B. (2020). Product-oriented direct cleavage of chemical linkages in lignin. *ChemSusChem* 13: 4367–4381.

34 Abu-Omar, M.M., Barta, K., Beckham, G.T. et al. (2021). Guidelines for performing lignin-first biorefining. *Energy Environ. Sci.* 14: 262–292.

35 Wong, S.S., Shu, R.Y., Zhang, J.G. et al. (2020). Downstream processing of lignin derived feedstock into end products. *Chem. Soc. Rev.* 49: 5510–5560.

36 Zhang, C.F. and Wang, F. (2017). Sell a dummy: Adjacent functional group modification strategy for the catalytic cleavage of lignin β-O-4 linkage. *Chin. J. Catal.* 38: 1102–1107.

37 Zhang, C.F. and Wang, F. (2020). Catalytic lignin depolymerization to aromatic chemicals. *Acc. Chem. Res.* 53: 470–484.

38 Wang, M. and Wang, F. (2019). Lignin: Catalytic scissoring of lignin into aryl monomers. In: *Adv. Mater*, vol. 50. John Wiley & Sons, Inc.

Part II

Lignin Introduction

2

Lignin Biosynthesis and Structure

2.1 Lignin Biosynthesis

As one important component of lignocellulose, lignin is primarily synthesized and deposited in the secondary cell wall of specialized plant cells such as xylem vessels, tracheids, and fibers. It is also deposited in minor amounts in the periderm [1]. The biological function of lignin in the plants includes four key roles [2] as follows: (1) Lignin gives rigidity to cell walls by cementing and fixing polymers in the woody plant cell walls, in close association with polysaccharides, making the fibers relatively stiff and suitable to provide mechanical support for stems and branches; (2) Lignin glues different cells together in woody tissues, in which the middle lamella consists mainly of lignin that works as an efficient and resistant glue that keeps different cells together; (3) Lignin makes the cell wall hydrophobic and strong as it inhibits the swelling of the cell walls in water and thereby water leakage from a woody cell wall in addition to this being the main function of lignin in nonwoody plants; and (4) Lignin serves as a compact barrier and protects the plant cells against microbial degradation by preventing the penetration of the polysaccharide-degrading proteins excreted by microorganisms and is therefore a necessary component for plants.

For lignin biosynthesis, which differs from cellulose and hemicellulose that have a simple, well-studied structure involving the connection of C_6 and C_5 sugar monomers, lignin is produced mainly by the oxidative coupling of three monolignols: *p*-coumaryl alcohol (H-unit), coniferyl alcohol (G-unit), and sinapyl alcohol (S-unit) [3–5] (Figure 2.1). In the field of lignin research, the process of lignin formation with the polymerization of various monolignol is also called lignification. Based on wood science research, plant lignification is a tightly regulated complex cellular process that occurs via three sequential steps: (1) The synthesis of monolignols from carbohydrates generated from the photosynthesis; (2) The transport of lignin monomeric precursors across the plasma membrane to the cells controlling the lignin synthesis in the plant; and (3) The oxidative polymerization of monolignols to form lignin macromolecules within the cell wall.

Lignin Conversion Catalysis: Transformation to Aromatic Chemicals, First Edition.
Chaofeng Zhang and Feng Wang.

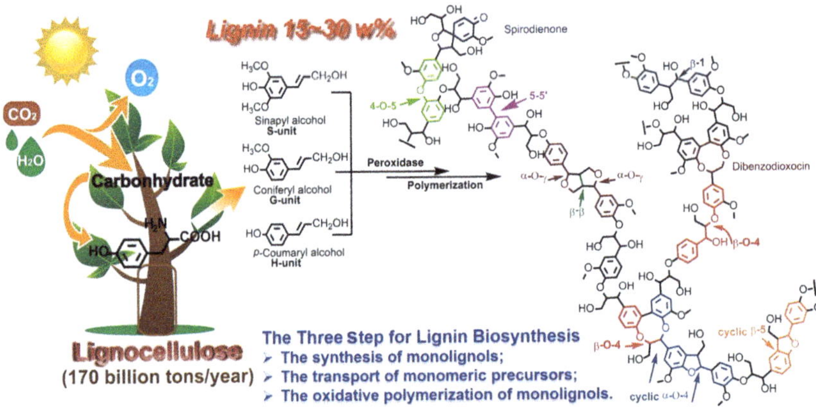

Figure 2.1 The Three-step for Lignin Biosynthesis. *Source*: Permission from Zhang et al. [3], © 2020, American Chemical Society.

Until now, the corresponding research has provided some reasonable understanding of monolignol biosynthesis and the transport of lignin monomeric precursors among plant cells, but some aspects of lignin assembly remain elusive [6]. Two major hypotheses exist for a unified mechanism of lignin biosynthesis. The first one is the combinatorial random coupling [1, 7] between monolignol radicals. This is the most widely accepted mechanism for lignin polymerization, but there is a second hypothesis based on protein-directed synthesis involving dirigent proteins [8, 9]. The lack of knowledge on the detailed lignification process is mainly because of the difficulty of its study. In the rest of Chapter 2.1, we will briefly introduce the biosynthesis of the lignin from monomer generation to their combination/polymerization to form the lignin molecule.

2.1.1 The Generation of Monolignols

In the plants, lignin biosynthesis requires a long sequence of reactions involving the following processes [10] to produce various monolignols and their precursors (Figure 2.2): (1) The shikimate pathway (shikimic acid pathway), which produces L-phenylalanine and L-tyrosine from phosphoenolpyruvate (PEP) and erythrose 4-phosphate (E-4-P) generated from the metabolism of glucose (Chapter 2.1.1.1); (2) The phenylpropanoid common pathway from L-phenylalanine (and/or L-tyrosine) to precursors of C_γ acyl-coenzyme A such as *p*-coumaroyl-CoA, feruloyl-CoA, and sinapoyl-CoA (Chapter 2.1.1.2); and (3) The monolignol specific pathway, which leads to the transformation of various acyl-coenzyme A compounds to *p*-coumaryl alcohol, coniferyl alcohol, and sinapyl alcohol (Chapter 2.1.1.2); and (4) The oxidation of monolignols to add a phenol group to yield the corresponding 4-*O*-phenoxy radicals, which polymerize spontaneously to give a growing lignin polymer. Steps 1-3 are primarily important in the generation of monolignols, and occur before the lignin monomeric precursors cross the plasma membrane to the cells controlling the lignin synthesis. Part 2.1.1 will briefly introduce the generation of three typical monolignols: *p*-coumaryl alcohol (H-unit), coniferyl alcohol (G-unit), and sinapyl alcohol (S-unit) as well as and other monolignols.

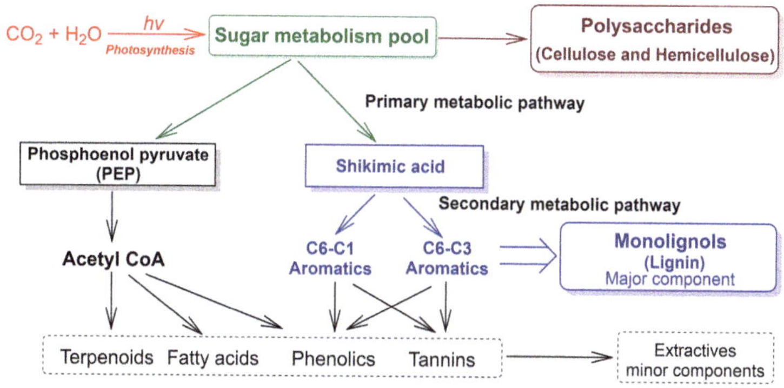

Figure 2.2 Primary and Secondary Metabolic Pathways Leading to the Biosynthesis of Lignin and Other Wood Components. *Source*: Calvo-Flores et al. [10] with Permission, © 2015 Wiley-VCH.

2.1.1.1 The Shikimic Acid Pathway

The shikimate pathway, present in bacteria, yeasts, and plants, is the entry pathway toward an overabundance of phenolic compounds. In seven enzymatic steps (Figure 2.3), the glycolytic intermediate phosphoenol pyruvate (PEP) and the pentose phosphate pathway intermediate erythrose-4-phosphate (E-4-P) are metabolized into chorismate via shikimate as an intermediate [11].

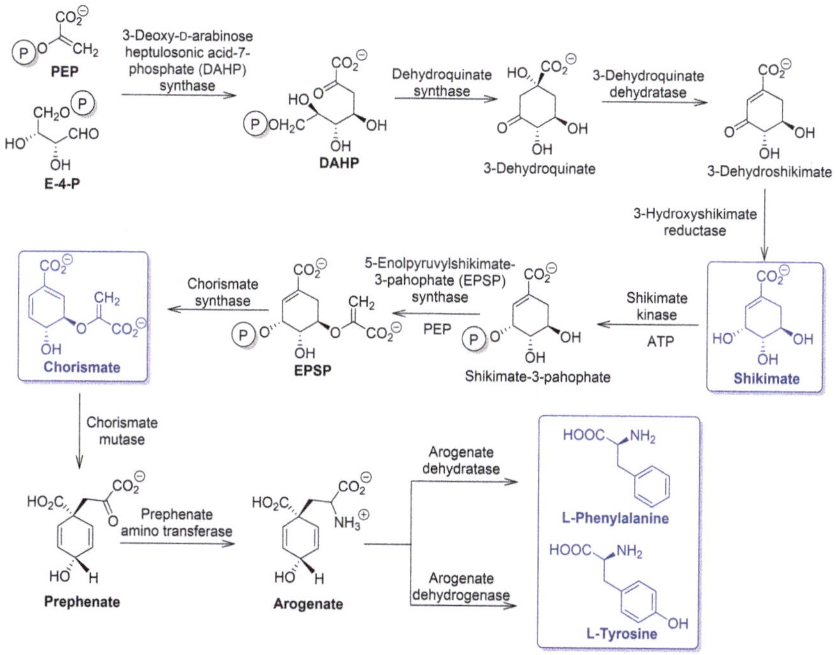

Figure 2.3 The Shikimate–Chorismate Pathway (from PEP/E-4-P to Chorismate via Shikimate) and the Biosynthetic Pathway from Chorismate to L-tyrosine and L-phenylalanine. *Source*: Calvo-Flores et al. [10] with Permission, © 2015 Wiley-VCH.

Next, chorismate can serve as the precursor for *p*-aminobenzoate, which is an intermediate in the biosynthesis of aromatic amino acids phenylalanine, tyrosine, and tryptophan [12]. L-Phenylalanine and L-tyrosine are produced from chorismate via arogenate by the action of three enzymes: (1) chorismate mutase; (2) prephenate aminotransferase; and (3) arogenate dehydrogenase or arogenate dehydratase (Figure 2.3).

2.1.1.2 The Common Phenylpropanoid Pathway and Monolignol-Specific Pathway

In 1971, Higuchi demonstrated that lignin is synthesized from L-phenylalanine and cinnamic acids [5, 13]. These acids are derived from carbohydrates through the shikimic and cinnamic acid pathways, which was further confirmed by administering radioactive ^{14}C-labeled glucose to plants that can produce labeled shikimic acid [14], and radioactive lignins [15, 16]. Beginning with L-tyrosine and L-phenylalanine, Figure 2.4a provides an overview of the common phenylpropanoid pathway for the generation of monolignol precursors of C_γ acyl-coenzyme A (*p*-coumaroyl-CoA, reruloyl-CoA, and sinapoyl-CoA).

The general phenylpropanoid pathway uses phenylalanine as an entry substrate and yields *p*-coumaroyl-CoA, feruloyl-CoA, and sinapoyl-CoA after several enzymatic steps [17]. The deamination process of phenylalanine to cinnamic acid/cinnamate is catalyzed by the key enzyme L-phenylalanine ammonia-lyase (PAL) [15, 16, 18]. In grasses, a PAL isozyme catalyzing the deamination of both phenylalanine (PAL activity) and tyrosine (tyrosine ammonia-lyase activity) might be involved in the direct conversion of tyrosine to *p*-coumarate [19]. An additional enzyme found only in grasses, tyrosine ammonia-lyase (TAL), can catalyze the formation of *p*-coumaric acid from L-tyrosine [20–22]. Hydroxylation of the aromatic ring of the cinnamic acid leads to *p*-coumarate through a reaction catalyzed by cinnamate 4-hydroxylase (C4H). Activation of the acid to a thioester by 4-coumarate CoA ligase (4CL) yields *p*-coumaroyl-CoA. As lignification proceeds, cinnamic acid is hydroxylated to *p*-coumaric and caffeic acids by specific hydroxylase enzymes (C4H and C3H) [23, 24], after which 4CL may convert caffeic acid into caffeoyl-CoA [25]. The caffeic acid can be further methylated to ferulic acid by O-methyltransferase (COMT) [26–28]. Ferulic acid can be transformed to feruloyl-CoA with the ligase 4CL or undergo hydroxylation with ferulate 5-hydroxylase (F5H) to yield 5-hydroxyferulic acid. 5-Hydroxyferulic acid can be further transformed to 5-hydroxyferuloyl-CoA with the assistance of the ligase 4CL or can be methylated with the enzyme COMT to yield sinapic acid. Sinapic acid then interacts with 4CL to form sinapoyl-CoA.

The monolignol-specific pathway involves the transformation of *p*-coumaroyl-CoA to feruloyl-CoA via caffeoyl-CoA and transformation of 5-hydroxyferuloyl-CoA to sinapoyl-CoA followed by further transformation of these CoA-esters to the final monolignol cinnamyl alcohol coniferyl alcohol and sinapyl alcohol (Figure 2.4b). The 3-hydroxylation of *p*-coumaroyl-CoA to caffeoyl-CoA involves three enzymatic steps, at least in dicots. First, *p*-coumaroyl-CoA is transesterified to its quinic or shikimic acid ester derivative by hydroxycinnamoyl-CoA: shikimate/quinate hydroxycinnamoyltransferase (HCT). *p*-Coumaroyl shikimate or quinate is

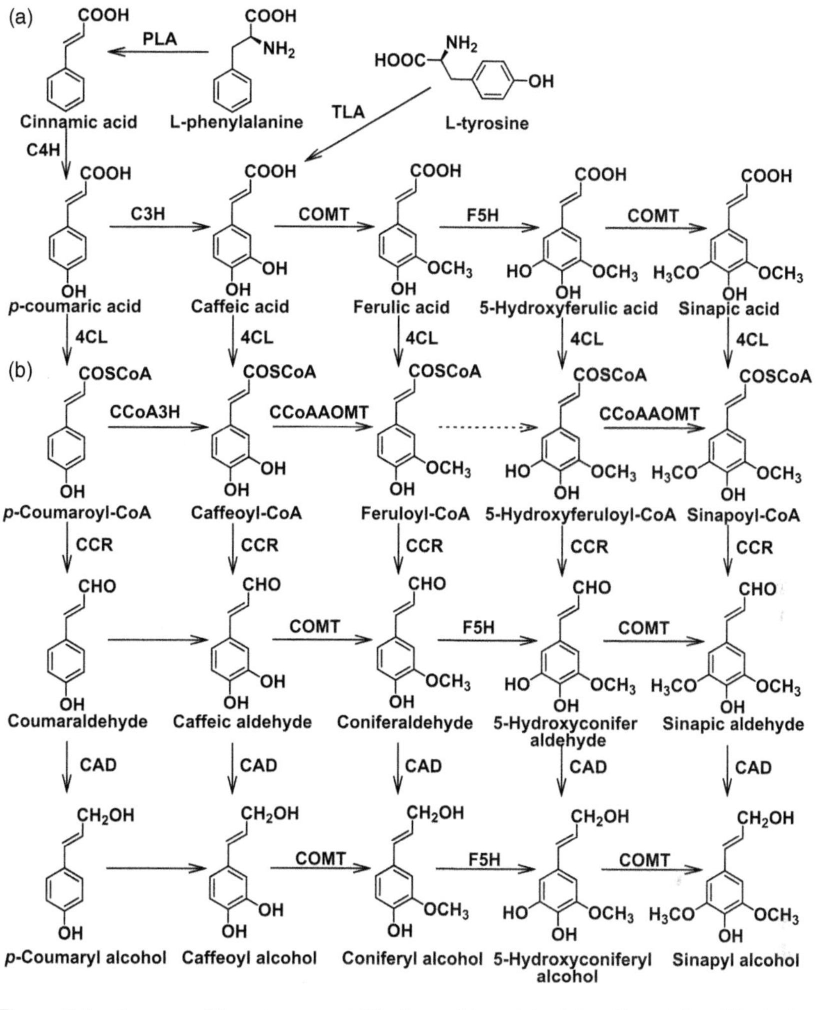

Figure 2.4 Common Phenylpropanoid Pathway (a) and the Monolignol-Specific Pathway (b) *Note*: PAL, Phenylalanine ammonia-lyase; TAL, tyrosine ammonia-lyase; C3H, emphp-coumarate 3-hydroxylase; C4H, cinnamic acid 4-hydroxylase; COMT, CoA O-methyltransferase; F5H, ferulate 5-hydroxylase; CCoA3H, 4-hydroxycinnamoyl-CoA 3-hydroxylase; CCoAOTM, caffeoyl-CoA O-methyltransferase; 4CL, emphp-coumaroyl-CoA ligase; CCR, cinnamoyl-CoA reductase; CAD, cinnamyl alcohol dehydrogenase.

then hydroxylated by *p*-coumarate 3-hydroxylase (C3H) and then transesterified again by HCT to caffeoyl-CoA. Further methylation of the 3-hydroxyl group by caffeoyl-CoA *O*-methyltransferase (CCoAOMT) yields feruloyl-CoA. Methylation of 5-hydroxyferuloyl-CoA can yield sinapoyl-CoA with the assistance of CCoAOMT. Next, the various CoA-esters can be transformed to the corresponding aldehydes with the assistances of enzyme cinnamoyl-CoA reductase (CCR) and the aldehydes undergo selective -CHO reduction to alcohols with the assistance of cinnamyl alcohol dehydrogenase (CAD), which provides the final *p*-coumaryl alcohol, coniferyl alcohol, and sinapyl alcohol. In addition to the above routes, the aldehydes and alcohols described can occur via hydroxylation with F5H and methylation with COMT. For detailed mechanisms of monolignol generation, one can check woods science references.

2.1.1.3 The Biosynthesis of Other Monolignols

In addition to the alcohol-based monolignols discussed above, plants can also generate some specific monomers such as dihydro-hydroxycinnamyl alcohols, hydroxybenzaldehydes, hydroxycinnamic acids, and other products such as hydroxycinnamaldehydes from an incomplete monolignol biosynthesis (Figure 2.5). Traditional monolignols are often acylated at their C_{γ}-position with acetate, *p*-hydroxybenzoate, or *p*-coumarate groups [29–35]. Some of these monomers can even be highly abundant in certain plants and may constitute 50% or more of the polymer units [36].

2.1.2 The Transport of Monolignols

p-Coumaryl alcohol, coniferyl alcohol, and sinapyl alcohol are toxic and unstable to plants, so they are often stored in the cambium cells of plants in the form of their glycosides (glycosylation of monolignols), namely as *p*-coumaryl alcohol glucoside, coniferyl alcohol glucoside, and sinapyl alcohol glucoside. This is a mechanism to protect plants themselves from the intrinsic toxicity of monolignols and to provide efficient transport as glycosides are easily transported to the cell via the cell membrane. The coniferyl alcohol glucoside is stored in ducts. Glucosyl alcohol glucoside accumulates in the differentiated xylem and inner phloem and represents about 1% of the fresh wood weight [19]. It is easy to separate coniferyl alcohol glucoside, coumarol glucoside, and sinapyl alcohol glucoside from the sap of many coniferous plant cambia. However, the amount of these substances in hardwood plant tissues is less, which could due to the fact that these plants do not accumulate these glycosides and may have other transport mechanisms. Additionally, monomeric

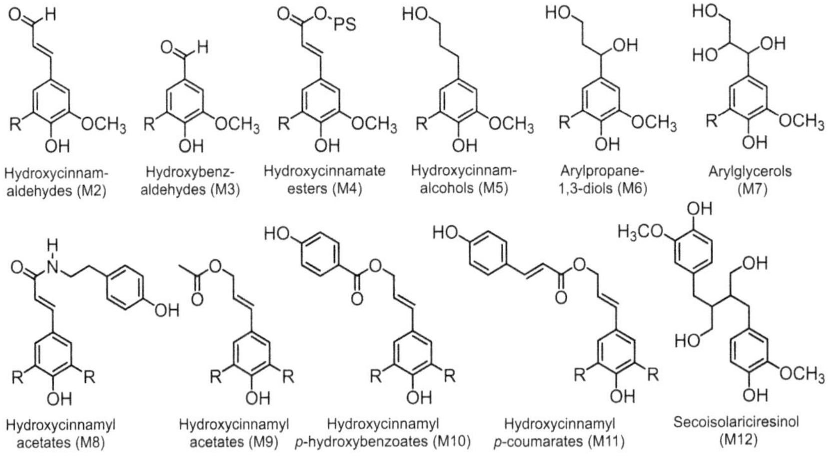

Figure 2.5 Chemical Formulas of the Other Monomers of Lignin. *Note:* M1monolignols are *p*-coumaryl alcohol, coniferyl alcohol, and sinapyl alcohol.

glycosides exist in the xylem vessel cells and cell walls of differentiated Pinaceae plants, indicating that monomeric glycosides are transported from the vascular tissue to the cell wall during the specific period of lignin partialization. During the lignification process of plants, when lignin needs to be synthesized, these monomer glycosides are transported to the lignin synthesis site of the cell wall and released and then the final lignin macromolecules are generated through polymerization [37]. During this complex process, monolignol glucoconjugates are generated by cytosolic glucosyl transferases (GT) and need to be deglucosylated by the cell wall-associated glucosylhydrolases (GH) before incorporation into lignin polymers. For a more detailed description of the monolignol transport mechanism, one can check the corresponding references [37–41].

2.1.3 The Polymerization of Lignin Monolignols

2.1.3.1 The Dehydrogenation of the Precursors

Upon entering the cell wall matrix, the polymerization of lignin monolignols begins with the dehydrogenation process of monolignols to produce the corresponding radicals. This process could be attributed to enzymes of several classes such as peroxidases (a group of enzymes containing the heme group, which uses H_2O_2 for activation), laccases (copper-containing enzymes that use molecular oxygen for activation), polyphenol oxidases, and coniferyl alcohol oxidases. These enzymes, which act alone or in combination, are responsible for the dehydrogenation of monolignols in plants. Usually, peroxidases are considered the main enzymes involved in the polymerization step to form lignin because both have a spatial and temporal correlation with cell wall lignification. However, there is a resurgence of theories in which other enzymes, such as laccases, participate in the polymerization pathway [42, 43]. A review of the literature regarding plant laccases and selected fungal laccases and their role during lignin biosynthesis has been made by Dean and Eriksson [44]. Additionally, two laccases involved in lignification have been identified [45].

These enzymes are capable of abstracting a proton from the phenolic hydroxyls of the precursor molecules, creating resonance-stabilized free radicals that can be detected and accurately quantified using ^{31}P NMR when the radicals react with a nitroxide phosphorus compound [46, 47]. Peroxidase induces one-electron oxidation of the phenol group, allowing the delocalization of the unpaired electron through resonance forms (Figure 2.6).

2.1.3.2 The Radical Polymerization

As discussed above, lignins are polymerized oxidation products of monolignols that form beginning with enzymatic dehydrogenation that yields reactive monolignol species with different free radical structures (Figure 2.6) that can couple with each other by generating new C–C or C–O bonds. Subsequent nucleophilic attacks by water, alcohols, or phenolic hydroxyl groups on the benzyl carbon of the quinone methide intermediate will restore the aromaticity of the benzene ring. The generated dilignols will then undergo further polymerization to generate the more complex lignin chains. Most of the links in the generated polymer are ether bonds between the central C_β of one methoxylated phenylpropane monomer and the

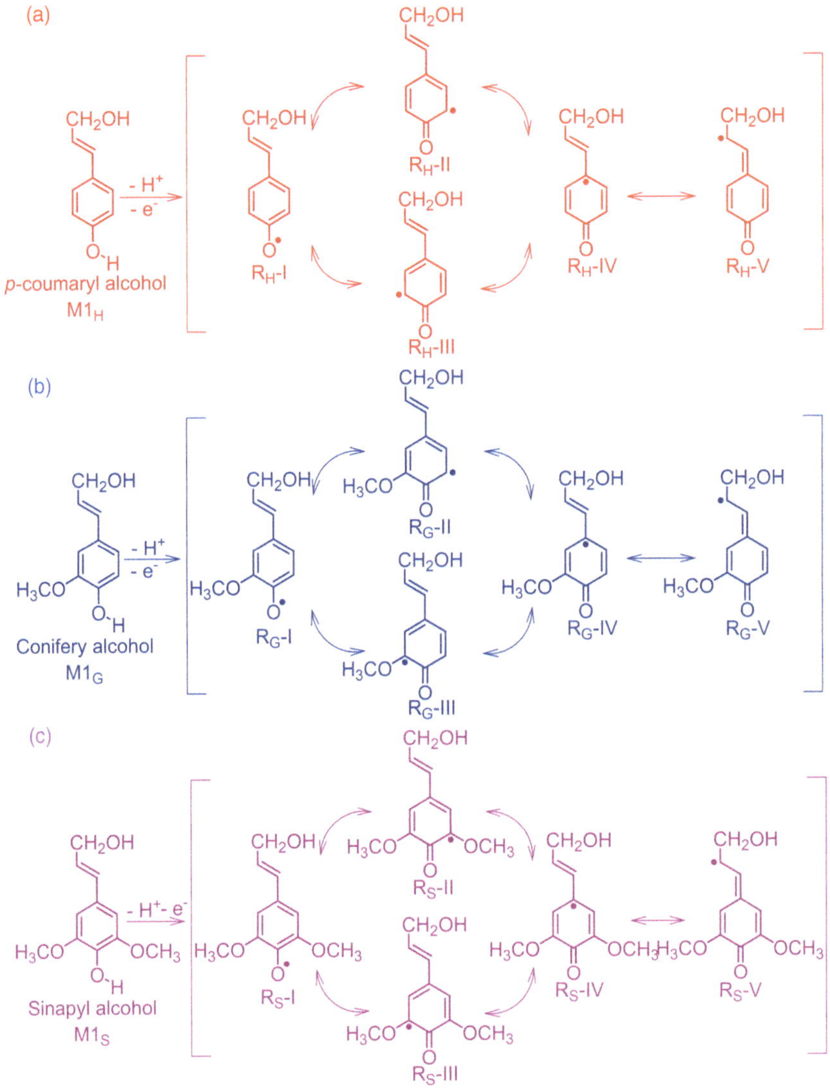

Figure 2.6 Resonance Structures of Monolignols Radicals. *Note*: R_H-II ≈ R_H-III and R_S-II ≈ R_S-III; R_G-III, R_S-II, and R_S-III participate less in lignin synthesis due to the steric effect of the methoxy group.

phenolic group of the other one. Although the exact structure of protolignin is still unknown and different from plant to plant, previous research has shown that lignin includes β-O-4, β-5, α-O-4, β-β, 5-5, 4-O-5, β-1, dibenzodioxocin, and other linkages and the approximate linkage abundance of these molecules in softwood/hardwood are listed in the table of Figure 2.7 [3].

The polymerization step is the last step in the biosynthesis of lignins from cinnamyl alcohol. After the oxidation of the monomeric alcohols to phenoxy radicals, the reaction changes dramatically. The reactions are no longer subjected to enzymatic

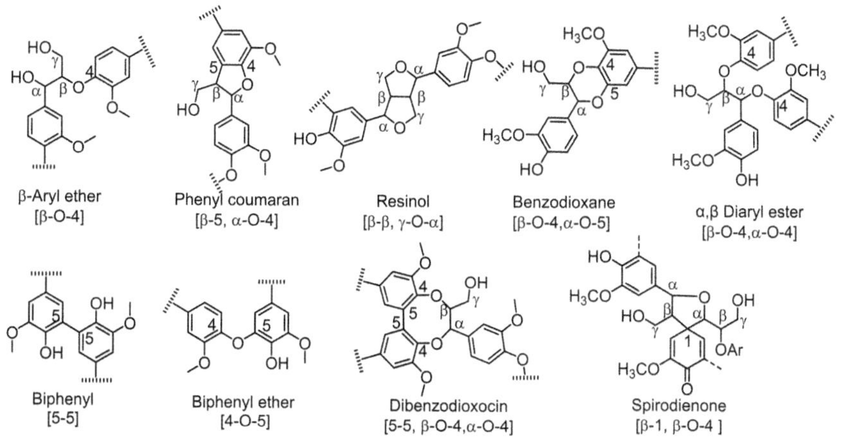

Linkages	Number /100 ppu		Functional groups	Number /100 ppu	
	Softwood	Hardwood		Softwood	Hardwood
β-O-4	43-50	50-65	$-OCH_3$	92-96	132-146
5-5′	10-25	4-10	−OH (phenolic)	20-28	9-20
β-5	9-12	4-6	−OH (benzyl)	16	-
α-O-4	6-8	4-8	−OH (aliphatic)	120	-
4-O-5	4	4-7	Carbonyl	20	3-7
β-1	3-7	5-7	Carboxyl	-	11-13
β-β	2-4	3-7			
Others	16	7-8			

Figure 2.7 Major Linkages Present in Lignins and Frequencies of Linkages and Main Functional Groups in Softwood and Hardwood.

control, but to a random polymerization process in which any phenolic molecule having the proper chemical kinetic and thermodynamic radical-generation and cross-coupling propensities can couple into the lignin polymer.

About 50% of the interunit linkages in lignins are β-aryl ethers arising from the β-O-4 coupling of a monolignol at its β-position to the 4-O-position of another monolignol or the growing oligomer, enabling the resulting 4-*O*-phenolic function produced after rearomatization of the quinone methide intermediate with H_2O attack at C_α (Figure 2.8, with the combination of two coniferyl alcohols as the example).

The generated C_β-radical of the monolignol can also combine with another monolignol or the growing oligomer at its 5-C-position to the β-5 linkage or the β-1 linkage after the rearomatization of the quinone methide intermediate with H_2O (Figure 2.9). Additionally, the carbonyl group of the C_5-aromatic ring after the generation of the β-5 precursor (quinone methide intermediate) can return to the

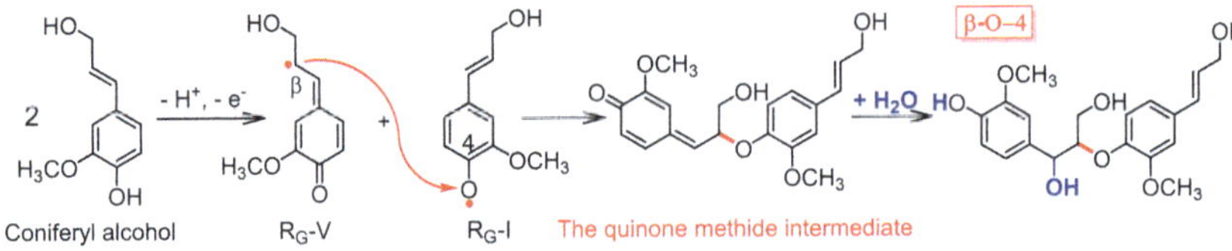

Figure 2.8 Dehydrogenation of Coniferyl Alcohol and Its Mesomeric Radicals Combination to the Lignin β-O-4 Linkage. *Note:* Similar combinations can be also found between two H units, S units, and the random combination between H, G, and S units.

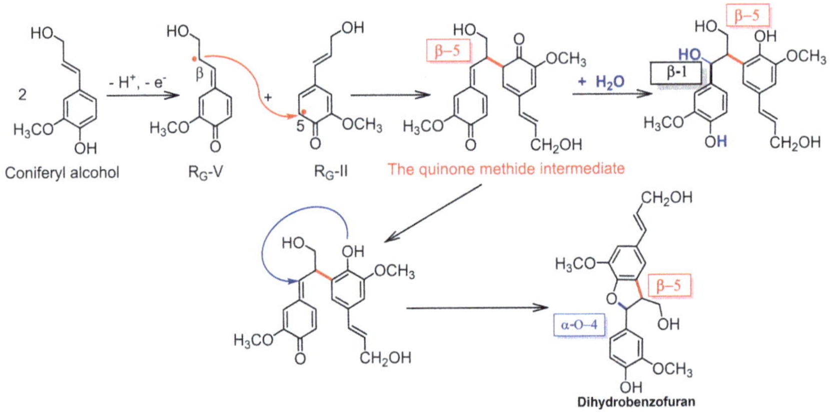

Figure 2.9 Dehydrogenation of Coniferyl Alcohol and Its Mesomeric Radicals Combination to the Lignin β-5 (β-1 Called in Some References), and α-O-4 Linkage in the Dihydrobenzofuran Structure. *Note:* Similar combinations can be also found between two H units and the combination between the H and G units. The hydration to provide the C_α–OH can also happen after the generation of α-O-4 linkage.

phenolic group via the keto-enol tautomerism, which can further attack the C_α of the rest quinone methide intermediate and provide a dihydrobenzofuran structure with the α-O-4 linkage generation.

Next, two generated C_β-radicals of the monolignol can bond together and provide a β-β linkage with the generation of a new C–C bond. Then, the C_γ–OH of the double quinone methide intermediate can attack the C_α of another half of the intermediate and achieve the rearomatization and provides a resinol structure with two conjoint tetrahydrofuran rings (Figure 2.10).

Besides the combination with the C_β-radical, the phenolic radical can bond with the C_5-radical to provide a 4-O-5 linkage after the rearomatization of the intermediate via the keto-enol tautomerism (Figure 2.11). Two C_5-radicals can bond together and provide the 5-5 linkage after the rearomatization of the intermediate via the keto-enol tautomerism (Figure 2.12). Neither 5-5 nor 4-O-5-linkages can be formed between S units or with the S unit for the methoxy groups at C_3 and C_5 positions in the S unit.

Following the 5-5-coupling of two phenolic end-units (see Figure 2.12), the next endwise coupling reaction with a monolignol (reacting at its favored β-position, and

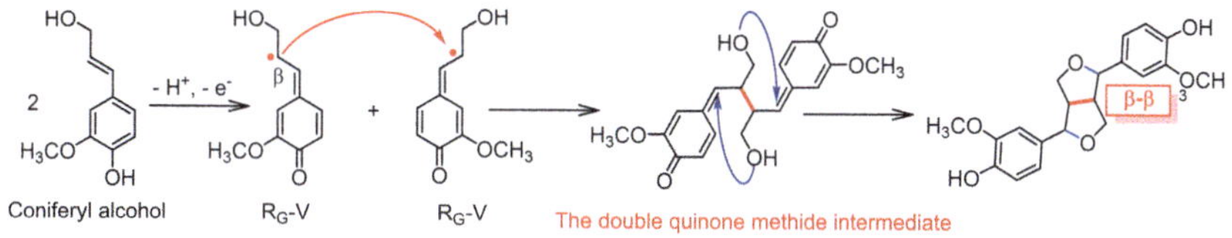

Figure 2.10 Dehydrogenation of Coniferyl Alcohol and Its Mesomeric Radicals Combination to the Lignin β-β Linkage in the Resinol Structure. *Note:* Similar combinations can be also found between two H units, S units, and the random combination between two of the H, G, and S units.

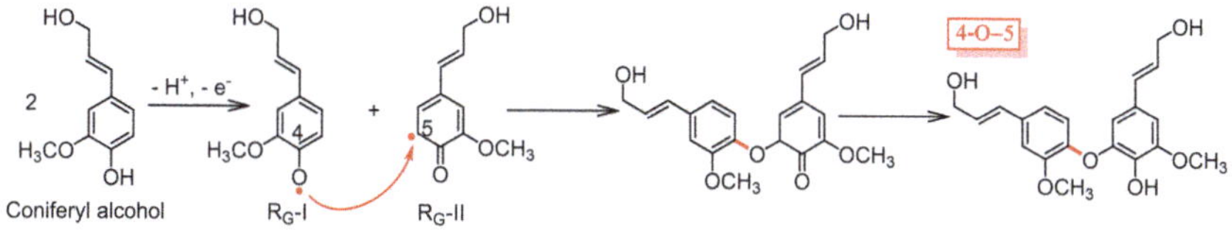

Figure 2.11 Dehydrogenation of Coniferyl Alcohol and Its Mesomeric Radicals Combination to the Lignin 4-O-5 Linkage. *Note:* Similar combinations can be also found between two H units and the combination between H and G units.

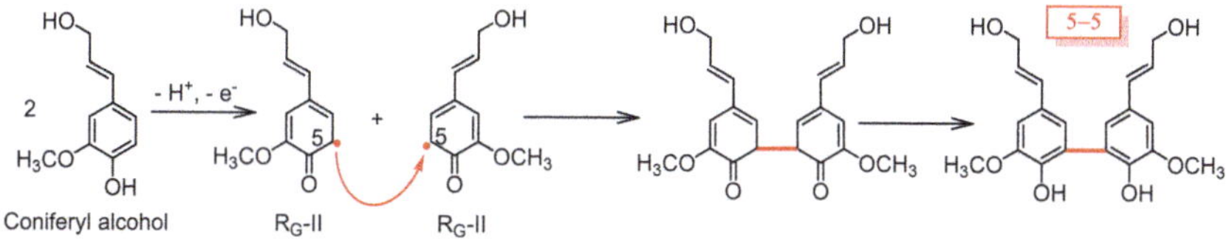

Figure 2.12 Dehydrogenation of Coniferyl Alcohol and Its Mesomeric Radicals Combination to the Lignin 5-5 Linkage. *Note:* Similar combinations can be also found between two H units and the combination between H and G units.

coupling to the only available position, the 4-O-position, of one of the phenolic moieties) produces the normal quinone methide intermediate (see Figure 2.13). This quinone methide is internally trapped by the other phenol in the 5-5-moiety, producing an 8-membered ring, a dibenzodioxocin structure. Such structures are strikingly prevalent in high-guaiacyl lignins [48].

Although the methoxy groups can reduce the activity of the adjacent carbon radical in lignin polymerization with other monolignol radicals, the C1 radical (R_H-IV, R_G-IV, and R_S-IV in Figure 2.6), connecting with an alkyl side chain, can participate in the radical combination and provide a real β-1cross-coupling linkage. The β-1 cross-coupling of a monolignol (at its favored β-position) with a preformed β-O-4 ether unit produces the usual quinone methide intermediate (Figure 2.14). The quinone methide may be trapped by water (Pathway A) to form a dienone. Alternatively, the quinone methide intermediate may be trapped by the C_α–OH of the β-O-4 ether unit to form a spirodienone unit (Pathway B) or the dienone may also occur dehydration between two C_α–OH groups to the spirodienone (Pathway D). Finally, the dienone or spirodienone structures may generate the conventional β–1-unit after

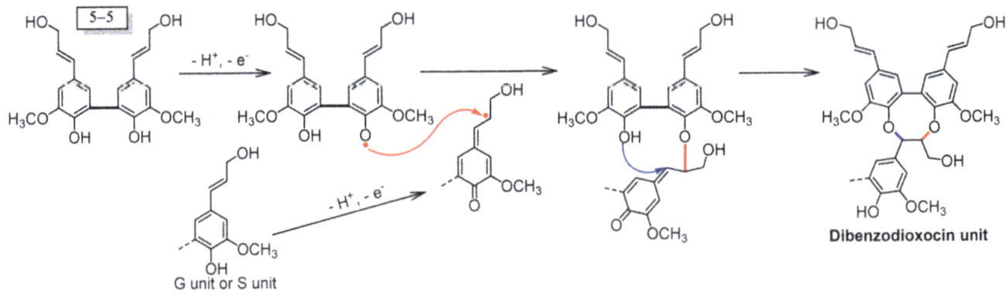

Figure 2.13 The Formation of Dibenzodioxocin Unit from the 5-5 Linkage.

the cleavage of the C_1–C_α bond (Pathway C). Furthermore, the spirodienone unit in the acid-catalyzed reaction can produce arylisochroman rearrangement products, with the C_1–C_α bond cleavage and the C_2–C_α generation (pathway E). Arylisochromans can be found at low levels in isolated lignins. Whether the conventional β–1-product, or aryl isochromanes, are present in the native lignins as such (from rearrangement during or post lignification) or if they are released from dienone structures during lignin isolation is unknown [48].

Besides the three normal monolignols, *p*-coumaryl alcohol, coniferyl alcohol, and sinapyl alcohol generated from the perfect enzyme catalysis system (Figure 2.4), the corresponding enzyme defects in certain kinds of plants (e.g., F5H-deficient, CAD-deficient, COMT-deficient, and so on) can provide some other monolignols listed in Figure 2.5, that can also participate in the biosynthesis of lignin and provide different linkage units, such as the 5-hydroxyguaiacyl coupling and coniferaldehyde coupling.

During the 5-hydroxyguaiacyl coupling (Figure 2.15), the phenolic radical $P_{5H}^{\bullet}$ generated from 5-hydroxyconiferyl alcohol (M5H) reacts with the coniferyl alcohol radical (R_G-V) at C_β to generate a β-O-4 quinone methide intermediate. In addition

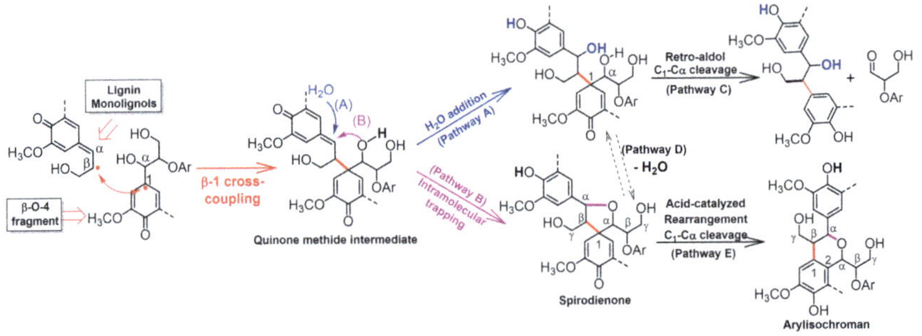

Figure 2.14 The β-1 Cross-coupling Mechanisms and the Corresponding Lignin Units.

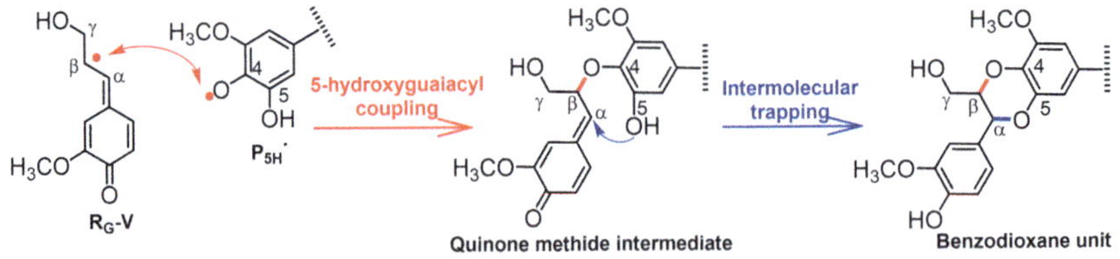

Figure 2.15 Mechanism of 5-Hydroxyconiferyl Alcohol Incorporation into a Guaiacyl Lignin to Produce Benzodioxane Units.

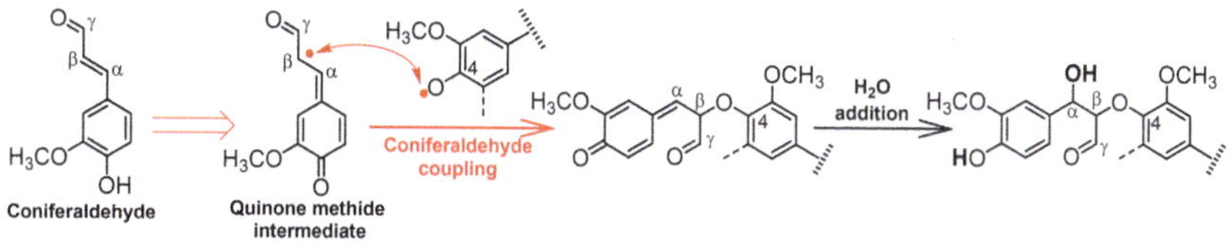

Figure 2.16 The Coniferaldehyde Coupling with Lignin Fragment to the β-O-4 Linkage.

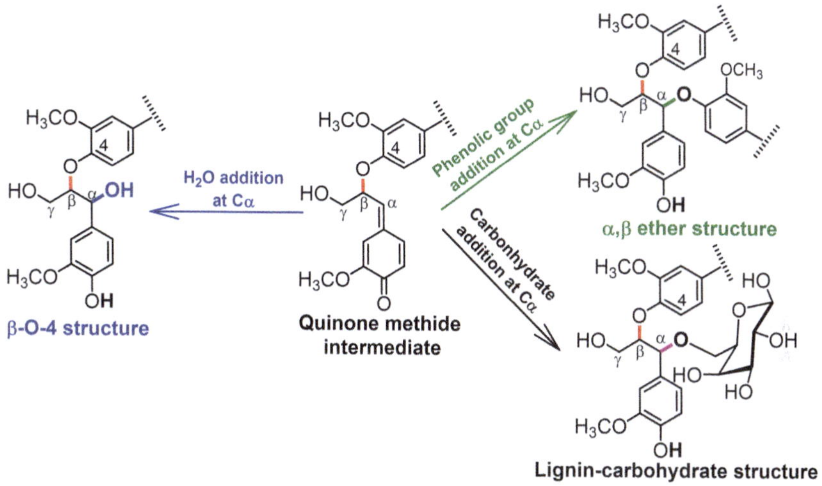

Figure 2.17 The Generation of α,β Ester Structure from the Quinone Methide Intermediate.

to the H_2O attack at the C_α, this time, the quinone methide can be internally trapped by the 5-OH of M5H, forming a new α-O-5 bond and creating the benzodioxane ring unit. The presence of benzodioxane structures in COMT-deficient plants and some transgenic plants is diagnostically revealed by NMR. The coniferaldehyde coniferaldehyde coupling is like the normal β-O-4 coupling but with a -C_γHO in the final product (Figure 2.16).

Furthermore, besides the first generation of the C_β-centered esters and because of the preferred C_β-radicals generation rather than the Cα-radicals from the basic monolignol, the phenol groups from other monolignols can compete with H_2O and some hydroxyl-containing carbohydrates to react with the quinone methide intermediate and provide an α,β ester structure (Figure 2.17). For a more detailed discussion about lignification, one can check the previous reviews [48].

2.2 Lignin Structure

For many years, researchers have been looking for an accurate model to describe native lignin (called protolignin in some literature and meaning a lignin isolated in such a way that the solvent does not react with or alter the lignin in any way). However, the random radicals combination synthesis process, the complexity of the spatial structure caused by the chiral centers, the complex functional group distribution, and the inherent difficulties in analysis have made it difficult to identify a complete structure of a lignin molecule, hampering the development of an accurate model. At the beginning of the study of lignin, the most common methodology employed was based on the hydrolysis of lignin and the subsequent analysis of the fragments obtained from the degradation process. Although the methods used for the detection and isolation of the fragments and the diversity of native lignin depending on the vegetable species contribute to the uncertainty concerning its structure, several models have been proposed over the years that have been revisited when new analytical methods are implemented. For many decades, lignin has been considered a cross-linked network polymer, but this perception has slightly changed recently. We will next summarize the structure models of protolignins.

2.2.1 Structure Models of Hardwood Lignin

Botanically, hardwoods are angiosperms. Anatomically, hardwoods are porous; that is, they contain vessel elements. Typically, hardwoods are plants with broad leaves that lose their leaves in autumn or winter. Hardwood lignin contains a higher proportion of sinapyl units, which results in a considerable percentage of unevenly distributed linear lignin. The typical composition of hardwood lignin is about a 100:70:7 ratio of coniferyl alcohol (G), sinapyl alcohol (S), and *p*-coumaryl alcohol (H), respectively. In 1974, Nimz proposed the first models of hardwood lignin for beech wood (Figure 2.18) [49]. Then, in 2003, Boerjan et al. [50] suggested a model for hardwood lignin based on poplar wood, which contains 20 units of monolignols and fewer branches. On the other hand, several acylated units have been detected in many hardwoods, for example, *p*-hydroxybenzoate (PB) or acetylated units in poplars, reported by Ralph in 2012 (Figure 2.19) [51].

2.2.2 Structure Models of Softwood Lignin

Botanically, softwoods are gymnosperms or conifers. Anatomically, softwoods are nonporous and do not contain vessels. Typically, softwoods are cone-bearing plants with needle- or scale-like evergreen leaves. In 1968, Neish [52] proposed the first softwood lignin model of spruce wood, which involved 18 units of monolignols. Later, in 1977, Adler [53] described a new softwood lignin model with 16 monolignols units derived from the oxidative degradation of spruce lignin. It was a partially limited model because certain units and linkages were not considered exactly due to the arbitrary choice of the sequence of units, meaning that proportions of certain structural details were not accounted for in a complete way. Despite this limitation, Adler's model is among the most extensive model of lignin structure (Figure 2.20). In 1980, Sakakibara [54] proposed a structural model of softwood lignin (Figure 2.21) based on degradation products resulting from hydrolysis with dioxane/water and catalytic hydrogenolysis. In this work, 39 hydrolysis products were identified.

In 1995, Karhunen et al. [55, 56] reported a new eight-membered ring (dibenzodioxocin) linkage in softwood lignin. This linkage was found using 2D NMR

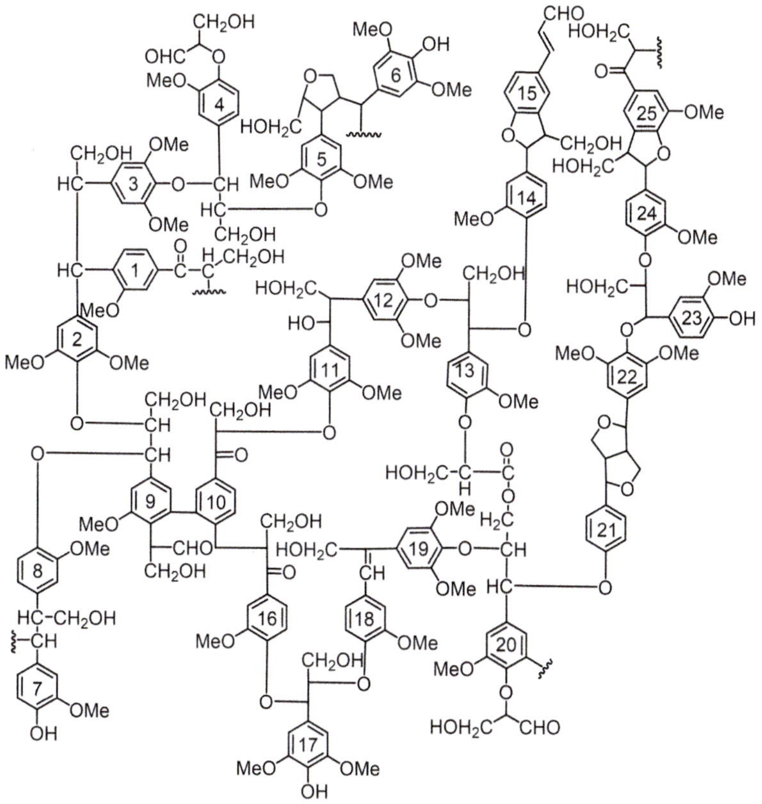

Figure 2.18 Structure Model of Beech Wood Lignin. *Source*: Nimz [52] with Permission, © 1974 Wiley-VCH.

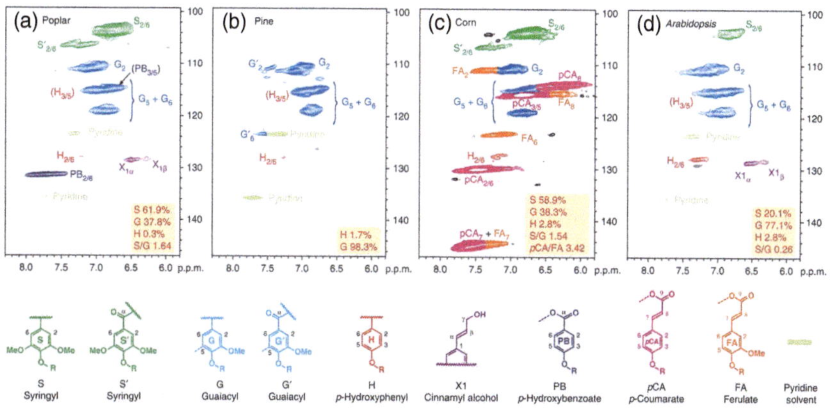

Figure 2.19 2D NMR Spectra Revealing Lignin Unit Compositions. Partial Short-range $^{13}C-^{1}H$ (HSQC) Correlation Spectra (Aromatic Regions Only) of Cell Wall Gels in 4:1 (vol/vol) DMSO-d_6/pyridine-d_5 from (a) Poplar Wood, (b) Pine Wood, (c) Corn Stalks and (d) Senesced Arabidopsis Inflorescence Stems. *Source*: Mansfield et al. [51] with Permission, © 2012 Springer Nature.

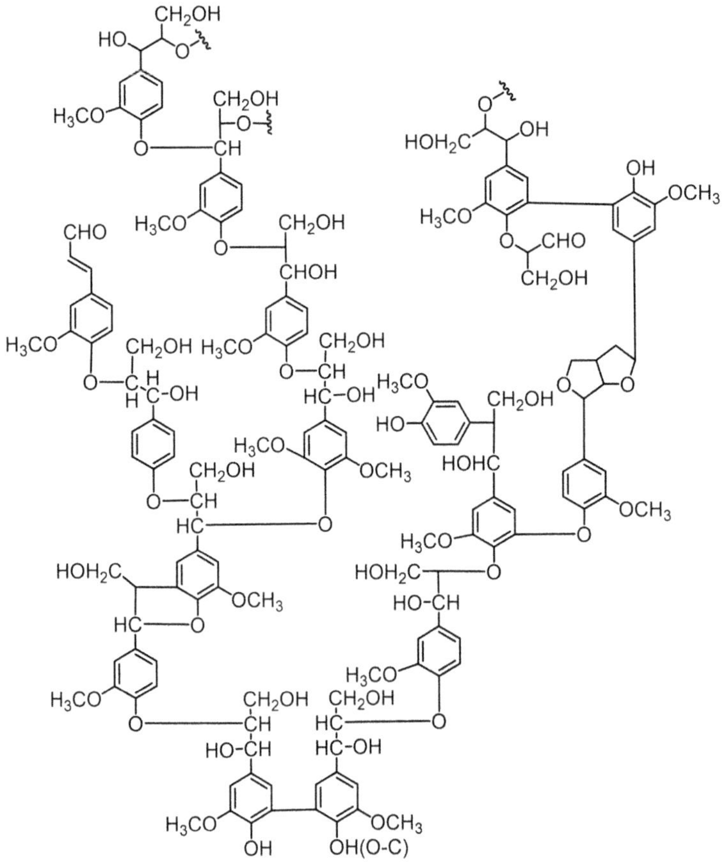

Figure 2.20 The Spruce Lignin Model. *Source*: Adler [53] with Permission, © 1977 Springer-Verlag.

techniques and is now proposed to be the main branching point in softwood lignin [57, 58]. Then, in 2001, Brunow developed a softwood-lignin model based on spruce wood data and formed with 25 units of monolignols (Figure 2.22). This model attempts to define more than the primary structure related to the coupling sequence, the idea of the main linkages, and the average of them. Softwood lignin is relatively branched by 4-*O*-5′ units and dibenzodioxocin units. This model also includes glycerol and cinnamaldehyde end units. For the stereochemical structure of the lignin, in 2011, Crestini et al. [59] proposed a model of milled softwood lignin based on supramolecular aggregates of linear oligomers rather than a network polymer (Figure 2.23).

Native lignin is an achiral material, but the structure shown in Figure 2.23 has 46 chiral centers and the optical centers must be generated randomly from the 17 billion physically distinct isomers due to the relative stereochemistries of pairs of centers in ring structures such as phenylcoumarans, resinols, and dibenzodioxocins. Therefore, it is impossible to propose an accurate model of lignin.

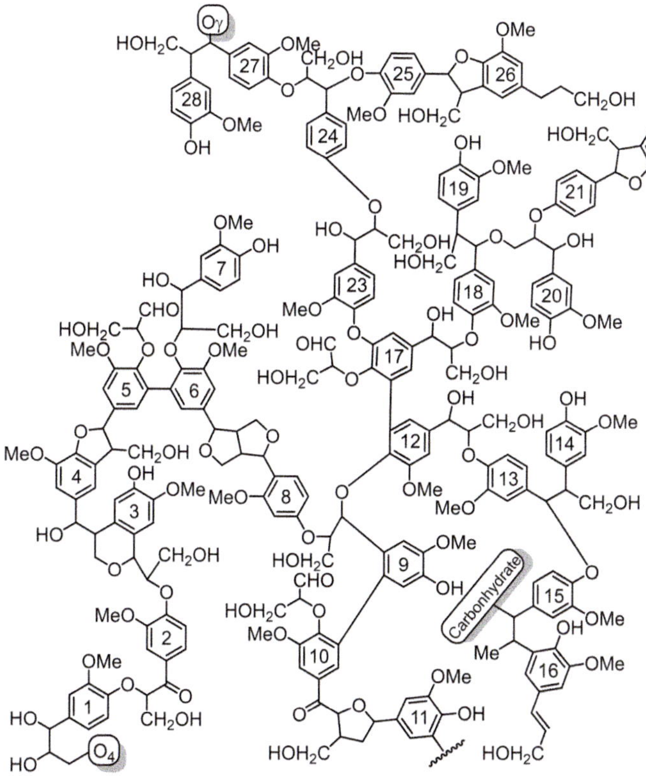

Figure 2.21 Structural Model for Softwood Lignin. *Source*: Sakakibara [54] with Permission, © 1980 Springer-Verlag.

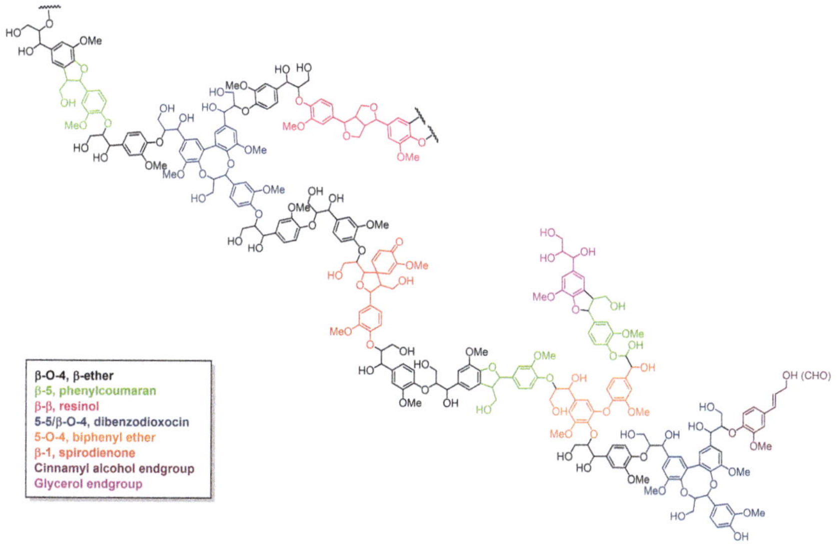

Figure 2.22 The Spruce Lignin Model. *Source*: Permission from Brunow [60], © 2001 John Wiley & Sons, Ltd.

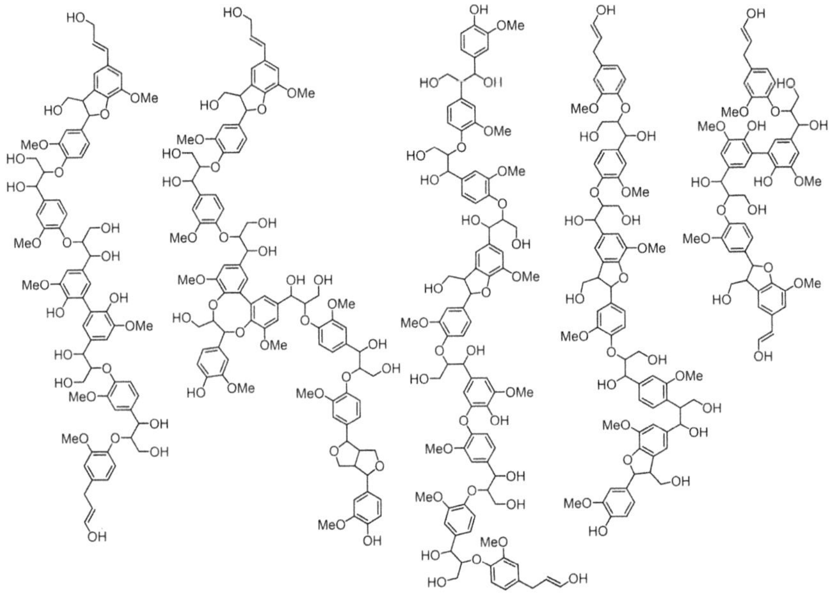

Figure 2.23 The Milled Softwood-lignin Model. *Source*: Crestini et al. [59] with Permission, © 2011 American Chemical Society.

2.2.3 Structure Models of Herbaceous Plant Lignin

In addition to the hardwood lignin and softwood lignin in trees, the herbaceous plants (such as what, rice, and reed) also contain lignin. The studies on lignin in herbaceous plants have shown that it presents greater variability by species, the part of the plant, and the isolation method.

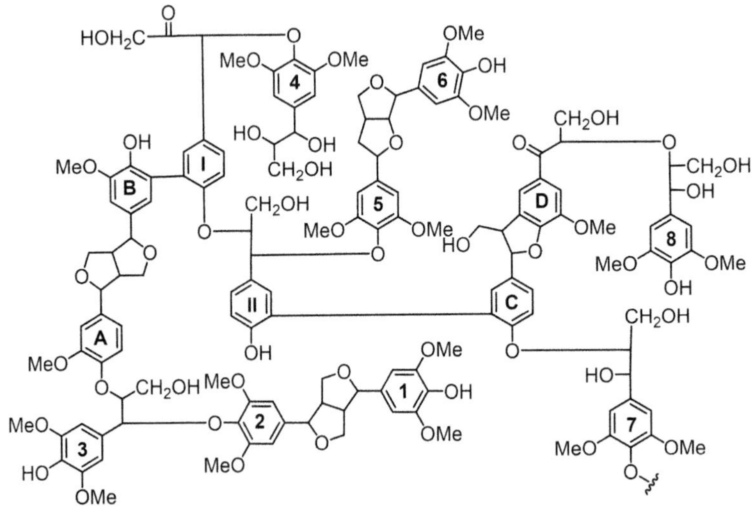

Figure 2.24 Structure Model Reed Stem Lignin.

In 1970, Simionescu [61] proposed a structural model of reed lignin, which consists of 14 phenylpropane units (Figure 2.24). The main type of bond present was alkyl-aryl ether (α-O-4 and β-O-4 bonds). In addition, one difference of this type of lignin from many herbaceous plants such as sisal, kenaf, abaca, or curaua is that these types of lignins are extensively acylated, especially at the C_γ of the lignin side chain (up to 80% acylation), with acetate and/or *p*-coumarate groups and preferentially over syringyl units. The structure of these acetylated lignins can be essentially regarded as syringyl units linked mostly through β-O-4 ether bonds. The lignin polymer for these herbaceous plants is rather linear and unbranched [62].

In conclusion, unlike for softwood and hardwood lignins, there is no commonly accepted structural model that accurately describes the lignins of herbaceous plants mainly due to the aforementioned diversity of lignin types in herbaceous plants. The most detailed structural data comes from composition analysis of the fragments and the study of some isolated fragments.

2.2.4 Lignin-Carbohydrate Complex

As we discussed in Chapter 1, the lignin in the lignocellulose is not self-existent but connects with cellulose, hemicellulose, or polysaccharides that can be called lignin–carbohydrate complex (LCC) in the study of lignin. The term lignin–carbohydrate complex was first introduced by Björkman in 1954 and is used to denominate the stable aggregate of lignin and polysaccharides interacting in vascular plant cell walls [63]. Three factors [64] determine the formation of LCC linkages: (1) The functional groups present in lignin and carbohydrate polymers (the primary and secondary alcohols, carboxyl, and carbonyl groups); (2) The enzymes involved in cell wall architecture (mainly glycosyl-transferases, laccases, and peroxidases); and (3) The free radicals or quinone methide intermediates formed in lignin polymerization are so reactive that they can react with carbohydrates (with the hydroxy group as the nucleophile) to the LCC. Although there are many possibilities for the formation of LCC linkages in the cell walls, investigation on wood yield of lignin fragments covalently bound to sugars [65–69] and the experimental findings support that LCC linkages in wood are mainly phenyl glycoside bonds [70], phenyl propyl glycoside bonds [71], benzyl esters [66], benzyl ethers [72], and acetal bonds [73]. Furthermore, carbohydrates from LCC are basically hemicellulose constituents (Figure 2.25). In previous publications, it was reported that, among the major three components of lignocellulose, the structure of hemicellulose and the linkage between hemicellulose and the other two (cellulose and lignin) was relatively less

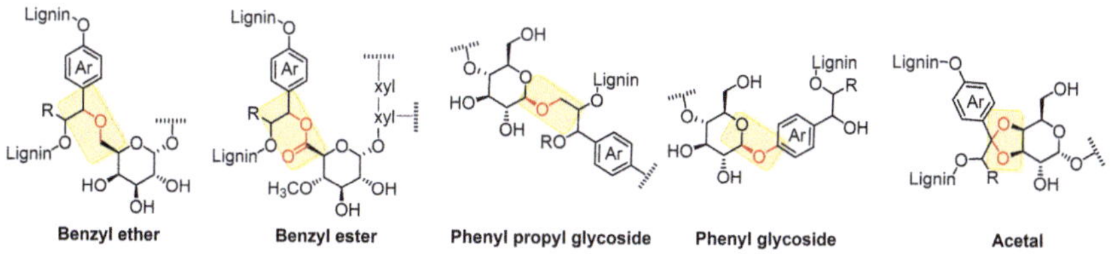

Figure 2.25 Five Main Types of LCC Linkages in the Wood.

stable and rigid [74, 75]. Hemicellulose is a heteropolymer formed by pentoses, mainly xylose and arabinose, and hexoses such as glucose, galactose, mannose, and sugar acid derivatives. Different from cellulose, hemicellulose is easily hydrolyzed by a diluted acid or base as well as by the hemicellulase enzyme. Cellulose is crystalline, strong, and resistant to hydrolysis, whereas hemicellulose has a random, amorphous structure with little strength. During native lignin depolymerization, the phenyl glycoside bonds can be also cleaved by the Lewis acids, which can release the C_α–OH for the following transformation (Chapter 6). For more information about the location of the lignin and its distribution in the plant, one can check a wood science or botany book.

2.3 Chapter Summary

In Chapter 2, we briefly introduce the biosynthesis of the lignin from the monolignols generation to their transport and combination/polymerization to the lignin molecule. We then give an introduction of the lignin structure considering lignin linkages, the mechanisms that generate these linkages, structural models of different kinds of protolignin, and the lignin-carbohydrate complex. Five keywords for this chapter should be Lignin biosynthesis, Monolignols, Lignification, Lignin linkages, and Lignin-carbohydrate complex.

References

1 Sarkanen, K. and Ludwig, C. (1971). *Lignins: occurrence, formation, structure and reactions*. New York: Wiley Interscience.

2 Ek, M., Gellerstedt, G., and Henriksson, G. (2009). *Wood Chemistry and Wood Biotechnology (Pulp and Paper Chemistry and Technology Book 1)*. New York: de Gruyter: Berlin.

3 Zhang, C.F. and Wang, F. (2020). Catalytic lignin depolymerization to aromatic chemicals. *Acc. Chem. Res.* 53: 470–484.

4 Zhang, C. and Wang, F. (2021). Chapter Six-Catalytic cleavage of lignin C–O and C–C bonds. *Adv. Inorg. Chem.* 77: 175–218.

5 Higuchi, T. (1990). Lignin biochemistry: Biosynthesis and biodegradation. *Wood Sci. Technol.* 24: 23–63.

6 Liu, C.J. (2012). Deciphering the enigma of lignification: Precursor transport, oxidation, and the topochemistry of lignin assembly. *Mol. Plant.* 5: 304–317.

7 Freudenberg, K. and Neish, A.C. (1968). *Constitution and Biosynthesis of Lignin*. Berlin: Springer-Verlag.

8 Davin, L.B., Wang, H.B., Crowell, A.L. et al. (1997). Stereoselective bimolecular phenoxy radical coupling by an auxiliary (dirigent) protein without an active center. *Science* 275: 362–366.

9 Davin, L.B. and Lewis, N.G. (2005). Lignin primary structures and dirigent sites. *Curr. Opin. Biotech.* 16: 407–415.

10 Calvo-Flores, F.G., Dobado, J.A., Isac-García, J., and Martín-MartíNez, F.J. (2015). *Lignin and lignans as renewable raw materials: Chemistry, technology and applications*. Hoboken: John Wiley and Sons, Ltd.

11 Herrmann, K.M. and Weaver, L.M. (1999). The shikimate pathway. *Annu Rev Plant Phys.* 50: 473–503.

12 Knaggs, A.R. (2003). The biosynthesis of shikimate metabolites. *Nat. Prod. Rep.* 20: 119–136.

13 Higuchi, T. (1971). Formation and biological degradation of lignins. *Advances in Enzymology and Related Areas of Molecular Biology* 207–283.

14 Hasegawa, M. and Higuchi, T. (1960). Formation of lignin from glucose in eucalyptus tree. *Nippon Rin Gakkaishi* 42: 305–308.

15 Kratzl, K. and Zauner, J. (1962). Incorporation of radioactivity in wood studied by means of xylitol-C14. *Holzforsch. Holzverwert.* 14: 108–111.

16 Schubert, W.J. and Acerbo, S.N. (1959). Conversion of D-glucose into lignin in Norwegian spruce. A preliminary report. *Arch. Biochem. Biophys.* 83: 178–182.

17 Humphreys, J.M. and Chapple, C. (2002). Rewriting the lignin roadmap. *Curr. Opin. Plant. Biol.* 5: 224–229.

18 Young, M.R., Towers, G.H.N., and Neish, A.C. (1966). Taxonomic distribution of ammonia lyases for L-phenylalanine and L-tyrosine in relation to lignification. *Can. J. Bot.* 44: 341–349.

19 Roesler, J., Krekel, F., Amrhein, N., and Schmid, J. (1997). Maize phenylalanine ammonia-lyase has tyrosine ammonia-lyase activity. *Plant Physiol.* 113: 175–179.

20 Higuchi, T., Ito, Y., and Kawamura, I. (1967). p-Hydroxyphenylpropane component of grass lignin and role of tyrosine ammonia-lyase in its formation. *Phytochemistry* 6: 875–881.

21 Neish, A.C. (1961). Formation of *m*- and *p*-coumaric acids by enzymic deamination of the corresponding isomers of tyrosine. *Phytochemistry* 1: 1–24.

22 Brown, S.A. (1961). Studies of lignin biosynthesis using isotopic carbon. IX. Taxonomic distribution of the ability to utilize tyrosine in lignification. *Can. J. Bot* 39: 253–258.

23 Vaughan, P.F.T. and Butt, V.S. (1970). Action of o-dihydric phenols in the hydroxylation of p-coumaric acid by a phenolase from leaves of spinach beet (Beta vulgaris). *Biochem. J.* 119: 89–94.

24 Russell, D.W. (1971). The metabolism of aromatic compounds in higher plants. *J. Biol. Chem.* 246: 3870–3878.

25 Chen, H.-C., Li, Q., Shuford, C.M. et al. (2011). Membrane protein complexes catalyze both 4- and 3-hydroxylation of cinnamic acid derivatives in monolignol biosynthesis. *Proc. Natl. Acad. Sci. USA* 108: 21253–21258. S21253/1-S21253/8.

26 Kutsuki, H. and Higuchi, T. (1981). Activities of some enzymes of lignin formation in reaction wood of *Thuja orientalis*, *Metasequoia glyptostroboides* and *Robinia pseudoacacia*. *Planta* 152: 365–368.

27 Kutsuki, H., Shimada, M., and Higuchi, T. (1982). Regulatory role of cinnamyl alcohol-dehydrogenase in the formation of guaiacyl and syringyl lignins. *Phytochemistry* 21: 19–23.

28 Luederitz, T., Schatz, G., and Grisebach, H. (1982). Enzymic synthesis of lignin precursors. Purification and properties of 4-coumarate: CoA ligase from cambial sap of spruce (*Picea abies* L.). *Eur. J. Biochem.* 123: 583–586.

29 Lu, F.C. and Ralph, J. (1999). Detection and determination of *p*-coumaroylated units in lignins. *J. Agr. Food. Chem.* 47: 1988–1992.

30 Lu, F.C. and Ralph, J. (2002). Preliminary evidence for sinapyl acetate as a lignin monomer in kenaf. *Chem. Commun.* 90–91.

31 Lu, F., Ralph, J., Morreel, K. et al. (2004). Preparation and relevance of a cross-coupling product between sinapyl alcohol and sinapyl p-hydroxybenzoate. *Org. Biomol. Chem.* 2: 2888–2890.

32 Morreel, K., Ralph, J., Kim, H. et al. (2004). Profiling of oligolignols reveals monolignol coupling conditions in lignifying poplar xylem. *Plant. Physiol.* 136: 3537–3549.

33 Del Rio, J.C., Marques, G., Rencoret, J. et al. (2007). Occurrence of naturally acetylated lignin units. *J. Agr. Food. Chem.* 55: 5461–5468.

34 Lu, F.C. and Ralph, J. (2008). Novel tetrahydrofuran structures derived from β-β-coupling reactions involving sinapyl acetate in Kenaf lignins. *Org. Biomol. Chem.* 6: 3681–3694.

35 Withers, S., Lu, F.C., Kim, H. et al. (2012). Identification of grass-specific enzyme that acylates monolignols with *p*-coumarate. *J. Biol. Chem.* 287: 8347–8355.

36 Ralph, J. (1996). An unusual lignin from kenaf. *J. Nat. Prod.* 59: 341–342.

37 Sibout, R. and Hofte, H. (2012). Plant cell biology: The ABC of monolignol transport. *Curr. Biol.* 22: R533–R535.

38 Miao, Y.C. and Liu, C.J. (2010). ATP-binding cassette-like transporters are involved in the transport of lignin precursors across plasma and vacuolar membranes. *Proc. Natl. Acad. Sci. U.S.A.* 107: 22728–22733.

39 Whetten, R. and Sederoff, R. (1995). Lignin Biosynthesis. *Plant Cell* 7: 1001–1013.

40 Steeves, V., Forster, H., Pommer, U., and Savidge, R. (2001). Coniferyl alcohol metabolism in conifers-I. Glucosidic turnover of cinnamyl aldehydes by UDPG: coniferyl alcohol glucosyltransferase from pine cambium. *Phytochemistry* 57: 1085–1093.

41 Dharmawardhana, D.P., Ellis, B.E., and Carlson, J.E. (1995). A β-glucosidase from lodgepole pine xylem specific for the lignin precursor coniferin. *Plant. Physiol.* 107: 331–339.

42 Dean, J.F.D. and Eriksson, K.E.L. (1994). Laccase and the deposition of lignin in vascular plants. *Holzforschung* 48: 21–33.

43 Omalley, D.M., Whetten, R., Bao, W.L. et al. (1993). The role of laccase in lignification. *Plant J.* 4: 751–757.

44 Dean, J.F.D., Sterjiades, R., and Eriksson, K.-E.L. (1994). Purification and characterization of an anionic peroxidase from sycamore maple (Acer pseudoplatanus) cell suspension cultures. *Physiol. Plant.* 92: 233–240.

45 Berthet, S., Demont-Caulet, N., Pollet, B. et al. (2011). Disruption of LACCASE4 and 17 results in tissue-specific alterations to lignification of *Arabidopsis thaliana* stems. *Plant Cell* 23: 1124–1137.

46 Zoia, L., Perazzini, R., Crestini, C., and Argyropoulos, D.S. (2011). Understanding the radical mechanism of lipoxygenases using ^{31}P NMR spin trapping. *Bioorg. Med. Chem.* 19: 3022–3028.

47 Zoia, L. and Argyropoulos, D.S. (2009). Phenoxy radical detection using ^{31}P NMR spin trapping. *J. Phys. Org. Chem.* 22: 1070–1077.

48 Ralph, J., Lundquist, K., Brunow, G. et al. (2004). Lignins: Natural polymers from oxidative coupling of 4-hydroxyphenylpropanoids. *Phytochem. Rev.* 3: 29–60.

49 Nimz, H. (1974). Beech lignin-Proposal of a constitutional scheme. *Angew. Chem. Int. Ed.* 13: 313–321.

50 Boerjan, W., Ralph, J., and Baucher, M. (2003). Lignin biosynthesis. *Annu. Rev. Plant Biol.* 54: 519–546.

51 Mansfield, S.D., Kim, H., Lu, F., and Ralph, J. (2012). Whole plant cell wall characterization using solution-state 2D NMR. *Nat. Protoc.* 7: 1579–1589.

52 Neish, A.C. (1968). Monomeric intermediates in the biosynthesis of lignin. In: *Constitution and Biosynthesis of Lignin, Molecular Biology, Biochemistry and Biophysics*, vol. 2 (ed. K. Freudenberg and A.C. Neish), 1–43. Springer-Verlag: Berlin.

53 Adler, E. (1977). Lignin chemistry-past, present and future. *Wood Sci. Technol.* 11: 169–218.

54 Sakakibara, A. (1980). A structural model of softwood lignin. *Wood Sci. Technol.* 14: 89–100.

55 Karhunen, P., Rummakko, P., Sipila, J. et al. (1995). Dibenzodioxocins; a novel type of linkage in softwood lignins. *Tetrahedron Lett.* 36: 169–170.

56 Karhunen, P., Rummakko, P., Sipila, J. et al. (1995). The formation of dibenzodioxocin structures by oxidative coupling. a model reaction for lignin biosynthesis. *Tetrahedron Lett* 36: 4501–4504.

57 Karhunen, P., Mikkola, J., Pajunen, A., and Brunow, G. (1999). The behavior of dibenzodioxocin structures in lignin during alkaline pulping processes. *Nord. Pulp Pap. Res. J.* 14: 123–128.

58 Argyropoulos, D.S., Jurasek, L., Kristofova, L. et al. (2002). Abundance and reactivity of dibenzodioxocins in softwood lignin. *J. Agr. Food. Chem.* 50: 658–666.

59 Crestini, C., Melone, F., Sette, M., and Saladino, R. (2011). Milled wood lignin: A linear oligomer. *Biomacromolecules* 12: 3928–3935.

60 Brunow, G. (2001). Methods to reveal the structure of lignin. In: *Biopolymers: Lignin, Humic Substances and Coal*, vol. 1 (ed. M. Hofrichter and A. Steinbuchel), 89–116. John Wiley and Sons, Ltd: Hoboken.

61 Simionescu, C.I. and Anton, I. (1970). A study of some representative types of reed lignin. *Cellul. Chem. Technol.* 4: 589–611.

62 del Rio, J.C., Rencoret, J., Marques, G. et al. (2008). Highly acylated (acetylated and/or *p*-coumaroylated) native lignins from diverse herbaceous plants. *J. Agric. Food Chem.* 56: 9525–9534.

63 Yuan, T.-Q., Sun, S.-N., Xu, F., and Sun, R.-C. (2011). Characterization of lignin structures and lignin-carbohydrate complex (LCC) linkages by quantitative ^{13}C and 2D HSQC NMR spectroscopy. *J. Agric. Food Chem.* 59: 10604–10614.

64 Cornu, A., Besle, J.M., Mosoni, P., and Grenet, E. (1994). Lignin-carbohydrate complexes in forages: Structure and consequences in the ruminal degradation of cell-wall carbohydrates. *Reprod. Nutr. Dev.* 34: 385–398.

65 Bjorkman, A. (1954). Isolation of lignin from finely divided wood with neutral solvents. *Nature* 174: 1057–1058.

66 Koshijima, T., Watanabe, T., and Azuma, J. (1737-1740). Existence of benzylated carbohydrate moiety in lignin-carbohydrate complex from pinewood. *Chem. Lett.* 1984.

67 Azuma, J., Takahashi, N., and Koshijima, T. (1981). Isolation and characterization of lignin-carbohydrate complexes from the milled-wood lignin fraction of pinus-densiflora sieb et zucc. *Carbohydr. Res.* 93: 91–104.

68 Lawoko, M., Henriksson, G., and Gellerstedt, G. (2006). Characterisation of lignin-carbohydrate complexes (LCCs) of spruce wood (Picea abies L.) isolated with two methods. *Holzforschung* 60: 156–161.

69 Lawoko, M., Henriksson, G., and Gellerstedt, G. (2006). Characterization of lignin-carbohydrate complexes from spruce sulfite pulp. *Holzforschung* 60: 162–165.

70 Kondo, R., Sako, T., Iimori, T., and Imamura, H. (1990). Formation of glycosidic lignin-carbohydrate complex in the enzymic dehydrogenative polymerization of coniferyl alcohol. *Mokuzai Gakkaishi* 36: 332–338.

71 Jiang, T. (2009). *Muzhisu*. Beijing: Chemical Industry Press.

72 Watanabe, T. and Koshijima, T. (1988). Evidence for an ester linkage between lignin and glucuronic-acid in lignin carbohydrate complexes by DDQ-oxidation. *Agr. Biol. Chem.* 52: 2953–2955.

73 Xie, Y., Yasuda, S., Wu, H., and Liu, H. (2000). Analysis of the structure of lignin-carbohydrate complexes by the specific ^{13}C tracer method. *J. Wood Sci.* 46: 130–136.

74 Rinaldi, R. (2014). Plant biomass fractionation meets catalysis. *Angew. Chem. Int. Ed.* 53: 8559–8560.

75 Melero, J.A., Iglesias, J., and Garcia, A. (2012). Biomass as renewable feedstock in standard refinery units. Feasibility, opportunities and challenges. *Energy Environ. Sci* 5: 7393–7420.

3

Lignin Isolation, Physicochemistry Properties, and Chemical Properties

3.1 Lignin Polymer Physical Properties

3.1.1 Lignin General Physical Properties

The physical properties of lignin are not only related to plant species, structure, location, growth period, and other factors, but also related to the method of separation and extraction from the lignocellulose. The main part of the original lignin in plants is closely connected with sugars and forms a huge spatial network structure. This lignin-sugar complex is insoluble in all solvents. Therefore, the so-called physical properties of lignin are usually the physical properties of the low molecular weight lignin fragments extracted/separated by some methods rather than the physical properties of the original lignin.

Relative Density and Caloricity Given the fact that the relative density of different woods produced by different preparation methods varies, the relative density of solid lignin is between 1.35 and 1.50 g/cm^3. Given its energy content, lignin is also be used as fuel. The combustion heat value of lignin is relatively high, which is one reason for the recovery of papermaking black liquor for the solid/liquid fuel. For example, the heat of combustion of spruce hydrochloric acid is 110 kJ/g.

Lignin Color Original lignin is a white or nearly colorless substance. However, lignin has carbonyl, vinyl, and other unsaturated double bonds connecting with the benzene ring to form a conjugate system. The π-electrons in this large conjugate system are quite active with low excitation energy. Lignin can easily absorb the longer light waves, making the absorption spectrum of lignin move from the ultraviolet region to the visible light region to display colors. This conjugate system is a chromophore group. Additionally, the presence of auxochrome groups such as -OH (hydroxyl), -NR (amino), -OR (ether bond), -COOH (carboxyl), and -X (halogen) can deepen the color of lignin. The color of the outer skin of the trees is darker because the lignin is more susceptible to discoloration by sunlight. Closer to the inside of the tree core, less light is irradiated and therefore the color of the tree core is extremely light. In addition, the phenolic hydroxyl group in the structural unit of lignin is easily oxidized by oxygen in the air and becomes a quinone-colored

Lignin Conversion Catalysis: Transformation to Aromatic Chemicals, First Edition.
Chaofeng Zhang and Feng Wang.

substance, thereby discoloring the lignin. The lignin that we usually see is colored, which is caused by the separation and preparation process. With different separation and preparation methods, lignin shows different shades of color. The quality of the separated Brauns spruce lignin is changed to cream color whereaas the colors of acid lignin, copper-ammonia lignin, and periodate lignin are darker in light yellow from brown to dark brown. In some situations, it needs to turn lignin into white or a light color, as when the application of lignin as a material is expanded and when using lignin in photocatalytic depolymerization. Lignin bleaching with H_2O_2, hypochlorite or lignin methylation with diazomethane can also make the color of lignin lighter [1].

Optical Properties Although the lignin contains chiral carbons in its skeleton, the lignin has unobvious optical activity. Besides, as one kind of fiber material, the refractive index of the lignin is about 1.5~1.7, such as the value of the copper-ammonia spruce lignin is 1.61. Lignin can be modified and used as an optical material. Given the distribution anisotropy of lignin because of the wood structure with porous channels in the vertical direction, not the horizontal direction, Hu et. al [2]. introduced the polymer material with a similar refractive index into the wood channels and provide a transparent wood.

Lignin Dissolvability Lignin is a type of aggregate with many polar groups in its structure, producing strong intra- and intermolecular hydrogen bonds and making native lignin insoluble in water and any solvents. Because of the structure condensation or degradation in isolated lignin, many physical properties and solubility properties of treated lignin are changed. Both soluble and insoluble forms of lignin exist. In general, the former is an amorphous structure and the morphological structure of the latter is the raw material fiber. Whether the separated lignin is soluble in a solvent is affected by the solubility parameters of the solvent, the hydrogen bond formation energy, and the functional groups of the lignin. The presence of phenolic hydroxyl and carboxylic acid groups can make the lignin dissolve in a strong alkali solution. In addition, some lignin dissolution in the organic solvent requires a certain amount of water. Brauns lignin and organic solvent lignins are soluble in dioxane, pyridine, methanol, ethanol, acetone, but, interestingly, the process requires the addition of a few drops of water to these solvents or it is almost insoluble. Therefore, ethanol is usually formulated into a 50% aqueous solution for use in lignin extraction. After dissolution in dioxane, alkali lignin and sulfur lignin are like colloidal solutions. Alkali lignin can be dissolved in dilute alkali water or acidic or neutral polarity solvents. Lignosulfonates are soluble in water in true colloidal solutions. Brauns lignin, phenol lignin, and many organic solvent lignins are clear after dissolving in dioxane, like a true solution. Acid lignin does not dissolve well in any solvents. The principle of the chemical pulping method is using alkali to hydrolyze the primary lignin to produce a large number of phenolic hydroxyl sodium salts or introducing sulfonic acid groups into the lignin macromolecules to form soluble lignin fragments, achieving the separation of lignin and cellulose.

Electrochemical Properties Given the existence of phenolic groups and carboxyl groups in the structure of lignin, lignin can be regarded as an anionic electrolyte with some electrical properties. Lignin can move toward the anode in an electric field. Using this feature, electrophoresis with glass fiber pile filter paper can be used to study the connection or disconnection between lignin and carbohydrates and to separate lignin and carbohydrates. Electrophoresis and electrodialysis can also be used to separate lignin and lignosulfonate from black liquor in the papermaking industry.

3.1.2 Lignin Polymer Physical Properties

Weight-average Molar Mass (M_W) and Number-average Molar Mass (M_n) For the determination of lignin structures, the most commonly used parameters are the weight-average molar mass (M_W) and number-average molar mass (M_n). The M_W of isolated lignin has been determined by usual techniques. The main problem in measuring the solution properties of isolated lignin is solubility in organic solvents. The M_W of the soluble portion of lignin is smaller than that of the truly isolated sample. The different techniques suggest that the M_W of isolated lignin is in the approximate range of $10^3$to 10^5, depending on the plant species, the processing method, and the method of measuring. The average M_W for softwood is greater than 2×10^4 and less for hardwoods. For lignosulfonates, the measured M_W is as high as 10^6.

Dispersity Index (Đ) The polydispersity index (Đ) is a measure of the width of molecular weight distributions (MWD) and is expressed as $Đ=M_W/M_n$ [3]. A considerable amount of experimental data supports the idea that soluble lignins and their derivatives show remarkable polydispersity. The Đ may vary from 2.5–3.0 to over 10–11 for different lignins [4], indicating that lignins are extremely polydisperse materials.

Determination of Molecular Weight For determining the lignin molecular weight, the values of number-average molecular weight (M_n) and average molecular weight (M_W) are not universally accepted at present because different analytical techniques yield discordant results. As previously mentioned, the major problem in measuring the physical properties of isolated lignin is its solubility in organic solvents. It is thought that unavoidable uncertainty results from the fact that the molecular mass of the soluble portion of lignin is comparably smaller than that of a truly isolated sample. This point becomes more complex because the solution properties of lignin are thought to depend not only on the plant species but also on the isolation process [5]. Whereas M_n is mainly determined by the measurement of colligative properties (e.g., using cryoscopy, ebulliometry, and membrane and vapor-pressure osmometry), the light scattering, ultra-high-speed centrifugation, and sedimentation balance methods can be used to determine the M_W.

However, the most commonly used method to obtain the molecular weight of lignin is by gel-permeation chromatography (GPC), also termed gel-filtration chromatography (GFC) or size-exclusion chromatography (SEC). GPC fractionates macromolecules according to molecular size [6]. The stationary phase used involves cross-linked polydextran gels having varying pore sizes. Such gels in water form a 3D network that acts as a molecular sieve. When an aqueous solution of macromolecules is allowed to move through a column containing the gel, the molecules of low molecular size penetrate the gel particle pores but large molecules are excluded and pass directly through the column to be eluted first. Detection is usually accomplished using UV. The application of this method requires that the lignin sample be completely soluble in THF; consequently, the vast majority of lignin preparations should be undertaken only after appropriate derivatization (acetylation or methylation). As technology has improved, the introduction of cross-linked, semirigid polyethylene (PE) gels of uniform particle size has made it possible to perform GPC in high-pressure systems using small columns. With such systems, the experimental time can be substantially reduced and the resolution power enhanced. A variety of solvents can be used. This technique is termed high-performance size-exclusion chromatography (HPSEC) [7]. For an insoluble lignin sample, the M_w can be calculated using the thermal softening method, which is based on the linear relation between log M_w and the thermal softening temperature (Ts).

Thermal Stability of Lignin Polymer The glass transition temperature (Tg) is one of the most widely used thermal properties for describing amorphous polymers. Tg curves for most lignins have indicated that mass decreases slightly at temperatures between 0 and 150 °C. This loss can be attributed to the vaporization of residual water in the samples undergoing testing. Because lignin has an aromatic structure and there are many intra- and intermolecular hydrogen bonds, most lignin samples are stable to heating. As an example of the evaluation of thermal stability, it has been found that the thermal decomposition of samples starts at approximately 200 °C. The extrapolated decomposition temperatures for Kraft lignin, alkaline lignin, and lignins from a sawdust sample, have been observed at approximately 321 °C, 294 °C, 324 °C, and 331 °C, respectively. By contrast, Tg of wood powder was observed at approximately 370 °C and 390 °C, which could be attributed to cellulose at 370 °C and lignin at 390 °C. When the temperature is increased to 760 °C, the residue of various lignin is approximately 20 to 30%.

3.2 Lignin Isolation from Lignocellulose and Technical Lignins

As discussed in Chapter 2, lignocellulose is a complex 3D polymer consisting of cellulose, hemicellulose, and lignin via complex linkages. For the utilization of lignocellulose or lignin with abundant aromatic structures, lignin isolation is a nontrivial initial step that determines the scale-up and optimization of many of the industrial processes concerning lignocellulose. The lignocellulose pretreatment is an essential technique to provide a robust and reproducible process because it can overcome some physical and chemical barriers that make lignin resistant to the following conversion. In 2009, Costa Sousa and co-workers [8] divided various pretreatment technologies into four categories: (1) physical pretreatment (e.g., ball milling); (2) solvent fractionation (including the organosolv process, together with phosphoric acid fractionation and the use of ionic liquids); (3) chemical pretreatment (acidic, alkaline, and oxidative); and (4) biological treatment (using mainly fungi or enzyme-like cellulase). Although many methods are currently available for the isolation and purification of lignin, none of them provides a 100% yield and structural authenticity. One reason for the lack of perfect isolation is that aggressive treatment is often performed during the process, which can cause the lignin linkages or fragments to occur condensation or repolymerization with stable bonds generation [9]. All obtained lignin from these preparations is distinct from so-called protolignin, which is a term used for the material in the plant tissue.

In addition, in a general perspective, lignin isolation methods can be classified into two main groups according to the scale of the process: (1) procedures for laboratory purposes, targeting lignin isolation for academic study; and (2) procedures for commercial application with a larger scale, during which lignin is usually produced as by-products such as in the pulping and bioethanol production processes. The laboratory-scale procedures can be further classified into two strategies: (1) lignin isolation from lignocellulose by first lignin dissolution (LBD); and (2) lignin isolation from lignocellulose with lignin as the residue (LAR).

Near the beginning of Chapter 3 (Section 3.2), the two above strategies for lignin isolation from lignocellulose are firstly summarized (Table 3.1) and discussed with typical examples. Next, the preparations for production and the structure of the typical technical/commercial lignins will be introduced.

3.2.1 Remove Lignin First by Dissolution

Brauns Native Lignin (BNL) In 1939, Brauns reported a method to prepare lignin in a mild condition. Water and ethanol were respectively used to extract approximately 100 to 200 mesh wood powders, after which complete extraction was performed using 95% ethanol and then the concentrated solution after extraction was poured into water. The crude lignin was then precipitated, after which is is dissolved in 1,4-dioxane that was further precipitated with ether, filtered, washed, and dried to obtain the refined lignin. This procedure has the drawback of a very low yield (2–10% of the lignin present in wood) and contamination with 2~3% xylan and some polyphenol impurities. Because this method is carried out at a low temperature with no use of acid, organic solvent molecules do not react with the lignin. Therefore, the extraction method will not cause structural changes of lignin and the product is called natural lignin. However, this kind of natural lignin is just the low molecular weight fraction of lignin, which is still very different from lignin in wood. To obtain a more refined natural lignin, use acetone-water (9:1) mixed solvent instead of ethanol as the extraction reagent and then use cold ethyl acetate to remove sugars and polyphenols and other hybrids.

Milled-Wood Lignin (MWL) In 1954, Björkman introduced the so-called MWL method to isolate the lignin [10], which uses a preliminary, extensive grinding, with the subsequent solvent extraction. It is the most commonly used method and involves thorough milling of the plant material, followed by extraction with dioxane/water. Briefly, 40 to 60 meshes wood powders are firstly pre-extracted by using ether or other organic solvents to remove rosin, fat, wax, and other substances at room temperature. After drying to remove the solvent, the milling is carried out either in a nonswelling medium such as toluene or in a dry state for 48 h or longer.

Table 3.1 Typical Method Used for the Lignin Isolation.

Procedures	Reagent/Solvent	Lignin Name	Lignin Yield (%)	Classification
Mechanical disintegration and cellulolytic enzymes	Aqueous-dioxane	Brauns	<10%	LBD
	Dioxane/H_2O	Björkman (MWL)	20~30	LBD
	Dioxane/H_2O	Pew (CEL)	80	LBD
	Ionic liquids	—	<80	LBD
	Enzymes	EMAL	60	LBD
Mild acid hydrolysis	Alcohol/dioxane, phenol, acetic acid with acid catalysts, alcohols	Organosolv	25~50	LBD
Steam hydrolysis	Steam/acetic acid	SEL	50~90	LBD
Alkaline hydrolysis	NaOH/EtOH or H_2O	Alkali (soda)	> 80	LBD
	NaOH/Na_2S	Kraft	90~95	LBD
	NaOH/O_2 or H_2O_2		>80	LBD
Sulfonation	HSO_3^-/SO_3^{2-} and base	Sulfite	>80	LBD
Biological treatment	EtOH or dioxane/H_2O	Rot fungi	20	LBD
Acid hydrolysis of polysaccharides	H_2SO_4	Klason	>80	LAR
	HCl	Willstäter	>80	LAR
	HCl/H_3PO_4	Urban	>80	LAR
	HCl/H_2SO_4	Halse	>80	LAR
	HF	Fredenhagen	>80	LAR
	HBr/H_2SO_4	Runkel	>80	LAR
	CF_3COOH	Fengel	>80	LAR
Polysaccharides oxidation	$Na_3H_2IO_6$	Purves	>80	LAR
	$Cu(NH_3)_4(OH)_2$	Freudenberg/ Cuoxam	>80	LAR

Note: LBD means removing lignin (from lignocellulose) by dissolution, LAR means removing lignin as residue.

As shown in Figure 3.1, ball milling can cause the structure modification or fragmentation reaction of the lignin structure [11]. After the extraction with the 1,4-dioxane solvent containing 10% water and removing the solvent, the crude MWL is obtained, which can account for the half amount of lignin in the wood.

Hardwood MWLs contain more carbohydrates than softwood MWLs. Many purification procedures have been used to reduce contaminants in isolated lignins. The first, developed by Björkman in 1954 for MWL includes two precipitations [12]: (1) Crude MWL dissolved in 90% acetic acid was precipitated into the water; and (2) The precipitation product was then dissolved in 1, 2-dichloroethane/EtOH (2:1) and

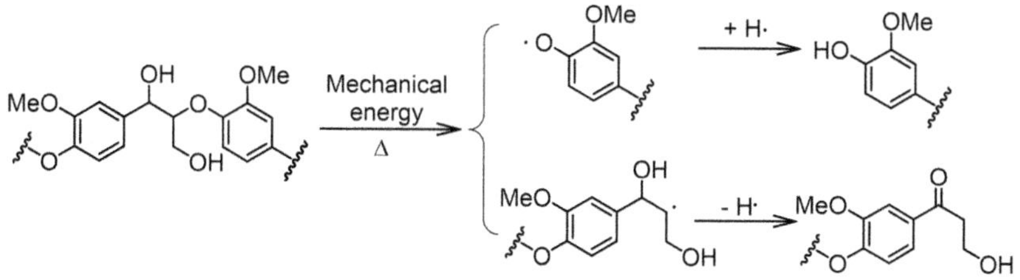

Figure 3.1 Lignin Fragmentation Reaction Taking Place during the Milling of Wood.

precipitated into ethyl ether. The final refined lignin is made of tan non-crystal powder with approximately 2 to 8 saccharides. To further reduce the saccharides, the crude MWL can be dissolved in the mixed solvent with a ratio of pyridine: acetic acid: H_2O (9:1:4) followed by the addition of chloroform and shaking to make the lignin enter the organic phase. After removing the water phase and removing the solvent from the organic phase, the MWL with low sugar content is obtained. When the ratio between chloroform and pyridine ratio is 6:1, the purified MWL with 0.05% sugar content can be obtained [13].

Dioxane (Acidolysis) Lignins (DL) In 1950, Stumpf and Freudenberg [14] showed that lignin could be extracted with dioxane/HCl (9:1) or with tetrahydrofuran and HCl from wood swollen by water. Briefly, after the pre-extraction with benzene-ethanol, the defatted wood powders are added to the 8 times amount of dioxane-water (9:1) solvent, and hydrochloric acid is then added to make the concentration of HCl reach 0.2 mol/L. After refluxing at 90~95 °C for 30 min to 48 h according to the requirement, the extract is concentrated and poured into water to precipitate the lignin. After the washing and drying, dioxane lignin is obtained with a sugar content of 1.6~7.5%. In addition, the defatted wood powders can be added into a mixed solvent of dioxane: water (1:1), reflux at 175 °C for 2 h. After the concentration, precipitation, washing, and drying treatment, a partially hydrolyzed, minimally condensed dioxane lignin is obtained. Spruce wood can dissolve 40% lignin and birch wood can dissolve 60% lignin [15].

Organosolv Lignin (OSL) The principle of the organic solvent method for lignin isolation requires selection of a suitable organic solvent to dissolve the lignin in the plant raw materials under a certain temperature and pressure leaving a remainder of cellulose that is further filtered, dried, and washed to obtain pulp [16]. The filtrate is concentrated by evaporation and finally provides the organosolv lignin. Organic solvents used for lignin extraction include alcohols (methanol, ethanol, n-butanol, 1,4-butanediol), acids (formic acid, acetic acid), ether, acetone, ethyl acetate, phenol, dimethyl sulfoxide, and others. Eight procedures for the commercial production of organosolv lignins have been the subject of major research efforts (Table 3.2), which summarize the main methods for lignin isolation with organic solvents.

One of the popular procedures is the Alcell© method [17]. Lignin is extracted with a mixture of water and a low-carbon alcohol solvent, such as EtOH. In this case, biomass is treated with a 1:1 mixture of water and EtOH at 175–195 °C for 1~2 h and

Table 3.2 Different Commercial Organosolv Lignins.

Lignin Name	Solvent for Isolation
Alcell©	$EtOH/H_2O$
acetocell	Acetic acid/H_2O
acetosolv	Acetic acid/HCl
ASAM	Alkaline sulfite/anthraquinone/MeOH
Battelle-Geneva phenol	Phenol/acid/H_2O
Formacell	Acetic acid/formic acid/H_2O
Milox	formic acid/H_2O_2
Organocell	MeOH pulping, followed by MeOH, MeOH, and anthraquinone pulping

a liquid/solid biomass ratio of about 4–7. Then, the filtrate is squeezed out and the mixed solvent is evaporated to obtain lignin.

The acetocell [18] and acetosolv [19] procedures are quite similar and both use a mixture of acetic acid/water to isolate lignin. In the case of the acetosolv method, a mineral acid is added as a catalyst, and thus 1.5 L of 95% aqueous acetic acid, containing 0.1 or 0.16% HCl, are required for 300 g of wood chips. For optimal results, the mixture must be refluxed for 3–5 h. Afterward, the excess acetic acid is decanted and the residue is extracted with hot (105 °C) 95% acetic acid in a column with water. Acetic and HCl can be recovered by evaporation of the extract.

ASAM pulping [20] is a method to produce lignin from wood chips. Starting wood chips are pre-steamed for 15 min and then sodium hydroxide, sodium sulfite, MeOH, and anthraquinone catalyst are added. The mixture is cooked for 90–180 min at 170–180 °C.

The Battelle-Geneva process [21] uses a mixture of phenol and water for the treatment of biomass using phenol as approximately 40% of the total amount of solvent. The pulping is carried out at temperatures near 100 °C and atmospheric pressure. The phase separation of the homogeneous mixture of phenol/water, at the pulping reaction condition, takes place by cooling down the spent liquor. In some cases, acids or alkalis can be added to water to facilitate the degradation of the lignin.

The Formacell pulping process [22] employs a mixture of water, formic acid, and acetic acid. The presence of formic acid in the pulping liquor provides an easier cleavage of lignin α-ether linkages and so lignin fragments are dissolved at lower temperatures and generally at atmospheric pressure.

The Milox procedure uses formic acid and H_2O_2 to produce in situ performic acid. The method is performed using three pathways with H_2O_2 added in the first and last ones. The temperature for the first and last stages is about 80 °C, whereas the temperature is increased to 100 °C for the middle pathway. The method can be applied to softwood, hardwood, and nonwoody plants and the lignin produced by this procedure has a lower M_W than others.

The Organocell process [23, 24] is a procedure based on the treatment of biomass with an alkali and an organic solvent. In the first stage, the wood chips are initially impregnated with a mixture of alcohol (MeOH or EtOH) and water at a temperature of approximately 120 °C. The wood chips are softened and then transported into the digester and treated with alkali and catalytic amounts of anthraquinone. The pulping liquor is heated to 165–170 °C, under pressure of approximately 13 bar for a specific period. The amount of alkali used in this method is related to the final degree of delignification of the wood chips.

Other Methods In addition to the above procedures for the commercial production of organosolv lignins, the first lignin dissolution can be also achieved via other methods such as the high-boiling-point solvent (HBS) method using glycols, 1,3-butanediol, or 1,4-butanediol as the solvent to extract the lignin from wood at approximately 190 to 220 °C. In addition, supercritical fluid, ionic liquids (IL), and deep eutectic solvent (DES) are widely used in lignin (in situ) isolation.

Complete Lignocellulose Solubilization In 2003, Lu and Ralph [25] suggested a method for the complete solubilization of milled wood by acetylation and dissolving the residue in *N*-methylimidazole/DMSO. The preparation obtained was called acetylated cell wall (Ac-CW). This allowed the characterization of wood components by NMR.

3.2.2 Remove Cellulose and Hemicellulose First Leaving Lignin Residue

3.2.2.1 Promoting Carbohydrate Hydrolysis with Mineral Acid

Klason Lignin (Sulfuric Acid) The most widely applied variation of the Klason (sulfuric acid) method employs 72% of H_2SO_4 to promote carbohydrate hydrolysis. Next, it is diluted with water an heated at high temperature (100–125 °C) to promote the second hydrolysis. Finally, the lignin is filtered, washed, and dried to get the Klason lignin [26].

Willstätter Lignin (Fuming Hydrochloric Acid) In 1913, Willstätter and Zechmeister [27] recorded that fuming HCl (42–43% at 15 °C) rather than H_2SO_4 will completely hydrolyze cellulose in the wood meal at low temperature, yielding an insoluble lignin residue.

Fredenhagen Lignin (Hydrogen Fluoride Acid) Furthermore, the hydrogen fluoride can also achieve the almost quantitative and rapid (30 min, rt) solubilization of the carbohydrates from the lignocellulosic matrix of cellulose [28]. In addition, the insoluble lignin was analyzed indicating that aryl ether linkages are not cleaved.

Lignin Isolation with the Mixed Mineral Acid Urban lignin is obtained by treating the wood powders with a mixture consisting of HCl and phosphoric acid (3:1, v/v). The residue that does not dissolve after heating wood meal with HCl and H_2SO_4 is called Halse lignin. The residue that does not dissolve after heating with

HBr and H_2SO_4 is named Runkel lignin. The residue that does not dissolve after heating with trifluoroacetic acid is named Fengel lignin [29].

3.2.2.2 Promoting Carbohydrate Hydrolysis or Decomposition with Metal Ion and Oxidant

Freudenberg Lignin (Cuoxam or Cuproxam Lignin) In this method, lignin is isolated by alternately hydrolyzing wood with boiling 1% H_2SO_4 and cuprammonium hydroxide solution (Schweizer's reagent, tetraamine copper(II)) [30]. All of the carbohydrate components in an extracted wood meal may be dissolved by complexation, giving high yields of the so-called Freudenberg, cuoxam, or cuproxam lignins. This lignin is totally insoluble in organic solvents and is known to retain the morphological features of wood [31]. with cuoxam lignin being the most suitable for model delignification experiments.

Purves Lignin (Periodate Lignin) Mild oxidation of extracted wood meal with periodic acid (HIO_3) can convert monosaccharide units in polysaccharide to dialdehydes (Figure 3.2), which are susceptible to hydrolysis by boiling water [32]. There is less alteration in lignin structure. Degraded carbohydrates are dissolved, finally giving an insoluble periodate or Purves lignin. This method avoids the condensation processes that cause changes in the acid lignins, but oxidative modification occurs instead, notably in the units with free phenolic hydroxyl groups, which are converted into o-quinones and then into muconic acid structures [33].

3.2.2.3 Promoting Carbohydrate Hydrolysis or Decomposition with Bio-Enzyme

Cellulolytic Enzyme Lignins (CEL) In addition to the acid-catalyzed hydrolysis of the polysaccharides and the oxidization decomposition of the monosaccharide structure, treatment of the finely ground wood meal with cellulolytic enzymes prior to solvent extraction removes part of the polysaccharides and increases the yield of lignin [34]. However, the crude extract still contains some carbohydrates after extraction with dioxane/water. To overcome this problem, one useful method is to reduce the connection between and hydrocarbons and lignin. Chen et al. [35] developed the swelled enzyme lignin (SEL) method. Swelling loosens and reduces the associations between lignin and hemicelluloses and therefore not only shortens the digestion time but also leads to a more intensive enzymatic degradation of the polysaccharides. Jung et al. [36] developed a lignin isolation method from wheat straw by ball milling (8 days) and enzyme treatment (4 days). Extraction with 50% dioxane/water produced lignin with higher yield compared with lignin extracted by standard methods (96% dioxane). Imai et al. reported a boost of 200% in the reaction

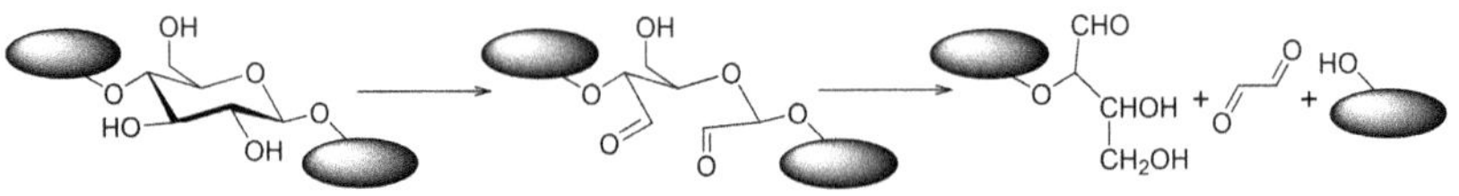

Figure 3.2 Degradation of Monosaccharide to Dialdehydes in the Periodate Lignin Preparation.

rate of the enzymatic hydrolysis after previous ultrasonic pretreatment of the sample [37]. The CEL is an efficient method to isolate the lignin amd Hu et al. [38] were able to obtain CEL preparations with yields of up to 86%. In addition, CEL is probably more representative of the total lignin in wood than is MWL, with higher yields and fewer chemical changes, but its preparation procedure is more tedious [36].

Enzymatic Mild Acidolysis Lignin (EMAL) To further increase the efficiency of carbohydrate hydrolysis, Wu et al. [39] proposed a procedure using the combination of enzymatic and mild acidolysis of lignin. Guerra et al. [40, 41] and reported that the yield of EMAL after the sufficient ball-milling treatment was about twofold greater than the CEL method isolated from the same batch of milled wood.

Brown-Rot Lignin (Schubert-Nord, Enzymatically Released Lignin) To degrade the carbohydrate material, the addition of fungi is a biological solution [42, 43] beside the use of the cellulolytic enzymes. Advantages of biological pretreatment (with white-, brown-, and soft-rot fungi) are a low-energy requirement and mild operation conditions. Nevertheless, the rate of biological hydrolysis is very low and requires long pretreatment times.

3.2.3 Technical Lignins

Every year, the paper industry can produce more than 50 million metric tons of lignin as a by-product, which is currently primarily discharged as black liquor or used as a fuel for the same factory. Depending on the pulping procedure, these commercial lignins can be divided into two major categories. The first category comprises sulfur-containing lignins, including the Kraft lignin and lignosulfonates. The second category comprises non-sulfur lignins, which include soda lignins, organosolv lignins, steam-explosion lignins, steam-explosion hydrolysis lignins, oxygen delignification lignins, and lignins prepared by other methods. As they are important substrates for the lignin depolymerization research, a short introduction about the most common commercial lignins will be addressed in the following paragraphs, including the preparation process and structural characterizations.

3.2.3.1 Kraft Lignin

Most of the paper pulp produced worldwide is generated by the Kraft process [44, 45], which converts wood into wood pulp when the wood is treated with Na_2S/NaOH solution (white liquor) at a temperature in the range of 155–175 °C for several hours. The method can dissolve about 90–95% of all lignin present in the starting material. Thus, cellulose is separated as a solid from a rich lignin fluid fraction called black liquor. Lignin remains in this solution as a fragmented polymer.

Kraft lignin (thiolignin, sulfate lignin) is isolated from black liquor in the so-called delignification process, involving precipitation and neutralization with an acid solution (pH=1-2) based on the low solubility of Kraft lignin in acidic aqueous media. Several methods have been carried out to obtain acid media to precipitate Kraft lignin from black liquors (CO_2, acetic acid, H_3PO_4, H_2SO_4, and HCl). The precipitation step is primarily influenced by pH, ion strength, and temperature. Additionally, there is an association between lower pH and higher lignin recovery.

Furthermore, the composition of the precipitated solid is almost independent of the conditions used in the precipitation and subsequent separation stage. This procedure yields a polymer with a lower M_w, but which is highly modified from the native lignin.

During digestion, besides the cleavage of most of the lignin–carbohydrate linkages, depolymerization of the lignin occurs as well as reactions with HO^- and HS^- ions and new polymerization reactions of the fragments (Figure 3.3) [46–50]. Delignification is controlled by two main reactions: degradation and condensation. The prevalent degradation reactions are the cleavage of the α- and β-aryl ether bonds [49]. Condensation reactions proceed via Michael addition of external (sulfide and hydroxy anions) and internal (carbanions from phenolic structures) ions to an acceptor (quinone methides or formaldehyde) (more discussion about the base-catalyzed condensation is provided in Chapter 11.1.2).

3.2.3.2 Sulfite Lignin

The sulfite process is a pulping method [51] normally carried out at 140–170 °C with an aqueous solution of a sulfite or bisulfite salt of sodium, ammonium, magnesium,

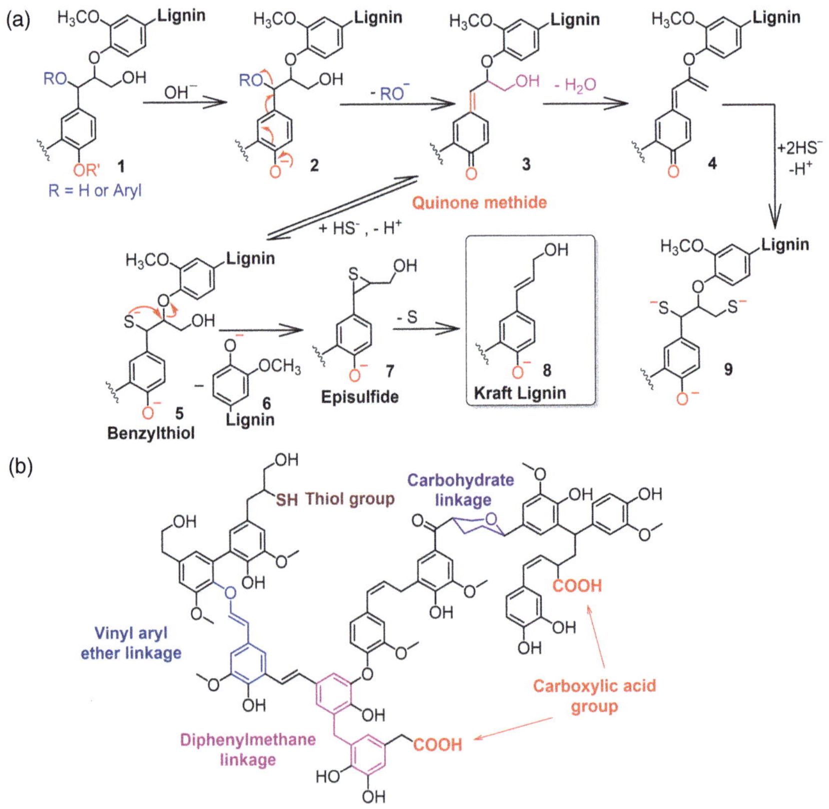

Figure 3.3 Main Reactions Involved in the Formation of Kraft Lignin during Pulping and the Structure Model of Kraft Lignin.

or calcium for wood digestion. The type of salt determines the pH of the digestion and, depending on the combination of salts employed, there are different versions of the method: the cooking may be acidic, neutral, or alkaline (Table 3.3) [52].

Several chemical transformations take place during the digestion process of wood, such as the cleavage of linkages between the lignin and the cellulose and other carbohydrates and some linkages of the lignin polymer as ether bonds. In addition, the sulfonation of the lignin aliphatic chain takes place in the Cα-positions via the carbocation intermediate [53]. The waste liquid formed during the separation of cellulose, generally by the treatment of softwood, is rich in lignins. These lignins, which cannot be isolated by precipitation from this pulping system and which contain about 4–8% sulfur, are called lignosulfonates. Lignosulfonates are water-soluble and chemically modified with sulfonate groups, carbohydrates, and small amounts of wood extracts and inorganic compounds (Figure 3.4) [50]. Lignin isolation from the sulfite pulping liquor can be carried out by diverse techniques: alcoholic fermentation of sugars in the sulfite liquor, sugar removal by chemical destruction, ultrafiltration to remove sugars, and precipitation. The last technique is performed in three steps as follows: (1) Treatment with a long-chain alkyl amine (e.g., octyl amine, N,N'-dimethylhexadecylamine); (2) Extraction with an organic solvent

Table 3.3 Sulfite-pulping Procedures Used to Extract Lignin from Wood.

Process	Reactive agent(s)	pH	Temperature (°C)
Acid sulfite	SO_2/HSO_3^-	1–2	125–145
Bisulfite	HSO_3^-	3–5	150–175
Neutral sulfite	HSO_3^-/SO_3^{2-}	6–7	150–175
Alkaline sulfite	$NaSO_3$	9–13	50–175

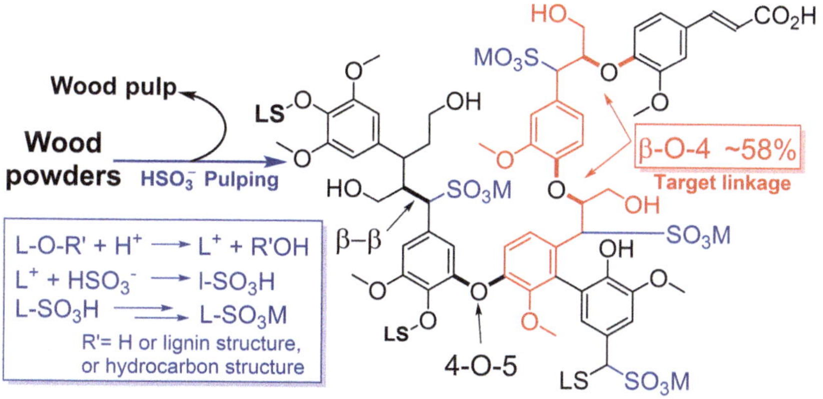

Figure 3.4 Structure Model of Lignosulfonate.

(for example, octan-1-ol); and (3) Fegeneration of sulfonate with alkaline water. The purity of isolated lignosulfonates can be measured by the amounts of reducing sugars and inorganic salts remaining in their solids. Lignosulfonates are commercially available mainly as calcium, magnesium, and ammonium salts. The M_W of the final product is higher than that of Kraft lignin, rendering a structure more similar to that of the original lignin.

3.2.3.3 Alkali Lignin

Alkali lignin is produced in the soda pulping process [49], industrialized since 1853, and traditionally used for nonwood fibers such as straw, sugarcane bagasse, and flax. Biomass is digested by the treatment with an aqueous solution of NaOH (or CaO, CaOH) at temperatures of approximately ≤ 160 °C. Lignin depolymerization occurs principally by the cleavage of α- and β-aryl ether bonds (Figure 3.5) and saponification of intermolecular ester bonds cross-linking hemicelluloses and lignin. Quinone methide structures are formed in soda pulping and contribute to depolymerization. Under these reaction conditions, stilbene or styryl aryl moieties are formed, which are alkali stable structures. The lignin is recovered lowering the pH until precipitation and the solid is isolated by decantation or filtration.

3.2.3.4 Steam-Exploded Lignin

Steam-exploded delignification is another simple and cheap method used in the separation of wood components [54, 55]. When wood is treated with steam at high temperature under pressure (i.e., about 1.38-3.00 MPa and 180-230 °C for 1-20 min), followed by a sudden decompression in the presence of some chemicals, the fibrous mass of wood or herbaceous plants actually explodes and lignin fibers are liberated. In addition, with this procedure, hemicellulose degradation and partial hydrolysis of lignin occur and the lignin itself has a somewhat reduced molecular mass because of certain acid hydrolysis reactions that take place. These conditions result in a water-insoluble lignin material with a low level of carbohydrate and wood extractive impurities. Once delignification occurs, the lignin fraction can be extracted, to a large extent, by treatment with either aqueous alkali or organic solvents. A residue highly enriched in cellulose remains separate from lignin [56, 57]. The general

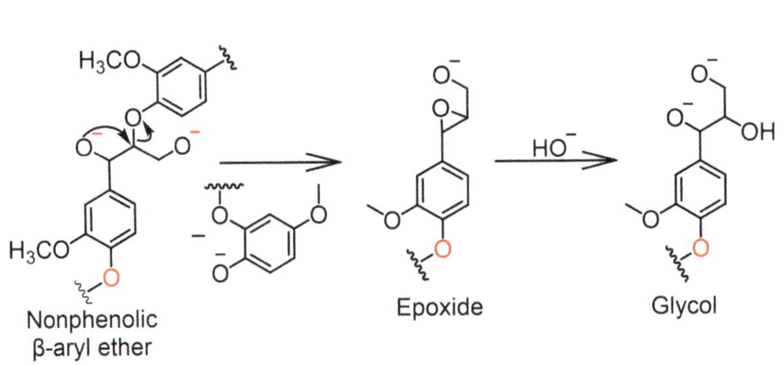

Figure 3.5 Main Reactions Involved in the Formation of Alkali Lignin.

advantages of steam-explosion processes compared to other pretreatment technologies for chemical utilization of lignocellulose are as follows: (1) No chemicals (organic solvent) are used except water; (2) Good yield of hemicelluloses is achieved with low degraded by-products; (3) Equipment corrosion is minimal due to a mild pH of reaction media when compared to acid hydrolysis processes; (D) Stages of acid handling and acid recycling are avoided; And (E) Disruption of the solid residues from bundles to individual fibers occurs due to the explosion effect.

In addition to the (water) steam-exploded lignin, several other explosion methods for pretreating and isolating lignins are available, including the ammonia fiber explosion (AFEX) process [58], and the CO_2 explosion (CDE) [59]. In the AFEX pretreatment, biomass is brought into contact with anhydrous liquid ammonia loading in a ratio of 1:1 to 1:2 for 10–60 min at 60–90 °C and pressures above 3 MPa. In a closed vessel under pressure, the biomass and ammonia mix is heated for about 30 min to the desired temperature. After holding the target temperature for about 5 min, the vent valve is opened rapidly to explosively relieve the pressure. The rapid release causes evaporation of the ammonia and a concomitant drop in the temperature of the system. The chemical effect of ammonia under pressure causes the cellulosic biomass to swell, thus increasing the accessible surface area while decrystalizing cellulose. A small amount of hemicellulose is solubilized in the oligomeric form during AFEX pretreatment. The lignin distribution in the biomass remains comparatively the same after AFEX pretreatment, but the lignin structure is rigorously altered resulting in increase water-holding capacity and digestibility. The CO_2 explosion involves the use of supercritical CO_2 (SC–CO_2) under pressure to enhance the digestibility of lignocellulosic biomass. In this process, the vessel with SC–CO_2 (1000–4000 psi) [60, 61] is heated to 200 °C and held for a designated length of time, usually several minutes [62]. CO_2 penetrates the biomass at high pressure and it is believed CO_2 forms carbonic acid once dissolved in water, improving the hydrolysis of hemicellulose. The release of pressurized gas results in the disruption of the native biomass structure increasing the accessible surface area [60]. CDE pretreatment is not effective on biomass with no moisture content, but enhances the hydrolytic yield of moisture-containing biomass with the yield increasing proportionately with the moisture content of the unprocessed feedstock [61]. The low cost of carbon dioxide as a pretreatment solvent, lack of generation of toxins, use of low temperatures, and high solid capacity are attractive features of CDE pretreatment. However, the high cost of equipment that can withstand high-pressure conditions of CDE pretreatment is a strong limitation to the application of this process on a large scale.

3.3 Lignin Spectroscopy Properties

Because lignin consists of atoms or groups of atoms that have characteristic states of motion, it can absorb electromagnetic waves with various energies and exhibit specific spectral properties. For lignin, Fourier transform infrared spectroscopy (FTIR), ultraviolet absorption spectroscopy (UV), and nuclear magnetic resonance spectroscopy (NMR) can provide considerable structural information that serves as the basis for structural analysis of lignin.

3.3.1 FT-IR Spectroscopy of Lignin

In infrared (IR) spectroscopy, a vibrational transition that involves a change in dipole moment results in the absorption of an IR photon. The energy of the absorbed photon is equal to the energy difference between two vibrational states of a molecule. From the 1950s, IR spectroscopy has been used often to characterize lignin because the technique is simple and the sample to be studied does not need to be dissolved in any solvent and is required only in very small quantities. In the 1980s, FTIR was used to study lignin. Compared with past scattering infrared spectroscopy, FTIR has the advantages of short analysis time, high sensitivity, good stability, convenience of use, and convenience of data processing. It makes the qualitative and quantitative analysis of the lignin sample more effective.

The characteristic groups in the molecular structure of lignin have clear characteristic peaks in the IR spectrum. Peaks at 1610~1600 cm^{-1} and 1520~1500 cm^{-1} result from the aromatic ring skeleton vibration. There are conjugated monkey groups at 1665~1670 cm^{-1}. The C–H bending vibrations of methyl and methylene appear between 1470 and^{-1} 1460 cm^{-1}. In the wavenumber range of these spectra, there are few other light bands and therefore these can be used to prove the existence of lignin in unknown samples whereas the aromatic ring vibrations of 1510 cm^{-1} and 1600 cm^{-1} can be used to quantitatively determine lignin. Table 3.4 lists the infrared absorption bands of characteristic groups in the molecular structure of lignin.

3.3.2 UV-vis Spectroscopy of Lignin

Given the existence of the aromatic ring structure and some conjugated carbonyl groups, lignin shows characteristic π-π^* transitions of the aromatic nuclei that make it a strong absorber of UV light, whereas sugars have almost no UV-vis absorption. Therefore, one can select identify the structure of lignin by using ultraviolet absorption spectroscopy in the presence of sugars. Usually, the ultraviolet and visible light range of 200~700 nm is used to study the absorption of lignin to obtain the UV-vis absorption spectra of lignin. The ε (molar absorption coefficient) value of the ultraviolet spectrum of lignin is relatively high and the repeatability is good, which is very suitable for the quantitative analysis of lignin.

The characteristic absorption peak of the UV-vis spectrum of lignin is first a strong absorption near 280 nm followed by absorption near 210 nm. In addition, there are weaker absorption peaks near 230 nm and 310~350 nm. The absorption near 210 nm is the absorption band of the conjugated ethylenic bond and the absorption near 230 nm is the absorption band of the aromatic ring. Softwood lignin has a maximum absorption value at 280 nm or a little lower. The maximum absorption value of hardwood lignin is 275~277 nm. This slight shift to the short-wavelength direction is because hardwood lignin contains more syringyl groups with a higher

Table 3.4 IR-absorption Bands of Lignins.

Structure and groups	Absorption Band (cm^{-1})	
	Softwood	**Hardwood**
OH stretching	3450–3400	
C–H stretching (in CH- and CH_2-)	2920	2940
C=O stretching (nonconjugated)	1715	1715–1710
C=O stretching (conjugated to aromatic ring)	1675–1660	1675–1660
Aromatic ring vibrations	1605	1595
Coumaran ring	1495	
C–H deformations, asymmetric	1470–1460	1470–1460
Aromatic ring skeletal vibrations	1430–1425	1425
C–H deformations, symmetric	1370	1370–1365
Syringyl ring breathing		1330–1325
Guaiacyl ring breathing	1270	1275
C–H inplane deformation in guaiacyl	1140	1145
C–H inplane deformation in syringyl		1130
C–O deformations in second alcohol and aliphatic ether	1085	1085
C–O deformations in primary alcohol	1035	1030

Source: Data were Abstracted from Refs. [63, 64].

symmetry. The absorption bands of the above two wavelengths of lignin can be used for the quantitative determination of trace lignin, which has the advantages of simplicity, speed, and directness. For soluble lignin, an appropriate solvent can be selected to prepare a solution of appropriate concentration in the determination cuvette whereas the solid sample compression method can be used for insoluble lignin. The sample is firstly ground with KB: fully combined, ground, and pressed into a translucent sheet with a thickness of 0.01 to 0.1 mm. Interference with the absorption of the lignin sample at 280 nm can be caused by the degradation products of glycans, such as aldehydes or hydroxymethyl furans, and these also have absorption maxima around 280 nm. In addition, the second absorption peak between 200 and 208 nm has a high extinction coefficient, so it has a large relationship with the measured solvent and can easily cause errors.

In addition to the use of the UV-vis to confirm lignin content with peak signals at 280 nm and 205 nm, [65–67] UV-vis spectroscopy can be used to study the bleaching pulp process and the lignification of wood during pulping with color change [68]. In addition, the ionization difference spectrum (ΔEi spectrum) can be used to conveniently determine the phenolic hydroxyl content in the lignin structure [69, 70]. The phenolic hydroxyl group in the lignin structure dissociates into phenolate ions under alkalinity, shifting the characteristic absorption peak to the long-wavelength direction and enhancing the absorption intensity. There is great absorption at 250 nm and 300 nm. The absorption band near 350 nm is caused by the aromatic ring structure with the conjugated double bond or carbonyl group on the side chain or by the biphenyl type structure. Therefore, the ΔEi spectrum can be obtained by subtracting the spectrum of the neutral solution from the spectrum of the alkaline solution of lignin (above pH 13), and the ΔEi spectrum of an appropriate model can be used as a standard to quantify the content of phenolic hydroxyl groups in lignin.

3.3.3 Nuclear Magnetic Resonance (NMR) Spectroscopy of Lignin

In addition to FTIR and UV-vis spectroscopy, NMR is also an important method to study the structure of lignin. The application of NMR spectroscopy to the characterization, classification, and detailed structural elucidation of lignins has seen widespread utility because it gives information on the lignin molecule that cannot be gained by chemical analysis. Commonly used nuclear magnetic resonance methods are hydrogen spectroscopy (^{1}H-NMR), carbon spectroscopy (^{13}C-NMR), and the corresponding 2D (or nD) NMR methods. In recent years, ^{29}Si-NMR, ^{31}P-NMR, and ^{19}F-NMR methods have been also used in lignin characterization.

3.3.3.1 ^{1}H-NMR

According to the chemical shift (δ) and spin coupling of the H protons in lignin, the ^{1}H-NMR measurement can provide the chemical structure information of lignin in-depth and the methoxy group, phenolic hydroxyl group, and other groups can be quantitatively analyzed (Table 3.5) [71, 72]. ^{1}H-NMR is only used to analyze low molecular weight lignin and it is usually carried out by dissolving lignin into a solution in a solvent. Commonly used solvents are heavy water (D_2O) or deuterated chloroform ($CDCl_3$), with tetramethylsilane (TMS) as the internal standard. In order to dissolve lignin in the solvent and increase the signal of H protons, pyridine and acetic anhydride (1:2, volume ratio) reagents are commonly used to acetylate lignin so that all -OH groups are converted into CH_3CO-, and one H proton becomes 3 H protons.

Besides the one-dimensional method, lignin ^{1}H NMR signals have been assigned based on model compound, decoupling, and 2D NMR experiments such as ^{1}H–^{1}H correlation spectroscopy (COSY). Therefore, ^{1}H-NMR can quantify many notable lignin structural features. Taking the spectra of spruce MWL and beech MWL as examples, the ^{1}H-NMR signals from acetylated lignins are summarized in Table 3.5. Nevertheless, there are some essential limitations to the 1H-NMR spectroscopy of lignins. These include the rather limited range of chemical shifts (12 ppm), extensive signal overlapping, and proton-coupling effects. Therefore, the NMR characterization of the lignin structure requires other methods.

3.3.3.2 Carbon Spectra (^{13}C NMR)

The ^{13}C-NMR has shown great power in molecular structure determination, isomer discrimination, conformation analysis, reaction mechanism research, and biosynthesis of natural organic compounds and therefore has become an indispensable tool in the research field of natural organic compounds. Compared with ^{1}H-NMR,

Table 3.5 Main ^1H Chemical Shifts (δ) of Acetylated Spruce MWL and Beech MWL.

Structure Assignment	δ (ppm)
Hydrocarbon contaminant	1.26
Aliphatic acetate	1.95, 2.01, 2.02
Aromatic acetate	2.28–2.29
H_β in β-β	3.08
OCH_3	3.76–3.81
H_γ	4.18, 4.27–4.28
H_γ primarily in β-*O*-4 (erythro) and β-5	4.39
H_γ	4.43
H_β in β-*O*-4 (birch)	4.60
H_β in β-*O*-4 (spruce) including CH_2 in cinnamyl alcohol	4.65
H_α in β-β (birch) including CH_2 in cinnamyl alcohol	4.70
H_α in β-5 and noncyclic benzyl aryl ethers (birch)	5.44
H_α in β-5 and noncyclic benzyl aryl ethers and H_β in aryloxypropiophenones (spruce)	5.49
H_α in β-*O*-4 and β-1 and vinyl protons	6.01–6.06
Aromatic and vinyl protons	6.93–6.94
Aromatics in benzaldehyde, vinyl protons on the carbon adjacent to aromatic rings in cinnamaldehyde(spruce)	7.41
Aromatics ortho to C=O (birch)	7.50
Aromatics ortho to C=O (spruce)	7.53
Formyl protons in cinnamaldehyde	9.64
Formyl protons in benzaldehyde	9.84–9.86

the advantage of ^{13}C-NMR is that the chemical shift (δ) of different ^{13}C-NMR varies widely (0~220) and ^{13}C nuclei with slightly different chemical environments have different chemical shift values that can be easily and directly analyzed the carbon atom skeleton of lignin. Convenience, speed, and a wealth of even quantitative information [73–78] have contributed to the widespread use of ^{13}C NMR for the structural analysis of lignins. Despite all of these advantages, some drawbacks remain concerning the efficient use of ^{13}C-NMR. These stem from the fact that the ^{13}C nucleus is not sensitive enough (1/5800 compared to the proton) and that it is only 1.1% naturally abundant. Consequently, relatively large sample sizes and long acquisition times are essential. However, a variety of newly developed pulse sequences such as distortionless enhancement by polarization transfer (DEPT) [79], incredible natural abundance double quantum transfer experiment (INADEQUATE) [80], heteronuclear multiple-quantum correlation (HMQC) [81], and acquisitions on ^{13}C-enriched lignin samples [78, 82, 83] have improved our understanding of lignin structure and reactivity. The ^{13}C chemical shifts of signals from acetylated

Table 3.6 Main ^{13}C Chemical Shifts (δ, ppm) Assignment of Lignin.

Assignment	δ (ppm)	Assignment	δ (ppm)
p-hydroxyphenyl C_4	167.8	p-hydroxyphenyl C_3, C_5	117
Syringyl C_3, C_5	153.6	Guaiacyl C_5	115.8
Guaiacyl C_4	148.6	Syringyl C_2, C_6	105.1
Guaiacyl C_3	145.5	Cβ	86.8
Syringyl C_4	138.5	Cα	73
Syringyl C_1	135.0	C_γ	65-61.7
p-hydroxyphenyl C_2, C_6	131	$-OCH_3$	56.6
Guaiacyl C_α=O	126.5		

lignins are summarized in Table 3.6, which can be divided into the following four regions: (A) Carbonyl carbons appear at 200–160 ppm; (B) C_1-C_6 aromatic carbons and carbons of double bonds on the side chain appear at 160–100 ppm; (C) Carbons on saturated side chains appear at 90 to 20 ppm; and (D) methoxyl carbons always appear in the narrow range of 56.3 ± 0.2 ppm.

3.3.3.3 Two-dimensional NMR Spectroscopy and Solid NMR Spectroscopy

Besides the ^1H NMR and ^{13}C NMR methods, the 2D NMR spectroscopy methods, such as ^{13}C-^1H HSQC (heteronuclear single-quantum correlation spectrum) are widely applied in the detailed structural elucidation of lignins. The 2D (and nD) NMR methods are superior to 1D ^{13}C-NMR and ^1H-NMR methods. These approaches can separate the side chain carbon of the lignin structural unit from the residual sugar carbon and overcome the interference of residual sugar carbon in ^{13}C-NMR. They can also separate the syringyl, guaiacyl groups, and certain kinds of lignin linkages without interference from other groups. In this book, we provide only a brief introduction to NMR methods and their application in lignin characterization. For more information about detailed NMR experiment operation and data analysis, one can check the work from Ralph [84–86], Sun [87], and other researchers.

In addition, solvent-soluble lignin samples are generally prepared into acetylated lignin and the relative molecular mass is small, so this cannot accurately reflect the real situation. When standard techniques of ^{13}C-NMR are applied, solid samples usually yield weak and broad signals. These problems can be overcome using cross-polarization pulse sequences and spinning samples at the magic angle of 54.7 °C relative to the applied magnetic field [88, 89]. The so-called CP/MAS is a combination of cross-polarization sequence (CP) and magic angle transformation technology (MAS); the former can overcome the shortcomings of weak solid-state NMR spectrum signals and the latter uses the magic angle (54.7 °) to rotate the sample to make broad signal peaks become sharp peaks. The solid ^{13}C-NMR CP/MAS method does not use any chemical or mechanical treatment and it is thus possible to gain some information about the lignin within a sample without its prior isolation [90–93]. With this method, it is possible to obtain information on special interactions and molecular dynamics in solid samples, so it can be directly used to determine the content of lignin in wood or pulp and the structural information of lignin [94].

3.3.3.4 ^{31}P-NMR

For the detection of active groups in lignin structure, such as different hydroxyl groups, that cannot be distinguished well using traditional ^{13}C NMR and ^{1}H NMR methods, one solution is to utilize the reaction differences between various functional groups and certain derivatization reagents to introduce the NMR signal element or groups and improve the discrimination of NMR signals between different structural groups by the signal strength and chemical shift that represent the specific chemical environment. In addition, by considering the abundance of NMR signal elements in the selection of derivative reagents, the strength and sensitivity of NMR signals that reflect the detailed lignin structural information will be intensified.

^{31}P-NMR technology is one of the methods that has provided efficient analytical tools to detect a variety of functional groups in isolated lignins. This approach detects the hydroxyl, carboxyl, carbonyl, and quinone groups of lignin with high sensitivity [95–98] as well as the two diastereomeric forms of β-O-4 structures [99]. These functional groups of lignin can be labeled by reacting with phosphorus reagents (Figure 3.6). Commonly used phosphorus reagents are 1,3,2-dioxaphospholanyl chloride and 2-chloro-4,4,5,5 tetramethyl-1,3,2-dioxaphospholane (TMDP) [100, 101].

Generally speaking, the ortho-substitution in the benzene ring has a greater influence on the ^{31}P-NMR chemical shift whereas the para/meta-substitution has less influence on the ^{31}P-NMR. There are guaiacyl, syringyl, and *p*-hydroxyphenyl, phenolic hydroxyl groups in the lignin. When the hydroxyphenyl groups form derivatives with the phosphating reagent 2-chloro-1,3,2-dioxaphospholane, their chemical shifts will appear around 130, 132, and 128, respectively, because the ortho substituents of the phenyl ring have a greater influence on the ^{31}P-NMR chemical shift. This feature can be used to accurately and quantitatively determine the content of various phenolic hydroxyl groups.

The ^{31}P-NMR chemical shifts of phosphorus derivatives of various alcoholic hydroxyl groups are in the range of 131.5~136.2 ppm. The ^{31}P-NMR chemical shifts of phosphorus derivatives of primary alcohols are shifted to a lower field by 2 ppm than that of phenolic hydroxyl groups. The ^{31}P-NMR chemical shift of phosphorus derivatives of secondary alcohols is 0.5-2.5 ppm lower than that of primary alcohols. The ^{31}P-NMR signal of the phosphorus derivative of the carboxyl group in lignin appears at 127 ppm, and the ^{31}P-NMR signal of the phosphorus indicator of phthalic acid or α-hydroxy acid appears as two peaks with a 7 ppm displacement difference. In addition, the ^{31}P-NMR chemical shift of the derivative formed by the compound

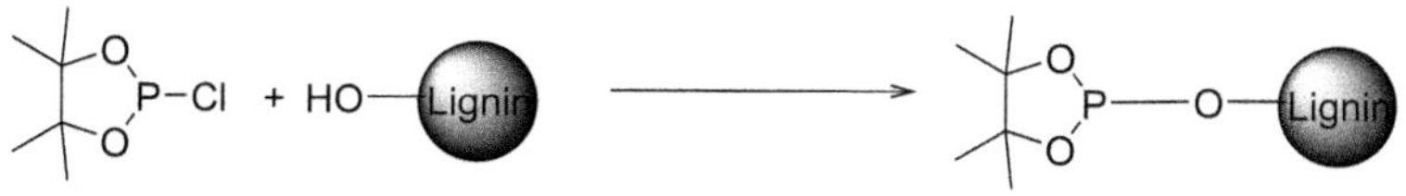

Figure 3.6 Derivatization of Lignin Hydroxylic Structures with TMPD.

aldehyde group and the phosphating reagent is in the range of 130~137 ppm, and the ketone group does not react with the phosphating reagent. This method will not show the signal of the ketone group in the lignin. In addition, trimethyl phosphite can react with aldehyde groups, o-quinone or p-quinone, α, β-unsaturated carboxyl groups, and carboxyl groups in lignin to generate phosphate derivatives and its solid ^{31}P-NMR CP/MAS has characteristic chemical shifts that can be used to determine the o-quinone group [96, 100–102].

3.3.3.5 ^{29}Si-NMR

Besides the phosphorization shown above organic compounds with hydroxyl, amino, or carboxyl groups are often siliconized in gas chromatography to reduce their vaporization temperature and this silanization technology has also been introduced into the study of NMR [103, 104]. The hydroxyl, carboxyl, and enolizable carboxyl groups in the lignin structure can be silanized so that Si atoms can be introduced and ^{29}Si-NMR can be used to analyze the structure of lignin. To determine the functional groups of trimethylsilyl ether derivatives of various lignin models, the silylation reagents used are bis(trimethylsilyl)acetamide with trimethylchlorosilane, and hexamethyldisilazane (HMDS) with trimethylchlorosilane (Figure 3.7) [104] and the quantitative standard could be N-benzoyl-N-phenylhydroxylamine that can further promote trimethylsilanization with an ^{29}Si resonance line of its trimethylsilylated derivative (at 32.9 ppm) tht is far enough from the resonance pattern of the trimethylsilylated lignin. The sample preparation, as listed by Nieminen, has four steps [104]: (1) weighing of the sample and the quantitative standard and dissolving them in pyridine-d_5, (2) introducing the silylation reagent (HMDS), (3) heating the reaction mixture under an N_2-flow at 50°C for 30 minutes; and (4) adding to the cooled reaction mixture the relaxation reagent tris(acetylacetonato)chromium(III), $Cr(acac)_3$, to a final concentration of 0.05 M and the internal standard tetramethylsilane (TMS). The ^{29}Si-NMR of different groups has different chemical shift values and the spectral lines are easily separated. The δ of trimethylsilyl ether on the aromatic ring is 22.2~19.8 ppm and the δ of trimethylsilyl ether on the alkyl group is 19.8~16.8 ppm. The δ of trimethylsilyl ether of guaiacyl is 20.2~20.9 ppm. The δ of trimethylsilyl ether of diphenol structure in 5-5 linkage is 19.1 ppm. For aromatic vanillyl with α-carbonyl structure, the δ of trimethylsilyl ether is 22.0~22.2 ppm.

3.3.3.6 ^{19}F-NMR

Besides the NMR methods shown above, ^{19}F-NMR has many advantages for the study of lignin structure: (A) The fluorinated derivatives of lignin are very stable; and (B) The position of the fluorine atom on the aromatic ring is very sensitive to chemical shifts, so it is easy to obtain various hydroxyl group signals, especially for

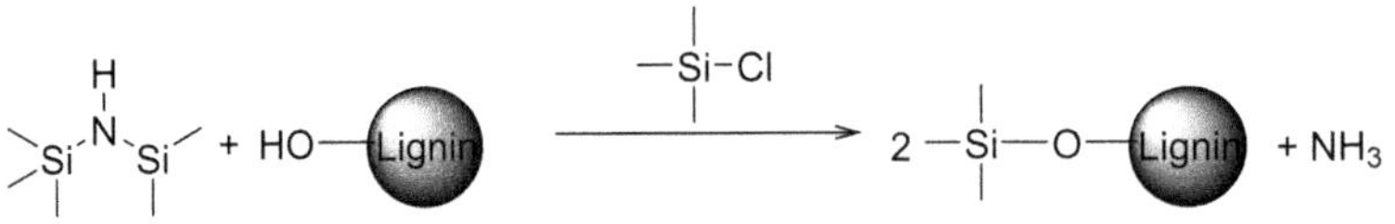

Figure 3.7 Derivatization of Lignin Hydroxylic Structures with HMDS and Me_3SiCl.

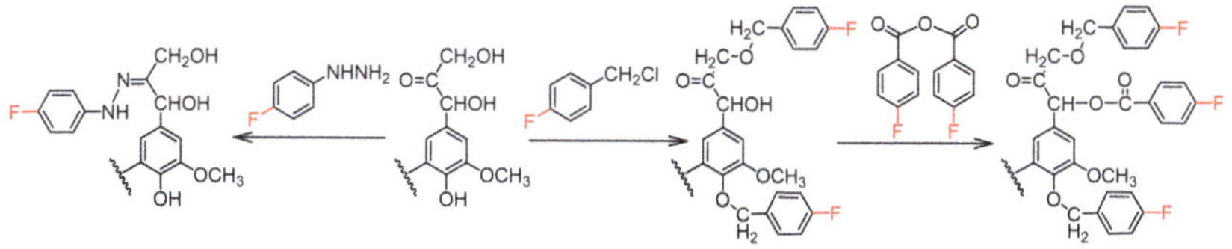

Figure 3.8 Derivatization of Lignin Hydroxylic Structures with Fluorine Reagents.

the quantification of the carboxyl group (Figure 3.8) [105] that is difficult to achieve with other methods. Frequently used fluorine reagents include 4-fluorobenzyl chloride, 4-fluorophenacyl bromide, 4-fluorobenzoic acid anhydride, and trifluoromethylphenylhydrazine, among others [105–110].

3.3.4 Electron Spin Resonance (ESR) Absorption Spectroscopy

If compounds with unpaired electrons or free radicals in their structure are placed in a magnetic field, they can absorb certain electromagnetic waves and produce electron spin resonance absorption (ESR). The spectrum generated by ESR absorption can be used to determine organic, inorganic, or coordination compounds. Free radicals are generated when lignin is subjected to light, heat, grinding, and chemical treatment. Therefore, ESR absorption spectroscopy can be used to study the free radicals in lignin [111–115]. The spin concentrations of radicals in spruce milled lignin, Brauns natural lignin, and Klasson lignin are 10×10^{16}, 5×10^{16}, and 4×10^{16} for one gram of each, respectively. The lignin location in the plant and the following treatment or isolation will affect the ESR spectrum of the lignin. In 1995, Jin et al. [116] calculated the g-factor and free radical spin concentration of larch lignin based on the electron spin resonance spectrum. They found that the g factors of several dioxane lignins are the same (between 2.0028 and 2.0029), indicating that these lignins have the same free-radical types. The larch heartwood lignin has a larger spin concentration than the value of the surface larch wood lignin, which indicates that the content of free radicals in heartwood larch lignin is high and it is easy to cause free radical reactions by light, heat, or initiators. The spin concentration of the heartwood lignin extracted with 0.1 mol/L NaOH is more than half lower than that of the larch heartwood lignin without alkali extraction, indicating that the free radical reaction activity of lignin may be inhibited by the dilute alkali treatment. The above conclusions may guide the optimization of the free radical oxidation process of target lignin. In addition, the ESR technique could be used to monitor the cation radical and other types of radical formation during the radical-induced breakdown of model compounds [111].

3.4 Lignin Chemical Properties

There are various active groups, such as aromatic groups, phenolic hydroxyl groups, alcoholic hydroxyl groups, carbonyl groups, methoxy groups, carboxyl groups, and conjugated double bonds in the molecular structure of lignin that can undergo oxidation, reduction, hydrogenation, acidolysis, acylation, sulfonation, alkylation, halogenation, nitration, polycondensation graft copolymerization reactions, and other reactions. Unlike complete degradation and the discussion about lignin fragmentation and repolymerization, we will focus in this section on introducing the chemical properties of lignin of value for these purposes. These corresponding processes can be called "lignin pre-chemical modification".

3.4.1 Oxidation

In the structure of lignin, many groups can undergo oxidation reactions to provide complex products. Peracetic acid, periodate, chlorine dioxide, hypochlorite, hydrogen peroxide, and oxygen in air can all oxidize the lignin with or without the assistance of redox metals.

Lignin Nitrobenzene Oxidation. Treating the wood powder or lignin sample with nitrobenzene in 2 mol/L NaOH at 180 °C for 2 h can generate a large amount of vanillin and other aromatic compounds. The main process is as follows (Figure 3.9): (1) Under the action of alkali, the β-O-4 structure first releases a formaldehyde at C_γ and the following dehydration produces Cα=Cβ; and (2) The side chain oxidation occurs again to form vanillin. The phenyl coumaran structure is oxidized by removing formaldehyde and generating a stilbene structure. After formation of the free phenolic hydroxyl group via the cleavage of the ether bond in the alkaline medium, the methylene quinone intermediate is produced and the side chain is easily oxidized to form vanilla, aldehydes, and syringaldehyde. Because it is possible to determine the proportion of H-,G-, and S-units in wood by measuring the amounts of vanillin, syringaldehyde, and *p*-hydroxybenzaldehyde in the lignin nitrobenzene oxidation under alkaline conditions, nitrobenzene oxidation of lignin

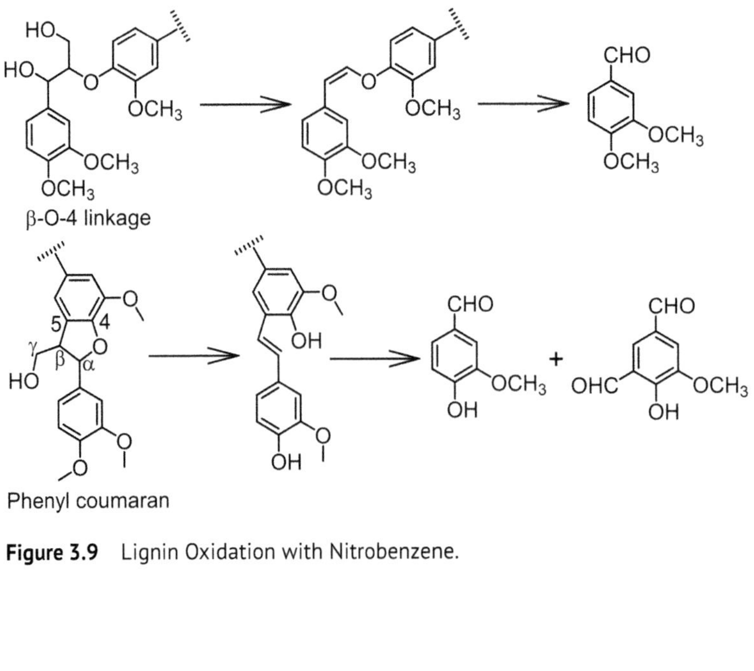

Figure 3.9 Lignin Oxidation with Nitrobenzene.

is an important first step to study the structure of lignin. We also add discussion about catalytic lignin oxidation with nitrobenzene in Chapter 7.2.

Lignin Cupric Oxide Oxidation. In 1942, Pearl [117] reported that heating sulfite liquor solids containing 40% of lignosulfonate with a 4 M NaOH solution in the presence of cupric sulfate at 160 °C for several hours could provide a 20% yield of vanillin. Cupric sulfate could be replaced by cupric hydroxide [118]. The optimal conditions for performing this cupric oxidation involve heating a mixture of lignin, cupric oxide, and NaOH (1.5–0.2 M) in a ratio of 1:6:3 in an autoclave at 170 °C for 2–5 h [119]. Other metal oxides that have been used as oxidants include Cu, Hg, Ag, and Co oxides. The main oxidation products are aromatic aldehydes or aromatic carboxylic acids, depending on the oxidant used. Silver oxide gives acids, cupric oxide yields aldehydes, and Hg and Co oxides produce a mixture of both acid and aldehyde.

Lignin Oxidation with O_2 and O_3. In general, O_2 cannot directly oxidize lignin. However, under alkaline conditions, the phenolic hydroxyl group of the lignin phenol structure can be dissociated by the O_2-NaOH, and electrons can be given to make O_2 generate free super anions radicals ($O_2^{-\bullet}$) that can cause lignin radical reactions [120, 121]. In other words, O_2-NaOH can only oxidize lignin with a phenolic structure to form a quinone structure or open the benzene ring to form muconic acid, which is the mechanism of the oxygen-alkali bleaching reaction without chlorine. Unlike O_2, O_3 has a large π bond formed by 3 atoms and 4 electrons and therefore has a strong electrophilic attack ability. O_3 can undergo an electrophilic substitution reaction with aromatic rings, double bonds at side-chains, alcohol hydroxy groups, ethers, and aldehyde groups in phenolic and non-phenolic lignin structures. [122] In general, aromatic rings react more slowly with ozone than do alkenes. Nevertheless, phenols and phenol ethers react rather rapidly with O_3. The reactivity follows the following order: syringyl > guaiacyl > 4-hydroxyphenyl [123, 124]. Nakano et al. [125] discovered that extensive ozonolysis of lignin destroyed the aromatic moieties while leaving the side chains intact. Ozonolysis provides the best opportunity for gathering information on the frequency of the different side chains or for characterizing their stereochemistry. When > 30 moles of O_3 per mole of the substrate is used, all aromatic rings are completely destroyed [125–127] and the β-O-4 linkages can be transformed to the corresponding threonic and erythronic acids (Figure 3.10) [128].

Lignin Oxidation with H_2O_2. In an alkaline medium, H_2O_2 can disintegrate the benzene ring and side chain of lignin and release the fragments. Hydrogen peroxide is an excellent pulp bleaching agent and its dissociation reaction is $H_2O_2 \rightarrow H^+ + HO_2^-$. As the solution pH value increases, the concentration of HO_2^- increases and the H_2O_2 oxidation activity increases. The aromatic ring of the structural unit of lignin is colorless. After being oxidized to form various quinoid structures, this ring becomes a colored solid. During the hydrogen peroxide bleaching process, these quinoid structures are destroyed and the colored structure becomes colorless. Other structures, even the benzene ring, are broken down into low-molecular aliphatic compounds (Figure 3.11). As a result of the hydrogen peroxide oxidation reaction of lignin, the lignin side chain is broken and the aromatic ring is cracked to produce a

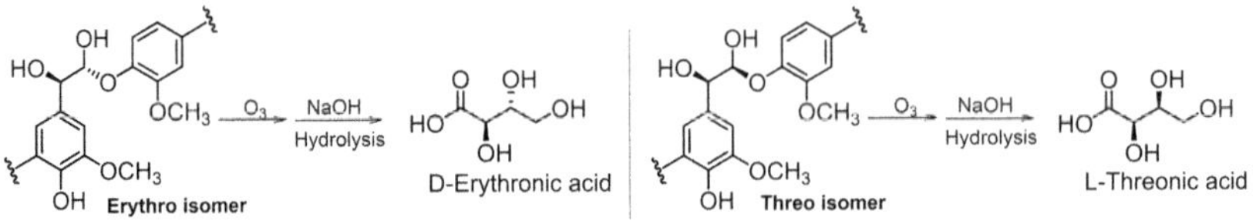

Figure 3.10 Formation of Low-molecular Aliphatic Acids from Ozonolysis of β-O-4 Linkages.

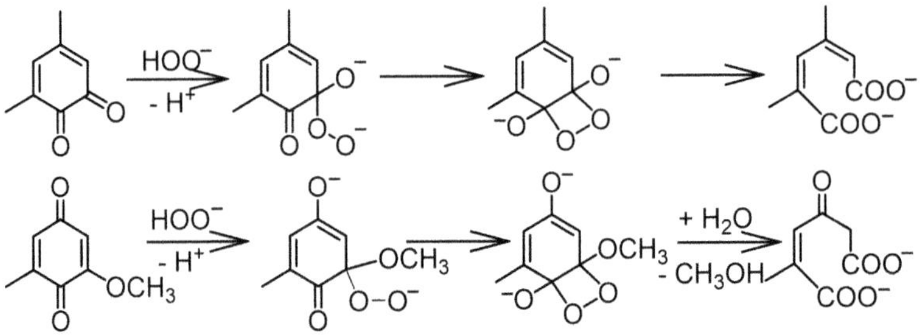

Figure 3.11 Lignin H_2O_2 Oxidation with Broken Down of the Aromatic Ring.

series of dibasic fatty acids and aromatic acids. At the same time, the demethylation reaction occurs on the aromatic ring, and some chromophoric groups of lignin are destroyed and become colorless groups, (the mechanism of alkaline lignin hydrogen peroxide bleaching) [122]. At the same time, H_2O_2 can also oxidize with the different structures of the side chain of lignin. We list some examples of catalytic BV oxidation of pre-oxidized lignin in Chapter 9.4.

Lignin Oxidation with ClO_2. ClO_2 used to be a good lignin bleaching agent. It is easy for ClO_2 to attack the phenolic hydroxyl of lignin to turn it into free radicals and then undergo a series of oxidation reactions. The phenolic structure in the guaiacyl- and syringyl-lignin can be oxidized by ClO_2 to form muconic acid derivatives, *o*-benzoquinone, or *p*-benzoquinone structure products. When there is a C=C bond in the lignin side chain, the methyl group or methylene group attached to the C=C bond is oxidized to a carbonyl group. The aromatic ring and the side chain of lignin can finally be oxidized to oxalic acid, chloroacetic acid, fumaric acid, and other compounds.

Lignin Oxidation with Hypochlorite. The reaction between hypochlorite and lignin is a nucleophilic reaction. When hypochlorite and hypochlorous acid are decomposed, they produce radicals and active chlorine, which will cause the lignin radical reaction as well as the electrophilic reaction between active chlorine and lignin. Hypochlorite mainly attacks the benzoquinone structure of the lignin aromatic ring and the conjugated C=C bond of the side chain. First, the electrophilic substitution reaction of the phenolic structural unit generates chlorinated lignin, which then occurs demethylation to form catechols. The phenols are further oxidized to *o*-benzoquinone or be converted into hydroxyquinone in an alkaline medium. The further oxidation by the hypochlorite can break down the aromatic structure and provide low-molecular organic acids and carbon dioxide.

Others Oxidants. In addition, different hydroxyl groups in lignin can be determined using an oxidative process. Free phenolic hydroxyl groups can be oxidized into quinone with metaperiodate (IO_4^-) [129] or nitrosodisulfonate (Fremy's salt) [130]. Lignin benzyl alcohols can be converted into aryl ketones by DDQ [131].

3.4.2 Hydrogenation

Another procedure that has significantly contributed to our understanding of the lignin structure is hydrogenolysis. The main reaction in the hydrogenolysis of lignins is the cleavage of ether linkages connecting the carbon atoms of the side chain and the para-position of a phenolic ring in an adjacent unit. Depending on the reaction conditions, C–C linkages may also be broken during hydrogenolysis. Moreover, hydrogenation accompanies hydrogenolysis. The factors affecting hydrogenolysis are catalyst activity, catalyst amount, hydrogen pressure, solvent, and pH. Various catalysts have been used in the hydrogenolysis of lignin: Raney nickel, copper chromite, cobalt octacarbonyl, and palladium, rhodium, and ruthenium on different supporting surfaces. Catalysts with high-hydrogenation activity convert lignin mainly into cyclohexyl derivatives, whereas catalysts having lower activity give rise to aromatic derivatives. In Part IV of this book, we review the lignin catalytic hydrogenation methods.

3.4.3 Esterification/Acylation

The structure of lignin contains alcoholic and phenolic hydroxyl groups, which can react with acylating reagents. The most commonly used acylation systems are acetic anhydride-pyridine, acetic anhydride-sulfuric acid, acetyl bromide, and others. The acetylation treatment of lignin can not only improve the solubility of lignin in organic solvents, but also can combine with nuclear magnetic and other characterization methods to study the alcohol and phenolic hydroxyl groups in the lignin structure.

3.4.4 Etherification

The structure of lignin contains alcoholic and phenolic hydroxyl groups, which can react with various alcohols to provide the corresponding esters. The most commonly used etherification system contains mineral acid or Lewis species as the catalysts. In the recent lignin conversion study, the etherification pretreatment of lignin became a useful method to restrain the linin fragment condensation by preventing the generation of the active C_α cation from the C_α–OH. More discussion can be found in Chapter 11.

3.4.5 Hydrolysis

Treating lignin with aqueous NaOH solution at high temperatures leads to phenols and derivatives. The phenolic α-aryl ether is most prone to alkaline hydrolysis because NaOH promotes the rearrangement of the phenolic structure and eliminates the α-aryl substituent to form a methylene quinone structure (Chapter 8). However, because etherification prevents the formation of the methylene quinone structure, non-phenolic α-aryl ethers are stable to alkalis. The non-phenolic α-aryl ethers mainly undergo β-proton elimination reaction and the β-formaldehyde elimination reaction. Most potential alkaline hydrolysis of phenolic β-aryl ethers reactions do not occur. Only a small amount of the connection can be hydrolyzed when an epoxy compound is formed through the nucleophilic attack of HO^- on the C_α atom. However, because HS^- has a stronger nucleophilic ability than HO^-, episulfide compounds are formed more smoothly in kraft pulping, which promotes the hydrolysis and cleavage of β-aryl ether bonds. For the alkaline hydrolysis of non-phenolic β-aryl ethers, only two cases can occur: (1) Non-phenolic β-aryl ethers with C_α–OH can undergo alkaline hydrolysis; and (2) Non-phenolic β-aryl ethers with α-carbonyl can undergo sulfate hydrolysis but must be able to form epoxides or episulfides between C_α and C_β, or C_β and C_γ.

During the acid condition, the α-aryl ether bond and the β-aryl ether bond are both unstable to acid and are prone to hydrolysis. The α-aryl ether bond occurs through the cleavage of C_α–OAr bond via an oxonium intermediate. For the β-aryl ether, the first activation C_α–OH can generate a C_α cation that can further transform to the vinyl ether intermediate C_α=C_β–OAr, which can further occur hydrolysis with C_β–OAr bond cleavage. For a more detailed mechanism discussion, one can check the discussion provided in Chapter 8.

3.4.6 Alkylation and Phenolation

The hydroxyl, carboxyl, and carbonyl groups of lignin can be alkylated. The most researched alkylation reaction is methylation (Figure 3.12). Commonly used methylation reagents include methanol-HCl, diazomethane, methyl iodide-silver oxide, dimethyl sulfate-sodium hydroxide, and others. The type of methylation reaction varies with the reagents used. Using the methanol-hydrochloric acid system, the benzyl hydroxyl, carbonyl, and carboxyl groups of the lignin side chain are all methylated. When diazomethane is used, the carboxyl, phenolic hydroxyl, and enol hydroxyl are methylated. Methyl iodide and dimethyl sulfate methylate all hydroxyl groups. Combined with nuclear magnetism and other means, the number of hydroxyl groups in the lignin molecule can be measured by increasing the methoxy group.

The phenolation process involves the treatment of lignins with phenol in an acidic medium with heating at approximately 70 °C, the boiling point of the common organic solvents (Figure 3.13). Under these conditions, phenol reacts with

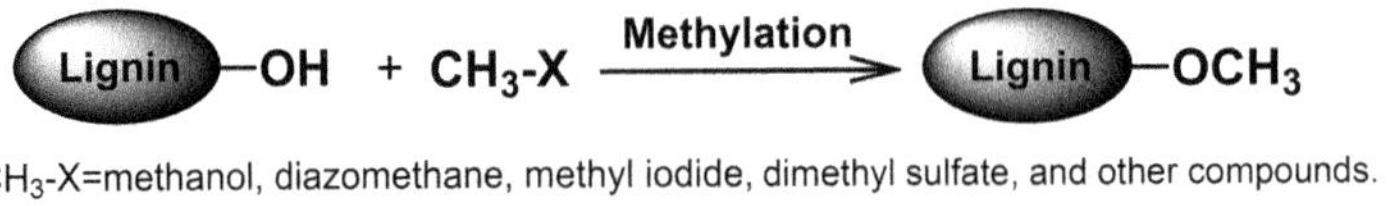

Figure 3.12 Methylation Reaction in Lignin.

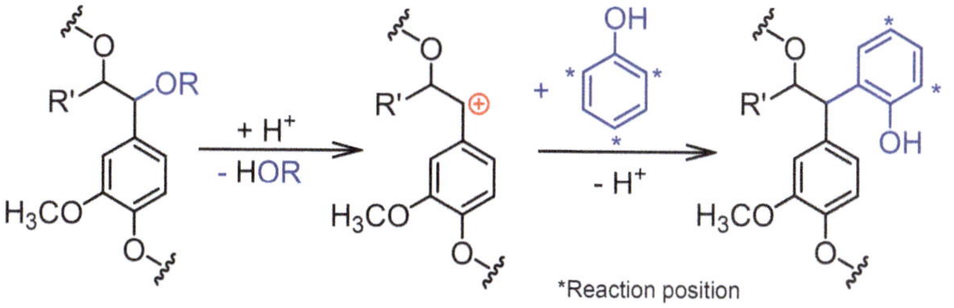

Figure 3.13 Phenolization Reaction in Lignin.

lignin aromatic rings and side chains. During the process, some ether bonds are cleaved so that a new sort of lignin is formed, decreasing the M_w of the molecule.

3.4.7 Demethylation

Demethylation of lignin means the conversion of methoxy groups on the aromatic ring to phenolic hydroxyl groups, forming a catechol structure in the lignin structure (Figure 3.14). The treatment of MeO-containing lignin linkages with concentrated HI or with HBr solution at refluxing temperature leads to the formation of demethylation and release of the phenolic hydroxyl group. For the demethylation of sulfate lignin [15], add 4~5% sulfur to 50% water-containing sulfate lignin, stir and treat for 30 min at 225~235 °C, then quickly cool to room temperature, dilute with water, and acidify with sulfuric acid. Next, perform extraction with ethyl acetate. After centrifugation, an aqueous phase and an organic phase separate. The organic phase and organic solvent can be removed and then dried under vacuum at room temperature to obtain a brown demethylated lignin powder.

3.4.8 Nitration

Lignin can react with nitric acid to produce nitrated lignin. In the lignin nitration reaction, besides the existence of an electrophilic substitution reaction, the process also contains the demethoxy reaction at the aromatic rings and the C–C bonds scission at the side-chain. When lignin is treated with dilute nitric acid, the reaction of lignin is more complex. Besides the reactions mentioned above, it also involves hydrolysis and reduction, as well as oxidation and other reactions. In these reactions, the scission reaction is the main one, besides the generation of nitroguaiacols, there is also α-O-(2-methoxyphenyl) glyceraldehyde. When dioxane lignin was treated with nitrous acid at pH=2.0 and 100 °C, 4-nitroguaiacol and 2,4-dinitroguaiacol were produced. Glycol-lignin was readily nitrated by adding it to a mixture

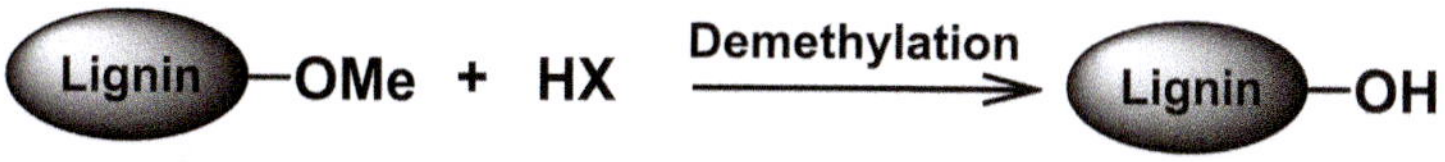

Figure 3.14 Demethylation Reaction in Lignin.

consisting of acetic anhydride/fuming HNO_3 (4:1) at a temperature below 0 °C. Acetylation occurred simultaneously. The nitrated lignin could be reduced with sodium amalgam in an alkaline solution.

3.4.9 Halogenation

The lignin halogenation reaction occurs primarily on its aromatic ring, which can be carried out at or below room temperature. Under mild conditions, the substitution reaction of chlorine mainly occurs on the aromatic ring through substitution of chlorine at the C_5 and C_6 positions with rare generation of rings with chlorine at C_2. Chlorine substitution will also occur at the C_1 position, but it is accompanied by the scission of the side chain at the same time. Under stronger conditions, the substitution reaction will also occur on the side chain. The more intense the reaction conditions, the greater proportion of chlorine that bonds to the side chain. Although the chlorination reaction of lignin occurs, the hydrolysis of the ether bond may also occur (e.g., the demethylation reaction, the rupture of the phenol ether bond, or other reactions) and the oxidation of the aromatic ring may also occur, even between the aromatic rings. A slight amount of dehydrogenation condensation reaction also occurs. Lignosulfonate can remove 75% of the methyl group when it is chlorinated. The bromination and iodination reactions of lignin are similar to chlorination, but the reactions are weaker than that of chlorination. Bromination should be carried out in an acidic medium. The stronger the acidity, the faster the bromination reaction. In addition, reduced pressure conditions can also accelerate the speed of the bromination reaction.

3.4.10 Sulfonation

When wood is heated with a solution of sulfurous acid and acid sulfites, the lignin is solubilized because of the introduction of the hydrophilic sulfonic acid groups, leaving cellulose in a more or less pure condition. This is the basis of the sulfite process for pulping wood. Phenolic lignin can be converted into methylene quinone structure under acidic, neutral or alkaline conditions, but the methylene quinone structure in an acidic medium can quickly form carbocations on C_α, which can facilitate the sulfonation of acid sulfite cooking.

Under the conditions of neutral sulfite cooking, there is no epoxidation reaction and episulfide reaction in the sulfonation of the methylene, but the neutral SO_3^{2-} ion has strong nucleophilicity to C_α, leading to the sulfonation reaction of the β-aryl ether with β-aryl ether bond cleavage. For non-phenolic lignin structural units, under alkaline conditions, α-aryl ether is stable, whereas β-aryl ether bonds are broken through epoxidation. Under neutral conditions, both α-aryl ether and β-aryl ether bonds are stable and do not react with neutral sulfite. Under acidic conditions, carbocations can still be formed on C_α, which is conducive to the progress of the sulfonation reaction with neutral sulfite. Under acidic conditions, both phenolic and non-phenolic lignin structural units can occur in favor of the sulfonation reaction on the C_α carbon, and sometimes it may also proceed on C_γ, but rarely on C_β.

During acidic bisulfite cooking, the -OR group on the C_α of phenolic and non-phenolic lignin structural unit will be quickly removed to form C_α carbocations, so the nucleophile HSO_3^- ion can attack at the C_α to achieve the sulfonation reaction. If the sulfonation reaction cannot occur in time, the carbocation will quickly undergo a condensation reaction. It must be pointed out that the nucleophilicity of this acidic HSO_3^- ion is not as strong as that of the neutral SO_3^{2-} ion, so the β-aryl ether bonds do not break.

3.4.11 Hydroxymethylation

The lignin hydroxymethylation reaction is the reaction between lignin and formaldehyde under alkaline conditions to generate hydroxymethyl groups on vacancies of the lignin aromatic ring and the side chain (Figure 3.15). When the hydroxymethylation reaction is carried out under high temperature and strong alkali conditions, the formaldehyde is prone to occur the Cannizzaro reaction, during which formaldehyde is converted into formic acid and methanol. Therefore, the high temperature and strong alkali conditions are not suited for the hydroxymethylation reaction. The optimal pH for the hydroxymethylation of sulfurized lignin is 8 and the temperature is 40 °C. The optimal pH of sulfate lignin is 12~12.5, and the reaction is at room temperature for 3 days [15].

3.4.12 Mannich Reaction

Lignin can undergo the Mannich reaction with formaldehyde and dimethylamine (Figure 3.16). The substitution of the amine methyl group mainly occurs at the C_5 position of the phenolic lignin structural unit. This is caused by the activation of the phenolic hydroxyl group on the C_5 position [132]. The reaction at room temperature

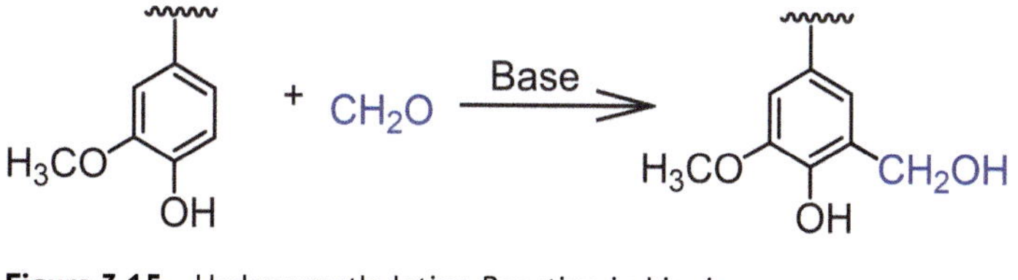

Figure 3.15 Hydroxymethylation Reaction in Lignin.

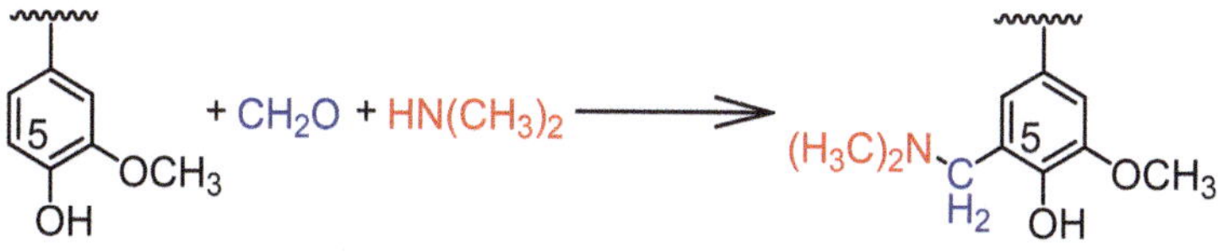

Figure 3.16 The lignin Mannich Reaction.

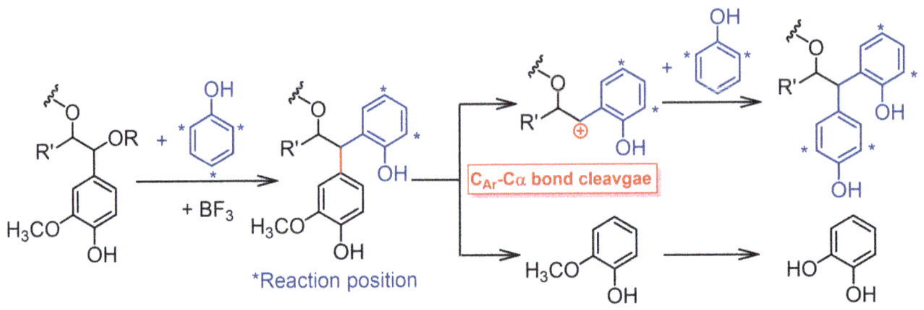

Figure 3.17 Nucleus-Exchange Reaction in Lignin.

for 1~2 days can introduce 2% nitrogen content, which is equivalent to adding about 0.35 nitrogen atoms for each $-OCH_3$. If there is a group on the C_5 position, or the phenolic hydroxyl group is etherified, the Mannich reaction will not occur. If the para-position of the phenolic hydroxyl group is a carboxyl group, the carboxyl group will be decomposed in the form of CO_2, and a dimethylaminomethyl group will be introduced at this position. If the para-position of the phenolic hydroxyl group is a benzyl alcohol group, a dimethylaminomethyl group is introduced at this position [15].

3.4.13 Nucleus-Exchange Reaction

In 1978, Funaoka [133] developed a chemical method to degrade lignin in a medium consisting of boron trifluoride and excess phenol. In this method, C_α-aryl carbon-carbon linkages in phenylpropane units are selectively and quantitatively cleaved, following phenolation at the C_α position, releasing phenyl nuclei typified by guaiacol from softwood lignins. Guaiacol subsequently is partially demethylated to catechol. The key step in this sequence, the release of a phenyl nucleus on reaction with phenol, was termed nucleus exchange (NE) by the original authors (Figure 3.17) [134]. Given the reaction mechanism, NE applied to softwood lignin can lead to the quantitative conversion of noncondensed guaiacyl nuclei to guaiacol and catechol because the sum of guaiacol and catechol represents the number of noncondensed phenyl nuclei in softwood lignins. However, the NE method is not suitable for grass lignins, which contain a significant amount of *p*-hydroxyphenyl units that can yield phenol in the NE reaction.

3.5 Chapter Summary

In Chapter 3, after a brief introduction about the lignin polymer physical properties and lignin isolation methods from lignocellulose, a basic introduction about the spectroscopy properties for the lignin structure characterization was provided. At the end of Chapter 3, the lignin chemical properties that are connected with the pre-chemical modification are discussed. The keywords of this chapter could be Lignin polymer, Lignin isolation, Lignin physicochemistry properties, and Lignin pre-chemical modification.

References

1 Xia, Q., Chen, C., Yao, Y. et al. (2021). In situ lignin modification toward photonic wood. *Adv. Mater.* 33: 2001588.

2 Zhu, M., Song, J., Li, T. et al. (2016). Highly anisotropic, highly transparent wood composites. *Adv. Mater.* 28: 5181–5187.

3 Rogosic, M., Mencer, H.J., and Gomzi, Z. (1996). Polydispersity index and molecular weight distributions of polymers. *Eur. Polym. J.* 32: 1337–1344.

4 Glasser, W.G., Dave, V., and Frazier, C.E. (1993). Molecular-weight distribution of (semi-) commercial lignin derivatives. *J. Wood. Chem. Technol.* 13: 545–559.

5 Benghedalia, D. and Yosef, E. (1994). Effect of osolation procedure on molecular-weight distribution of wheat-straw lignins. *J. Agr. Food. Chem.* 42: 649–652.

6 Forss, K., Kokkonen, R., and Sågfors, P. (1989). Determination of the molar mass distribution of lignins by gel permeation chromatography. In: *Lignin* (ed. W. Glasser and S. Sarkanen), 124–133. Washington, D.C.: American Chemical Society.

7 Striegel, A., Yau, W., Kirkland, J., and Bly, D. (2009). *Modern Size-Exclusion Liquid Chromatography: Practice of Gel Permeation and Gel Filtration Chromatography, Second Edition*. Hoboken: John Wiley & Sons, Inc.

8 da Costa Sousa, L., Chundawat, S.P.S., Balan, V., and Dale, B.E. (2009). 'Cradle-to-grave' assessment of existing lignocellulose pretreatment technologies. *Curr. Opin. Biotech.* 20: 339–347.

9 Zhang, C., Li, H., Lu, J. et al. (2017). Promoting lignin depolymerization and restraining the condensation via an oxidation-hydrogenation strategy. *ACS Catal.* 7: 3419–3429.

10 Björkman, A. (1954). Isolation of lignin from finely divided wood with neutral solvents. *Nature* 174: 1057–1058.

11 Ikeda, T., Holtman, K., Kadla, J.F. et al. (2002). Studies on the effect of ball milling on lignin structure using a modified DFRC method. *J. Agr. Food. Chem.* 50: 129–135.

12 Björkman, A.F. and divided wood. I. (1956). Extraction of lignin with neutral solvents. *Sven. Papperstidn.* 59: 477–485.

13 Lundquist, K. and Simonson, R. (1975). Lignin preparations with very low carbohydrate content. *Sven. Papperstidn.* 78: 390–390.

14 Stumpf, W. and Freudenberg, K. (1950). Lösliches lignin aus fichten-und buchenholz. *Angew. Chem. Int. Ed.* 62: 537–537.

15 Jiang, T. (2009). *Muzhisu*. Beijing: Chemical Industry Press.

16 Mcdonough, T.J. (1993). The chemistry of oOrganosolv delignification. *Tappi J.* 76: 186–193.

17 Pye, E.K. and Lora, J.H. (1991). The Alcell process-a proven alternative to kraft pulping. *Tappi J.* 74: 113–118.

18 Neumann, N. and Balser, K. (1993). Acetocell - an innovative process for pulping, totally free from sulfur and chlorine. *Papier* 47: V16–V23.

19 Parajo, J.C., Alonso, J.L., Vazquez, D., and Santos, V. (1993). Optimization of catalyzed acetosolv fractionation of pine wood. *Holzforschung* 47: 188–196.

20 Black, N.P. (1991). Asam alkaline sulfite pulping process shows potential for large-scale application. *Tappi J.* 74: 87–93.
21 Johansson, A., Aaltonen, O., and Ylinen, P. (1987). Organosolv pulping-methods and pulp properties. *Biomass* 13: 45–65.
22 Saake, B., Lummitsch, S., Mormanee, R. et al. (1995). Production of pulps using the formacell process. *Papier* 49: V1–V7.
23 Lindner, A. and Wegener, G. (1988). Characterization of cignins from organosolv pulping according to the organocell process. 1. Elemental analysis, nonlignin portions and functional-groups. *J. Wood. Chem. Technol* 8: 323–340.
24 Lindner, A. and Wegener, G. (1989). Characterization of lignins from organosolv pulping according to the organocell process. 2. Residual lignins. *J. Wood. Chem. Technol* 9: 443–465.
25 Lu, F.C. and Ralph, J. (2003). Non-degradative dissolution and acetylation of ball-milled plant cell walls: high-resolution solution-state NMR. *Plant J.* 35: 535–544.
26 Klason, P. (1910). Determination of lignin in sulfite wood pulp. *Papierfabrikant* 8: 1285–1286.
27 Willstätter, R. and Zechmeister, L. (1913). Zur Kenntnis der Hydrolyse von Cellulose I. *Ber. dtsch. chem. Ges.* 46: 2401–2412.
28 Fredenhagen, K. and Cadenback, G. (1933). The decomposition of cellulose by hydrofluoric acid and a new method of wood saccharification by means of highly concentrated hydrofluoric acid. *Angew. Chem.* 46: 113–117.
29 Calvo-Flores, F.G., Dobado, J.A., Isac-García, J., and Martín-MartíNez, F.J. (2015). *Lignin and lignans as renewable raw materials: Chemistry, technology and applications.* Hoboken: John Wiley & Sons, Inc.
30 Freudenberg, K., Harder, M., and Markert, L. (1928). Bemerkungen zur chemie des lignins. (VII. Mitteilung über lignin und cellulose). *Ber. dtsch. chem. Ges.* 61: 1760–1765.
31 Fengel, D. (1976). Fractionation experiments with the alkali extract from spruce holocellulose. Part 2. Optimizing the ion exchange chromatographic fractionation. *Holzforschung* 30: 143–148.
32 Wald, W.J., Ritchie, P.F., and Purves, C.B. (1947). The elementary composition of lignin in Northern Pine and Black Spruce Woods, and of the isolated klason and periodate lignins. *J. Am. Chem. Soc.* 69: 1371–1377.
33 Adler, E. and Hernestam, S. (1955). Estimation of phenolic hydroxyl croups in lignin. 1. Periodate oxidation of guaiacol compounds. *Acta. Chem. Scand.* 9: 319–334.
34 Chang, H.M., Cowling, E.B., Brown, W. et al. (1975). Comparative studies on cellulolytic enzyme lignin and milled wood lignin of wweetgum and spruce. *Holzforschung* 29: 153–159.
35 Chen, J.Y., Shimizu, Y., Takai, M., and Hayashi, J. (1995). A method for isolation of milled-wood lignin involving solvent swelling prior to enzyme treatment. *Wood Sci. Technol.* 29: 295–306.
36 Jung, H.J.G. and Himmelsbach, D.S. (1989). Isolation and characterization of wheat straw lignin. *J. Agr. Food. Chem.* 37: 81–87.

37 Imai, M., Ikari, K., and Suzuki, I. (2004). High-performance hydrolysis of cellulose using mixed cellulase species and ultrasonication pretreatment. *Biochem. Eng. J.* 17: 79–83.

38 Hu, Z.J., Yeh, T.F., Chang, H.M. et al. (2006). Elucidation of the structure of cellulolytic enzyme lignin. *Holzforschung* 60: 389–397.

39 Wu, S. and Argyropoulos, D.S. (2003). An improved method for isolating lignin in high yield and purity. *J. Pulp. Pap. Sci.* 29: 235–240.

40 Guerra, A., Filpponen, I., Lucia, L.A. et al. (2006). Toward a better understanding of the lignin isolation process from wood. *J. Agr. Food. Chem.* 54: 5939–5947.

41 Guerra, A., Filpponen, I., Lucia, L.A., and Argyropoulos, D.S. (2006). Comparative evaluation of three lignin isolation protocols for various wood species. *J. Agr. Food. Chem.* 54: 9696–9705.

42 Nord, F.F. and Schubert, W.J. (1951). Enzymatic studies on cellulose, lignin and the mechanism of lignification. *Holzforschung* 5: 1–9.

43 Ishikawa, H., Schubert, W.J., and Nord, F.F. (1963). Investigations on lignins and lignification 27. Enzymic degradation of softwood lignin by white-rot fungi. *Arch. Biochem. Biophys.* 100: 131–139.

44 Gierer, J. (1980). Chemical aspects of kraft pulping. *Wood Sci. Technol.* 14: 241–266.

45 Chakar, F.S. and Ragauskas, A.J. (2004). Review of current and future softwood kraft lignin process chemistry. *Ind. Crops Prod.* 20: 131–141.

46 Gierer, J. and Ljunggren, S. (1979). Reactions of lignins during sulfate pulping 16. Kinetics of the cleavage of beta-aryl ether linkages in structures containing carbonyl groups. *Sven. Papperstidn.* 82: 71–81.

47 Gierer, J. and Ljunggren, S. (1979). Reactions of lignin during sulfate pulping 17. Kinetic treatment of the formation and competing reactions of quinone methide intermediates. *Sven. Papperstidn.* 82: 503–512.

48 Gierer, J., Ljunggren, S., Ljungquist, P., and Noren, I. (1980). Reactions of lignin during sulfate pulping 18. Significance of alpha-carbonyl groups for the cleavage of beta-aryl ether structures. *Sven. Papperstidn.* 83: 75–82.

49 Gierer, J. (1985). Chemistry of delignification 1. General concept and reactions during pulping. *Wood Sci. Technol.* 19: 289–312.

50 Zakzeski, J., Bruijnincx, P.C.A., Jongerius, A.L., and Weckhuysen, B.M. (2010). The catalytic valorization of lignin for the production of renewable chemicals. *Chem. Rev.* 110: 3552–3599.

51 Sridach, W. (2010). The environmentally benign pulping process of non-wood fibers. *Suranaree J. Sci. Technol.* 17: 105–123.

52 Calvo-Flores, F.G. and Dobado, J.A. (2010). Lignin as renewable raw material. *ChemSusChem* 3: 1227–1235.

53 Lora, J. (2008). Chapter 10 - Industrial Commercial Lignins: Sources, Properties and Applications. In: *Monomers, Polymers and Composites from Renewable Resources* (ed. M.N. Belgacem and A. Gandini), 225–241. Amsterdam: Elsevier.

54 Bauer, A., Lizasoain, J., Theuretzbacher, F. et al. (2014). Steam explosion pretreatment for enhancing biogas production of late harvested hay. *Bioresour. Technol.* 166: 403–410.

55 Fu, F., Lin, L., and Xu, E. (2017). Chapter 4 Functional pretreatments of natural raw materials. In. In: *Advanced High Strength Natural Fibre Composites in Construction* (ed. M. Fan and F. Fu), 87–114. Woodhead Publishing.

56 Josefsson, T., Lennholm, H., and Gellerstedt, G. (2002). Steam explosion of aspen wood. *Characterisation of reaction products. Holzforschung* 56: 289–297.

57 Li, J.B., Gellerstedt, G., and Toven, K. (2009). Steam explosion lignins; their extraction, structure and potential as feedstock for biodiesel and chemicals. *Bioresour. Technol.* 100: 2556–2561.

58 Teymouri, F., Laureano-Perez, L., Alizadeh, H., and Dale, B.E. (2005). Optimization of the ammonia fiber explosion (AFEX) treatment parameters for enzymatic hydrolysis of corn stover. *Bioresour. Technol.* 96: 2014–2018.

59 Agbor, V.B., Cicek, N., Sparling, R. et al. (2011). Biomass pretreatment: Fundamentals toward application. *Biotechnol. Adv.* 29: 675–685.

60 Zheng, Y.Z., Lin, H.M., Wen, J.Q. et al. (1995). Supercritical carbon-dioxide explosion as a pretreatment for cellulose hydrolysis. *Biotechnol. Lett.* 17: 845–850.

61 Kim, K.H. and Hong, J. (2001). Supercritical CO_2 pretreatment of lignocellulose enhances enzymatic cellulose hydrolysis. *Bioresour. Technol.* 77: 139–144.

62 Hendriks, A.T.W.M. and Zeeman, G. (2009). Pretreatments to enhance the digestibility of lignocellulosic biomass. *Bioresour. Technol.* 100: 10–18.

63 Hergert, H. (1971). Infrared spectra. In: *Lignins: occurrence, formation, structure and reactions* (ed. K. Sarkanen and C. Ludwig), 267–297. New York: Wiley-Interscience.

64 Monteil-Rivera, F., Phuong, M., Ye, M.W. et al. (2013). Isolation and characterization of herbaceous lignins for applications in biomaterials. *Ind. Crops Prod.* 41: 356–364.

65 Alibert, G., Boudet, A., and Lignification of poplar. I (1979). Demonstration of method for the determination and monomeric analysis of lignins. *Physiol. Veg.* 17: 67–74.

66 Janshekar, H., Brown, C., and Fiechter, A. (1981). Determination of biodegraded lignin by ultraviolet spectrophotometry. *Anal. Chim. Acta.* 130: 81–91.

67 Steinitz, Y.L. (1981). Microbial desulfonation of lignosulfonate-a new approach. *Eur. J. Appl. Microbiol.* 13: 216–221.

68 Wardrop, A.B. (1957). The phase of lignification in the differentiation of wood fibers. *Tappi* 40: 225–243.

69 Aulin-Erdtman, G. and Hegbom, L. (1957). Spectrographic contributions to lignin chemistry. VII. The ultraviolet absorption and ionization δε-curves of some phenols. *Sven. Papperstidn.* 60: 671–681.

70 Aulin-Erdtman, G. and Sanden, R. (1968). Spectrographic contributions to lignin chemistry. IX. Absorption properties of some 4-hydroxyphenyl guaiacyl and 4-hydroxy-3,5-dimethoxyphenyl type model compounds for hardwood lignins. *Acta. Chem. Scand.* 22: 1187–1209.

71 Ludwig, C.H., Nist, B.J., McCarthy, J., and Lignin, L. (1964). XIII. The high resolution nuclear magnetic resonance spectroscopy of protons in acetylated lignins. *J. Am. Chem. Soc.* 86: 1196–1202.

72 Li, S. and Lundquist, K. A new method for the analysis of phenolic groups in lignins by"1H NMR spectrometry(1994). *Nord. Pulp. Pap. Res. J.* 9: 191–195.
73 Lüdemann, H.D. and Nimz, H.H. (1973). ^{13}C nuclear magnetic resonance spectra of lignins. *Biochem. Biophys. Res. Commun.* 52: 1162–1169.
74 Nimz, H.H., Mogharab, I., and Lüdemann, H.D. (1974). ^{13}C NMR spectra of lignins. 3. Comparison of spruce lignin with synthetic lignin according to freudenberg. *Makromol. Chem.* 175: 2563–2575.
75 Nimz, H.H. and Lüdemann, H.D. (1976). ^{13}C NMR spectra of lignins. 6. Lignin and DHP acetates. *Holzforschung* 30: 33–40.
76 Landucci, L.L. (1985). Quantitative ^{13}C NMR characterization of lignin 1. *A methodology for high precision. Holzforschung* 39: 355–359.
77 Obst, J.R. and Landucci, L.L. (1986). Quantitative ^{13}C NMR of lignins-methoxyl:aryl ratio. *Holzforschung* 40: 87–92.
78 Bardet, M., Gagnaire, D., Nardin, R. et al. (1986). Use of ^{13}C enriched wood for structural NMR investigation of wood and wood components, cellulose and lignin, in solid and in solution. *Holzforschung* 40: 17–24.
79 Bardet, M., Foray, M.F., and Robert, D. (1985). Use of the DEPT pulse sequence to facilitate the ^{13}C NMR structural analysis of lignins. *Macromol. Chem. Phys.* 186: 1495–1504.
80 Ralph, J. (1988). NMR of lignin model compounds; application of long-range C-H correlations through oxygen. *Holzforschung* 42: 273–275.
81 Ede, R.M. and Kilpeläinen, I. (1995). Homo- and hetero-nuclear 2D NMR techniques: Umambiguous structural probes for non-cyclic benzyl aryl ethers in soluble lignin samples. *Res. Chem. Intermediat.* 21: 313–328.
82 Ellwardt, P.C., Haider, K., and Ernst, L. (1981). Investigation of microbial lignin degradation by ^{13}C NMR spectroscopy of specifically carbon-13-enriched DHP-lignin from coniferyl alcohol. *Holzforschung* 35: 103–109.
83 Lewis, N.G., Newman, J.P., Just, G., and Ripmeister, J. (1987). Determination of bonding patterns of carbon-13 specifically enriched dehydrogenatively polymerized lignin in solution and solid state. *Macromolecules* 20: 1752–1756.
84 Kim, H., Ralph, J., and Akiyama, T. (2008). Solution-state 2D NMR of Ball-milled Plant Cell Wall Gels in DMSO-d_6. *Bioenerg. Res.* 1: 56–66.
85 Ralph, J. and Landucci, L.L. (2010). Chapter 5 NMR of Lignins. In: *Lignin and Lignans: Advances in Chemistry* (ed. C. Heitner, D. Dimmel and J. Schmidt), 137–244. CRC Press.
86 Mansfield, S.D., Kim, H., Lu, F., and Ralph, J. (2012). Whole plant cell wall characterization using solution-state 2D NMR. *Nat. Protoc.* 7: 1579–1589.
87 Wen, J.L., Sun, S.L., Xue, B.L., and Sun, R.C. (2013). Recent advances in characterization of lignin polymer by solution-state nuclear magnetic resonance (NMR) methodology. *Materials* 6: 359–391.
88 Schaefer, J. and Stejskal, E.O. (1976). Carbon-13 nuclear magnetic resonance of polymers spinning at the magic angle. *J. Am. Chem. Soc.* 98: 1031–1032.
89 Zhang, M. and Maciel, G.E. (1990). Enhanced signal-to-noise ratios in the nuclear magnetic resonance analysis of solids, using large-sample magic-angle spinners. *Anal. Chem.* 62: 633–638.

90 Maciel, G.E., O'Donnell, D.J., Ackerman, J.J.H. et al. (1981). A 1^{3}C NMR study of four lignins in the solid and solution states. *Makromol. Chem.* 182: 2297–2304.

91 Haw, J.F. and Maciel, G.E. (1985). Nuclear magnetic resonance study of autohydrolyzed and organosolv-treated lodgepole pinewood using carbon-13 with cross polarization and magic-angle spinning. *Holzforschung* 39: 99–107.

92 Leary, G.J. and Newman, R.H. (1992). Cross Polarization/Magic Angle Spinning Nuclear Magnetic Resonance (CP/MAS NMR) Spectroscopy. In: *Methods in Lignin Chemistry* (ed. S.Y. Lin and C.W. Dence), 146–161. Berlin, Heidelberg: Springer Berlin Heidelberg.

93 Argyropoulos, D.S. and Morin, F.G. (1995). Probing the macromolecular structure of wood and pulps with proton spin-lattice relaxation time measurements in the solid state. *Wood Sci. Technol.* 29: 19–30.

94 Kimura, T., Kimura, F., Argyropoulos, D.S., and Gray, D.G. (1992). ^{13}C CP/MAS NMR study of photodegraded stoneground wood pulp. *Holzforschung* 46: 331–336.

95 Argyropoulos, D.S., Heitner, C., and Morin, F.G. (1992). ^{31}P- NMR spectroscopy in wood chemistry. 3. solid-state ^{31}P NMRof trimethyl phosphite derivatives of chromophores in mechanical pulp. *Holzforschung* 46: 211–218.

96 Argyropoulos, D.S. (1994). Quantitative Phosphorus-31 NMR analysis of lignins, a new tool for the lignin chemist. *J. Wood. Chem. Technol.* 14: 45–63.

97 Argyropoulos, D.S. (1995). ^{31}P NMR in wood chemistry: A review of recent progress. *Res. Chem. Intermediat.* 21: 373.

98 Argyropoulos, D.S., Heitner, C., and Schmidt, J.A. (1995). Observation of quinonoid groups during the light-induced yellowing of softwood mechanical pulp. *Res. Chem. Intermediat.* 21: 263.

99 Saake, B., Argyropoulos, D.S., Beinhoff, O., and Faix, O. (1996). A comparison of lignin polymer models (DHPs) and lignins by ^{31}P NMR spectroscopy. *Phytochemistry* 43: 499–507.

100 Granata, A. and Argyropoulos, D.S. (1995). 2-chloro-4,4,5,5-tetramethyl-1,3,2-dioxaphospholane, a reagent for the accurate determination of the uncondensed and condensed phenolic moieties in lignins. *J. Agr. Food. Chem.* 43: 1538–1544.

101 Jiang, Z.-H., Argyropoulos, D.S., and Granata, A. (1995). Correlation analysis of ^{31}P NMR chemical shifts with substituent effects of phenols. *Magn. Reson. Chem.* 33: 375–382.

102 Argyropoulos, D.S. (1994). Quantitative Phosphorus-31 NMR analysis of six soluble lignins. *J. Wood. Chem. Technol.* 14: 65–82.

103 Brežný, R. and Schraml, J. (1987). Silicon-29 NMR spectral studies of kraft lignin and related model compounds. *Holzforschung* 41: 293–298.

104 Nieminen, M.O., Pulkkinen, E., and Rahkamaa, E. (1989). Determination of hydroxyl groups in kraft pine lignin by silicon-29 NMR spectroscopy. *Holzforschung* 43: 303–307.

105 Sevillano, R.M., Mortha, G., Barrelle, M., and Lachenal, D. (2001). ^{19}F NMR spectroscopy for the quantitative analysis of carbonyl groups in lignins. *Holzforschung* 55: 286–295.

106 Barrelle, M., Fernandes, J.C., Froment, P., and Lachenal, D. (1992). An approach to the determination of functional groups in oxidized lignins by ^{19}F NMR. *J. Wood. Chem. Technol.* 12: 413–424.

107 Barrelle, M. (1993). A new method for the quantitative ^{19}F NMR spectroscopic analysis of hydroxyl groups in lignins. *Holzforschung* 47: 261–267.

108 Barrelle, M. (1995). Improvements in the structural investigation of lignins by ^{19}F NMR spectroscopy. *J. Wood. Chem. Technol.* 15: 179–188.

109 Ahvazi, B.C. and Argyropoulos, D.S. (1996). Quantitative trifluoromethylation of carbonyl-containing lignin model compounds. *J. Fluorine Chem.* 78: 195–198.

110 Ahvazi, B.C. and Argyropoulos, D.S. (1996). ^{19}F Nuclear magnetic resonance spectroscopy for the elucidation of carbonyl groups in lignins. 1. Model compounds. *J. Agr. Food. Chem.* 44: 2167–2175.

111 Hatakeyama, H. and Nakano, J. (1970). Electron spin resonance studies on lignin and lignin model compounds. *Cellul. Chem. Technol.* 4: 281–291.

112 Hon, D.N.-S. (1992, Vol. Springer Series in Wood Science). Electron Spin Resonance (ESR) Spectroscopy. In: *Methods in Lignin Chemistry* (ed. S.Y. Lin and C.W. Dence), 274–286. Berlin, Heidelberg: Springer Berlin Heidelberg.

113 Kalyanaraman, B. (1995). Radical intermediates during degradation of lignin-model compounds and environmental pollutants: an electron spin resonance study. *Xenobiotica* 25: 667–675.

114 Sun, Y.-P., Yates, B., Abbot, J., and Chen, C.-L. (1996). ESR study of lignin model compounds irradiated by UV (254 nm Photons) in the presence of hydrogen peroxide. *Holzforschung* 50: 233–236.

115 Wan, J.K.S. and Depew, M.C. (1998). Applications of ESR and CIDEP to mechanistic studies of lignin chemistry. *Res. Chem. Intermediat.* 24: 831.

116 Jin, Y., Wu, M., Tai, D., and Li, Z. (1995). The study on the chemical structure characterizatics of Dahuiian larch lignin. *J. Cellulose Sci. Technol.* 4: 28–37.

117 Pearl, I.A. (1942). Vanillin from Lignin Materials. *J. Am. Chem. Soc.* 64: 1429–1431.

118 Pearl, I.A. and Beyer, D.L. (1950). Lignin and related products. VI. The oxidation of liquor with cupric oxide under pressure. *Tappi* 33: 544–548.

119 Pepper, J.M., Casselman, B.W., and Karapally, J.C. (1967). Lignin oxidation. Preferential use of cupric oxide. *Can. J. Chem.* 45: 3009–3012.

120 Gierer, J. (1990). Basic principles of bleaching. 2. Anionic processes. *Holzforschung* 44: 395–400.

121 Gierer, J. (1990). Basic principles of bleaching. 1. Cationic and radical processes. *Holzforschung* 44: 387–394.

122 Gierer, J. (1986). Chemistry of delignification. 2. Reactions of lignins during bleaching. *Wood Sci. Technol.* 20: 1–33.

123 TanaHASHI, M., Nakatsubo, F., and Higuchi, T. (1975). Structural elucidation of bamboo lignin by acidolysis and ozonolysis I. *Wood. Res. (Kyoto, Jpn.)* 58: 1–11.

124 Haluk, J.P. and Metche, M. (1986). Chemical and spectrographic characterization of poplar lignin by acidolysis and ozonolysis. *Cellul. Chem. Technol.* 20: 31–50.

125 Matsumoto, Y., Ishizu, A., and Nakano, J. (1986). Studies on chemical structure of lignin by ozonation. *Holzforschung* 40: 81–85.

126 Matsumoto, Y., Ishizu, A., and Nakano, J. (1984). Determination of glyceraldehyde-2-aryl ether type structure in lignin [from Picea jezoensis and Fagus crenata] by the use of ozonolysis. *Mokuzai Gakkaishi.*

127 Tsutsumi, Y., Islam, A., Anderson, C.D., and Sarkanen, K.V. (1990). Acidic permanganate oxidations of lignin and model compounds - comparison with zonolysis. *Holzforschung* 44: 59–66.

128 Akiyama, T., Sugimoto, T., Matsumoto, Y., and Meshitsuka, G. (2002). Erythro/threo ratio of β-O-4 structures as an important structural characteristic of lignin. I: Improvement of ozonation method for the quantitative analysis of lignin side-chain structure. *J. Wood. Sci.* 48: 210–215.

129 Adler, E., Falkehag, I., and Smith, B. (1962). Periodate oxidation of phenols. VIII. ^{18}O-Studies of the mechanism of oxidative cleavage of the monoether of pyrocatechol and of hydroquinone. *Acta. Chem. Scand.* 16: 529–540.

130 Adler, E. and Lundquist, K. (1961). Estimation of "uncondensed" phenolic units in spruce lignin. *Acta. Chem. Scand.* 15: 223–224.

131 Beck, H.D. and Adle, E. (1961). *Oxidation with quinones. Acta. Chem. Scand.* 15: 218–219.

132 Li, J.J. (2009). *Name Reactions*. Berlin: Springer-Verlag.

133 Funaoka, M. and Abe, I. (1978). The reaction of lignin under the presence of phenol and boron trifluoride. I. On the formation of catechol from MWL, dioxane lignin and Kraft lignin. *Mokuzai Gakkaishi* 24: 256–261.

134 Funaoka, M., Abe, I., and Chiang, V.L. (1992). Nucleus Exchange Reaction. In: *Methods in Lignin Chemistry* (ed. S.Y. Lin and C.W. Dence), 369–386. Berlin, Heidelberg: Springer Berlin Heidelberg.

Part III

Lignin Depolymerization: Scientific Questions, Challenges, and Current Progress

4

Scientific Questions for Lignin Conversion and a Brief Summary of Methods for Lignin Depolymerization

4.1 Opportunity and Challenges of New Biorefinery Approaches for Lignin Valorization

As we mentioned in Chapter 1, lignocellulose (LC) is the most abundant biomass form with an annual production of around 170 billion tons [1]. Traditionally, most large-scale industrial processes that use plant polysaccharides have burned lignin to generate the power to support the corresponding biomass transformation. Among the three main components of lignocellulose, lignin (15-30 w% of LC) is a complex, three-dimensional amorphous polymer, consisting of various methoxylated phenylpropanoid units, that can be regarded as the potential renewable resource to replace fossil resources to directly prepare functionalized aromatic chemicals and non-aromatic hydrocarbon stocks/fuels [2–4]. Therefore, the direct conversion of lignin in lignocellulose is an attractive process. In addition, the advent of biorefineries that convert cellulosic biomass into liquid transportation fuels will generate substantially more lignin than necessary to power the operation and therefore the potential use of non-edible lignocellulosic biomass for the production of value-added chemicals could provide an attractive alternative for fossil-based processes to prepare chemicals and fuels, thus being of great significance. However, the current paper industry and, more recently, biorefineries produce large quantities of lignin that today is almost considered a by-product, not to mention the transformation of native lignin. Not only from an economic standpoint but also from a sustainability perspective, the misuse or even nonuse of lignin appears to be a colossal mistake. Lignin, in its different versions available whether as native lignin from plants or partially transformed by industrial separation procedures, is a complex material with great potential by itself or as a source of chemicals [5].

For the potential practical utilization of lignin to chemicals, maintaining the aromatic structure and preparing high-value aromatic chemicals from renewable lignin seems more profitable compared with lignin conversion to liquid fuels with extra H_2 consumption (Figure 4.1). In the past decade, research on lignin depolymerization [2–9] and the transformation of its downstream products [10] have drawn

Lignin Conversion Catalysis: Transformation to Aromatic Chemicals, First Edition.
Chaofeng Zhang and Feng Wang.

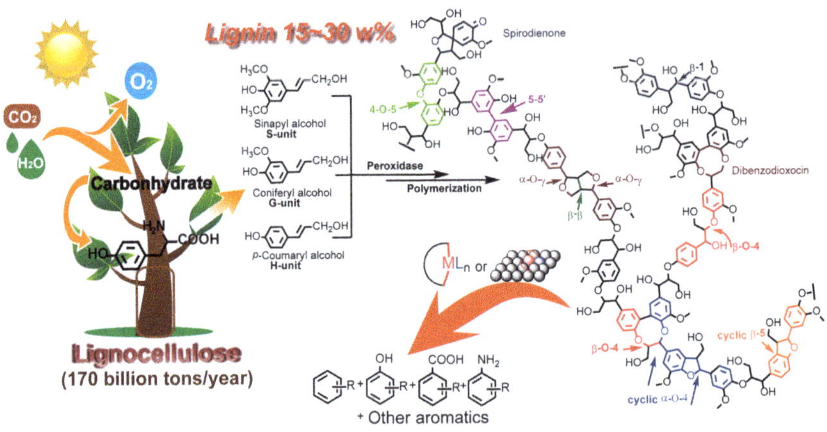

Figure 4.1 Lignin Depolymerization to Aromatic Chemicals. *Source:* Reprinted from Zhang and Wang [6] with Permission, © 2020, American Chemical Society.

enormous attention from academia to industry worldwide aimed at harvesting aromatic compounds from this abundant and renewable biomass resource. Lignin conversion efforts trace back to the 1920s and various non-catalytic and catalytic methods have been explored to depolymerize lignin the direct lignin conversion or study of lignin model conversions. However, the complexity of lignin structure, its various linkages, the high stability of lignin bonds, and its diverse fragments from the condensation process mean that lignin depolymerization to monomers, and selectively to aromatics, is still a highly challenging task. Therefore, lignin depolymerization to easy-to-handle aromatic monomers with acceptable conversion and selectivity is of great importance. Although different from cellulose, hemicellulose, and other bio-platform molecules with a relatively straightforward structures, lignin structures are much more complex and therefore the key issue for lignin transformation lies in the development of strategies with highly selective and active catalysts to effectively cleave the ubiquitous C–O/C–C bonds while leaving the aromatic benzene rings unconverted.

The detailed reasons making lignin depolymerization to monomers difficult include the highly complex mechanism caused by disturbing the lignin structure with various targeted linkages as well as factors such as transformations of the lignin structure and the varying condensation modes of lignin fragments or depolymerized monomers. To better make use of lignin to prepare aromatic chemicals, it is necessary to summarize the existing catalysts and processes. Based on current knowledge regarding lignin conversion, we can update the catalysts and develop new routes for efficient lignin conversion. As described in Chapter 1 about the book structure, the following chapters aim to provide an overview of key advances in the field of lignin depolymerization to aromatics. Unlike previous review works [2, 9, 11–16] that classified research by the lignin substrate (native lignin, kraft lignin, organosolv lignin, etc.), cleavage methods (oxidation, hydrogenation, etc.), or main products (arenes, phenol, acid, etc.), this book will provide a brief introduction of past technologies and new catalytic methods or strategies in Chapter 4

(Part III) and then Parts IV and V will pay more attention to the relationships among the strategy/method, catalyst, and mechanism viewed from the cleavage order and type of corresponding chemical bonds in various catalytic systems. First, however, this chapter will present the critical scientific questions involved in lignin depolymerization and the foundation of strategies (Chapter 4.2) as well as two different approaches for the foundation of lignin depolymerization strategies: direct lignin conversion and the bottom-up approach (Chapter 4.3).

4.2 Scientific Questions Involved in Lignin Depolymerization and the Foundation of Strategies

Before working on lignin depolymerization, it is necessary to first make clear the scientific questions involved as this is beneficial to understanding the foundation of catalytic strategies for efficient lignin utilization. For the depolymerization of lignin polymer with a complex 3D structure (or 2D structure in some genetically modified plants) to easily handled aromatic monomers with acceptable conversion and selectivity, reactions should firstly focus on the target substrate structure to be cleaved. Although the exact structure of protolignin is quite complex and different from plant to plant, *the target is clear that the key issue for lignin utilization to preparing aromatic chemicals depends on the efficient strategies to cleave the ubiquitous C–C or C–O linkages bonds while leaving the aromatic rings intact* (Figure 4.2).

Compared with other linkages, *the β-O-4 linkage is dominant and accounts for more than half of interunit linkage in lignin. Its C–O–Ar and C_α–C_β bonds are the bonds chiefly targeted to break during the delignification process. In addition, the C–O/C–C bonds in β-5, α-O-4, β-β, 5-5, 4-O-5, and β-1 linkages should be also cleaved to increase the depolymerization efficiency*. As we discussed above, the key issues for the lignin transformation lie in the development of highly selective and active catalysts to effectively cleave the ubiquitous C–C or C–O linkages bonds, while leaving the aromatic benzene rings unconverted. The next step is founding effective depolymerization strategies. For lignin depolymerization to aromatic monomers, unlike other organic reactions, two factors in particular should be considered before catalyst screening: *(1) Promoting the lignin depolymerization via selective cleavage of the ubiquitous lignin C–O/C–C bonds, increasing the conversion; and (2) Restraining the lignin fragments recondensation with more stable chemical bonds generation or multimers formation, increasing the selectivity and final yield of monomers*. A detailed summary discussion about lignin catalytic conversion based on mechanism classification is provided in Chapter 5 of Part IV.

To obtain higher lignin depolymerization conversion and more useful monomers, in addition to directly using the various isolated lignins as the substrate, recent lignin conversion research has paid more attention to the effect of lignin structure change of the pretreatment of lignocellulose (the lignin linkages transformation and fragment condensation). In recent reviews of new lignin biorefinery reported by Sels et. al. [17, 18], it is noted that *new biorefinery methods need first to actively*

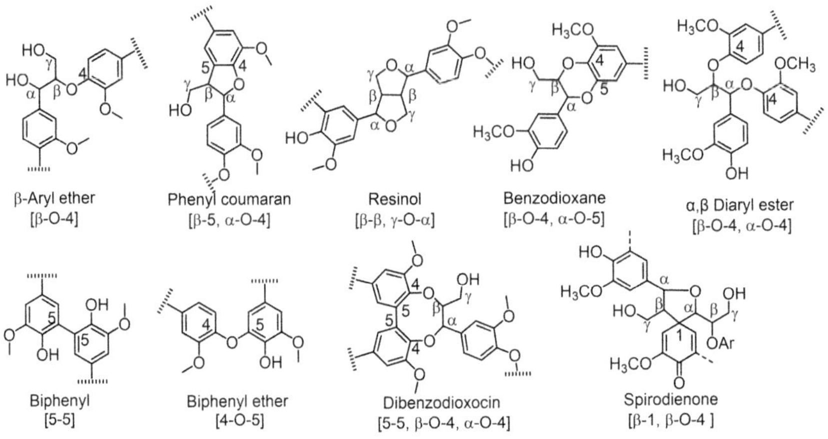

Linkages	Number /100 ppu		Linkages	Number /100 ppu	
	Softwood	Hardwood		Softwood	Hardwood
β-O-4	43-50	50-65	4-O-5	4	4-7
5-5'	10-25	4-10	β-1	3-7	5-7
β-5	9-12	4-6	β-β	2-4	3-7
α-O-4	6-8	4-8	Others	16	7-8

Figure 4.2 Frequencies of MajorLignin Linkages.

preserve dominant β*-O-4 bonds.* During organic synthesis, one can keep the selectivity of the reaction via the initial protection of sensitive groups that do not participate in the conversion. For lignin conversion, the similar β-O-4 bond protection mode means that one can first protect/transform the active functional groups of lignin at the beginning of pretreatment, which could preserve the β-O-4 structure for further controllable conversion and restrain the possible condensation of lignin fragments. Recently, the pre-protection methods (discussed in Chapter 11) have drawn much attention to lignocellulose hydrogenation approaches such as the formaldehyde stabilization method and its derivative methods, which involve the first protection of C_α–OH and C_γ–OH in the β-O-4 structure [19]. Some etherification/esterification pretreatment at C_α–OH and C_γ–OH in the β-O-4 structure with protonic polar solvents or regents can weaken or replace hydrogen bonds to promote lignin release from lignocellulose and the etherification/esterification with alcohol/acid can further increase the dissolving capacity of lignin for the following depolymerization. Furthermore, as another promotion effect, the first oxidation of C_α–OH to C_α=O could not only restrain the generation of the C_α carbocation (which can cause the lignin fragments condensations) but also can promote the hydrogenation cleavage of β-O-4 linkage by weakening the C_β–OAr bond (discussed in Chapter 9). In addition, tandem depolymerization–stabilization of native lignin depolymerization fragments/monomers is another method to prevent unwanted condensation (discussed in Chapter 11). These cascaded models usually combine two different methods to achieve the in-situ lignin fragments depolymerization to monomers and instant monomers stabilization or transformation. Compared with the single transformation method, the cascaded transformation represented by the acidolysis+acetal formation, acidolysis+hydrogenation, acidolysis+decarbonylation [20], and alkali catalysis+hydrogenation/oxidation methods can restrain the possible condensation and make the conversion process more controllable. More introduction to these approaches will be presented in Part IV.

Furthermore, for the practical conversion of the complex lignin polymer, *the first fragmentation of the huge lignin molecule to the one with a lower molecular weight (2~15 aromatic rings) can promote the reaction rate by increasing the activity of reaction collisions between molecule target bonds and catalytic centers to achieve the cleavage of linkage bonds, during which the solvent molecules and additives play a critical role.*

As one example of the study of lignin fragmentation from 2013, Wang and Xu [21] found a fragmentation–hydrogenolysis process for protolignin depolymerization into monomeric phenols in alcoholic solvents during which alcohols (such as methanol, ethanol, and ethylene glycol) not only worked as solvents and hydrogen donors but also acted as the nucleophilic reagent for cleaving ether linkages to achieve the preliminary fragmentation of birch wood into smaller and soluble species consisting of several aromatic rings. The Ni/C then catalyzed the hydrogenolysis of resultant fragments to phenols. The best selectivity toward monomeric phenols at 200 °C was 90% with a 54% birch lignin conversion in methanol. In addition to the study of lignin fragments conversion, they focused on the fragmentation mechanism and the solvolysis effect of various solvents, especially for alcohols, on the degree of fragmentation. They first performed lignin dissolubility experiments in various solvents monitored by UV-Vis spectroscopy (Figure 4.3), in which the absorption band at ca. 275 nm was used to quantify the amount of dissolved lignin according to the guaiacol standard

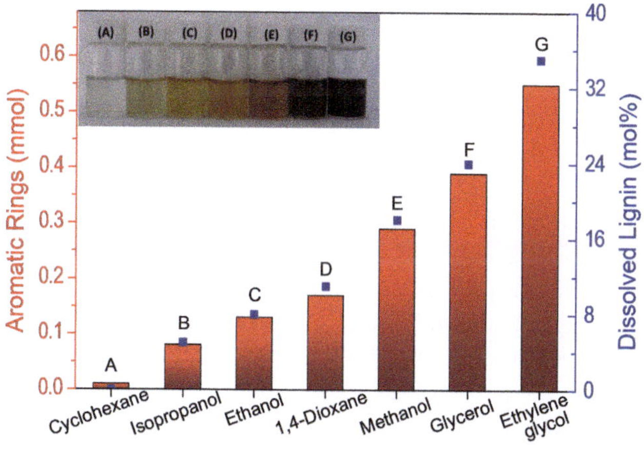

Figure 4.3 The Solubility of Birch Lignin in Various Solvents Characterized by the UV-Vis Spectrometer. Reaction conditions: Birch Sawdust (1.0 g), Solvent (10 mL), Ar, 200 °C, 30 min. *Source:* Reprinted from Zhang and Wang [6] with Permission, © 2020, American Chemical Society.

curve [22]. Two points were noted: (1) protonic polar solvents (alcohols) exhibit higher dissolving capacity than that of nonpolar solvents; and (2) the solvents with more hydroxyl groups are better for lignin dissolution because the hydroxyl groups can weaken or replace hydrogen bonds to promote lignin release from lignocellulose and the following etherification with alcohol can further increase the dissolving capacity.

The lignin dissolubility experiment can explain some solvent effects on lignin conversion [23, 24], but does not provide the whole story. The MALDI-TOF technique was employed to monitor the molecular weight change of fragmented lignin or the degree of lignin fragmentation (Figure 4.4 A-C). The reaction between lignin and methanol/ethanol made alcohol solvents different from other solvents without hydroxyl groups like 1,4-dioxane. The 1,4-dioxane-dissolved lignin fragments stretch beyond 4000 (M_w=5558, M_n=5049), but alcohol-dissolved lignin fragments were

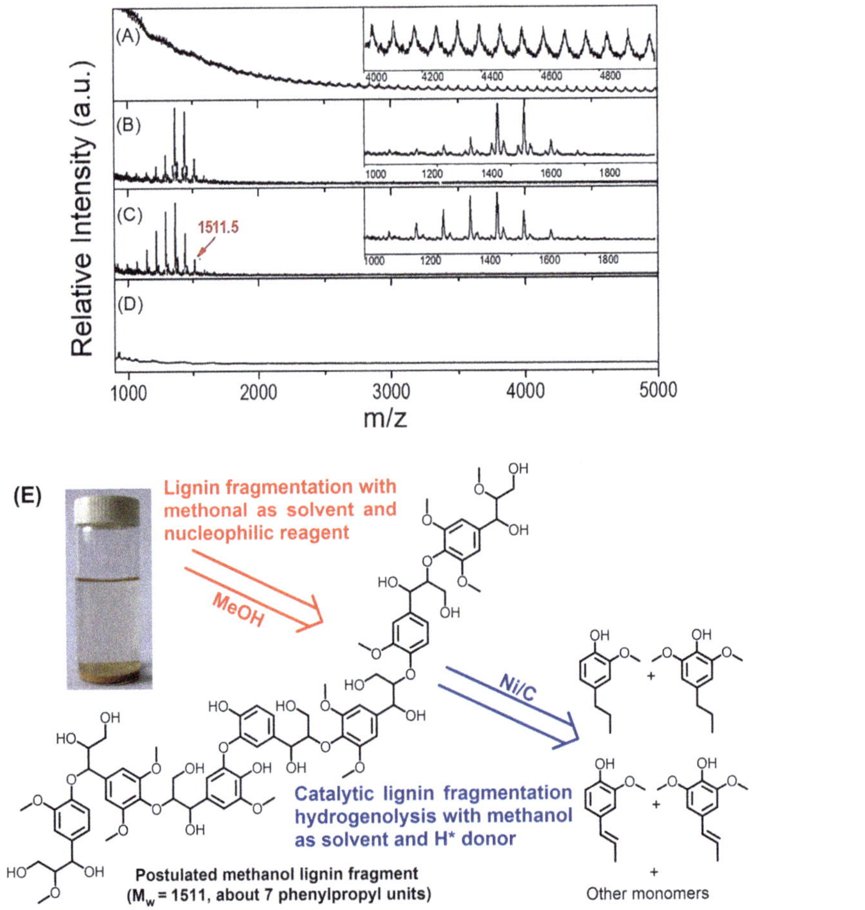

Figure 4.4 MALDI-TOF Profiles of Dissolved Birch Lignin in 1,4-Dioxane (A), Ethanol (B), Methanol (C), Sample C after Reaction over Ni/C (D), and the Fragmentation–hydrogenolysis Process with a Postulated Fragment m/z=1511 (E). *Source:* Reprinted from Zhang and Wang [6] with Permission, © 2020, American Chemical Society.

much smaller. It was challenging to give an exact structure of the corresponding fragment due to its complexity. They found that a typical fragment from methanol soluble lignin (M_w=1309, M_n=1292) is m/z=1511.5, which was calculated to contain seven phenylpropyl rings (Figure 3E). The ethanol-soluble lignin has a similar pattern of molecular weight distribution (M_w=1406, M_n=1309) compared with the methanol-soluble lignin. The difference is that M_W of the ethanol-soluble lignin is 97 m/z larger than the methanol sample, approximately equal to seven -CH_2- groups. The difference suggests that each phenylpropyl unit of the fragment might consist of one alkoxyl group from solvent in the fragmentation step (Figure 4.4E). As shown in Figure 4.4D, methanol-lignin fragments were further cut into monomeric phenols over the Ni/C catalyst. Furthermore, this fragmentation–hydrogenolysis process understanding can provide a solution for the separation difficulty between heterogeneous catalyst particles and carbohydrate solid residua after lignin conversion, by separating the solvent fragmentation and hydrogenolysis/following conversion into two processes or trapping the solid catalysts in a basket [25, 26]. In addition, it can be concluded that *the dissolution of lignin fragments from solid lignocellulose, its solubility in solvents, its size/molecular weight, and its structural characteristics have important impacts on the subsequent depolymerization and conversion to monomers.*

4.3 Two Different Approaches for the Foundation of Lignin Depolymerization Strategies

For the practical utilization of a lignin polymer with a complex structure, the lignin should first be depolymerized to easily handled aromatic monomers with acceptable conversion and selectivity. However, the natural complexity and high stability of lignin bonds make lignin depolymerization a highly challenging task. To achieve this goal, various non-catalytic and catalytic methods have been explored to cleave the various lignin C–C or C–O linkages bonds via two approaches: (1) a 'direct plant powders or isolated lignin conversion' approach; and (2) a from typical models to lignin approach (bottom-up approach). These approaches place extra emphasis on the disciplines of direct lignin conversion or lignin model conversion, respectively. Although a vast difference exists between studies on model compounds and studies on technical/native lignin because of the complex structure of lignin, the bottom-up approach has presented many effective catalytic systems and strategies in the technical/native lignin conversion. Nevertheless, each approach has its advantages and limitations. Hence, both domains should be complementary in developing efficient methods of lignin conversion.

4.3.1 Direct Plant Powders or Isolated Lignin Conversion

At the beginning of lignin conversion research, lignin was regarded as a resource like oil and coal and the thoughts and equipment for oil refinery and coal transformation were employed in lignocellulose and lignin utilization. These included pyrolysis and gasification processes carried out above 400 °C. Although these

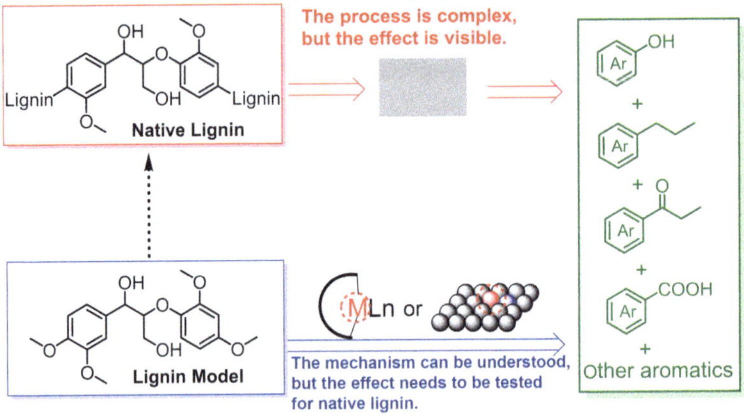

Figure 4.5 Two Different Approaches for the Foundation of Lignin Depolymerization Strategies.

processes can produce a certain amount of liquid product, known as bio-oil products, and a mixture of phenols, low-carbon alcohol, acid, ester, ether, and hydrocarbon as well as some biomass structures that can be transformed into CO, CO_2, H_2, and other gas products, the harsh conditions with high temperatures can easily cause the vast generation of carbon to deposit or to remain as unconverted carbon polymer from lignin carbon constituents.

To efficiently utilize biomass with different constituents, or lignins with different structures from different plants and isolated processes, the new biorefinery with a clear and reasonable step-by-step process is necessary and efficient [2, 5, 17]. Furthermore, to get a higher conversion and more simple but useful chemicals, besides directly using the various isolated lignins as the substrate, recent direct lignin conversion research paid more attention to the lignin structure change effect (linkage transform and fragments condensation) of the pretreatment of lignocellulose. In Sels' recent work [17], they point out that the new biorefinery paradigm includes two distinct strategies to actively prevent structural degradation during biomass fractionation, namely: (1) tandem depolymerization–stabilization of native lignin; and (2) active preservation of β-O-4 bonds. In addition, as shown in Figure 4.5, catalysis is regarded as a key enabling technology for efficiency and conversion selectivity of biomass conversion in general and for fulfilling the promise of lignin valorization in particular [2, 27]. Until the present, some catalytic systems have achieved the direct conversion of native lignin to aromatics [9, 15, 17], but the detailed process is very complex. More knowledge about the conversion disciplines is useful for further lignin direct conversion.

4.3.2 Bottom-up Approach: From Models Conversion to Lignin Depolymerization

Unlike direct depolymerization of native lignin, another approach termed "From typical models to lignin" or "Bottom-up" focuses on the mechanism of typical C–O or C–C bonds cleavage in lignin model compounds and then utilizes the discovered methods in the native lignin conversion. As shown in table of Figure 4.2, the β-O-4 ether bond is dominant and accounts for more than half of interunit linkage in lignin. This is therefore the chief targeted bond to break during the delignification process [2]. Besides β-O-4 linkage models and their derivatives, other common used lignin models contain 5-5, α-O-4, 4-O-5, β-1, β-1, and β-β models (Figure 4.2). As an obvious advantage, lignin model research elucidates the mechanism. According to the analysis of lignin conversion research in the past decade, more than half of the articles included the study of the conversion of lignin model compounds and then the catalysts, followed by the development of methods/strategies used in real lignin transformation. Unlike direct lignin conversion research, which evaluates the effect by the product yield and proposes the reaction process by the product distribution, the "Bottom-up" approach can give a more detailed study of the mechanism by focusing on the transformation of target bonds in the lignin model over the catalyst. This can strengthen the design of the process for lignin conversion and promote the upgrade of transformation technology. In addition, the conversion of lignin model compounds provides new tests and application outlets for the application of new chemical (catalytic) methods and proposes new research contents for other disciplines.

Although a vast difference exists between model conversion and native/technical lignin conversion because the complex lignin structure makes the latter conversion more challenging, the "bottom-up" approach has presented many effective catalytic systems and strategies in the technical/native lignin conversion. Nevertheless, each approach has its advantages and limitations. Hence, both domains should be complementary in establishing efficient methods in lignin conversion.

4.4 Classification of Lignin Conversion Methods by Reaction Types

With great efforts and contributions from many research groups, significant advances have been made in recent years in the upstream depolymerization of lignin into platform chemicals via various pathways. To understand the multitudinous methods for lignin depolymerization with different strategies and catalysts, it is extremely necessary and useful to first make a summary of catalytic/non-catalytic methods for lignin conversion. By the end of 2021, a number of excellent reviews focusing on lignin depolymerization technologies had been published. In 2010, Weckhuysen and co-workers [2] provided an elegant review regarding the valorization of lignin via the approaches of (hydro)cracking, lignin reduction, and oxidation reactions. In 2012, Dumesic and co-workers [28] thoroughly reviewed different catalytic systems for biomass upgrading into fuels and chemicals. In 2014, Ragauskas and co-workers [5] reviewed the biosynthesis, characterization, and recovery and briefly summarized the valorization of lignin as a whole. Luque and co-workers [3, 29] overviewed both existing and emerging lignin depolymerization strategies. In 2015, Li, Zhang, Huber, and co-workers [9] summarized the critical methods and strategies in lignin depolymerization. In 2016, Weckhuysen and co-workers [30] provided an in-depth analysis of the biosynthesis and structure of lignin, the structural differences between native and technical lignin, and the relationships between the lignin structure and catalyst performance in depolymerization. In 2018, Sels and co-workers [31] comprehensively discussed the strategies of lignocellulose fractionation and depolymerization. In 2018, Barta and co-workers [15] elegantly summarized chemical and biological pathways to depolymerize lignin into high yield monomers, and the functionalization and defunctionalization of lignin-derived compounds, with a particular emphasis on biobased polymers and pharmaceuticals. These reviews, together with others more than 100 review works by the end of 2021, viewing from different points [32–35], serve as guides for future developments in lignin utilization technologies.

On the basis of previous work, we herein make a brief summary in this chapter of methods for lignin depolymerization, which begins with the classification of lignin conversion methods by reaction types. First, on the basis of methods classification reported in previous reviews and new strategies in recent years, the methods for lignin depolymerization can be broadly classified into the following categories: (1) pyrolysis ([P]); (2) gasification ([G]); (3) liquid phase reforming ([LPR]); (4) chemical oxidation ([O]); (5) hydrogenation ([H]); (6) transfer hydrogenation ([TH]); (7) reduction ([R]); (8) solvolysis ([S]); (9) acidolysis ([A]); and (10) alkaline hydrolysis ([B]); as well as newly developed methods including (11) redox-neutral process ([RN]); (12) combinatorial strategies ([CS]); (13) pre-protection strategies ([PS]); and (14) self-decomposition ([SD]), among others. These approaches may also contain several submethods according to the classification of reaction formats, such as thermal, photo, and electro processes, photo-thermal/-electro processes, different biodegradation methods, and others.

During these methods, hydrogenation is well studied and widely used in the native lignin conversion. Hydroprocessing involves the thermal reduction with H_2 at temperatures typically ranging from 100 to 350 °C, which has the potential to produce simple bulk aromatic compounds such as phenols, benzene, toluene, and xylene, and even alkane fuels with the hydrogenation of aromatic rings. During lignin conversion under oxygen-free conditions and a high temperature, the lignin active -OH and -OMe groups and H-containing solvent can also work as the hydrogen donors and participate in the transfer hydrogenation, which can produce similar products like the hydroprocessing, but the ratio of alkane products with hydrogenated aromatic rings is usually lower. Additionally, the chemical oxidation methods occur at lower temperatures ranging from room temperature to 250 °C with O_2, H_2O_2, or other peroxides, which favors producing aromatic aldehydes, acids, and esters. Alternatively, the reduction with an active hydride (Metal–H, B–H, Si–H), thiol (R–SH), an active metal in acid solution ($Zn+NH_4Cl$) or alkali metal in liquid ammonia ($Na+NH_3$), or electrons from the cathode, photocatalysts, or biodegradation processes can achieve the lignin reduction primarily at a lower temperature (<100 °C). Furthermore, at the beginning of lignin conversion research, lignin was regarded as another resource like coal and oil and people tried to achieve the refining of lignin and its extractive via pyrolysis or gasification at the temperature of 450–700 °C, which produced a liquid product known as bio-oil" and complex gas products with synthesis gas (CO and H_2), CO_2, and others. Liquid-phase reforming (typically at 250–400 °C) has been attempted for lignin conversion that produced hydrogen and light gases from the lignin. Unlike liquid-phase reforming, solvolysis can be regarded as another relatively mild reforming method that can produce a certain amount of monomers but with obvious fragmentation of lignin polymer with supercritical CO_2, H_2O, CH_3OH, and other compounds. Supercritical alcoholysis or high boiling point alcohol solvolysis can efficiently fragment the lignin chain to multimers with a much lower molecular weight than the native lignin, which can be dissolved in the solvent. Cooperating with some transfer hydrogenation catalysts, these dissolved multimers can be cleaved into monomers. In addition, depolymerization reactions catalyzed by both acid (typically at 0–200 °C) and base (100–300 °C) can break the C–O or C–C linkages between lignin units to offer small segments including monomeric phenols.

Furthermore, besides the one-step or direct method to efficiently catalyze the lignin conversion, some combinatorial strategies (CS) contain two or more processes of [O], [H], [TH], [A], [B], and others have drawn much attention. To make the lignin conversion more efficient and reduce the condensation of fragments condensation, acidolysis and alkaline hydrolysis are usually combined with other catalytic methods such as hydrogenation and decarbonylation. Similarly, the combinatorial strategies can also begin with selective oxidation and dehydrogenation such as the following: (1) oxidation-Baeyer-Villiger oxidation-hydrolysis; (2) oxidation-Beckmann rearrangement-hydrolysis; (3) oxidation-reduction; (4) oxidation-hydrogenation; and (5) oxidation-hydrolysis. The combinatorial strategies are far more than those listed in Table 4.1.

In addition, some redox-neutral ([R-N]) methods have been presented for the C–C/C–O bonds cleavage of lignin and its model compounds via the first rearrangement of the corresponding atoms in the lignin structure, which produce an active intermediate and lead to the lignin linkages broken to monomers. The [R-N] methods herein are different from the [P], [G], [LPR], [A], [B], and [TH] methods. To achieve the efficient lignin depolymerization to monomers with restraint of the potential recondensation of fragments, some pre-protection strategies [PS] have been developed. These include the typical method with the first etherization of the C_α–OH. Furthermore, some researchers have also tried to achieve the self-decomposition [SD] of the lignin linkages, which requires the introduction of high energy such as strong light or even a high energy ray to the lignin linkage skeleton. Therefore, a more detailed summary of the main methods for catalytic conversion of lignin is presented in Table 4.1.

Table 4.1 A Brief Summary of Main Methods for Lignin Conversion by Reaction Types.

Methods	Subgroup	Temperature (°C)	Catalysts/Systems	Note
Pyrolysis [P] [32, 36]	—	450~700	None, zeolites, oxide/zeolite, metal NPs, etc.	Produce bio-oil (aromatic chemicals, and alkane), synthesis gas (CO and H_2), CO_2, and others
Gasification [G] [37]	—	480~700	None, noble metal salts, different forms of metals, etc.	
Liquid phase reformation [LPR]	—	200~450	None, Group VIII metals, noble metals, alloys, etc.	Produce H_2, light gases, and aromatic monomers
Oxidation [O] [12, 13, 38, 39]	Thermal oxidation [TO]	<300	(1) Organometallics and metal salt with O_2: Co [40–43], V [44–49], Cu [50–53], Mn [54], Re [38, 55], Ce [56], Fe [57], etc.; (2) High valence metal or metal with peroxide (SET mechanism): M=Fe, Cu, V, Mo, Co, Mn, Ag, Ce, etc. [58–68]; (3) Metal salt+organocatalytic system: benzoquinones, $R_2NO^•/R_2NOH$ (e.g. TEMPO, NHPI) [69], organoiodine [70], Cu+$PhNO_2$ [71–75], etc.; (4) Polyoxometalates [76]; (5) Ionic liquid (as solvent and catalyst) [77]; (6) Metal oxide and mixed metal oxide: CuO_x, [78] MnO_2 [79], $LaCoO_3$, $LaMnO_3$, $LaFeO_3$, $LaA_{1-x}B_xO_3$ [80–86], $CuCeO_x$ [87] etc.; (7) (Noble) metal/metal oxide (or SiO_2 as the support) [88]; (8) Metal-zeolite; (9) Carbon-based catalyst [89]; (10) M-CN_x catalyst: e.g. Co-CN_x [90–93]; etc.	Air, O_2, H_2O_2, ROOH, RCOOOH, NaClO, $K_2S_2O_8$, etc. are used as the oxidant. The phenolic acid/aldehyde, quinones are the main products. In some reactions, the aromatic ring can be broken with the generation of conjugate carboxylic acid.

Photo-oxidation [PO]	~r.t.	(1) Organometallics: V [94], etc.; (2) Organic photocatalyst: Acridinium catalyst [95, 96], DDQ [96, 97], DCA [95, 98, 99], Bengal Rose [100], etc.; (3) High valence matal species:Ce, Ru, Ir, etc. [101–103]; (4) Metal oxide or semiconductors: MnO_2 [104], $CuO_X/CeO_2/TiO_2$ [105], etc.; (5) CN-based photocatalysts: C_3N_4 [94], etc,; (6) Metal sulfide: e.g. $ZnIn_2S_4$ [106]; etc.	Air, O_2, H_2O_2, ROOH, or azo compounds are used as the final oxidant. The main products are the phenolic aldehyde/acid.
Electro-oxidation [EO]	~r.t.	(1) Anode surface oxidation: Ru, graphite, Ni, Pt, PbO_2, etc. [107–113]; (2) Anode oxidation with mediate: TEMPO/$TEMPO^+$ [114, 115], NHPI [116], Ar_3N [107], etc.; (3) Anode oxidation product as the oxidant: e.g. NaCl/NaClO [117], O_2/H_2O_2 [118], etc.	The anode catalyzes the oxidation of the substrate or the generation of oxidant/mediate in an oxidative state. The phenolic acid/aldehyde, quinones are the main products.
Enzymatic oxidation [EnO]	r.t.~50	(1) HAT processes: Laccase with a HAT cocatalyst [49, 119–124]; (2) SET processes: Lignin peroxidase [66, 125–127], Manganese peroxidase [128–130]; etc.	O_2, H_2O_2, or other peroxides are used as the final oxidant. The main products are phenolic acid/aldehyde, quinones.
Photo-electro/Photo-thermal Oxidation [PEO/PTO]	r.t.~200	The ingenious [TO] or [EO] systems are enhanced by photocatalysis [131, 132].	See above about the [TO], [PO], and [EO] systems.

(Continued)

Table 4.1 (Continued)

Methods	Subgroup	Temperature (°C)	Catalysts/Systems	Note
Hydrogenation [H] [34, 133–135]	—	100~350	(1) Organometallics: Ni [136], Ru, Rh [137], etc.; (2) Metal catalyst: Ni [138–142], Ru [143–145], Pt [146], Pd [146], Cu [147], Rh [148], etc.; (3) Metal alloy: [149, 150] NiW [151], B-modified alloy [152, 153] etc.; (4) Metal with homogeneous acid: Pd/C or Ru/C+ $M(OTf)_n$ [20, 154–158], etc.; (5) Metal or alloy NPs with mineral acid or in acid ionic liquid [159–162]; (6) Metal or alloy NPs with base [163, 164]; (7) Mixed metal oxide: CuCrO [165, 166], $ZnMoO_4$ [167], $NiAlO_x$ [168], etc.; (8) Metal sulfid: MoS_2, [169] $NiMoS_x$ [170], $CoMoS_x$ [171], etc.; (9) Metal carbide: MoC_x [172], WC_x [173] etc.; (10) Metal nitride: TiN [174–176] etc.; (11) Metal phosphide: MoP [169], FeMoP [177], Ni_2P [178], etc.; (12) Metal boride: NiB_x [179], etc.; (13) MOF [180]/Zeolite [181], and other porous catalysts [182] without metal NPs; (14) $M\text{-}CN_x$ catalyst: e.g. $Fe\text{-}CN_x$ [183]; etc.	The main products of these reactions are alkylphenol, alkylbenzene, substituted cyclohexanol or cyclohexane, etc.
Transfer hydrogenation [TH] [184, 185]	Intermolecular Transfer hydrogenation [ITH]	r.t.~250	(1) Organometallics; (2) Thermal catalytic systems: Pd/C [186], Pt/C [187] with HCOOH, Pd/C with 1,4-dioxane [188], etc.;	The H* species come from the solvent or extra H* donor reagents. The main products of these reactions are acetyl-/propionyl-phenols, alkylphenol, alkylbenzene, etc.

	Self-transfer Hydrogenation [STH]	r.t.~250	Systems with C_{α}/C_{γ}–OH as hydrogen dornor: (1) Homogeneous thermal catalytic system Organometallics Ru [189–192], Rh [193, 194], etc.; (2) Heterogeneous thermal catalytic system: Pd/C [188, 195, 196], Pt/C [197], Alloy [198], etc.; (3) Photocatalytic system: $ZnIn_2S_4$ [199], CdSe [200], Ag_2S@CdS [201], etc.; System with hydroxy group and MeO- as the hydrogen donor: (4) Ru-WO_x [202], M/$NiAl_2O_4$ (M=Pt, Pd, Ru) [203, 204], RuW/Zeolite HY_{30} [205]; etc.	The H* species come from the substrate hydroxy, methoxy groups, or the released CO after the water-gas shift reaction. The main products of these reactions are acetyl-/propionyl-phenols, alkylphenol, alkylbenzene, or even benzene, etc.
Reduction [R]	Thermal reduction [TR]	<100	(1) Hydride: B_xH_y, $NaBH_4$ [206–208], $LiAlH_4$ [206], R_3SiH [209, 210], systems with/without transition metals or I_2; (2) Active metal: e.g. Zn/NH_4Cl [211]; (3) Electron: e.g. Na in liquid NH_3 [212]; etc.	The main products of these reactions are phenols, arenes, or methyl phenyl ethers after the cleavage of the ether bond.
	Photoreduction [PR]	~r.t.	Substrate as the sacrificial agent: CdS [213, 214], $Zn_4In_2S_7$ [215], etc.	
	Electroreduction [ER]	~r.t.	Cathode surface reduction: Ni, PtNiB [216–219], etc.	
	Enzymatic reduction [EnR]	r.t~<50	LigDFG enzyme system [220], etc.	

(Continued)

Table 4.1 (Continued)

Methods	Subgroup	Temperature (°C)	Catalysts/Systems	Note
Solvolysis [S]	Supercritical solvolysis [SS][a]	> 200	With/without catalyst: water and other solvents	Produce H_2, light gases, and aromatic monomers.
	Alcoholysis [SA] [221]	> 200	(1) Without catalyst: MeOH [21], EtOH [24] etc.; (2) With catalyst: Ni/C [21], MoC_{1-x} [222, 223], $CuMgAlO_x$ [224, 225], MoC_{1-x}/Cu-MgAlOz [226], Mo_2N [227], Ru/C; [228] etc.	Alcoholysis provides lignin fragments that can be further cleaved by the catalyst to aromatic monomers.
Acidolysis [A] [16]	—	r.t.~ < 150	(1) Mineral acid (HCl [229, 230], H_2SO_4 [231–233], H_3PO_4 [234]) and organic acid [235–238]; (2) Lewis acid: MCl_n [239, 240], $M(OTf)_n$ [20, 157, 241–244], etc.; (3) Acid ionic liquid [245]; (4) Organic metal compound: methyltrioxorhenium (MTO) [246, 247], etc.; (5) Heteropolyacid [248]; (6) Solid super acid [249] and niobic acid [250, 251]; (7) Solid metal oxide [252, 253] and Zeolite [254–256]; etc.	The main products of these reactions are acetyl-/propionyl-phenols.
Alkaline hydrolysis [B]	—	r.t.~ < 300	(1) Homogeneous base: LiOH, NaOH, KOH, Cs_2CO_3, and ^{t}BuOLi [257–260], etc.; (2) Solid metal oxide: MgO, CaO, etc.; (3) Hydrotalcite (HT) and hydroxyapatite (HAP) [261]; (4) Zeolitic catalyst: NaX, NaY, NaP [262], etc.; (5) Organic base: 1,5,7-Triazabicyclo [4.4.0] dec-5-ene (TBD) [263], etc.; (6) Basic ionic liquid; etc.	The main products of these reactions are conjugated unsaturated phenols and methoxy phenols.

Combinatorial strategies [CS]	Processes contains two or more steps of [O], [H], [TH], [A], [B], [R] and other reaction steps.	r.t.~ < 200	(1) Oxidation- Baeyer-Villiger oxidation-Hydrolysis [264–269]; (2) Oxidation-Beckmann Rearrangement-Hydrolysis [270, 271]; (3) Other oxidation-Rearrangement with a N atom insertion [272, 273]; (4) Oxidation-Reduction: Thermal reduction with: Zn/NH_4Cl [211], thiol [274]; Photocatalytic reduction catalyzed by (i) Ir catalysts [275–277]; (ii) Organic molecule: perylene diimide (PDI) [278], etc.; (iii) Semiconductore: TiO_2 [106], etc.; (iv) Solid organic photocatalysts (OPCs) [279]; etc.; (5) Oxidation-Hydrogenation [69, 280]; (6) Oxidation-Hydrolysis (HCOOH/Na) [281, 282]; (7) Muitiple oxidation [283, 284]; (8) Acidolysis-hydrogenation: $HOT_f/M(OTf)_n$+Ru/C [20, 242], etc.; (9) Acidolysis-transfer hydrogenation: RuW/HY zeolite [205], etc.; (10) Acidolysis-decarbonylation: HOTf+Ir [20, 242], etc.; (11) Solvolysis-hydrogenation [285]; (12) Hydrogenation-(oxidation)-dealkylation [286–289]; (13) Acidolysis-dealkylation: HY_{30} zeolite [290]; $La(OTf_3)$ [291]; etc.	The main products of these reactions are based on the process but all of them focus on the generation of aromatic monomers.
Redox-neutral process [R-N]	Thermal process [TR-N]	r.t.~ < 200	(1) Molecule self-rearrangement with C–C bond cleavage: Ru (Retro-aldol reaction) [292], Ir (Retro-aldol reaction) [293], Ir-(PCET) [101, 103], V-(LMCT) [294]; (2) Molecule self-rearrangement with C–OAr bond cleavage: V [295], Ir [296]; etc.	Processes do not belong to [P], [G], [LPR], [TH], [A] or [B].
	Photo process [PR-N]	r.t.~ < 200	Exogenous molecules/free radicals promote structural rearrangement: (1) Acridinium-catalyzed the acidolysis diaryl ether with ArCOOH [297]; (2) UO_2^{2+}-catalyzed hydrolysis of diaryl ethers bond [298]; etc.	The main products are based on a detailed process.

(Continued)

Table 4.1 (Continued)

Methods	Subgroup	Temperature (°C)	Catalysts/Systems	Note
Pre-protection strategies [PS] [299]	—	< 250	(1) Formaldehyde stabilization strategy with following hydrogenation [19, 300] and the modified strategies with pre-protection of the C_α–OH and C_γ–OH [301]; (2) Pre-etherification of the C_α–OH with following hydrogenation [19, 301–303]; (3) Pre-etherification of the C_α–OH with following with oxidation [304]; etc.	Processes begin with the pre-protection of lignin active groups, which make the following lignin transformation more controllable.
Self-decomposition [SD]	—	r.t.~ < 200	(1) Irradiation decomposition under high energy UV [305, 306] with γ-rays [307, 308] (2) Self-catalysis process with active fragments under irradiation [309]; etc.	Extra energy introduction promotes lignin depolymerization to produce monomers or low molecular fragments that accelerate the reaction.

Table Note: The abbreviation of each process is put in the bracket; [a] Supercritical solvolysis [SS] in some research can be Gasification [G] or Liquid phase reformation[LPR].

$$\text{Lignin Conv.} = \frac{\text{The total weight of detected monomer products}}{\text{The weight of lignin in added lignin}} \times 100\%$$

$$\text{Selec.}(i) = \frac{\text{The revised GC area of product } (i)}{\text{The intergrated GC area of all detected products}} \times 100\% \qquad \text{(A)}$$

$$\text{Lignin Conv.} = \frac{\text{The total amount of detected monomeric aromatic rings}}{\text{The amount of aromatics rings in added lignin}} \times 100\%$$

$$\text{Selec.}(i) = \frac{\text{The amount of aromatic ring in product } (i)}{\text{The intergrated amount of aromatic rings in all monomers}} \times 100\% \qquad \text{(B)}$$

Figure 4.6 Definitions of Conversion and Product Yield for Lignin (A) and Lignin models (B).

4.5 Brief Index of Progress of Native/Technical Lignin Conversion

4.5.1 Definitions of Lignin Conversion and Product Yield

Before checking the recent progress of lignin conversion and comparing the catalytic systems, it is useful to first make clear the definitions of lignin conversion and product yields. Given the fact that the complex structure of the lignin can occur changes with or without the generation of monomers, the common definition of lignin conversion usually ignores changes without monomers generation and this differs from the conversion definition of lignin models. Therefore, the definition of lignin conversion just focuses on the generation of the monomers. Besides this definition, some researchers also use the weight ratio between the oil product and the weight of lignin substrate to calculate the final conversion. In addition, given the complex process to accurately calculate the change of the lignin depolymerization to the final monomers (e.g., the cleavage of the lignin ether bonds needs to add two H atoms in the monomers, acidolysis/alkaline hydrolysis of lignin may lead to the release of the C_γ structure, and lignin β-O-4 oxidation to carboxylic acids can add an extra O atom to the C_α but can lose the C_β and C_γ in the phenolic products), the common method for calculating the lignin conversion just refers to the weight of total detected monomers products (Figure 4.6A) [21]. To some extent, the weight calculation method is not the most accurate approach but is still a useful method. Alternatively, one can calculate the total amount of aromatic rings in the lignin substrate, using the methods such as the UV-vis that we have discussed in Chapter 3 or the standard nitrobenzene oxidation method to calculate the aromatic rings, and then calculate the lignin product conversion with the total aromatic rings of the monomers or monocyclic hydrocarbons [310] (Figure 4.6B). In addition, for the lignocellulose direct conversion without lignin extraction/or separation, the lignin content in wood (without bark) can be measured by the TAPPI standard method [311].

4.5.2 Recent Process of Native/Technical Lignin Conversion

Table 4.2 provides a brief list of some representative systems for native/technical lignin conversion reported in the last 10 years, focusing on the hydrogenation,

Table 4.2 A Brief List of Some Representative Systems for Native/Technical Lignin Conversion to Monomers.

Part 1 Hydrogenation, Transfer Hydrogenation, and Reduction				
Part 1.1 Hydrogenation without Pretreatment				
Entry Lignin/ Plan Information		**Reaction Description and Reference**		
		Catalytic System and Reference	**Notes for the Reaction Condition**	**Products Information**
1	Lignin sulfonate	Ni/AC 2012, Song [312]	Sodium lignin sulfonate 2.0 g, 10 w% Ni/AC 0.4 g, ethylene glycol 140 mL, H_2 5 MPa, 200 C, 6 h	The yield of phenolic monomers is 90 w%.
2	Birch wood	4%Ni-30%W_2C/AC 2012, Li [313]	Birch wood 1.0 g, Ni-W_2C/AC 0.4 g, H_2O 100 mL, H_2 6 MPa, 235 °C, 4 h.	The yield of phenolic monomers is 46.5 w%.
3	Birch sawdust	Ru/C 2015, Sels [4]	Birch sawdust (2.0 g), Ru/C(5 w%, 0.3 g), methanol (40 mL), H_2 3 MPa, 250 °C, 6.0 h.	The yield of phenolic monomers is 52 w%.
4	Organosolv pine lignin	Ru/C+NaOH 2015, Ma [163]	1.0 g lignin, 40 mL CH_3OH, 4 MPa H_2, 0.030 mol NaOH/0.20 g Ru/C (5 w%) at 260 °C for 4 h.	The yield of phenolic monomers is 12.7 w%.
5	Woods	Pd/C and Yb(OTf)$_3$ 2016, Hensen [156]	Wood feedstock 2.0 g, Pd/C (5 wt% Pd loading, 200 mg), metal triflate (0.0322 mmol), methanol (40 mL), H_2 30 bar, 180 °C, 2 h.	This process provides aromatic monomer yields of 36–48 wt% from different woods.
6	Wood lignin	Ni-Mo_2C/C 2016, Jiang [314]	Lignin (1.0 g), Ni-Mo_2C/C (0.25 g), isopropanol (100 mL), H_2 2 MPa, 250 °C, 2 h.	The yield of phenolic monomers is 31.9 w%.
7	EMALs	MoO_x/CNT 2017, Song [315]	Birch enzymatic mild acidolysis lignin (EMAL) 50 mg, MoO_x/CNT (5 mg), MeOH (10 mL), H_2 (3 MPa), 260 °C, 4 h.	The yield of phenolic monomers is 47 w%. Other lignins range from13~40 w%.

8	Birch Lignin	Ru/Nb_2O_5 2017, Wang [316]	Birch 100 mg, 2 w% Ru/Nb_2O_5 200 mg, H_2O 15mL, H_2 0.7 MPa, 250 °C, 20 h.	The yield of C7–C9 monomer is 35.5 wt% and the selectivity of arenes is 71 wt%.
9	Kraft Lignin	$Ru/NbOPO_4$ 2019, Wang [310]	Lignin 0.1 g, 5 w% $Ru/NbOPO_4$ 0.2 g, dodecane 5 mL, H_2 0.5 MPa, 310 °C, 40 h	$Ru/NbOPO_4$ can cleave C–C and C–O bonds to yield 124%–153% of monomers [317].
10	Birch wood	$Pd_{0.25}/CN_x$ 2020, Kim [318]	Biomass 3 g, 0.45 g of $Pd_{0.25}/CN_x$, 60 mL of methanol, H_2 3.0 MPa, 250 °C, 3 h	The yield of phenolic monomers is 52.7 w%
11	C-lignin	M/C (M=Pt, Pd, Ru) 2018, Ralph [146]	C-lignin 200 mg, 5 w% M/C 100 mg, methanol 30 mL, H_2 40 bar, 200 °C, 15 h.	The Pd/C provides an 89 w% yield of 4-(3-hydroxypropyl)-catechol.
12	C-lignin	Ru/ZnO/C 2021, Song [144]	C-lignin 50 mg, Ru/ZnO/C 30 mg, methanol 10 mL, H_2 3 MPa, 200 °C, 4 h.	The C-lignin depolymerization provides a 77 w% yield to propenylcatechol.

Part 1.2 Hydrogenation with Pretreatment

Entry	Lignin/Plan Information	Reaction Description and Reference: Catalytic System and Reference	Notes for the Reaction Condition	Products Information
1	Transgenic Poplar (high-S)	Step 1: Formaldehyde stabilization Step 2: $Ru/C + H_2$ 2016, Shuai [19]	(1) Wood particles 1.0 g, 1,4-dioxane 9 mL, 37% HCl solution 420 μL, 36.5% FA solution 1 mL, 80 °C, 5 h. After the neutralization and separation, the lignin was dissolved in 20 mL THF.; (2) Lignin-FA-THF 20 mL, 5 w% Ru/C 100 mg, H_2 40 bar, 250 °C, 6 h.	With F5H-poplar as the substrate, yields of 78 w% lignin monomers were obtained.

(Continued)

Table 4.2 (Continued)

Part 1.2 Hydrogenation with Pretreatment				
Entry Lignin/ Plan Information		Reaction Description and Reference		
		Catalytic System and Reference	Notes for the Reaction Condition	Products Information
2	Alkali lignin	Step 1. Methylation $Fe_2(SO_4)_3$(microwave) Step2. Hydrogenation $Pd/C + H_2$ 2016, Ouyang [302]	(1) Lignin (0.3 g), $Al_2(SO_4)_3 \cdot 16H_2O$ (0.03 g), and methanol (20 mL) were added into the microwave vessel. The pretreatment was carried out at 160 °C for 20 min under microwave irradiation with an initial power of 400 W; (2) Above mixture, 10 w%Pd/C (0.03 g), HCOOH (1.0 mL), N_2, 280 °C, 2 h.	The yield of phenolic monomers via the two-step method is 16.8 wt%.
4	Raw Birch Powder	Step 1: Oxidation $NaNO_2$/DDQ/NHPI+O_2 Step 2: Hydrogenation $NiMoS_x + H_2$ 2017, Zhang [69]	(1) Birch powder 100 mg, acetonitrile 5.0 mL, $NaNO_2$ 8 mg, DDQ 12 mg, NHPI 12 mg, O_2 1.0 MPa, 90 °C, 4 h; (2) Preoxidative birch powder 50 mg, methanol 5.0 mL, NiMo sulfide 50 mg, H_2 1.0 MPa, 200 °C for 4 h.	The preoxidized birch powders were transformed to phenolic monomers with a 32 w% yield. The unoxidized lignin provided a 3.2 w% yield.
5	Birch Lignin	Step 1: Oxidation DDQ+ ^{t}BuONO+O_2 Step 2: Hydrogenation Ni/MgAlO-C + H_2 2018, Wang [280]	(1) Abstracted birch lignin 1.20 g, 2-methoxyethanol/1, 2-dimethoxyethane (2:3, 17 mL) 5.0 mL, ^{t}BuONO 105 μL, DDQ 120 mg, O_2 1atm, 80°C, 14 h; (2) Preoxidative birch powder 50 mg, methanol 2.0 mL, Ni/MgAlO-C 20 mg, H_2 1.0 MPa, 200 °C for 6 h.	The preoxidized birch lignin was transformed to phenolic monomers with a 22 w% yield.
6	Soda lignin	Step 1:Oxidation AcNH-TEMPO/HNO_3/O_2 Step 2: Hydrogenation Cu-TiN +H_2 2018, Rashidi [176]	(1) Lignin 2.2 g, R-TEMPO 0.1 g, HNO_3 (0.1 mL, 67%), HCl (0.01 mL, 37%), CH_3CN 10 mL, O_2 2 atm, 65 °C, 48 h; (2) 0.05 M of lignin with 10 wt % of Cu-TiN dissolved in solvent dodecane, 10 bar H_2, 250 °C, 12 h	The pre-oxidized lignin produced a 56.7 wt % yield of aromatic monomers. The unoxidized lignin provided an 11.8 w% yield.

Part 1.3. Transfer Hydrogenation with Extra Hydrogen Donor

Entry	Lignin/ Plan Information	Reaction Description and Reference		
		Catalytic System and Reference	Notes for the Reaction Condition	Products information
1	Birch sawdust	Ni/AC 2013, Song [21]	Birch sawdust (2.0 g), Ni/AC catalyst (0.1 g), methanol (40 mL), 1 atm Ar, 220 °C, 6 h.	The yield of aromatic monomers is 60 w%.
2	Birch wood	Pd/C 2014, Samec [311]	Untreated sawdust (10.00 g), Pd/C (5 wt%, 1.3 g), $EtOH/H_2O$ (v/v=1:1, 100 mL), Ar 4 bar, 195 °C, 1 h.	Birch wood provided a 49 w% yield of 2, 6-dimethoxy-4-(prop-1-enyl)-phenol.
3	Soda lignin	$CuMgAlO_x$ 2014, Hensen [224]	Lignin1.0 g, $CuMgAlO_x$ 0.5 mg, ethanol 20 mL, N_2, 300 °C, 8 h.	The yield of aromatic monomers is 23 w%.
4	Kraft lignin	α-MoC_{1-x}/AC 2014, Li [222]	Kraft lignin (1.0 g), catalyst (0.5 g), ethanol 100 mL, N_2, 280 °C, 6 h.	The complete ethanolysis of Kraft lignin was obtained [319].
5	Kraft lignin	MoC_{1-x}/Cu-MgAlOz 2017, Li [226]	Kraft lignin 1 g, MoC_{1-x}/Cu-MgAlOz 0.5 g, ethanol 100 mL, 330 °C, 5 h.	The highest yield of aromatic compounds is 575 mg/g lignin (57.5 w%), at 330 °C

Part 1.4. Thermal Reforming with Intramolecular Hydrogen Transfer

Entry	Lignin/ Plan Information	Reaction Description and Reference		
		Catalytic System and Reference	Notes for the Reaction Condition	Products information
1	Alkaline lignin	Binuclear Rh 2019, Li [193]	Basswood lignin (100 mg), binuclear Rh (4 mg), NaOH (80 mg), and H_2O (1 mL), Ar, 110 °C, 24 h.	An 11 w% yield of aromatic ketones was obtained.

(Continued)

Table 4.2 (Continued)

Entry Lignin/ Plan Information		Reaction Description and Reference		
Part 1.4. Thermal Reforming with Intramolecular Hydrogen Transfer				
		Catalytic System and Reference	**Notes for the Reaction Condition**	**Products information**
2	Birch lignin	$Pt/NiAl_2O_4$ 2020, Wang [203, 204]	Birch lignin 0.102 g, $M/NiAl_2O_4$ (10 w% Pt) 0.201 g, H_2O 15.1 mL, N_2 2 MPa, 280 °C, 20 h. (M=Pt, Pd, Ru)	A 17.3 wt% yield of 4-alkylphenols can be obtained from birch lignin.
3	Pine lignin	$La(OTf)_3$ 2020, Hanl [291]	Organosolv Lignin 1.5 g, $La(OTf)_3$ 0.6 g, methanol 120 mL, H_2O 0.3 mL, Ar 1 atm, 270 °C, 24 h.	1.5 g organosolv lignin can provide 0.32 g of isolated guaiacol (21.2 wt%).
4	Poplar lignin	Zeolite HY_{30} 2020, Han [290]	Lignin 50.0 g, HY_{30} 50.0 g, H_2O 600 mL, Ar 0.5 MPa, 200 °C, 3 h.	A scale-up experiment with 50.0 g lignin can provide 4.1 g of pure phenol (8.2 w%).
5	Pine lignin	RuW/Zeolite HY_{30} 2021, Han [205]	Lignin 50.0 g, RuW/HY_{30} (3.5 wt% Ru, 20 wt% W) 50.0 g, H_2O 650 mL, N_2 0.1 MPa, 240 °C, 12 h.	A scale-up experiment with 50.0 g lignin can provide 8.5 g of pure benzene (17 w%)
Part 1.5 Photocatalytic Reforming with Intramolecular Hydrogen Transfer (Reduction)				
Entry Lignin/ Plan Information		**Reaction Description and Reference**		
		Catalytic System and Reference	**Notes for the Reaction Condition**	**Products information**
1	Dioxanesolv lignin	$ZnIn_2S_4$ 2017, Luo [199]	Dioxanesolv poplar lignin (30 mg), $ZnIn_2S_4$ (10 mg), acetone (0.8 mL), isopropanol (0.2 mL), Ar, blue LEDs (455 nm, 9.6 W), 42°C, 24 h.	This photocatalytic method provided a 10 w% yield of aromatic monomers.
2	Birch woodmeal	CdS quantum dots 2018, Wang [213]	Birch woodmeal (100 mg), CdS QDs (15 mg), CH_3OH/H_2O (5 mL), N_2, 300 W Xe lamp with a UV cut-off filter ($\lambda \geq 400$ nm), 8 h.	This photocatalytic method provided a 27 w% yield of aromatic monomers.

Part 2. Oxidation				
Part 2.1 Thermal oxidation				
Entry	**Lignin/ Plan Information**	**Reaction Description and Reference** — **Catalytic System and Reference**	**Notes for the Reaction Condition**	**Products information**
1	Lignin sulfonates	Cu/nitrobenzene 1999, Bjørsvik [75]	Lignosulfonate (110.6 g), NaOH (18 g dissolved in 16 mL of water), $CuSO_4 \cdot 5H_2O$ (1.08 g), $PhNO_2$ (2.0 mL), 190 °, 20 min.	The lignin sulfonate provided a 5.9 wt % yield of vanillin.
2	Raw Poplar Sawdust	$MoO_2(acac)_2$ 2019, Ma [320]	1.0 g poplar sawdust, 0.5 g $MoO_2(acac)_2$, 30 mL CH_3OH, 180 °C, 3 h, 1.0 MPa O_2.	It provides 27.8 w% of aromatic monomers from lignin and 25.8 w% of low-carbon organic acid esters from carbohydrate.
	Note: The aromatic monomers yields with other catalysts: (1) MoS_2 17.1 w%; (2) $VO(acac)_2$ 12.1 w%; (3) $VOSO_4$ 5.3 w%; (4) $Cu(OAc)_2$ 5.7 w%; (5) $MnPO_4$·3.7 w%;			
		$MoO_2(acac)_2$	1.0 g poplar, sawdust, 0.5 g $MoO_2(acac)_2$, 30 mL solvent, 180 °C, 6 h, 1.0 MPa O_2.	It provides 34.8 w% of aromatic monomers and 27.4 w% of organic acid esters.
3	Raw poplar sawdust	Co-N-C catalyst 2021, Luo [93]	0.1 g poplar sawdust, 10 wt% Co-N-C catalyst, 20 mL acetone, 190 °C, 35 bar 6% O_2 in N_2, 12 h.	It provides 15 w% of aromatic monomers.

(Continued)

Table 4.2 (Continued)

Part 2.2 Electrooxidation				
		Reaction Description and Reference		
Entry	Lignin/ Plan Information	Catalytic System and Reference	Notes for the Reaction Condition	Products information
1	Lignin sulfonate	Ni mesh electrode 2011, Utley [110]	A flow reactor (FM01) allowed up to 150 g to be electrolyzed at up to12 A, the electrolyser operated at 145 °C/500 kPa/3 M NaOH.	The electrolysis of lignin provided a 5-7 wt % yield of vanillin.
2	Kraft lignin	Ni-based electrode 2020, Waldvogel [112]	In the electrolysis cell, lignin (525 mg) was dissolved in aqueous NaOH (3 M, 85 g) under vigorous stirring. The electrolysis cell was sealed and heated in an oil bath up to 160 °C. Constant current electrolysis (10 mA/cm^2, 60 mA, applied voltage range: 2–3 V) was performed using planar Ni electrodes until a charge of 1418 C (2.7 C per mg lignin) had passed at 160 °C.	The high-temperature electrolysis of lignin provided a 4.5~7.0 wt % yield of vanillin.
3	Poplar Lignin	Step 1:Electrooxidation Anode/AcNH-TEMPO Step 2: Acid treatment H_2O/HCOOH 2019, Stahl [114]	(1) Lignin 100 mg, AcNH-TEMPO 0.23 mmol, 8 mL of H_2O/CH_3CN (70/30), $NaHCO_3/Na_2CO_3$ electrolyte pH=10, scan rate for cyclic voltammetry 100 mV s^{-1}, electrolysis at 0.80 V vs Ag/AgCl; (2) Oxidized lignin 50 mg, 5 mL of H_2O/HCO_2H (1:9), 100 °C, 12 h.	Treatment of the oxidized lignin under acidic conditions results in depolymerization of the material into characterized aromatic monomers in nearly a 30 wt% yield.

Part 3. Multi-step Process for Lignin Depolymerization

Entry	Lignin/ Plan Information	Reaction Description and Reference		
		Catalytic System and Reference	Notes for the Reaction Condition	Products Information
1	Aspen Lignin	Step 1:Oxidation AcNH-TEMPO/HNO_3/O_2 Step 2: Hydrolysis HCOOH/Na 2014, Stahl [281]	(1) Lignin 35 mg, AcNH-TEMPO 1 mg, HNO_3 (1.5 μL, 67%), HCl (1 μL, 37%), CH_3CN 10 mL, H_2O 100 μL, O_2 2 atm, 65 °C, 48 h; (2) Oxidized lignin, HCOONa (25 mg, 3 equiv.), 85~90% aqueous HCOOH 5.0 mL, 110 °C, 24 h.	The oxidized lignin under the mild condition in aqueous formic acid that results in a 61.2 wt% yield of low molecular-mass aromatics
	Note: The phenolic monomers yields from other biomass resources: (1) Poplar 52.6 w%; (2) Maize 24.2 w%; (3) Maple1 1.2 w% [321].			
2	Dioxasolv birch lignin	Step 1 Oxidation DDQ/^{t}BuONO/O_2 Step 2: Reduction Zn/NH_4Cl 2016, Westwood [211]	(1) Lignin (14 mL/g),10 w% DDQ/^{t}BuONO, 2-2-methoxyethanol /1,2-dimethoxyethane (v:v=2:3), O_2 (balloon), 80 °C, 14 h; (2) Oxidized lignin (600 mg), 2-methoxyethanol (8.4 mL), water (2.1 mL), NH_4Cl (740 mg), and zinc dust (900 mg), 80 °C for 1 h.	The oxidized lignin provided a 6.0 w% yield of aromatic monomers.
3	Dioxasolv pine lignin	Step 1: Electrooxidation NHPI/2,6-Lutidine Step 2:Photoreduction Ir + DIPEA 2017, Stephenson [276]	(1) Dioxasolv pine lignin, NHPI/2,6-Lutidine (10 mol%), MeCN, KPF_6, O_2, 0.64 V vs Fc*/Fc; (2) $[Ir(ppy)_2(dtbbpy)]PF_6$ (0.5 mol%), DIPEA (1.1 equiv.), HCOOH (10 mol%), blue LEDs, MeCN.	The electro-photo tandem process provided a 2.4 w% yield of aromatic monomers.

(Continued)

Table 4.2 (Continued)

Entry	Lignin/Plan Information	Reaction Description and Reference		
		Catalytic System and Reference	Notes for the Reaction Condition	Products Information
4	Dioxasolv poplar lignin	Step 1: Oxidation Step 2: N-modification Step 3: Photocatalysis 2019, Li [273]	(1) Poplar lignin (1.8 g), dioxane (90 mL), DDQ (3.6 g, 200 wt%), 80 °C, 6 h; (2) Oxidized lignin (0.8 g), 2-methoxyehtanol (30 mL), hydroxylamine hydrochloride (8 mmol), pyridine (12 mmol), 80 °C, 18 h; (3) N-modified lignin (30 mg), phenothiazine (photocatalyst, 1 mg), ascorbic acid (17.6 mg), $MeCN:H_2O:HCOOH=18:1:1$ (2.0 mL), 365 nm, Ar, 8 h.	This method can provide a 6.4 w%yield of aromatic monomers, Including 1.9 wt % primary arylamine monomers.

Part 4. Multi-step Cascaded Process for Single Aromatic Monomer Preparation

Entry	Lignin/ Plan Information	Reaction Description and Reference: Catalytic System and Reference	Notes for the Reaction Condition	Products information
1	Corn Stover Lignin	Step 1: Demethoxylation $MoO_x/AC/H_2$ Step 2: Carbonylation PdL_n/CO Step 3: Oxidation Co-Mn-Br/O_2 2019, Song [322]	(1) Demethoxylation: lignin oil 0.11 g, MoO_x/AC 0.2 g, toluene 3 mL, H_2=30 bar, T=320 °C, t=10 h; (2) Carbonylation: a) 4-propylphenol 10 mmol, DCM 50 mL, pyridine 20 mmol, triflic anhydride 12 mmol; b) 4-propylphenol triflate 2 mmol, $Pd(OAc)_2$ 5 mol%, ligand 20 mol%, KOAc 8 mmol, DMSO 25 mL, CO balloon, 60 °C, 24 h; (3) Oxidation: 4-propylbenzoic acid 2 mmol, $Co(OAc)_2$ 2.5 mol%, $Mn(OAc)_2$ 2.5 mol%, KBr 5.0 mol%, acetic acid 3 mL, O_2=10 bar, 12 h.	TPA (terephthalic acid) yield based on lignin content of corn stover could reach 15.5 wt% with a greater than 99% purity.
2	Poplar lignin	$Ru/CeO_2/H_2$-$CuCl_2$ 2018 Wang [286]	Poplar lignin 100 mg, 0.01 mmol of Ru (Ru/CeO_2, 5 w%), 2 mL of 0.25 M $CuCl_2$ aqueous solution, H_2 3 MPa, 200 °C, 6 h.	Phenol can be obtained from separated poplar lignin with a 13 wt % yield.
3	Birch wood	A new integrated biorefinery process 2020, Liao [289]	The integrated biorefinery process [289] for xylochemicals production from wood is discussed in Chapter 12.3.2. The critical dealkylation of the phenols is discussed in the references with zeolite catalyst [287, 288, 323, 324]	After the hydroprocessing and dealkylation, 20 wt % of phenol and 9 wt % of propylene were obtained.

Note: Some other one-pot methods for Single aromatic monomer preparation can be found in Part 1.4 Thermal Reforming with Intramolecular Hydrogen Transfer (Research from Han et. al.) or Part 1.1 Hydrogenation without Pretreatment with C-lignin as the substrate (Research from Ralph and Song, discussed in Chapter 12).

oxidation, reduction, and multi-step process. Given the content limitation of this book, one can check the corresponding references list in Tables 4.1 and 4.2 for more detailed information and some of this content will be discussed in the following chapters.

4.6 Chapter Summary

At the beginning of Chapter 4, a discussion about the opportunity and challenge for lignin valorization, especially for the aromatic chemicals preparation, is firstly provided (Chapter 4.1). Then, focusing on the topic of lignin (catalytic) depolymerization to aromatic monomers, the involved scientific questions are briefly summarized (Chapter 4.2) as follows: (1) promoting the lignin depolymerization via selective cleavage of the ubiquitous lignin C–O/C–C bonds, increasing the conversion; and (2) restraining the lignin fragments recondensation with more stable chemical bonds generation or multimers formation, increasing the selectivity and final monomer yield. Then, two different approaches for the foundation of lignin depolymerization strategies are introduced: (1) direct lignin conversion; and (2) the bottom-up approach (Chapter 4.3). To provide a macro understanding of the research field of lignin conversion, the classification of lignin conversion methods by reaction types is summarized (Chapter 4.4), during which the corresponding catalyst types are also briefly provided. Then, a brief list of some representative systems reported in the last 10 years for native/technical lignin conversion is provided (Chapter 4.5). Based on the brief introduction and summary of the lignin conversion research in this chapter, we hope to provide a basic understanding of the lignin conversion process, which can make the acceptance of the following mechanism discussion easier.

References

1 Amidon, T.E. and Liu, S. (2009). Water-based woody biorefinery. *Biotechnol. Adv.* 27: 542–550.

2 Zakzeski, J., Bruijnincx, P.C.A., Jongerius, A.L., and Weckhuysen, B.M. (2010). The catalytic valorization of lignin for the production of renewable chemicals. *Chem. Rev.* 110: 3552–3599.

3 Xu, C., Arancon, R.A.D., Labidi, J., and Luque, R. (2014). Lignin depolymerisation strategies: towards valuable chemicals and fuels. *Chem. Soc. Rev.* 43: 7485–7500.

4 Van den Bosch, S., Schutyser, W., Vanholme, R. et al. (2015). Reductive lignocellulose fractionation into soluble lignin-derived phenolic monomers and dimers and processable carbohydrate pulps. *Energy Environ. Sci.* 8: 1748–1763.

5 Ragauskas, A.J., Beckham, G.T., Biddy, M.J. et al. (2014). Lignin valorization: improving lignin processing in the biorefinery. *Science* 344: 1246843.

6 Zhang, C.F. and Wang, F. (2020). Catalytic lignin depolymerization to aromatic chemicals. *Acc. Chem. Res.* 53: 470–484.

7 Chen, X. and Yan, N. (2014). Novel catalytic systems to convert chitin and lignin into valuable chemicals. *Catal. Surv. Asia* 18: 164–176.

8 Besson, M., Gallezot, P., and Pinel, C. (2014). Conversion of biomass into chemicals over metal catalysts. *Chem. Rev.* 114: 1827–1870.

9 Li, C., Zhao, X., Wang, A. et al. (2015). Catalytic transformation of lignin for the production of chemicals and fuels. *Chem. Rev.* 115: 11559–11624.

10 Wong, S.S., Shu, R.Y., Zhang, J.G. et al. (2020). Downstream processing of lignin derived feedstock into end products. *Chem. Soc. Rev.* 49: 5510–5560.

11 Pandey, M.P. and Kim, C.S. (2011). Lignin depolymerization and conversion: A review of thermochemical methods. *Chemical Engineering & Technology* 34: 29–41.

12 Chatel, G. and Rogers, R.D. (2014). Review: Oxidation of lignin using ionic liquids-An innovative strategy to produce renewable chemicals. *ACS Sustainable Chem. Eng.* 2: 322–339.

13 Hanson, S.K. and Baker, R.T. (2015). Knocking on wood: Base metal complexes as catalysts for selective oxidation of lignin models and extracts. *Acc. Chem. Res.* 48: 2037–2048.

14 Zhang, J. (2018). Conversion of lignin models by photoredox catalysis. *ChemSusChem* 11: 3071–3080.

15 Sun, Z., Fridrich, B., de Santi, A. et al. (2018). Bright side of lignin depolymerization: Toward new platform chemicals. *Chem. Rev.* 118: 614–678.

16 Wu, Z., Hu, L., Jiang, Y. et al. (2020). Recent advances in the acid-catalyzed conversion of lignin. *Biomass Convers. Biorefin.* 1–21.

17 Renders, T., Van den Bosch, S., Koelewijn, S.F. et al. (2017). Lignin-first biomass fractionation: the advent of active stabilisation strategies. *Energy Environ. Sci.* 10: 1551–1557.

18 Abu-Omar, M.M., Barta, K., Beckham, G.T. et al. (2021). Guidelines for performing lignin-first biorefining. *Energy Environ. Sci.* 14: 262–292.

19 Shuai, L., Amiri, M.T., Questell-Santiago, Y.M. et al. (2016). Formaldehyde stabilization facilitates lignin monomer production during biomass depolymerization. *Science* 354: 329–333.

20 Deuss, P.J., Scott, M., Tran, F. et al. (2015). Aromatic monomers by in situ conversion of reactive intermediates in the acid-catalyzed depolymerization of lignin. *J. Am. Chem. Soc.* 137: 7456–7467.

21 Song, Q., Wang, F., Cai, J. et al. (2013). Lignin depolymerization (LDP) in alcohol over nickel-based catalysts via a fragmentation–hydrogenolysis process. *Energy Environ. Sci.* 6: 994–1007.

22 Balfour, W.J. (1985). The 275 nm absorption system of anisole. *I. Mol. Spectrosc.* 109: 60–72.

23 Schuerch, C. (1951). Solubility Effects in lignin alcoholysis reactions. *J. Am. Chem. Soc.* 73: 2385–2386.

24 Cramer, A.B., Hunter, M.J., and Hibbert, H. (1939). Studies on lignin and related compounds. XXXV. The ethanolysis of spruce wood. *J. Am. Chem. Soc.* 61: 509–516.

25 Anderson, E.M., Stone, M.L., Katahira, R. et al. (2017). Flowthrough reductive catalytic fractionation of biomass. *Joule* 1: 613–622.

26 Van den Bosch, S., Renders, T., Kennis, S. et al. (2017). Integrating lignin valorization and bio-ethanol production: on the role of Ni-Al2O3catalyst pellets during lignin-first fractionation. *Green Chem.* 19: 3313–3326.

27 Rinaldi, R. (2014). Plant Biomass Fractionation meets catalysis. *Angew. Chem. Int. Ed.* 53: 8559–8560.

28 Alonso, D.M., Wettstein, S.G., and Dumesic, J.A. (2012). Bimetallic catalysts for upgrading of biomass to fuels and chemicals. *Chem. Soc. Rev.* 41: 8075–8098.

29 Colmenares, J.C. and Luque, R. (2014). Heterogeneous photocatalytic nanomaterials: prospects and challenges in selective transformations of biomass-derived compounds. *Chem. Soc. Rev.* 43: 765–778.

30 Rinaldi, R., Jastrzebski, R., Clough, M.T. et al. (2016). Paving the way for lignin valorisation: Recent advances in bioengineering, Biorefining and Catalysis. *Angew. Chem. Int. Ed.* 55: 8164–8215.

31 Schutyser, W., Renders, T., Van den Bosch, S. et al. (2018). Chemicals from lignin: an interplay of lignocellulose fractionation, depolymerisation, and upgrading. *Chem. Soc. Rev.* 47: 852–908.

32 Liu, C.J., Wang, H.M., Karim, A.M. et al. (2014). Catalytic fast pyrolysis of lignocellulosic biomass. *Chem. Soc. Rev.* 43: 7594–7623.

33 Zhang, Z.R., Song, J.L., and Han, B.X. (2017). Catalytic Transformation of lignocellulose into chemicals and fuel products in ionic liquids. *Chem. Rev.* 117: 6834–6880.

34 Jing, Y. and Wang, Y. (2020). Catalytic hydrodeoxygenation of lignin-derived feedstock into arenes and phenolics. *Front. Chem. Eng.* 10.

35 Jing, Y.X., Guo, Y., Xia, Q.N. et al. (2019). Catalytic production of value-added chemicals and liquid fuels from lignocellulosic biomass. *Chem* 5: 2520–2546.

36 Liu, W.-J., Li, W.-W., Jiang, H., and Yu, H.-Q. (2017). Fates of chemical elements in biomass during its pyrolysis. *Chem. Rev.* 117: 6367–6398.

37 Yu, J., Guo, Q., Gong, Y. et al. (2021). A review of the effects of alkali and alkaline earth metal species on biomass gasification. *Fuel Process. Technol.* 214: 106723.

38 Lange, H., Decina, S., and Crestini, C. (2013). Oxidative upgrade of lignin-Recent routes reviewed. *Eur. Polym. J.* 49: 1151–1173.

39 Vangeel, T., Schutyser, W., Renders, T., and Sels, B.F. (2018). Perspective on lignin oxidation: Advances, challenges, and future directions. *Top. Curr. Chem.* 376.

40 Cedeno, D. and Bozell, J.J. (2012). Catalytic oxidation of para-substituted phenols with cobalt-Schiff base complexes/O_2-selective conversion of syringyl and guaiacyl lignin models to benzoquinones. *Tetrahedron Lett.* 53: 2380–2383.

41 Zhu, W.M. and Ford, W.T. (1993). Oxidation of lignin model compounds in water with dioxygen and hydrogen-peroxide catalyzed by metallophthalocyanines. *J. Mol. Catal.* 78: 367–378.

42 Cui, F.T. and Dolphin, D. (1995). Metallophthalocyanines as possible lignin peroxidase models. *Bioorg. Med. Chem.* 3: 471–477.

43 Barbat, A., Gloaguen, V., Sol, V., and Krausz, P. (2010). Aqueous extraction of glucuronoxylans from chestnut wood: New strategy for lignin oxidation using phthalocyanine or porphyrin/H_2O_2 system. *Bioresour. Technol.* 101: 6538–6544.

44 Hanson, S.K., Wu, R.L., and Silks, L.A. (2012). C—C or C—O bond cleavage in a phenolic lignin model compound: Selectivity depends on vanadium catalyst. *Angew. Chem. Int. Ed.* 51: 3410–3413.

45 Hanson, S.K., Baker, R.T., Gordon, J.C. et al. (2010). Aerobic oxidation of lignin models using a base metal vanadium catalyst. *Inorg. Chem.* 49: 5611–5618.

46 Chan, J.M.W., Bauer, S., Sorek, H. et al. (2013). Studies on the vanadium-catalyzed nonoxidative depolymerization of *Miscanthus giganteus*-derived lignin. *ACS Catal.* 3: 1369–1377.

47 Jiang, Y.Y., Yan, L., Yu, H.Z. et al. (2016). Mechanism of vanadium-catalyzed selective C–O and C–C cleavage of lignin model compound. *ACS Catal.* 6: 4399–4410.

48 Ma, Y.Y., Du, Z.T., Xia, F. et al. (2016). Mechanistic studies on the $VO(acac)_2$-catalyzed oxidative cleavage of lignin model compounds in acetic acid. *RSC Adv.* 6: 110229–110234.

49 Ma, Y.Y., Du, Z.T., Liu, J.X. et al. (2015). Selective oxidative C–C bond cleavage of a lignin model compound in the presence of acetic acid with a vanadium catalyst. *Green Chem.* 17: 4968–4973.

50 Zhou, Z.-z., Liu, M., and Li, C.-J. (2017). Selective copper–N-heterocyclic carbene (Copper-NHC)-catalyzed aerobic cleavage of β-1 lignin models to aldehydes. *ACS Catal.* 7: 3344–3348.

51 Sedai, B., Diaz-Urrutia, C., Baker, R.T. et al. (2013). Aerobic oxidation of β-1 lignin model compounds with copper and oxovanadium catalysts. *ACS Catal.* 3: 3111–3122.

52 Liu, H., Wang, M., Li, H. et al. (2017). New protocol of copper-catalyzed oxidative C(CO)—C bond cleavage of aryl and aliphatic ketones to organic acids using O_2 as the terminal oxidant. *J. Catal.* 346: 170–179.

53 Wang, M., Lu, J.M., Li, L.H. et al. (2017). Oxidative C(OH)—C bond cleavage of secondary alcohols to acids over a copper catalyst with molecular oxygen as the oxidant. *J. Catal.* 348: 160–167.

54 Riano, S., Fernandez, D., and Fadini, L. (2008). Oxidative cleavage of vic-diols catalyzed by manganese(III) complexes in ionic liquids. *Catal. Commun.* 9: 1282–1285.

55 Crestini, C., Pro, P., Neri, V., and Saladino, R. (2005). Methyltrioxorhenium: a new catalyst for the activation of hydrogen peroxide to the oxidation of lignin and lignin model compounds. *Bioorg. Med. Chem.* 13: 2569–2578.

56 Cho, D.W., Parthasarathi, R., Pimentel, A.S. et al. (2010). Nature and kinetic analysis of carbon-carbon bond fragmentation reactions of cation radicals derived from SET-oxidation of lignin model compounds. *J. Org. Chem.* 75: 6549–6562.

57 Mottweiler, J., Rinesch, T., Besson, C. et al. (2015). Iron-catalysed oxidative cleavage of lignin and β-O-4 lignin model compounds with peroxides in DMSO. *Green Chem.* 17: 5001–5008.

58 Shimada, M., Habe, T., Higuchi, T. et al. (1987). Biomimetic approach to lignin degradation. 2. The mechanism of oxidative C-C bond cleavage reactions of lignin model compounds with natural iron(III) porphyrin chloride as a heme-enzyme model system. *Holzforschung* 41: 277–285.

59 Heiba, E.A.I., Dessau, R.M., and Koehl, W.J. Jr. (1969). Oxidation by metal salts. V. Cobaltic acetate oxidation of alkylbenzenes. *J. Am. Chem. Soc.* 91: 6830–6837.

60 Hirose, S., Kasuga, K., and Hatakeyama, H. (1985). Alkali-oxygen degradation of lignin catalyzed by cobalt(III). *Tappi J.* 39: 1151–1156.

61 Huynh, V.-B. (1986). Biomimetic oxidation of lignin model compounds by simple inorganic complexes. *Biochem. Biophys. Res. Commun.* 139: 1104–1110.

62 Sedai, B., Diaz-Urrutia, C., Baker, R.T. et al. (2011). Comparison of copper and vanadium homogeneous catalysts for aerobic oxidation of lignin models. *ACS Catal.* 1: 794–804.

63 Sedai, B. and Baker, R.T. (2014). Copper catalysts for selective C–C bond cleavage of β-O-4 lignin model compounds. *Adv. Synth. Catal.* 356: 3563–3574.

64 Mottweiler, J., Rauber, C., Bolm, C. et al. (2015). Copper- and vanadium-catalyzed oxidative cleavage of lignin using dioxygen. *ChemSusChem* 8: 2106–2113.

65 Dicosimo, R. and Szabo, H.C. (1988). Oxidation of lignin model compounds using single-electron-transfer catalysts. *J. Org. Chem.* 53: 1673–1679.

66 Harvey, P.J., Schoemaker, H.E., Bowen, R.M., and Palmer, J.M. (1985). Single-electron transfer processes and the reaction mechanism of enzymic degradation of lignin. *FEBS Lett.* 183: 13–16.

67 Maggioni, P. and Minisci, F. (1979). Electron-transfer processes-selective syntheses of aromatic-aldehydes of industrial interest by catalyzed peroxydisulfate oxidation. *Chim. Ind.* 61: 101–105.

68 Minisci, F., Citterio, A., and Giordano, C. (1983). Electron-transfer processes: Peroxydisulfate, a useful and versatile reagent in organic-chemistry. *Acc. Chem. Res.* 16: 27–32.

69 Zhang, C.F., Li, H.J., Lu, J.M. et al. (2017). Promoting lignin depolymerization and restraining the condensation via an oxidation-hydrogenation strategy. *ACS Catal.* 7: 3419–3429.

70 Dabral, S., Hernandez, J.G., Kamer, P.C.J., and Bolm, C. (2017). Organocatalytic chemoselective primary alcohol oxidation and subsequent cleavage of lignin model compounds and lignin. *ChemSusChem* 10: 2707–2713.

71 Leopold, B. and Malmstrom, I.L. (1951). Constitution of resin phenols and their biogenetic relations. 15. Nitrobenzene oxidation of compounds of the lignan type. *Acta. Chem. Scand.* 5: 936–940.

72 Leopold, B. and Malmstrom, I.L. (1952). Studies on lignin. 4. Investigation on the nitrobenzene oxidation products of lignin from different woods by paper partition chromatography. *Acta. Chem. Scand.* 6: 49–54.

73 Leopold, B. (1952). Studies on lignin. 3. Oxidation of wood from Picea-Abies (L) Karst (Norway Spruce) with nitrobenzene and alkali. *Acta. Chem. Scand.* 6: 38–48.

74 Kavanagh, K.R. and Pepper, J.M. (1955). The alkaline nitrobenzene oxidation of aspen wood and lignin model substances. *Can. J. Chem.* 33: 24–30.

75 Bjørsvik, H.R. and Minisci, F. (1999). Fine chemicals from lignosulfonates. 1. Synthesis of vanillin by oxidation of lignosulfonates. *Org. Process. Res. Dev.* 3: 330–340.

76 Weinstock, I.A., Atalla, R.H., Reiner, R.S. et al. (1997). A new environmentally benign technology for transforming wood pulp into paper - Engineering polyoxometalates as catalysts for multiple processes. *J. Mol. Catal. A: Chem.* 116: 59–84.

77 Yang, Y.Y., Fan, H.L., Song, J.L. et al. (2015). Free radical reaction promoted by ionic liquid: a route for metal-free oxidation depolymerization of lignin model compound and lignin. *Chem. Commun.* 51: 4028–4031.

78 Qu, C., Ito, K., Katsuyama, I. et al. (2020). Directly Microwave-accelerated cleavage of C–C and C–O Bonds of lignin by copper oxide and H_2O_2. *ChemSusChem* 13: 4510–4518.

79 Kurek, B. and Gaudard, F. (2000). Oxidation of spruce wood sawdust by MnO_2 plus oxalate: A biochemical investigation. *J. Agr. Food. Chem.* 48: 3058–3062.

80 Deng, H., Lin, L., Sun, Y. et al. (2008). Perovskite-type oxide $LaMnO_3$: An efficient and recyclable heterogeneous catalyst for the wet aerobic oxidation of lignin to aromatic aldehydes. *Catal. Lett.* 126: 106–111.

81 Deng, H., Lin, L., Sun, Y. et al. (2008). Activity and stability of $LaFeO_3$ catalyst in lignin catalytic wet oxidation to aromatic aldehydes. *Chin. J. Catal.* 29: 753–757.

82 Deng, H., Lin, L., Sun, Y. et al. (2009). Activity and stability of perovskite-type oxide $LaCoO_3$ catalyst in lignin catalytic wet oxidation to aromatic aldehydes process. *Energy Fuel* 23: 19–24.

83 Zhang, J.H., Deng, H.B., and Lin, L. (2009). Wet aerobic oxidation of lignin into aromatic aldehydes catalysed by a perovskite-type oxide: $LaFe_{1-x}Cu_xO_3$ (x=0, 0.1, 0.2). *Molecules* 14: 2747–2757.

84 Deng, H., Lin, L., and Liu, S. (2010). Catalysis of Cu-doped Co-based perovskite-type oxide in wet oxidation of lignin to produce aromatic aldehydes. *Energy Fuel* 24: 4797–4802.

85 Gao, P., Li, N., Wang, A. et al. (2013). Perovskite $LaMnO_3$ hollow nanospheres: The synthesis and the application in catalytic wet air oxidation of phenol. *Mater. Lett.* 92: 173–176.

86 Gao, P., Li, C., Wang, H. et al. (2013). Perovskite hollow nanospheres for the catalytic wet air oxidation of lignin. *Chin. J. Catal.* 34: 1811–1815.

87 Li, L., Kong, J., Zhang, H. et al. (2020). Selective aerobic oxidative cleavage of lignin C—C bonds over novel hierarchical Ce-Cu/MFI nanosheets. *Appl. Catal., B* 119343.

88 Deng, W., Zhang, H., Wu, X. et al. (2015). Oxidative conversion of lignin and lignin model compounds catalyzed by CeO_2-supported Pd nanoparticles. *Green Chem.* 17: 5009–5018.

89 Blandez, J.F., Navalon, S., Alvaro, M., and Garcia, H. (2015). Graphenes as metal-free catalysts for the oxidative depolymerization of lignin models. *ChemCatChem* 7: 3020–3026.

90 Luo, H., Wang, L., Shang, S. et al. (2020). Cobalt nanoparticles-catalyzed widely applicable successive C—C bond cleavage in alcohols to access esters. *Angew. Chem. Int. Ed.* 59: 19268–19274.

91 Zhou, H., Hong, S., Zhang, H. et al. (2019). Toward biomass-based single-atom catalysts and plastics: Highly active single-atom Co on N-doped carbon for oxidative esterification of primary alcohols. *Appl. Catal., B* 256: 117767.

92 Liu, S., Bai, L., van Muyden, A.P. et al. (2019). Oxidative cleavage of β-O-4 bonds in lignin model compounds with a single-atom Co catalyst. *Green Chem.* 21: 1974–1981.

93 Luo, H., Weeda, E.P., Alherech, M. et al. (2021). Oxidative Catalytic Fractionation of lignocellulosic biomass under non-alkaline conditions. *J. Am. Chem. Soc.* 143: 15462–15470.

94 Liu, H., Li, H., Lu, J. et al. (2018). Photocatalytic cleavage of C–C bond in lignin models under visible light on mesoporous graphitic carbon nitride through π–π stacking interaction. *ACS Catal.* 8: 4761–4771.

95 Zhou, W., Nakahashi, J., Miura, T., and Murakami, M. (2018). Light/copper relay for aerobic fragmentation of lignin model compounds. *Asian. J. Org. Chem.* 7: 2431–2434.

96 Li, H.J., Subbotina, E., Bunrit, A. et al. (2019). Functionalized spirolactones by photoinduced dearomatization of biaryl compounds. *Chem. Sci.* 10: 3681–3686.

97 Mitchell, L.J. and Moody, C.J. (2014). Solar photochemical oxidation of alcohols using catalytic hydroquinone and copper nanoparticles under oxygen: Oxidative cleavage of lignin models. *J. Org. Chem.* 79: 11091–11100.

98 Cho, D.W., Latham, J.A., Park, H.J. et al. (2011). Regioselectivity of enzymatic and photochemical single electron transfer promoted carbon–carbon bond fragmentation reactions of tetrameric lignin model compounds. *J. Org. Chem.* 76: 2840–2852.

99 Lim, S.H., Lee, W.S., Kim, Y.I. et al. (2015). Photochemical and enzymatic SET promoted C–C bond cleavage reactions of lignin β-1 model compounds containing varying number of methoxy substituents on their arene rings. *Tetrahedron* 71: 4236–4247.

100 Crestini, C. and DAuria, M. (1997). Singlet oxygen in the photodegradation of lignin models. *Tetrahedron* 53: 7877–7888.

101 Wang, Y., Liu, Y., He, J., and Zhang, Y. (2019). Redox-neutral photocatalytic strategy for selective C–C bond cleavage of lignin and lignin models via PCET process. *Sci. Bull.* 64: 1658–1666.

102 Wang, Y., He, J., and Zhang, Y. (2020). $CeCl_3$-promoted simultaneous photocatalytic cleavage and amination of C_α–C_β bond in lignin model compounds and native lignin. *CCS Chem.* 2: 107–117.

103 Nguyen, S.T., Murray, P.R.D., and Knowles, R.R. (2020). Light-driven depolymerization of native lignin enabled by proton-coupled electron transfer. *ACS Catal.* 10: 800–805.

104 Dai, J.H., Patti, A.F., Styles, G.N. et al. (2019). Lignin oxidation by MnO_2 under the irradiation of blue light. *Green Chem.* 21: 2005–2014.

105 Hou, T., Luo, N., Li, H. et al. (2017). Yin and Yang dual characters of CuO_x clusters for C–C bond oxidation driven by visible light. *ACS Catal.* 3850–3859.

106 Luo, N., Wang, M., Li, H. et al. (2016). Photocatalytic oxidation–hydrogenolysis of lignin β-O-4 models via a dual light wavelength switching strategy. *ACS Catal.* 6: 7716–7721.

107 Pardini, V.L., Smith, C.Z., Utley, J.H.P. et al. (1991). Electroorganic reactions. 38. Mechanism of electrooxidative cleavage of lignin model dimers. *J. Org. Chem.* 56: 7305–7313.

108 Pardini, V.L., Vargas, R.R., Viertler, H., and Utley, J.H. (1992). Anodic cleavage of lignin model dimers in methanol. *Tetrahedron* 48: 7221–7228.

109 Deng, G., Huang, H., and Wei, H. (1995). Study on the eleetrochemical oxidation of lignosulfonates by different anodes. *J. Cellulose Sci. Technol.* 3: 26–32.

110 Smith, C.Z., Utley, J.H.P., and Hammond, J.K. (2011). Electro-organic reactions. Part 60 [1]. The electro-oxidative conversion at laboratory scale of a lignosulfonate into vanillin in an FM01 filter press flow reactor: preparative and mechanistic aspects. *J. Appl. Electrochem.* 41: 363–375.

111 Zhu, H., Wang, L., Chen, Y. et al. (2014). Electrochemical depolymerization of lignin into renewable aromatic compounds in a non-diaphragm electrolytic cell. *RSC Adv.* 4: 29917–29924.

112 Zirbes, M., Quadri, L.L., Breiner, M. et al. (2020). High-temperature electrolysis of kraft lignin for selective vanillin formation. *ACS Sustainable Chem. Eng.* 8: 7300–7307.

113 Cui, T., Ma, L., Wang, S. et al. (2021). Atomically dispersed Pt–N_3C_1 sites enabling efficient and selective electrocatalytic C–C bond cleavage in lignin models under ambient conditions. *J. Am. Chem. Soc.* 143: 9429–9439.

114 Rafiee, M., Alherech, M., Karlen, S.D., and Stahl, S.S. (2019). Electrochemical aminoxyl-mediated oxidation of primary alcohols in lignin to carboxylic acids: Polymer modification and depolymerization. *J. Am. Chem. Soc.* 141: 15266–15276.

115 Subbotina, E., Rukkijakan, T., Marquez-Medina, M.D. et al. (2021). Oxidative cleavage of C—C bonds in lignin. *Nat. Chem.* 1118–1125.

116 Shiraishi, T., Takano, T., Kamitakahara, H., and Nakatsubo, F. (2012). Studies on electro-oxidation of lignin and lignin model compounds. Part 2: N-Hydroxyphthalimide (NHPI)-mediated indirect electro-oxidation of non-phenolic lignin model compounds. *Holzforschung* 66: 311–315.

117 Tamburini, E., Bernardi, T., Castaldelli, G. et al. (2011). Green electrochemical approach for delignification of wheat straw in second-generation bioethanol production. *Energy Environ. Sci.* 4: 551–557.

118 Ko, M., Pham, L.T.M., Sa, Y.J. et al. (2019). Unassisted solar lignin valorisation using a compartmented photo-electro-biochemical cell. *Nat. Commun.* 10: 5123.

119 Lundquist, K. and Kristersson, P. (1985). Exhaustive laccase-catalysed oxidation of a lignin model compound (vanillyl glycol) produces methanol and polymeric quinoid products. *Biochem. J.* 229: 277–279.

120 Higuchi, T. (1990). Lignin Biochemistry - Biosynthesis and biodegradation. *Wood Sci. Technol.* 24: 23–63.

121 Kawai, S., Nakagawa, M., and Ohashi, H. (2002). Degradation mechanisms of a nonphenolic β-O-4 lignin model dimer by Trametes versicolor laccase in the presence of 1-hydroxybenzotriazole. *Enzyme. Microb. Tech.* 30: 482–489.

122 Reddy, G.V.B., Sridhar, M., and Gold, M.H. (2003). Cleavage of nonphenolic β-1 diarylpropane lignin model dimers by manganese peroxidase from Phanerochaete chrysosporium-Evidence for a hydrogen abstraction mechanism. *Eur. J. Biochem.* 270: 284–292.

123 Ahmad, M., Taylor, C.R., Pink, D. et al. (2010). Development of novel assays for lignin degradation: comparative analysis of bacterial and fungal lignin degraders. *Mol Biosyst* 6: 815–821.

124 Hilgers, R., van Dam, A., Zuilhof, H. et al. (2020). Controlling the competition: Boosting laccase/HBT-catalyzed cleavage of a β-O-4′ linked lignin model. *ACS Catal.* 10: 8650–8659.

125 Schoemaker, H.E., Harvey, P.J., Bowen, R.M., and Palmer, J.M. (1985). On the mechanism of enzymatic lignin breakdown. *FEBS Lett.* 183: 7–12.

126 Paterson, A. and Lundquist, K. (1985). Radical Breakdown of Lignin. *Nature* 316: 575–576.

127 Kersten, P.J., Tien, M., Kalyanaraman, B., and Kirk, T.K. (1985). The ligninase of phanerochaete-chrysosporium generates cation radicals from methoxybenzenes. *J. Biol. Chem.* 260: 2609–2612.

128 Glenn, J.K. and Gold, M.H. (1985). Purification and characterization of an extracellular Mn(II)-dependent peroxidase from the lignin-degrading basidiomycete. *Phanerochaete chrysosporium. Arch. Biochem. Biophys.* 242: 329–341.

129 Jensen, K.A., Bao, W.L., Kawai, S. et al. (1996). Manganese-dependent cleavage of nonphenolic lignin structures by Ceriporiopsis subvermispora in the absence of lignin peroxidase. *Appl. Environ. Microb.* 62: 3679–3686.

130 Hofrichter, M. (2002). Review: lignin conversion by manganese peroxidase (MnP). *Enzyme Microb. Technol.* 30: 454–466.

131 Li, T., Mo, J.Y., Weekes, D.M. et al. (2020). Photoelectrochemical decomposition of lignin model compound on a $BiVO_4$ photoanode. *ChemSusChem* 3622–3626.

132 Li, S., Kim, S., Davis, A.H. et al. (2021). Photocatalytic chemoselective C–C bond cleavage at room temperature in dye-sensitized photoelectrochemical cells. *ACS Catal.* 11: 3771–3781.

133 Jin, W., Pastor-Pérez, L., Shen, D. et al. (2019). Catalytic upgrading of biomass model compounds: Novel approaches and lessons learnt from traditional hydrodeoxygenation – a review. *ChemCatChem* 11: 924–960.

134 Zaheer, M. and Kempe, R. (2015). Catalytic hydrogenolysis of aryl ethers: A key step in lignin valorization to valuable chemicals. *ACS Catal.* 5: 1675–1684.

135 Sreenavya, A., Neethu, P., and Sakthivel, A. (2021). The role of group VIII metals in hydroconversion of lignin to value-added chemicals and biofuels. *Catalysis for Clean Energy and Environmental Sustainability: Biomass Conversion and Green Chemistry-Volume* 1: 739–765.

136 Sergeev, A.G. and Hartwig, J.F. (2011). Selective, nickel-catalyzed hydrogenolysis of aryl ethers. *Science* 332: 439–443.

137 Zhu, J., Wang, J.C., and Dong, G.B. (2019). Catalytic activation of unstrained C(aryl)–C(aryl) bonds in 2,2'-biphenols. *Nat. Chem.* 11: 45–51.

138 Saeman, J.F. and Harris, E.E. (1946). Hydrogenation of lignin over Raney nickel. *J. Am. Chem. Soc.* 68: 2507–2509.

139 He, J., Zhao, C., and Lercher, J.A. (2012). Ni-catalyzed cleavage of aryl ethers in the aqueous phase. *J. Am. Chem. Soc.* 134: 20768–20775.

140 Zaheer, M., Hermannsdorfer, J., Kretschmer, W.P. et al. (2014). Robust heterogeneous nickel catalysts with tailored porosity for the selective hydrogenolysis of aryl ethers. *ChemCatChem* 6: 91–95.

141 Stone, M.L., Anderson, E.M., Meek, K.M. et al. (2018). Reductive catalytic fractionation of C-lignin. *ACS Sustainable Chem. Eng.* 6: 11211–11218.

142 Lama, S.M.G., Pampel, J., Fellinger, T.-P. et al. (2017). Efficiency of Ni nanoparticles supported on hierarchical porous nitrogen-doped carbon for hydrogenolysis of kraft lignin in flow and batch Systems. *ACS Sustainable Chem. Eng.* 5: 2415–2420.

143 Zhang, W., Chen, J., Liu, R. et al. (2014). Hydrodeoxygenation of lignin-derived phenolic monomers and dimers to alkane fuels over bifunctional zeolite-supported metal catalysts. *ACS Sustainable Chem. Eng.* 2: 683–691.

144 Wang, S., Zhang, K., Li, H. et al. (2021). Selective hydrogenolysis of catechyl lignin into propenylcatechol over an atomically dispersed ruthenium catalyst. *Nat. Commun.* 12: 416.

145 Verziu, M., Tirsoaga, A., Cojocaru, B. et al. (2018). Hydrogenolysis of lignin over Ru-based catalysts: The role of the ruthenium in a lignin fragmentation process. *Mol Catal* 450: 65–76.

146 Li, Y., Shuai, L., Kim, H. et al. (2018). An "ideal lignin" facilitates full biomass utilization. *Sci. Adv.* 4: eaau2968.

147 Strassberger, Z., Alberts, A.H., Louwerse, M.J. et al. (2013). Catalytic cleavage of lignin β-O-4 link mimics using copper on alumina and magnesia–alumina. *Green Chem.* 15: 768–774.

148 Yan, N., Zhao, C., Dyson, P.J. et al. (2008). Selective degradation of wood lignin over noble-metal catalysts in a two-step process. *ChemSusChem* 1: 626–629.

149 Singh, A.K., Jang, S., Kim, J.Y. et al. (2015). One-pot defunctionalization of lignin-derived compounds by dual-functional $Pd_{50}Ag_{50}/Fe_3O_4$/N-rGO catalyst. *ACS Catal.* 5: 6964–6972.

150 Zhang, J., Teo, J., Chen, X. et al. (2014). A series of NiM (M = Ru, Rh, and Pd) bimetallic catalysts for effective lignin hydrogenolysis in water. *ACS Catal.* 4: 1574–1583.

151 Horacek, J., Mikkola, J.-P., Samikannu, A. et al. (2013). Studies on sodium lignosulfonate depolymerization over Al_2O_3 supported catalysts loaded with metals and metal oxides in a continuous flow reactor. *Top. Catal.* 56: 794–799.

152 Rajan, K., Mann, J., Regmi, Y. et al. (2018). Depolymerized lignin & acrylate-based renewable photopolymers. *Abstr. Pap. Am. Chem. S.* 255.

153 Wei, L., Bibi, R., Zheng, Y. et al. (2018). Promoting effect of boron on the stability and activity of Ni/Mo_2C catalyst for hydrogenation of alkali lignin. *Catal. Lett.* 148: 1856–1869.

154 Assary, R.S., Atesin, A.C., Li, Z. et al. (2013). Reaction pathways and energetics of etheric C–O bond cleavage catalyzed by lanthanide triflates. *ACS Catal.* 3: 1908–1914.

155 Lohr, T.L., Li, Z., and Marks, T.J. (2015). Selective ether/ester C—O cleavage of an acetylated lignin model via tandem aatalysis. *ACS Catal.* 5: 7004–7007.

156 Huang, X., Zhu, J., Koranyi, T.I. et al. (2016). Effective release of lignin fragments from lignocellulose by Lewis acid metal triflates in the lignin-first approach. *ChemSusChem* 9: 3262–3267.

157 Jastrzebski, R., Constant, S., Lancefield, C.S. et al. (2016). Tandem catalytic depolymerization of lignin by water-tolerant Lewis acids and rhodium complexes. *ChemSusChem* 9: 2074–2079.

158 Huang, X.M., Gonzalez, O.M.M., Zhu, J.D. et al. (2017). Reductive fractionation of woody biomass into lignin monomers and cellulose by tandem metal triflate and Pd/C catalysis. *Green Chem.* 19: 175–187.

159 Yan, N., Yuan, Y.A., Dykeman, R. et al. (2010). Hydrodeoxygenation of lignin-derived phenols into alkanes by using nanoparticle catalysts combined with Bronsted acidic ionic liquids. *Angew. Chem. Int. Ed.* 49: 5549–5553.

160 Chen, L., Xin, J.Y., Ni, L.L. et al. (2016). Conversion of lignin model compounds under mild conditions in pseudo-homogeneous systems. *Green Chem.* 18: 2341–2352.

161 Scott, M., Deuss, P.J., de Vries, J.G. et al. (2016). New insights into the catalytic cleavage of the lignin β-O-4 linkage in multifunctional ionic liquid media. *Catal. Sci. Technol.* 6: 1882–1891.

162 Liu, F., Liu, Q.Y., Wang, A.Q., and Zhang, T. (2016). Direct catalytic hydrogenolysis of kraft lignin to phenols in choline-derived ionic liquids. *ACS Sustainable Chem. Eng.* 4: 3850–3856.

163 Long, J.X., Xu, Y., Wang, T.J. et al. (2015). Efficient base-catalyzed decomposition and in situ hydrogenolysis process for lignin depolymerization and char elimination. *Applied Energy* 141: 70–79.

164 Konnerth, H., Zhang, J., Ma, D. et al. (2015). Base promoted hydrogenolysis of lignin model compounds and organosolv lignin over metal catalysts in water. *Chem. Eng. Sci.* 123: 155–163.

165 Harris, E.E., D'Ianni, J., and Adkins, H. (1938). Reaction of hardwood lignin with hydrogen. *J. Am. Chem. Soc.* 60: 1467–1470.

166 Deutsch, K.L. and Shanks, B.H. (2012). Hydrodeoxygenation of lignin model compounds over a copper chromite catalyst. *Appl. Catal., A* 447: 144–150.

167 Cao, Y., Ma, R., Wang, N. et al. (2018). Selective hydrodeoxygenation of lignin β-O-4 model compounds and aromatic ketones promoted by palladium chloride with acidic CO_2/MeOH system. *J. CO2 Util.* 24: 328–333.

168 Cui, X.J., Yuan, H.K., Junge, K. et al. (2017). A stable and practical nickel catalyst for the hydrogenolysis of C—O bonds. *Green Chem.* 19: 305–310.

169 Whiffen, V.M.L. and Smith, K.J. (2010). Hydrodeoxygenation of 4-methylphenol over unsupported MoP, MoS_2, and MoO_x Catalysts. *Energy Fuel* 24: 4728–4737.

170 Zhang, C., Lu, J., Zhang, X. et al. (2016). Cleavage of the lignin β-O-4 ether bond via a dehydroxylation-hydrogenation strategy over a NiMo sulfide catalyst. *Green Chem.* 18: 6545–6555.

171 Jongerius, A.L., Jastrzebski, R., Bruijnincx, P.C.A., and Weckhuysen, B.M. (2012). CoMo sulfide-catalyzed hydrodeoxygenation of lignin model compounds: An extended reaction network for the conversion of monomeric and dimeric substrates. *J. Catal.* 285: 315–323.

172 Cao, Z.W., Engelhardt, J., Dierks, M. et al. (2017). Catalysis meets nonthermal separation for the production of (alkyl)phenols and hydrocarbons from pyrolysis oil. *Angew. Chem. Int. Ed.* 56: 2334–2339.

173 Fang, H., Chen, W., Wu, L. et al. (2021). Stable and antisintering tungsten carbides with controllable active phase for selective cleavage of aryl ether C–O bonds. *ACS Appl. Mater. Inter.* 13: 8274–8284.

174 Molinari, V., Giordano, C., Antonietti, M., and Esposito, D. (2014). Titanium Nitride-Nickel Nanocomposite as heterogeneous catalyst for the hydrogenolysis of aryl ethers. *J. Am. Chem. Soc.* 136: 1758–1761.

175 Molinari, V., Clavel, G., Graglia, M. et al. (2016). Mild continuous hydrogenolysis of kraft lignin over titanium nitride-nickel catalyst. *ACS Catal.* 6: 1663–1670.

176 Rashidi, M., Konarova, M., Aslam, W., and Beltramini, J.N. (2018). TiN-Cu heterogeneous nanocatalysts for effective depolymerisation of oxidised lignin. *ChemistrySelect* 3: 3379–3385.

177 Rensel, D.J., Rouvimov, S., Gin, M.E., and Hicks, J.C. (2013). Highly selective bimetallic FeMoP catalyst for C—O bond cleavage of aryl ethers. *J. Catal.* 305: 256–263.

178 Huang, X., Atay, C., Zhu, J. et al. (2017). Catalytic depolymerization of lignin and woody biomass in supercritical ethanol: Influence of reaction temperature and feedstock. *ACS Sustainable Chem. Eng.* 5: 10864–10874.

179 Loubinoux, B., Heitz, M., Coudert, G., and Guillaumet, G. (1980). Hydrogenolysis of lignins: Nickel boride catalyst. *Tetrahedron Lett.* 21: 4991–4994.

180 Stavila, V., Parthasarathi, R., Davis, R.W. et al. (2016). MOF-based catalysts for selective hydrogenolysis of carbon-oxygen ether bonds. *ACS Catal.* 6: 55–59.

181 Singh, S.K. and Ekhe, J.D. (2015). Cu-Mo doped zeolite ZSM-5 catalyzed conversion of lignin to alkyl phenols with high selectivity. *Catal. Sci. Technol.* 5: 2117–2124.

182 Barta, K., Warner, G.R., Beach, E.S., and Anastas, P.T. (2014). Depolymerization of organosolv lignin to aromatic compounds over Cu-doped porous metal oxides. *Green Chem.* 16: 191–196.

183 Li, J., Sun, H., Liu, J.-x. et al. (2018). Selective reductive cleavage of C—O bond in lignin model compounds over nitrogen-doped carbon-supported iron catalysts. *Mol Catal* 452: 36–45.

184 Jing, Y., Dong, L., Guo, Y. et al. (2020). Chemicals from lignin: A review of catalytic conversion involving hydrogen. *ChemSusChem* 13: 4181–4198.

185 Zhang, H., Fu, S., Du, X., and Deng, Y. (2021). Advances in versatile nanoscale catalyst for the reductive catalytic fractionation of lignin. *ChemSusChem* 14: 2268–2294.

186 Galkin, M.V., Sawadjoon, S., Rohde, V. et al. (2014). Mild heterogeneous palladium-catalyzed cleavage of β-O-4'-ether linkages of lignin model compounds and native lignin in air. *ChemCatChem* 6: 179–184.

187 Xu, W.Y., Miller, S.J., Agrawal, P.K., and Jones, C.W. (2012). Depolymerization and hydrodeoxygenation of switchgrass lignin with formic acid. *ChemSusChem* 5: 667–675.

188 Zhou, X., Mitra, J., and Rauchfuss, T.B. (2014). Lignol cleavage by Pd/C under mild conditions and without hydrogen: a role for benzylic C–H activation? *ChemSusChem* 7: 1623–1626.

189 Nichols, J.M., Bishop, L.M., Bergman, R.G., and Ellman, J.A. (2010). Catalytic C–O bond cleavage of 2-aryloxy-1-arylethanols and its application to the depolymerization of lignin-related polymers. *J. Am. Chem. Soc.* 132: 12554–12555.

190 Wu, A., Patrick, B.O., Chung, E., and James, B.R. (2012). Hydrogenolysis of β-O-4 lignin model dimers by a ruthenium-xantphos catalyst. *Dalton Trans.* 41: 11093–11106.

191 Zhao, X., Yang, Y., Zhu, R. et al. (2019). Mechanistic picture of the redox-neutral C–C bond cleavage in 1,3-dilignol lignin model compound catalyzed by [Ru(Cl)(H)(PPh3)3]/triphos. *Mol Catal* 471: 77–84.

192 Chmely, S.C., Kim, S., Ciesielski, P.N. et al. (2013). Mechanistic study of a Ru-xantphos catalyst for tandem alcohol dehydrogenation and reductive aryl-ether cleavage. *ACS Catal.* 3: 963–974.

193 Liu, Y.X., Li, C.Z., Miao, W. et al. (2019). Mild redox-neutral depolymerization of lignin with a binuclear Rh complex in water. *ACS Catal.* 9: 4441–4447.

194 Liu, Y., Li, C., Miao, W. et al. (2020). Rhodium-terpyridine catalyzed redox-neutral depolymerization of lignin in water. *Green Chem.* 22: 33–38.

195 Lu, J., Wang, M., Zhang, X. et al. (2016). β-O-4 bond cleavage mechanism for lignin model compounds over Pd catalysts identified by combination of first-principles calculations and experiments. *ACS Catal.* 6: 5589–5598.

196 Gao, F., Webb, J.D., Sorek, H. et al. (2016). Fragmentation of lignin samples with commercial Pd/C under ambient pressure of hydrogen. *ACS Catal.* 6: 7385–7392.

197 Liu, X., Bouxin, F.P., Fan, J. et al. (2020). Microwave-assisted catalytic depolymerization of lignin from birch sawdust to produce phenolic monomers utilizing a hydrogen-free strategy. *J. Hazard. Mater.* 123490.

198 Zhang, J.W., Lu, G.P., and Cai, C. (2017). Self-hydrogen transfer hydrogenolysis of β-O-4 linkages in lignin catalyzed by MIL-100(Fe) supported Pd-Ni BMNPs. *Green Chem.* 19: 4538–4543.

199 Luo, N., Wang, M., Li, H. et al. (2017). Visible-light-driven self-hydrogen transfer hydrogenolysis of lignin models and extracts into phenolic products. *ACS Catal.* 7: 4571–4580.

200 Enright, M.J., Gilbert-Bass, K., Sarsito, H., and Cossairt, B.M. (2019). Photolytic C–O bond cleavage with quantum dots. *Chem. Mater.* 31: 2677–2682.

201 Yoo, H., Lee, M.-W., Lee, S. et al. (2020). Enhancing photocatalytic β-O-4 bond cleavage in lignin model compounds by silver-exchanged cadmium sulfide. *ACS Catal.* 10: 8465–8475.

202 Huang, Y.B., Yan, L., Chen, M.Y. et al. (2015). Selective hydrogenolysis of phenols and phenyl ethers to arenes through direct C–O cleavage over ruthenium-tungsten bifunctional catalysts. *Green Chem.* 17: 3010–3017.

203 Li, L., Dong, L., Li, D. et al. (2020). Hydrogen-free Production of 4-alkylphenols from lignin via self-reforming-driven depolymerization and hydrogenolysis. *ACS Catal.* 10: 15197–15206.

204 Li, L., Zhang, T., Guo, Z. et al. (2021). Unraveling the role of metal in $M/NiAl_2O_4$ (M= Pt, Pd, Ru) catalyst for the self-reforming-driven hydrogenolysis of lignin. *Ind. Eng. Chem. Res.* 60: 11699–11706.

205 Meng, Q., Yan, J., Wu, R. et al. (2021). Sustainable production of benzene from lignin. *Nat. Commun.* 12: 4534.

206 Ren, Y.L., Yan, M.J., Wang, J.J. et al. (2013). Selective reductive cleavage of inert aryl C—O bonds by an iron catalyst. *Angew. Chem. Int. Ed.* 52: 12674–12678.

207 Nagy, M., David, K., Britovsek, G.J.P., and Ragauskas, A.J. (2009). Catalytic hydrogenolysis of ethanol organosolv lignin. *Holzforschung* 63: 513–520.

208 Chouhan, M., Kumar, K., Sharma, R. et al. (2013). $NiCl_2 \cdot 6H_2O/NaBH_4$ in methanol: a mild and efficient strategy for chemoselective deallylation/ debenzylation of aryl ethers. *Tetrahedron Lett.* 54: 4540–4543.

209 Ohta, T., Michibata, T., Yamada, K. et al. (2003). Reductive cleavage of the C–O bond of acetals and orthoesters: reduction by silane in the presence of a $Rh\text{-}PPh_3$ complex. *Chem. Commun.* 1192–1193.

210 Fedorov, A., Toutov, A.A., Swisher, N.A., and Grubbs, R.H. (2013). Lewis-base silane activation: from reductive cleavage of aryl ethers to selective ortho-silylation. *Chem. Sci.* 4: 1640–1645.

211 Lancefield, C.S., Ojo, O.S., Tran, F., and Westwood, N.J. (2015). Isolation of functionalized phenolic monomers through selective oxidation and C–O bond cleavage of the β-O-4 linkages in lignin. *Angew. Chem. Int. Ed.* 54: 258–262.

212 Prinsen, P., Narani, A., Hartog, A.F. et al. (2017). Dissolving lignin in water through enzymatic sulfation with aryl sulfotransferase. *ChemSusChem* 10: 2267–2273.

213 Wu, X.J., Fan, X.T., Xie, S.J. et al. (2018). Solar energy-driven lignin-first approach to full utilization of lignocellulosic biomass under mild conditions. *Nat. Catal.* 1: 772–780.

214 Wu, X.J., Xie, S.J., Liu, C.X. et al. (2019). Ligand-controlled photocatalysis of CdS quantum dots for lignin valorization under visible light. *ACS Catal.* 9: 8443–8451.

215 Lin, J., Wu, X., Xie, S. et al. (2019). Visible-light-driven cleavage of C—O linkage for lignin valorization to functionalized aromatics. *ChemSusChem* 12: 5023–5031.

216 Bawareth, B., Di Marino, D., Nijhuis, T.A., and Wessling, M. (2018). Unravelling electrochemical lignin depolymerization. *ACS Sustainable Chem. Eng.* 6: 7565–7573.

217 Mahdavi, B., Lafrance, A., Martel, A. et al. (1997). Electrocatalytic hydrogenolysis of lignin model dimers at Raney nickel electrodes. *J. Appl. Electrochem.* 27: 605–611.

218 Wu, W.B. and Huang, J.M. (2014). Electrochemical cleavage of aryl ethers promoted by sodium borohydride. *J. Org. Chem.* 79: 10189–10195.

219 Zhou, Y.L., Gao, Y.J., Zhong, X. et al. (2019). Electrocatalytic upgrading of lignin-derived bio-oil based on surface-engineered PtNiB nanostructure. *Adv. Funct. Mater.* 29: 1807651.

220 Reiter, J., Strittmatter, H., Wiemann, L.O. et al. (2013). Enzymatic cleavage of lignin β-O-4 aryl ether bonds via net internal hydrogen transfer. *Green Chem.* 15: 1373–1381.

221 Zhu, S., Guo, J., Wang, X. et al. (2017). Alcoholysis: A promising technology for conversion of lignocellulose and platform chemicals. *ChemSusChem* 10: 2547–2559.

222 Ma, R., Hao, W., Ma, X. et al. (2014). Catalytic ethanolysis of kraft lignin into high-calue small-molecular chemicals over a nanostructured **α**-molybdenum carbide catalyst. *Angew. Chem. Int. Ed.* 53: 7310–7315.

223 Ma, X.L., Ma, R., Hao, W.Y. et al. (2015). Common pathways in ethanolysis of kraft lignin to platform chemicals over molybdenum-based catalysts. *ACS Catal.* 5: 4803–4813.

224 Huang, X., Koranyi, T.I., Boot, M.D., and Hensen, E.J.M. (2014). Catalytic depolymerization of lignin in wupercritical ethanol. *ChemSusChem* 7: 2276–2288.

225 Huang, X.M., Atay, C., Koranyi, T.I. et al. (2015). Role of Cu-Mg-Al mixed oxide catalysts in lignin depolymerization in supercritical ethanol. *ACS Catal.* 5: 7359–7370.

226 Yan, F., Ma, R., Ma, X.L. et al. (2017). Ethanolysis of kraft lignin to platform chemicals on a $MoC_{1-x}/Cu\text{-}MgAlO_z$, catalyst. *Appl. Catal., B* 202: 305–313.

227 Chen, M., Hao, W., Ma, R. et al. (2017). Catalytic ethanolysis of kraft lignin to small-molecular liquid products over an alumina supported molybdenum nitride catalyst. *Catal. Today* 298: 9–15.

228 Kim, J.-Y., Park, S.Y., Lee, J.H. et al. (2017). Sequential solvent fractionation of lignin for selective production of monoaromatics by Ru catalyzed ethanolysis. *RSC Adv.* 7: 53117–53125.

229 Phillips, M. and Goss, M.J. (1932). The hydrolysis of lignin with 12% hydrochloric acid. *J. Am. Chem. Soc.* 54: 3374–3377.

230 Zhang, G., Wen, Y., Liu, Z. et al. (2017). Acid-catalyzed hydrolysis of conifer lignosulfonate in black liquor for the production of value-added chemicals. *Appl. Catal., A* 542: 1–9.

231 Sturgeon, M.R., Kim, S., Lawrence, K. et al. (2014). A mechanistic investigation of acid-catalyzed cleavage of aryl-ether linkages: Implications for lignin depolymerization in acidic environments. *ACS Sustainable Chem. Eng.* 2: 472–485.

232 Zhang, L., Yan, L., Wang, Z. et al. (2015). Characterization of lignin derived from water-only and dilute acid flowthrough pretreatment of poplar wood at elevated temperatures. *Biotechnol. Biofuels* 8: 203.

233 Mahmood, N., Yuan, Z.S., Schmidt, J., and Xu, C.B. (2015). Hydrolytic depolymerization of hydrolysis lignin: Effects of catalysts and solvents. *Bioresour. Technol.* 190: 416–419.

234 Anderson, E.M., Katahira, R., Reed, M. et al. (2016). Reductive catalytic fractionation of corn stover lignin. *ACS Sustainable Chem. Eng.* 4: 6940–6950.

235 Holmelid, B., Barth, T., Brusletto, R., and Kleinert, M. (2017). Production of monomeric phenols by formic acid assisted hydrous liquefaction of lignin. *Biomass Bioenergy* 105: 298–309.

236 da Silva, S.H.F., Egues, I., and Labidi, J. (2019). Liquefaction of kraft lignin using polyhydric alcohols and organic acids as catalysts for sustainable polyols production. *Ind. Crops Prod.* 137: 687–693.

237 Huang, S.H., Mahmood, N., Zhang, Y.S. et al. (2017). Reductive de-polymerization of kraft lignin with formic acid at low temperatures using inexpensive supported Ni-based catalysts. *Fuel* 209: 579–586.

238 Wu, Z., Zhao, X.X., Zhang, J. et al. (2019). Ethanol/1,4-dioxane/formic acid as synergistic solvents for the conversion of lignin into high-value added phenolic monomers. *Bioresour. Technol.* 278: 187–194.

239 Schwiderski, M., Kruse, A., Grandl, R., and Dockendorf, D. (2014). Comparison of the influence of a Lewis acid $AlCl_3$ and a Bronsted acid HCl on the organosolv pulping of beech wood. *Green Chem.* 16: 1569–1578.

240 Wang, H.L., Zhang, L.B., Deng, T.S. et al. (2016). $ZnCl_2$ induced catalytic conversion of softwood lignin to aromatics and hydrocarbons. *Green Chem.* 18: 2802–2810.

241 Lahive, C.W., Deuss, P.J., Lancefield, C.S. et al. (2016). Advanced model compounds for understanding acid-catalyzed lignin depolymerization: Identification of renewable aromatics and a lignin-derived solvent. *J. Am. Chem. Soc.* 138: 8900–8911.

242 Deuss, P.J., Lahive, C.W., Lancefield, C.S. et al. (2016). Metal triflates for the production of aromatics from lignin. *ChemSusChem* 9: 2974–2981.

243 Güvenatam, B., Heeres, E.H.J., Pidko, E.A., and Hensen, E.J.M. (2016). Lewis-acid catalyzed depolymerization of Protobind lignin in supercritical water and ethanol. *Catal. Today* 259: 460–466.

244 Deuss, P.J., Lancefield, C.S., Narani, A. et al. (2017). Phenolic acetals from lignins of varying compositions via iron(III) triflate catalysed depolymerisation. *Green Chem.* 19: 2774–2782.

245 Jia, S.Y., Cox, B.J., Guo, X.W. et al. (2010). Cleaving the β-O-4 bonds of lignin model compounds in an acidic ionic liquid, 1-H-3-methylimidazolium chloride: An optional strategy for the degradation of lignin. *ChemSusChem* 3: 1078–1084.

246 Harms, R.G., Markovits, I.I., Drees, M. et al. (2014). Cleavage of C–O bonds in lignin model compounds catalyzed by methyldioxorhenium in homogeneous phase. *ChemSusChem* 7: 429–434.

247 Zhang, B., Li, C., Dai, T. et al. (2015). Microwave-assisted fast conversion of lignin model compounds and organosolv lignin over methyltrioxorhenium in ionic liquids. *RSC Adv.* 5: 84967–84973.

248 Liu, X., Zhao, G., Ren, S. et al. (2018). Zirconia-supported 12-tungstophosphoric acid: Characterization and catalytic behavior in degradation of alkali lignin. *Ind. Crops Prod.* 125: 496–504.

249 Wang, H., Li, W., Wang, J. et al. (2017). A ternary composite oxides $S_2O_8^{2-}$/ ZrO_2-TiO_2-SiO_2 as an efficient solid super acid catalyst for depolymerization of lignin. *RSC Adv.* 7: 50027–50034.

250 Jin, S.H., Xiao, Z.H., Chen, X. et al. (2015). Cleavage of lignin-derived 4-O-5 aryl ethers over nickel nanoparticles supported on niobic acid-activated carbon composites. *Ind. Eng. Chem. Res.* 54: 2302–2310.

251 Santana, J.A., Carvalho, W.S., and Ataide, C.H. (2018). Catalytic effect of ZSM-5 zeolite and HY-340 niobic acid on the pyrolysis of industrial kraft lignins. *Ind. Crops Prod.* 111: 126–132.

252 Deepa, A.K. and Dhepe, P.L. (2014). Solid acid catalyzed depolymerization of lignin into value added aromatic monomers. *RSC Adv.* 4: 12625–12629.

253 Deepa, A.K. and Dhepe, P.L. (2015). Lignin depolymerization into aromatic monomers over solid acid catalysts. *ACS Catal.* 5: 365–379.

254 Adhikari, S., Srinivasan, V., and Fasina, O. (2014). Catalytic pyrolysis of raw and thermally treated lignin using different acidic zeolites. *Energy Fuel* 28: 4532–4538.

255 Kim, J.-Y., Lee, J.H., Park, J. et al. (2015). Catalytic pyrolysis of lignin over HZSM-5 catalysts: Effect of various parameters on the production of aromatic hydrocarbon. *J. Anal. Appl. Pyrolysis* 114: 273–280.

256 Ma, Z., Troussard, E., and van Bokhoven, J.A. (2012). Controlling the selectivity to chemicals from lignin via catalytic fast pyrolysis. *Appl. Catal. A: Gen* 423-424: 130–136.

257 Roberts, V.M., Stein, V., Reiner, T. et al. (2011). Towards quantitative catalytic lignin depolymerization. *Chem. Eur. J.* 17: 5939–5948.

258 Dabral, S., Mottweiler, J., Rinesch, T., and Bolm, C. (2015). Base-catalysed cleavage of lignin β-O-4 model compounds in dimethyl carbonate. *Green Chem.* 17: 4908–4912.

259 Katahira, R., Mittal, A., McKinney, K. et al. (2016). Base-catalyzed depolymerization of biorefinery lignins. *ACS Sustainable Chem. Eng.* 4: 1474–1486.

260 Dabral, S., Engel, J., Mottweiler, J. et al. (2018). Mechanistic studies of base-catalysed lignin depolymerisation in dimethyl carbonate. *Green Chem.* 20: 170–182.

261 Toledano, A., Serrano, L., and Labidi, J. (2012). Organosolv lignin depolymerization with different base catalysts. *J. Chem. Technol. Biotechnol.* 87: 1593–1599.

262 Chaudhary, R. and Dhepe, P.L. (2017). Solid base catalyzed depolymerization of lignin into low molecular weight products. *Green Chem.* 19: 778–788.

263 Jia, S., Cox, B.J., Guo, X. et al. (2010). Decomposition of a phenolic lignin model compound over organic N-bases in an ionic liquid. *Holzforschung* 64: 577–580.

264 Rahimi, A., Azarpira, A., Kim, H. et al. (2013). Chemoselective metal-free aerobic alcohol oxidation in lignin. *J. Am. Chem. Soc.* 135: 6415–6418.

265 Patil, N.D., Yao, S.G., Meier, M.S. et al. (2015). Selective cleavage of the C_α–C_β linkage in lignin model compounds via Baeyer-Villiger oxidation. *Org. Biomol. Chem.* 13: 3243–3254.

266 Wang, Y.L., Wang, Q.Y., He, J.H., and Zhang, Y.T. (2017). Highly effective C–C bond cleavage of lignin model compounds. *Green Chem.* 19: 3135–3141.

267 Jennings, J.A., Parkin, S., Munson, E. et al. (2017). Regioselective Baeyer-Villiger oxidation of lignin model compounds with tin beta zeolite catalyst and hydrogen peroxide. *RSC Adv.* 7: 25987–25997.

268 Santos, W.C.C., Dias, K.A., Santos, L.P. et al. (2018). Evaluating gold and selenium chemistry for selective transformations of lignin model compounds. *Adv. Synth. Catal.* 360: 1376–1383.

269 Hao, K., Zhang, L.-L., Song, L. et al. (2020). Highly active Mo-V-based bifunctional catalysts for catalytic conversion of lignin dimer model compounds at room temperature. *Inorg. Chem. Commun.* 116: 107910.

270 Wang, Y., Du, Y., He, J., and Zhang, Y. (2018). Transformation of lignin model compounds to N-substituted aromatics via Beckmann rearrangement. *Green Chem.* 20: 3318–3326.
271 Li, H., Wang, M., Liu, H. et al. (2018). NH_2OH–mediated lignin conversion to isoxazole and nitrile. *ACS Sustainable Chem. Eng.* 6: 3748–3753.
272 Liu, J.Z., Qiu, X., Huang, X.Q. et al. (2019). From alkylarenes to anilines via site-directed carbon-carbon amination. *Nat. Chem.* 11: 71–77.
273 Li, H.J., Bunrit, A., Lu, J.M. et al. (2019). Photocatalytic cleavage of aryl ether in modified lignin to non-phenolic aromatics. *ACS Catal.* 9: 8843–8851.
274 Klinger, G.E., Zhou, Y., Hao, P. et al. (2019). Biomimetic reductive cleavage of keto aryl ether bonds by small-molecule thiols. *ChemSusChem* 12: 4775–4779.
275 Nguyen, J.D., Matsuura, B.S., and Stephenson, C.R. (2014). A photochemical strategy for lignin degradation at room temperature. *J. Am. Chem. Soc.* 136: 1218–1221.
276 Bosque, I., Magallanes, G., Rigoulet, M. et al. (2017). Redox catalysis facilitates lignin depolymerization. *ACS Cent. Sci.* 3: 621–628.
277 Magallanes, G., Kärkäs, M.D., Bosque, I. et al. (2019). Selective C–O bond cleavage of lignin systems and polymers enabled by sequential palladium-catalyzed aerobic oxidation and visible-light photoredox catalysis. *ACS Catal.* 9: 2252–2260.
278 Li, S., Hao, Z., Wang, K. et al. (2020). Visible light-enabled selective depolymerization of oxidized lignin by an organic photocatalyst. *Chem. Commun.* 56: 11243–11246.
279 Luo, J., Zhang, X., Lu, J., and Zhang, J. (2017). Fine tuning the redox potentials of carbazolic porous organic frameworks for visible-light photoredox catalytic degradation of lignin β-O-4 models. *ACS Catal.* 7: 5062–5070.
280 Wang, M., Zhang, X., Li, H. et al. (2018). Carbon modification of nickel catalyst for depolymerization of oxidized lignin to aromatics. *ACS Catal.* 8: 1614–1620.
281 Rahimi, A., Ulbrich, A., Coon, J.J., and Stahl, S.S. (2014). Formic-acid-induced depolymerization of oxidized lignin to aromatics. *Nature* 515: 249–252.
282 Gasser, C.A., Cvancarova, M., Ammann, E.M. et al. (2017). Sequential lignin depolymerization by combination of biocatalytic and formic acid/formate treatment steps. *Appl. Microbiol. Biotechnol.* 101: 2575–2588.
283 Wang, M., Lu, J.M., Zhang, X.C. et al. (2016). Two-step, catalytic C-C bond oxidative cleavage process converts lignin models and extracts to aromatic acids. *ACS Catal.* 6: 6086–6090.
284 Wang, M., Li, L.H., Lu, J.M. et al. (2017). Acid promoted C–C bond oxidative cleavage of β-O-4 and β-1 lignin models to esters over a copper catalyst. *Green Chem.* 19: 702–706.
285 Anderson, E.M., Stone, M.L., Hulsey, M.J. et al. (2018). Kinetic Studies of lignin solvolysis and reduction by reductive catalytic fractionation decoupled in flow-through reactors. *ACS Sustainable Chem. Eng.* 6: 7951–7959.
286 Wang, M., Liu, M., Li, H. et al. (2018). Dealkylation of lignin to phenol via oxidation-hydrogenation strategy. *ACS Catal.* 8: 6837–6843.
287 Liao, Y.H., Zhong, R.Y., Makshina, E. et al. (2018). Propylphenol to phenol and propylene over acidic zeolites: Role of shape selectivity and presence of steam. *ACS Catal.* 8: 7861–7878.

288 Verboekend, D., Liao, Y.H., Schutyser, W., and Sels, B.F. (2016). Alkylphenols to phenol and olefins by zeolite catalysis: A pathway to valorize raw and fossilized lignocellulose. *Green Chem.* 18: 297–306.

289 Liao, Y., Koelewijn, S.-F., Van den Bossche, G. et al. (2020). A sustainable wood biorefinery for low-carbon footprint chemicals production. *Science* 367: 1385–1390.

290 Yan, J., Meng, Q.L., Shen, X.J. et al. (2020). X. Selective valorization of lignin to phenol by direct transformation of C_{sp2}—C_{sp3} and C—O bonds. *Sci. Adv.* 6: eabd1951.

291 Shen, X., Meng, Q., Mei, Q. et al. (2020). Selective catalytic transformation of lignin with guaiacol as the only liquid product. *Chem. Sci.* 11: 1347–1352.

292 vom Stein, T., den Hartog, T., Buendia, J. et al. (2015). Ruthenium-Catalyzed C–C bond cleavage in lignin model substrates. *Angew. Chem. Int. Ed.* 54: 5859–5863.

293 Lancefield, C.S., Teunissen, L.W., Weckhuysen, B.M., and Bruijnincx, P.C.A. (2018). Iridium-catalysed primary alcohol oxidation and hydrogen shuttling for the depolymerisation of lignin. *Green Chem.* 20: 3214–3221.

294 Gazi, S., Ng, W.K.H., Ganguly, R. et al. (2015). Selective photocatalytic C—C bond cleavage under ambient conditions with earth abundant vanadium complexes. *Chem. Sci.* 6: 7130–7142.

295 Son, S. and Toste, F.D. (2010). Non-oxidative vanadium-catalyzed C—O bond cleavage: Application to degradation of lignin model compounds. *Angew. Chem. Int. Ed.* 49: 3791–3794.

296 Haibach, M.C., Lease, N., and Goldman, A.S. (2014). Catalytic cleavage of ether C—O bonds by pincer iridium complexes. *Angew. Chem. Int. Ed.* 53: 10160–10163.

297 Tan, F.-F., He, X.-Y., Tian, W.-F., and Li, Y. (2020). Visible-light photoredox-catalyzed C–O bond cleavage of diaryl ethers by acridinium photocatalysts at room temperature. *Nat. Commun.* 11: 1–11.

298 Zhou, Y., Hu, D., Li, D., and Jiang, X. (2021). Uranyl-photocatalyzed hydrolysis of diaryl ethers at ambient environment for the directional degradation of 4-O-5 Lignin. *JACS Au* 1: 1141–1146.

299 Questell-Santiago, Y.M., Galkin, M.V., Barta, K., and Luterbacher, J.S. (2020). Stabilization strategies in biomass depolymerization using chemical functionalization. *Nat. Rev. Chem.* 4: 311–330.

300 Karkas, M.D. (2017). Lignin Hydrogenolysis: Improving lignin disassembly through formaldehyde stabilization. *ChemSusChem* 10: 2111–2115.

301 Lan, W., Amiri, M.T., Hunston, C.M., and Luterbacher, J.S. (2018). Protection group effects during α,γ-diol lignin stabilization promote high-selectivity monomer production. *Angew. Chem. Int. Ed.* 57: 1356–1360.

302 Zhu, G., Qiu, X., Zhao, Y. et al. (2016). Depolymerization of lignin by microwave-assisted methylation of benzylic alcohols. *Bioresour. Technol.* 218: 718–722.

303 Chen, J., Lu, F., Si, X. et al. (2016). High yield production of natural phenolic alcohols from woody biomass using a nickel-based catalyst. *ChemSusChem* 9: 3353–3360.

304 Xiao, G., Montgomery, J.R.D., Lancefield, C.S. et al. (2020). Copper-mediated conversion of complex ethers to esters: Enabling biopolymer depolymerisation under mild conditions. *Chem. Eur. J.* 26: 12397–12402.

305 Wang, J.Y., Deng, Y.H., Qian, Y. et al. (2016). Reduction of lignin color via one-step UV irradiation. *Green Chem.* 18: 695–699.

306 Sun, Y.-P., Yates, B., Abbot, J., and Chen, C.-L. (1996). ESR study of lignin model compounds irradiated by UV (254 nm photons) in the presence of hydrogen peroxide. *Holzforschung* 50: 233–236.

307 Hatakeyama, H. and Nakano, J. (1970). Electron spin resonance studies on lignin and lignin model compounds. *Cellul. Chem. Technol.* 4: 281–291.

308 Ponomarev, A.V. and Ershov, B.G. (2018). Radiation-thermal decomposition of lignin: Products and the mechanism of their formation (review). *High Energy Chem.* 52: 58–70.

309 Zhou, H., Zhang, R.L., Zhan, W. et al. (2016). High biomass loadings of 40 wt% for efficient fractionation in biorefineries with an aqueous solvent system without adding adscititious catalyst. *Green Chem.* 18: 6108–6114.

310 Dong, L., Lin, L., Han, X. et al. (2019). Breaking the limit of lignin monomer production via cleavage of interunit carbon–carbon linkages. *Chem* 5: 1521–1536.

311 Galkin, M.V. and Samec, J.S.M. (2014). Selective route to 2-propenyl aryls directly from wood by a tandem organosolv and palladium-catalysed transfer hydrogenolysis. *ChemSusChem* 7: 2154–2158.

312 Song, Q., Wang, F., and Xu, J. (2012). Hydrogenolysis of lignosulfonate into phenols over heterogeneous nickel catalysts. *Chem. Commun.* 48: 7019–7021.

313 Li, C.Z., Zheng, M.Y., Wang, A.Q., and Zhang, T. (2012). One-pot catalytic hydrocracking of raw woody biomass into chemicals over supported carbide catalysts: simultaneous conversion of cellulose, hemicellulose and lignin. *Energy Environ. Sci.* 5: 6383–6390.

314 Wang, Y.-Y., Ling, L.-L., and Jiang, H. (2016). Selective hydrogenation of lignin to produce chemical commodities by using a biochar supported Ni–Mo_2C catalyst obtained from biomass. *Green Chem.* 18: 4032–4041.

315 Xiao, L.-P., Wang, S., Li, H. et al. (2017). Catalytic hydrogenolysis of lignins into phenolic compounds over carbon nanotube supported molybdenum oxide. *ACS Catal.* 7: 7535–7542.

316 Shao, Y., Xia, Q., Dong, L. et al. (2017). Selective production of arenes via direct lignin upgrading over a niobium-based catalyst. *Nat. Commun.* 8: 16104.

317 The yield calculation is obtained from the established nitrobenzene oxidation method, which is regarded as 100%.

318 Park, J., Cahyadi, H.S., Mushtaq, U. et al. (2020). Highly efficient reductive catalytic fractionation of lignocellulosic biomass over extremely low-loaded Pd catalysts. *ACS Catal.* 10: 12487–12506.

319 The mixture contains aromatic chemicals and low molecular weight liquid products.

320 Ma, Y.Y., Ma, J.P., Xia, F. et al. (2019). Molybdenum-catalyzed oxidative cleavage of raw poplar sawdust into mono-aromatics and organic acid esters. *Asian. J. Org. Chem.* 8: 1348–1353.

321 Das, A., Rahimi, A., Ulbrich, A. et al. (2018). Lignin conversion to low-molecular-weight aromatics via an aerobic oxidation-hydrolysissequence: Comparison of different lignin sources. *ACS Sustainable. Chem. Eng.* 6: 3367–3374.

322 Song, S., Zhang, J.G., Gozaydin, G., and Yan, N. (2019). Production of terephthalic acid from corn stover lignin. *Angew. Chem. Int. Ed.* 58: 4934–4937.

323 Liao, Y.H., Zhong, R.Y., d'Halluin, M. et al. (2020). Aromatics production from lignocellulosic biomass: Shape selective dealkylation of lignin-derived phenolics over hierarchical ZSM-5. *ACS Sustainable Chem. Eng.* 8: 8713–8722.

324 Liao, Y., d'Halluin, M., Makshina, E. et al. (2018). Shape selectivity vapor-phase conversion of lignin-derived 4-ethylphenol to phenol and ethylene over acidic aluminosilicates: Impact of acid properties and pore constraint. *Appl. Catal., B* 234: 117–129.

Part IV

Review on Lignin Linkages Cleavage Strategies and Mechanisms via an IDA Method

5

The Inverse Disassembly Analysis Method for Classifying Lignin Conversion Strategies

5.1 Introduction of Inverse Disassembly Analysis for Lignin Conversion

As discussed in Chapters 1 and 2, lignin is a three-dimensional amorphous polymer consisting of methoxylated phenylpropane structures that can be regarded as a promising resource to prepare aromatic chemicals. However, nature leaves us with challenging work to utilize this resource. We should firstly depolymerize lignin to aromatics monomers with a reasonable conversion and aromatic yields. The depolymerization process relies on the development of efficient strategies and catalysts to selectively cleavage the stable and ubiquitous C–O or C–C bonds and meanwhile to keep the aromatic benzene rings unconverted. Therefore, the lignin conversion can be regarded as a disassembly process of complex natural organic products, which is the opposite of the artificial synthesis of natural products, assembling smaller molecules to the target and complex molecule. Although these two processes are different, the strategies and thoughts used in the artificial synthesis of the natural product via a retrosynthetic analysis can give us some inspiration in classifying and understanding the various lignin conversion methods.

For the artificial synthesis of natural products, the disassembly analysis is an important process because it is necessary to know what kinds of reagents are necessary to synthesize the target molecule. In this respect, Prof. Corey gives the famous retrosynthetic analysis (RSA) method (Figure 5.1a, with Propindilactone G as the example) [1–3], and its core idea is to find the desired synthons or synthetic equivalents by breaking down the characteristic chemical bonds of complex organic substrates, basing on the reported synthesis processes, catalysis reactions, or newly discovered/developed reactions. Furthermore, appropriate substrates modification, group protection, or group transformation can efficiently promote the retrosynthetic analysis and provide more route choices for the target molecular synthesis. The RSA method inspired us to think about whether there is a similar analysis method to be developed used to further systematically classify the multifarious methods used in the lignin bonds cleavage.

Lignin Conversion Catalysis: Transformation to Aromatic Chemicals, First Edition.
Chaofeng Zhang and Feng Wang.

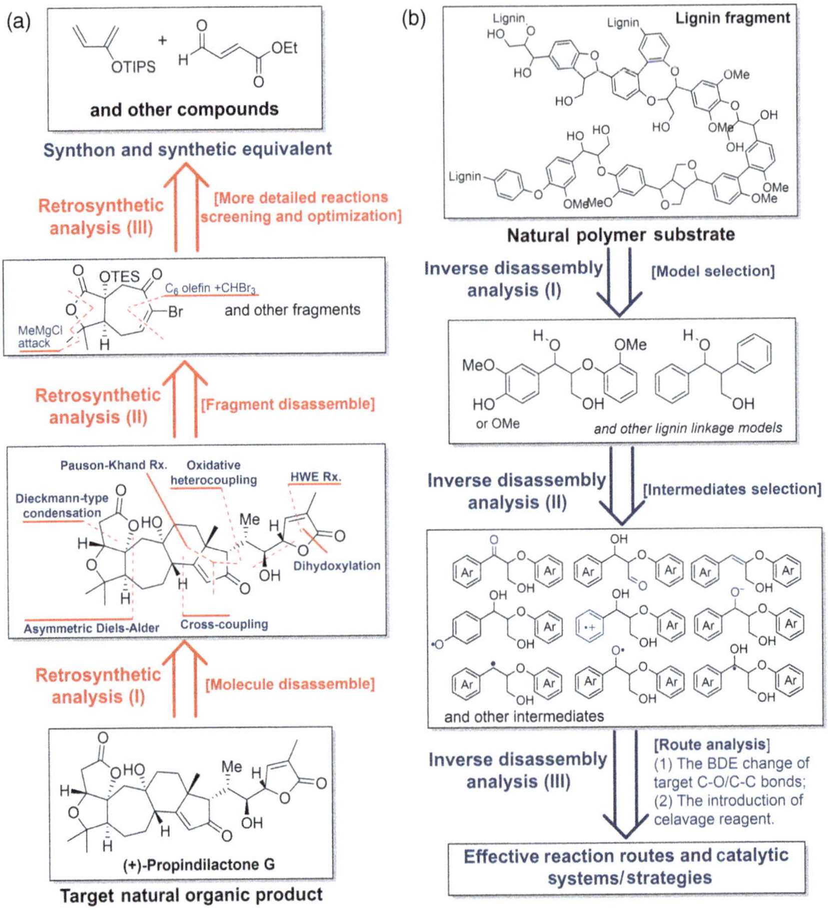

Figure 5.1 The Retrosynthetic Analysis of the Natural Product to Confirm the Synthons and Synthetic Equivalents (a, with Propindilactone G as the Example Reported by Yang [4]), and the Inverse Disassembly Analysis Methods to Explore the Efficient Strategies for Lignin Depolymerization (b, with β-O-4 as the Main Cleavage Model).

For the first step of lignin depolymerization, different catalytic routes produce different intermediates, and these intermediates will lead to variations in the lignin linkage cleavage. In an ideal route with cleavage of the β-O-4 bond and other linkages generates an activated intermediate after a series of steps with low activation energy, where the target bond has lower dissociation energy and easily breaks in the subsequent transformations. Therefore, the discussion or classification of the methods or strategies for lignin depolymerization based onthese generation strategies and steps following transformation of critical intermediates will be more systemic and comprehensive. Referring to the retrosynthetic analysis used in the synthesis of organic natural products, we herein decide to put forward an inverse disassembly analysis (IDA) method to summarize and understand the methods used in lignin and its model compounds transformation (Figure 5.1b). Different from the RSA method, which focuses on providing the optimal reaction route to recombine the corresponding synthons/synthetic equivalents to the natural product, the IDA method aims to find the potential lignin depolymerization routes via the study of the intermediates after the artificial pruning of the corresponding bonds like RSA method. The IDA can refer to the bond dissociation enthalpy (BDE) change of the target linkage bonds during the transformation, or refer to the introduction of an active chemical group or substrate site for linkage bonds cleavage.

Based on the determined reaction path with the generation of active intermediates and further controllable conversion/transformation, the nature of the catalyst will be explicit, which can improve the efficiency of catalysts screening and foundation of the method. Additionally, the IDA method can be used in the classification and discussion of the various methods for lignin conversion as viewed from the cleavage order and type of corresponding chemical bonds over various catalytic systems. In some situations, oxidation methods and hydrogenation methods will be put together for discussion, not based on the reaction types but based on a deeper connection of their mechanism, which may be more efficient than the lignin depolymerization summary study by reaction types, such as oxidation, hydrogenation, hydrolysis, and other approaches.

5.2 Different Analysis Modes for Lignin Depolymerization

As the important parameter for the following discussion of lignin bond cleavage mechanism, the bond dissociation enthalpy of typical bonds in lignin structure and its fragments are summarized. As shown in Figure 5.2, the bonddissociation enthalpy (BDE) for each linkage in lignin falls in the range 175.7-372.4 kJ mol^{-1} for

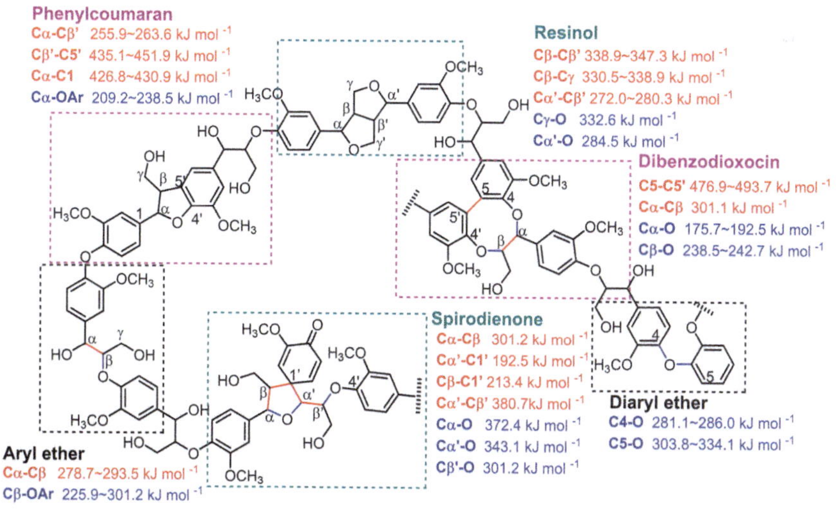

Figure 5.2 Bond Dissociation Enthalpy of Typical Bonds in Lignin Linkages. *Source*: The Data are Abstracted from to the Refs [5–10].

Table 5.1 Bond Dissociation Energies of the $C_β$–O and $C_α$–$C_β$ Bonds in β–O–4 Linkage Models.

Lignin model	$C_α$–$C_β$ bond BDE (kJ·mol^{-1})	$C_β$–O bond BDE (kJ·mol^{-1})	Lignin model	$C_α$–$C_β$ bond BDE (kJ·mol^{-1})
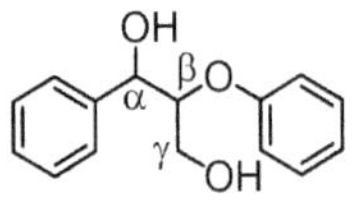	279.7	273.2	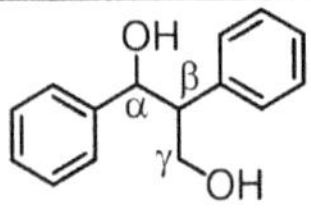	235.3
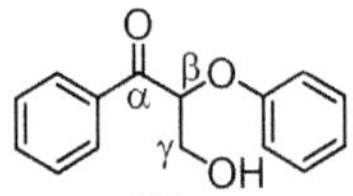	307.7	215.6	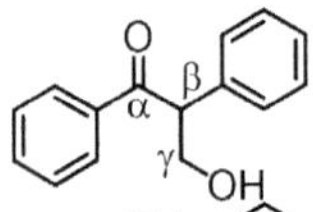	297.6
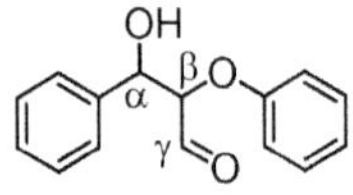	213.9	218.0	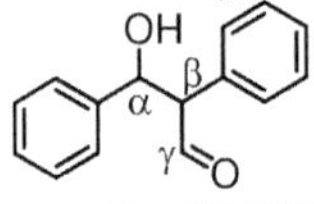	194.9

C–O linkages and 192.5–493.7 kJ mol^{-1} for C–C linkages, which indicates the difficulty of breaking all of the C–C/C–O linkages in purely thermal processes with the singlestep homolytic cleavage [5–10]. The calculation results show that the C–O linkage bond is generally weaker than the C–C bond and the average bond dissociation energy of $C_α$–O is the lowest, and that of $C_β$–O is second lowest. The order of the average BDE of the various linkage bonds is as follows: $C_α$–O < $C_β$–O < $C_α$–$C_β$ ($_β$–1) < $C_α$–$C_β$ ($_β$–O–4) < C_4–O < O–C_5 < $C_α$–C_1 < C_5–C_5. The most robust linkage bond should be the C_5–C_5 bond.

According to the DFT calculation results of six prevalent linkages (β-O-4, α-O-4, 4-O-5, β-1, α-1, and 5–5) [5], the substituent groups on both the aromatic and alkyl groups have different degrees of influence on the BDE of C–O and C–C bonds. The methoxyl groups on benzene ring connected to the $C_α$ have little influence on the BDE of C–O, but the methoxyl groups on benzene ring connected to the ether can increase the BDE of C–O. The substituent groups on the aromatic rings have a tiny influence on the BDE of the C–C bond.

To achieve the selective cleavage of lignin ether bonds and C–C bonds, one more feasible strategy is to first weaken the target bonds and then employ a catalytic system to facilitate the selective cleavage of weakened C–O/C–C bonds during the subsequent conversion. Compared with the direct cleavage of $C_β$–OAr or C–C bonds, more reported catalytic systems involve the first activation of its neighboring chemical bonds, which offer an activate intermediate (molecular or fragment) with a low BDE of linkages bonds for the following transformation [11]. As one important example, the carbonyls on alkyl groups ($C_α$ and $C_γ$) can reduce the BDE of $C_β$–O in β-O-4 linkage by 50-70 kJ mol^{-1} (Table 5.1). The carbonyls on the alkyl groups have different influences on the BDE of the $C_α$–$C_β$ bond. The $C_α$=O can slightly increase the BDE of $C_α$–$C_β$ bond by 24~30 kJ mol^{-1}, whereas the $C_γ$=O can reduce the BDE of $C_α$–$C_β$ bond by ~70 kJ mol^{-1} [5], which is the reaction driving force of the retrograde aldol condensation with $C_α$–$C_β$ bond cleavage. The same carbonyl effect on the BDE of the $C_α$–$C_β$ bond can be found in the β-1 linkage. In addition, besides the molecule intermediate, the lignin conversion usually contains various radical intermediates, such as the $C_\alpha^{\bullet}$ radical in the β-O-4 linkage conversion can obviously reduce the BDE of C_β–O by 200 kJ mol^{-1} [12], the $C_\alpha O^{\bullet}$ radical can efficiently weaken various C–C bond and induce its cleavage [13].

The different catalytic routes produce different intermediates, which can cause variations in the linkage bond cleavage. The dominating factor for the linkage bond cleavage should not only contain the target bond BDE of the original substrate but also one of the activated intermediates' phases. In an ideal route for cleavage of the lignin linkage bonds, an activated intermediate is generated after a series of activation steps with low activation energy with a target bond that has a lower energy and easily breaks in the subsequent transformations. Although weakening the target bond with BDE reducing can be regarded as one approach to design lignin conversion strategies, but it is not the whole story. In addition, some pretreatment/transformation with adjacent functional group modification (AFGM) [11] of the target linkage bonds can generate new substrate sites to introduce the cleavage reagent for further conversion, or the AFGM can prevent the recondensation and repolymerization of the lignin fragments intermediates, during which the BDE of target linkage bonds may increase. Therefore, to effectively break the lignin bonding structure or directly cleave the target chemical bonds, as we mentioned above, we can first transform the lignin linkage to the intermediate structure with weakened C–C/C–O bonds or with the substrate sites that can be used to introduce the cleavage reagents.

During the classification of lignin depolymerization methods with inverse disassembly analysis thought, to generalize as many lignin conversion methods with various intermediates as possible, four thinking modes for lignin conversion need be first emphasized, namely: (1) Tailoring mode; (2) Wedging mode; (3) Protection mode; and (4) Cascaded mode (vide infra).

Tailoring Mode. In the Tailoring mode for transformation of lignin or its model compounds, the lignin linkages are first tailored to small molecular or intermediate species over catalyst centers or with active reagents/solvent molecules. The intermediates then undergo the cleavage of lignin C–C and C–O bonds and the generated carbon- and oxygen-centered fragments react/combine with the activated reductive, oxidative, or nucleophilic species. Besides some processes beginning with the direct cleavage of target C–C or C–OAr bonds in the lignin models with a stable group such as an alkane structure, recent research shows that adjacent functional groups, such as hydroxyl groups or the consequently generated carbonyl, alkenyl, and aldehyde groups, play a critical role in the lignin linkage activation and following cleavage [11]. Detailed mechanism studies confirmed that these efficient processes tend to begin with the adjacent functional group modification (AFGM, Figure 5.3) [11], which can directly decrease the BDE of targeted bonds in the molecule or fragment intermediates for the lignin selective depolymerization to monomers. For example, initial C_α–OH oxidation/dehydrogenation to C_α=O or dehydroxylation to a C_α-centred radical in β-O-4 linkage can facilitate the cleavage of C_β–OAr bond by reducing the BDE of C_β–OAr, and the generated carbon- or oxygen-centered radicals can be quenched by the H* during the oxidation-hydrogenation mechanism or dehydroxylation-hydrogenation mechanism.

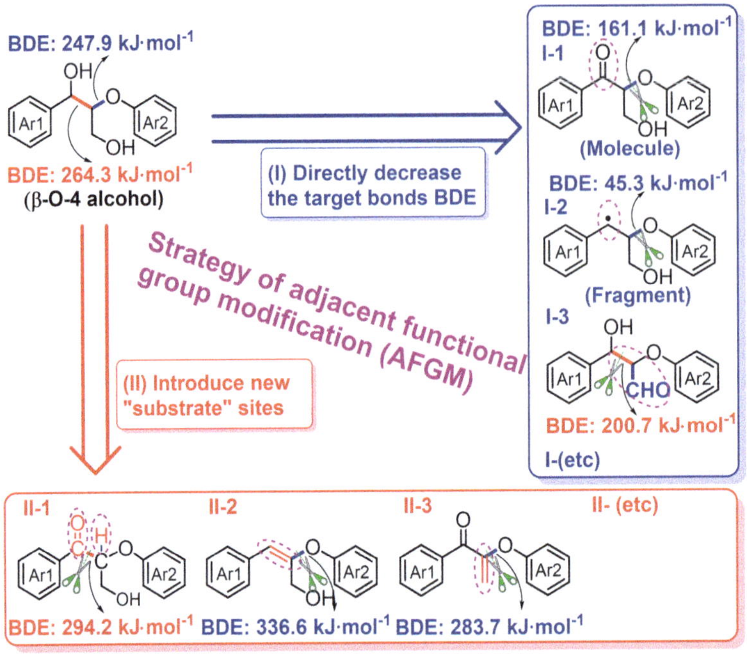

Figure 5.3 Adjacent Functional Group Modification (AFGM) Strategy for the Cleavage of Basic Lignin $C_\alpha-C_\beta$ and C_β–OAr Bonds in β-O-4 Lignin Linkage. *Note:* BDE values are based on the Ar1=Ar2=Ph-; kcal·mol^{-1} = 4.18 kJ·mol^{-1}. *Source:* Reprinted from Zhang and Wang [14] with Permission, © 2020 American Chemical Society.

Wedging Mode. As a daily skill, to efficiently cleave the wood stake into pieces, people can add a wedge in the original crevice. At the molecule level, this Wedging mode could also be also a good idea to cleave the stable lignin molecule to aromatic monomers. Wedging mode needs the lignin conversion can first generate unstable sites or organic groups for introducing lignin cleavage reagents. For the cleavage of the lignin C–C bonds, the Baeyer-Villiger oxidation [15] or Beckmann rearrangement [16] of the lignin C_α ketone intermediate could be the representative methods of Wedging mode. The C_α=O can directly react with NH_2OH or peroxide, and the N and O atoms can insert in lignin C–C bonds to generate C(=O)–O and C(=O)–N bonds in the generated ester and amide that could be easily cleaved by the subsequent hydrolysis. In addition, although the first C_α–OH oxidation to C_α=O can cause the increase of $C_\alpha-C_\beta$ BED by 30 kJ·mol^{-1}, C_β–H can be easily abstracted as H$^\bullet$ or H^+ and the rest C_β can work as the "substrate site" to react with active O_2* or other O* species, causing the cleavage of $C_\alpha-C_\beta$ bond [17]. The more stable $C_\alpha=C_\beta$ in lignin alkenyl ether can work with the adjacent phenolic O atom as the "substrate site" to introduce the proton and promote the hydrolysis of C_β–OAr bond. The most stable C(aryl)–C(aryl) bonds in the 2,2'-biphenol linkage can be cleaved by inserting the active Rh center [18]. And there are many other examples, which will be discussed in the following chapters [19–21]. In addition, as another field, some researchers tried to wedge some chemically labile linkages into the lignin backbone via the gene engineering method [22], which target to make the lignin depolymerization to small fragments more easily. One can read the corresponding kinds of literature and reviews.

Protection Mode. During organic synthesis, one can keep the selectivity of the reaction via the first protection of sensitive groups that do not participate in the conversion. For lignocellulose conversion, the similar "β-O-4 bond protection" mode suggests that one can first protect/transform the active functional groups of lignin at the beginning of pretreatment, which could preserve the β-O-4 structure for further controllable conversion and restrain the recondensation of possible lignin fragments [23]. Recently, the pre-protection methods have drawn much attention to lignocellulose hydrogenation, such as the formaldehyde stabilization method and its derivative methods which involve the initial protection of C_α–OH and C_γ–OH in the β-O-4 structure [24]. Some etherification/esterification pretreatment at C_α–OH and C_γ–OH in β-O-4 structure with protonic polar solvents or reagents can weaken or replace hydrogen bonds to promote lignin release from lignocellulose, and the etherification/esterification with alcohol/acid can further increase the dissolving capacity of lignin for the following depolymerization. Furthermore, the first oxidation of the C_α–OH to C_α=O could not only restrain the generation of the C_α carbocation which can cause the lignin fragments condensation but also can promote the hydrogenation cleavage of β-O-4 linkage by weakening the C_β–OAr bond [25].

Cascaded Mode. Cascaded mode is another efficient strategy used in lignin depolymerization, which usually combines two or more different methods to achieve the in-situ lignin fragments depolymerization to monomers and instant monomers stabilization or transformation. Compared with the single transformation method, the Cascaded mode, represented by the acidolysis + acetal formation, acidolysis + hydrogenation, acidolysis + decarbonylation [26], and alkali catalysis + hydrogenation/oxidation methods, can restrain the recondensation of fragments and make the conversion process more controllable and efficient with higher monomers yields.

5.3 IDA Catalogue of Lignin Conversion Methods Discussed in the Following Chapters

As developed from the 1920s, the catalytic methods for lignin conversion/depolymerization are multitudinous (Chapter 4). Furthermore, during the conversion of lignin or its model compounds, the catalysts characterization and substrate structure have effects on the reaction mechanism. Besides the final cleavage of the target bonds, these processes usually contain various activation steps involving the transformation of several bonds among lignin C_α–H, C_α–O_αH, O_α–H_o, C_β–H, C_β–C_γ, C_γ–O_γH bonds and/or aromatic substituent groups (Figure 5.4, with β-O-4 as the example).

Regarding β-O-4 linkage conversion as the main lignin depolymerization model, as well as other lignin linkages, the critical summary of the following Chapter 6 - 11 aims to provide an overview of key advances in the field of lignin depolymerization.

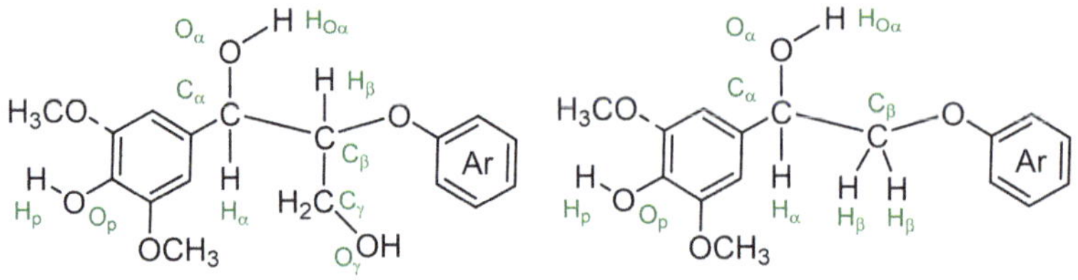

Figure 5.4 The β-O-4 Linkage Model with Atoms Denomination. *Note:* The phenolic -O_pH in lignin model could also be -OCH_3.

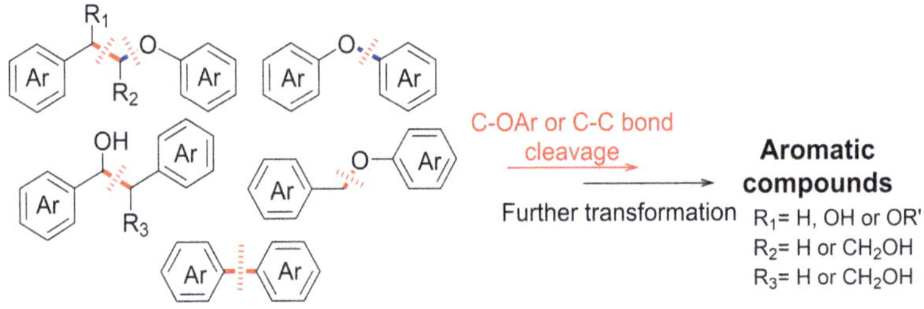

Figure 5.5 Direct Lignin C–OAr, ArO–Ar or C–Ar Bonds Cleavage without First Activation of the Adjacent Chemical Bonds.

We will employ the inverse disassembly analysis (IDA) thought and four thinking modes to summarize and understand the mechanism of lignin linkages cleavage, viewed from the cleavage order and type of corresponding chemical bonds. In this subsection, an IDA catalogue of lignin conversion methods discussed in the following chapters will be briefly summarized.

The direct cleavage of C_β–OAr, C_α–OAr or ArO–Ar ether bonds, and C–Ar bonds is the simplest mode to tailor the β-O-4 linkage and other lignin linkages (Figure 5.5). In Chapter 6, we will first review the stoichiometric reactions and catalytic systems for the direct cleavage of ether bonds with strong nucleophile and electrophilic reagents. Then, as the modification and update of the above methods, Brönsted/Lewis acid + metal systems, base/organometallic systems, and other heterogeneous catalysts used in the direct hydrogenative cleavage of C–OAr ether bonds will be systemically introduced. In addition, for the direct reductive cleavage of ether bond, Chapter 6 will also give a brief review of the lignin conversion methods with hydride reagents, and e^- species.

The first activation/transformation of the neighboring lignin phenolic hydroxyl group or aromatic rings to the active radicals can also cause the cleavage of lignin C–C and C–O bonds (Figure 5.6). Chapter 7 first summarizes two kinds of mechanisms for lignin C_{Ar}–C bonds cleavage via the first phenolic hydroxyl group dehydrogenation to the phenolic radical (Figure 5.6a) and mechanism *via* the first single-electron transfer from the aromatic ring (Figure 5.6b). Furthermore, as an extension of these methods, the first transformation of one aromatic ring to the active intermediate, such as semi-hydrogenation/addition to a cyclohexenyl ether (Figure 5.6c), or first inserting a cleavage reagent between the target bond atoms

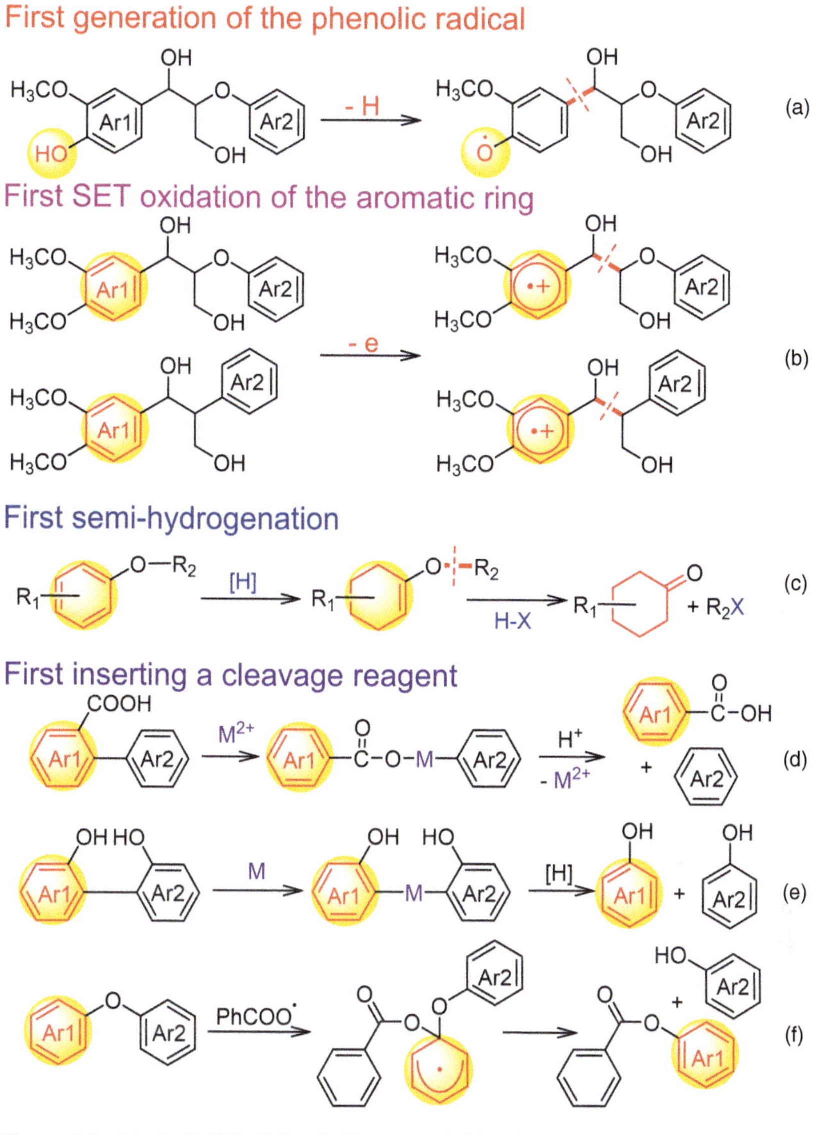

Figure 5.6 Lignin C–C/C–O Bonds Cleavage via First Phenolic Hydroxyl Group Dehydrogenation or First Aromatic Rings Activation.

with the assistance of the aromatic rings conjoint groups, metal catalyst or extra radical (Figure 5.6d, e and f) could also provide some useful methods.

In Chapter 8, the main topic is about the strategies for lignin linkages cleavage beginning with C_αO–H/ArO–H heterolysis or C_α–OH bond heterolysis, which is mainly focusing on the lignin alkaline hydrolysis, and acidolysis. Chapter 8.1 will give a summary of base-catalyzed cleavage of C_β–OAr bond beginning with C_αO–H or ArO–H heterolysis to oxygen anion, which further induces the cleavage of the lignin C_β–OAr bond (Figure 5.7a and b). Given the fact that the appropriate base pretreatment can induce the lignin polymerization fragmentation, the tandem base-promoted oxidation can further cause the cleavage of lignin C–C bonds in the fragments. Chapter 8.2 summarizes the various acid-catalyzed methods for the C_β–OAr bond cleavage beginning with C_α–OH heterolysis, which further induce the C_β–OAr bond cleavage via a C_α=C_β–OAr intermediate (Figure 5.7c).

In Chapter 9, strategies for lignin catalytic depolymerization which begin with the activation of C_α–H, C_α–OH, or C_αO–H bonds via a non-ionized route are mainly summarized. As shown in Figure 5.8, the first cleavage of C_αO–H and C_α–H bonds can generate a β-O-4 ketone intermediate, which can weaken the C_β–OAr bond and make a little stronger C_α–C_β bond but with an active C_β–H. Focusing on this molecule intermediate, the lignin depolymerization strategies of catalytic transfer hydrogenation (Chapter 9.1), dehydrogenation/oxidation-hydrogenation (reduction) (Chapter 9.2), and the multiple oxidation process (Chapter 9.3) are first summarized. Given the Baeyer-Villiger oxidation and Beckmann rearrangement can be used in the cleavage of C–C bonds in lignin linkages that contain a ketone substrate site, the corresponding strategies, and their derivative methods are summarized in Chapter 9.4. In addition, the first oxidation of C_α–OH can promote the lignin hydrolysis, an embellishing β-O-4 linkages hydrolysis method involving the C_α–OH first oxidation and C_γ–OH transformation is discussed in Chapter 9.5. Besides the first cleavage of C_αO–H and C_α–H bonds to C_α=O at the beginning of the β-O-4 model transformation, some catalytic strategies begin with the C_α–H or C_αO–H activation to the corresponding active radical intermediates can also lead to the cleavage of the lignin fragments. The first activation of C_α–OH(–H) to a $C_\alpha^\bullet$ radical intermediate can obviously reduce the BDE of the C_β–OAr bond for the following self-cleavage of β-O-4 linkage (Chapter 9.6). Besides, the first activation of C_αO–H bond to the $C_\alpha O^\bullet$ via a HAT, PCET oxidation or redox-neutral LMCT process can lead to the selective cleavage of C_α–C_β bonds (Chapter 9.7).

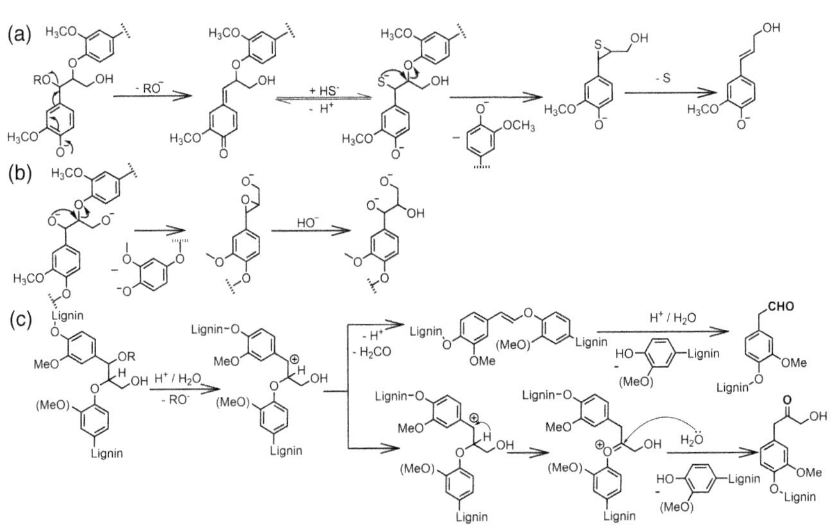

Figure 5.7 Lignin Linkages Cleavage Beginning with C_αO–H/ArO–H Heterolysis or C_α–OH Bond Heterolysis.

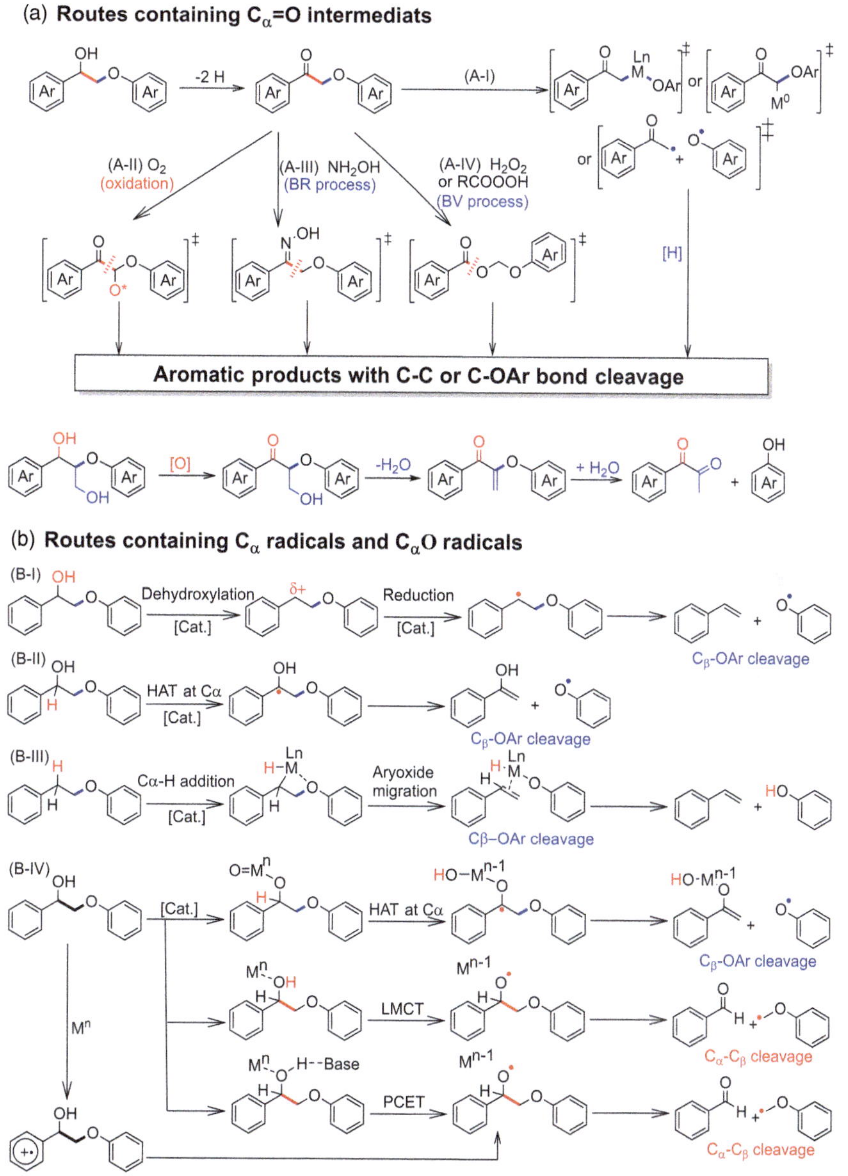

Figure 5.8 Lignin Linkages Cleavage Beginning with the Activation of C_α–H, C_α–OH, or C_αO–H Bonds via a Non-ionized Route. *Note:* O*: Active oxgen species; BV: Baeyer-Villiger Oxidation; BR: Backmann Rearrangment.

In Chapter 10, the lignin linkages cleavage strategies beginning with C_β–H, C_γ–H, or C_γO–H direct activation are summarized (Figure 5.9). Chapter 10.1 will first describe some photocatalytic strategies with direct activation of C_β–H bonds that introduce the cleavage agent O*, leading to the cleavage of the C_α–C_β bonds in β-1 and β-O-4 linkages. Then, it is a similar C_β–H activation process but with the direct generation of C_β–OH to provide a hemiacetal structure in the oxidized β-O-4

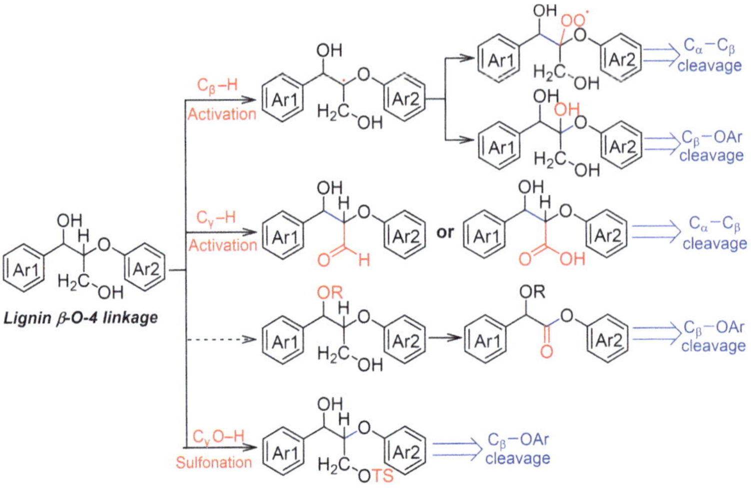

Figure 5.9 Lignin Linkages Cleavage Beginning with C_β–H, C_γ–H, or C_γO–H Direct Activation.

linkage, which can easily occur hydrolysis to cleave the C_β(OH)–OAr bond. Chapter 10.2 will firstly introduce some Retro-aldol strategies with first -$C_\gamma H_2OH$ activation to -C_γHO/-C_γOOR (R=H or alkyl). Then, as the modification of the direct-$C_\gamma H_2OH$ activation, the first-etherification of the $C_\alpha H_2OH$ with following -C_βH(OAr)–$C_\gamma H_2OH$ oxidation to -C_β(=O)–OAr is also discussed in the chapter 10.2. Chapter 10.3 will introduce some strategies for the lignin C–OAr bond cleavage with first –C_γHOH sulphonation.

As the important manifestation of the protection mode for lignin depolymerization, in Chapter 11, the main topic is promoting the lignin linkages cleavage with considering the potential lignin fragments condensation. Chapter 11.1 will give a brief summary of potential lignin fragments condensation mechanism, including acid-catalyzed condensation, base-catalyzed condensation, radical fragments condensation, and other lignin linkages transformations. When condensation mechanisms are made clear, Chapter 11.2 will briefly introduce the developed methods for restraining the potential condensation of lignin fragments. The methods are divided into six categories as follows: (1) Pre-protection of the active groups to inhibit the condensation; (2) First transformation of the active groups to promote the depolymerization; (3) In-situ converting the active intermediates and scavenging the unwanted species; (4) Modifying and designing the catalysts; (5) Intensifying reaction systems; (6) Employing an "ideal lignin" as the substrates. The corresponding examples will be provided and discussed to show the importance of restraining the fragments' condensation or repolymerization.

In the following chapters, we will provide a detailed discussion about the above strategies with reported researches in lignin depolymerization, which may put more attention on the catalyst and mechanism research with the corresponding C–C/C–O bonds transformation rules under the IDA thoughts.

5.4 Chapter Summary

As discussed in Chapter 5, multitudinous methods have been developed for lignin depolymerization to provide monomers. It is hard work to systemically summarize all of the strategies, just as using catalyst or reaction types as the standard of classification. Facing this question, in Chapter 5.1, inspired by the retrosynthetic analysis of the natural product to confirm the synthons and synthetic equivalents, we herein put forward an inverse disassembly analysis (IDA) method to explore the efficient strategies for lignin depolymerization. Besides, the IDA method can be also used in the classification and discussion of the various methods for lignin conversion, viewing from the cleavage order and type of corresponding chemical bonds over various catalytic systems. Then, to generalize as many lignin conversion methods with various routes and intermediates as possible, four thinking modes based on the IDA for lignin conversion are emphasized in Chapter 5.2, namely: (1) Tailoring mode; (2) Wedging mode; (3) Protection mode; and (4) Cascaded mode. In Chapter 5.3, with β-O-4 linkage as the main lignin mode, we briefly introduce the contents of the following chapters with the IDA thoughts, which can be regarded as the outline of the following chapters of Part IV. The keywords of this chapter could be Lignin depolymerization strategies, Inverse disassembly analysis, β-O-4 linkage, BDE of lignin linkages.

References

1 Corey, E.J. and Cheng, X.M. (1995). *The Logic of Chemical Synthesis*. Hoboken: John Wiley & Sons, Ltd.

2 Corey, E.J. (1988). Retrosynthetic thinking-Essentials and examples. *Chem. Soc. Rev.* 17: 111–133.

3 Corey, E.J. (2010). The logic of chemical synthesis: Multistep synthesis of complex carbogenic molecules (Nobel Lecture). *Angew. Chem. Int. Ed.* 30: 455–465.

4 You, L., Liang, X.T., Xu, L.M. et al. (2015). Asymmetric total synthesis of propindilactone G. *J. Am. Chem. Soc.* 137: 10120–10123.

5 Huang, J.-b., Wu, S.-b., Cheng, H. et al. (2015). Theoretical study of bond dissociation energies for lignin model compounds. *J. Fuel. Chem. Technol.* 43: 429–436.

6 Elder, T. (2014). Bond dissociation enthalpies of a pinoresinol lignin model compound. *Energy Fuel* 28: 1175–1182.

7 Younker, J.M., Beste, A., and Buchanan, A.C. (2012). Computational study of bond dissociation enthalpies for lignin model compounds: β-5 Arylcoumaran. *Chem. Phys. Lett.* 545: 100–106.

8 Kim, S., Chmely, S.C., Nimos, M.R. et al. (2011). Computational study of bond dissociation enthalpies for a large range of native and modified lignins. *J. Phys. Chem. Lett.* 2: 2846–2852.

9 Elder, T., Berstis, L., Beckham, G.T., and Crowley, M.F. (2017). Density functional theory study of spirodienone stereoisomers in lignin. *ACS Sustainable Chem. Eng.* 5: 7188–7194.

10 Elder, T. (2013). Bond dissociation enthalpies of a dibenzodioxocin lignin model compound. *Energy Fuel* 27: 4785–4790.
11 Zhang, C.F. and Wang, F. (2017). Sell a dummy: Adjacent functional group modification strategy for the catalytic cleavage of lignin β-O-4 linkage. *Chin. J. Catal.* 38: 1102–1107.
12 Zhang, C., Lu, J., Zhang, X. et al. (2016). Cleavage of the lignin β-O-4 ether bond via a dehydroxylation-hydrogenation strategy over a NiMo sulfide catalyst. *Green Chem.* 18: 6545–6555.
13 Gazi, S., Ng, W.K.H., Ganguly, R. et al. (2015). Selective photocatalytic C–C bond cleavage under ambient conditions with earth abundant vanadium complexes. *Chem. Sci.* 6: 7130–7142.
14 Zhang, C.F. and Wang, F. (2020). Catalytic lignin depolymerization to aromatic chemicals. *Acc. Chem. Res.* 53: 470–484.
15 Rahimi, A., Azarpira, A., Kim, H. et al. (2013). Chemoselective metal-free aerobic alcohol oxidation in lignin. *J. Am. Chem. Soc.* 135: 6415–6418.
16 Li, H., Wang, M., Liu, H. et al. (2018). NH2OH–mediated lignin conversion to isoxazole and nitrile. *ACS Sustainable Chem. Eng.* 6: 3748–3753.
17 Wang, M., Li, L.H., Lu, J.M. et al. (2017). Acid promoted C–C bond oxidative cleavage of β-O-4 and β-1 lignin models to esters over a copper catalyst. *Green Chem.* 19: 702–706.
18 Zhu, J., Wang, J.C., and Dong, G.B. (2019). Catalytic activation of unstrained C(aryl)–C(aryl) bonds in 2,2′-biphenols. *Nat. Chem.* 11: 45–51.
19 Liu, J.Z., Qiu, X., Huang, X.Q. et al. (2019). From alkylarenes to anilines via site-directed carbon-carbon amination. *Nat. Chem.* 11: 71–77.
20 Li, H.J., Bunrit, A., Lu, J.M. et al. (2019). Photocatalytic cleavage of aryl ether in modified lignin to non-phenolic aromatics. *ACS Catal.* 9: 8843–8851.
21 Tan, F.-F., He, X.-Y., Tian, W.-F., and Li, Y. (2020). Visible-light photoredox-catalyzed C–O bond cleavage of diaryl ethers by acridinium photocatalysts at room temperature. *Nat. Commun.* 11: 1–11.
22 Wilkerson, C.G., Mansfield, S.D., Lu, F. et al. (2014). Monolignol ferulate transferase introduces chemically labile linkages into the lignin backbone. *Science* 344: 90–93.
23 Questell-Santiago, Y.M., Galkin, M.V., Barta, K., and Luterbacher, J.S. (2020). Stabilization strategies in biomass depolymerization using chemical functionalization. *Nat. Rev. Chem.* 4: 311–330.
24 Shuai, L., Amiri, M.T., Questell-Santiago, Y.M. et al. (2016). Formaldehyde stabilization facilitates lignin monomer production during biomass depolymerization. *Science* 354: 329–333.
25 Zhang, C.F., Li, H.J., Lu, J.M. et al. (2017). Promoting lignin depolymerization and restraining the condensation via an oxidation-hydrogenation strategy. *ACS Catal.* 7: 3419–3429.
26 Deuss, P.J., Scott, M., Tran, F. et al. (2015). Aromatic monomers by in situ conversion of reactive intermediates in the acid-catalyzed depolymerization of lignin. *J. Am. Chem. Soc.* 137: 7456–7467.

6

Direct Lignin C–OAr, ArO–Ar or C–Ar Bonds Cleavage without First Activation of the Adjacent Chemical Bonds

6.1 Brönsted/Lewis Acid + Metal Systems for the Direct Hydrogenative Cleavage of Ether Bonds and C–Ar Bonds

At the initial exploration of lignin conversion, the lignin models usually do not contain hydroxyl groups at aromatic rings or branches. Therefore, previous strategies and methods for lignin conversion focused on the cleavage of various ether bonds. Although the organic alkali metal compounds, alkali metals, or other compounds with strong nucleophilicity [1] can promote the cleavage of ether bond (Figure 6-1a), given the fact that the ether bond can be activated more easily by the Brönsted or Lewis acid reagents, stoichiometric Brönsted acids (HBr, HI, etc.) or Lewis acids (BBr_3, TMSI, metal halides, etc.) were then used in the deprotection of alkyl/aryl ethers [2], leading to the generation of phenols and halohydrocarbons (Figure 6.1b). Besides, thioacidolysis is another acid-promoted method for the cleavage of the ether bonds [3]. This acidic solvolysis uses boron trifluoride etherate, a potent Lewis acid, together with ethanethiol in dioxane to cleave aryl ether bonds to provide thiolated monomers and phenols (Figure 6.1b).

Then, various acid-catalyzed systems for the ether bond cleavage to esters and halohydrocarbons with acyl chlorides were founded, which involved Lewis acid catalysts, such as $ZnCl_2$, [4] $FeCl_3$, [5] $SnBr_2$, [6] $CoCl_2$, [7–9] $AlCl_3$, [10] YCl_3 [11], $LnCl_3$, [12] Al+I_2 [13], or other acyl chloride-based systems with Al [14], Zn [15], graphite [16], $Mo(CO)_6$ [17], and $ArMo(CO)_3$ [18]. It is generally believed that these catalytic reactions proceed *via* a cationic mechanism in which the acid chloride reacts with the catalyst first to generate a RCO^+ ion (Figure 6.1c–I). The addition of ether gives an oxonium salt ($RCO_2{}^+R'_2$), followed by the cleavage of the C–O bond to give an ester [17, 18]. While, Takahashi and co-workers [19] reported another mechanism for the reaction of acyclic ethers and acid chlorides in the presence of $MCl_{5/6}$ (M = Mo, W, Nb or Ta). As shown in Figure 6.1c-II, when $MoCl_5$ was used as the catalyst, a stoichiometric reaction of the dioctyl ether with the $MoCl_5$ afforded 1-chlorooctane and Cl_4MoOBu. Then benzoyl chloride provides Cl^- to Cl_4MoOR to

Lignin Conversion Catalysis: Transformation to Aromatic Chemicals, First Edition.
Chaofeng Zhang and Feng Wang.

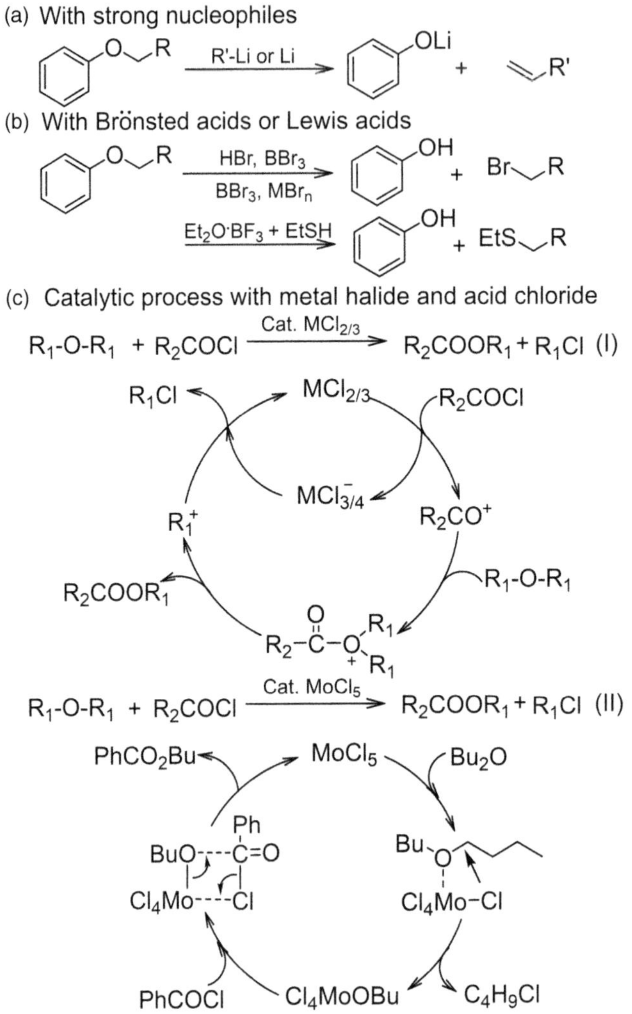

Figure 6.1 Stoichiometric Reactions for C–O Ether Bonds Cleavage (A and B), Proposed Reaction Mechanism for Catalytic Acylative Cleavage of Ethers Bonds with Acyl Chloride (C).

regenerate $MoCl_5$ and esters. As the substitution of acyl chlorides, the organic anhydrides can be also used in the above processes with Lewis acid like $FeCl_3$ [20].

However, the generated organic chlorides need to be further transformed, which usually releases a molecular HX and the process was not atom-economic [21, 22]. In 2012–2015, Marks and co-workers [23] developed an efficient and green tandem catalytic system for the hydrogenolysis of ethers, which comprised of a high-valent metal triflate Lewis acid and a supported Pd catalyst. According to the choice of metal triflate, the saturated alcohol [24, 25], or hydrocarbon [26] was obtained as the major products respectively. In the proposed catalytic cycles (Figure 6.2), the high-valent metal triflate Lewis acid ($Ln(OTf)_3$, Ln= Yb or Hf) first promote the C–O scission to yield alkenols (Cycle A), which is further

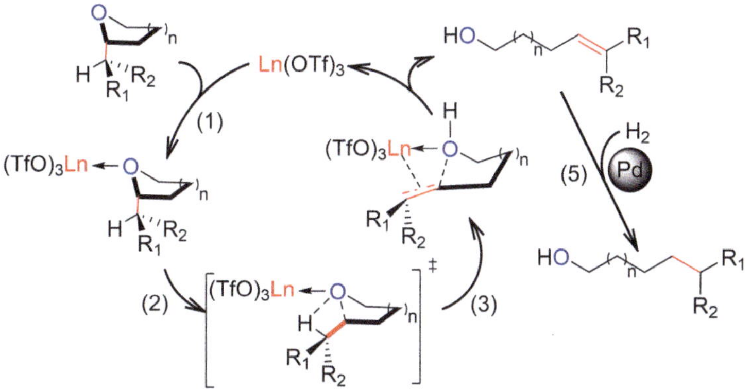

Figure 6.2 Acidolysis-hydrogenation Tandem Pathway for Lanthanide Triflate/Pd-mediated Ether Bond Cleavage. *Source:* Atesin et al. [24] with Permission, © 2012 American Chemical Society.

hydrogenated over the heterogeneous Pd catalysts (Cycle B). Mechanistic studies and DFT-level computation reveal that lanthanides with smaller ionic radii/greater electrophilicity, i.e., higher effective charge density, exhibit greater catalytic activity than protonic acids [27, 28].

In 2015, the Marks group [29] further applied this tandem system to cleaving the β-O-4 aryl-ether bonds of lignin models. They first modified the hydroxyl groups via acetylation, which served to (1) solubilize both lignin and lignin models and to (2) modify the reactivity of pendant hydroxyl groups to promote more selective C–OAr cleavage. As shown in the proposed mechanism (Figure 6.2), the $M(OTf)_n$ first promotes the elimination of a CH_3COOH between C_α and C_β, leading to the generation of an olefin intermediate (*i*). Then the C=C bond is hydrogenated to saturated bond (*ii*), and the C–OAr ether bonds in the generated intermediates (2a,b in Figure 6.3) further go through the acidolysis-hydrogenation tandem cleavage with $M(OTf)_n$ + Pd/C system mentioned above [26].

Then, in 2016 and 2017, Hensen and co-workers [30, 31] presented a catalytic process for the upgrading of woody biomass into mono-aromatics with homogeneous Lewis acid metal triflates and Pd/C in methanol. As shown in Figure 6.4, lignin fragments were firstly extracted from the lignocellulosic matrix by cleavage of ester and ether linkages between lignin and carbohydrates (See LCC, discussed in Chapter 2) by the catalytic action of homogeneous Lewis acid, metal triflates, in methanol. The released lignin fragments were then converted into lignin monomers by the combined catalytic action of Pd/C and metal triflates with hydrogen to cleave the ether bonds. Under the optimized condition, 55 wt% aromatic monomers, mainly alkylmethoxyphenols, were obtained from the lignin fraction (23.8 wt%) of birch wood over the Pd/C + Al(III)-triflate system.

Besides the Pd/C catalyst, in 2017, Beller and co-workers [32] demonstrated the reductive hydrogenolysis of a wide variety of ethers including diaryl, aryl–alkyl, and aryl–benzyl derivatives catalyzed by a stable heterogeneous $NiAlO_x$ catalyst in the presence of a Lewis acid (LA). Different above Lewis acid/Pd systems, the special

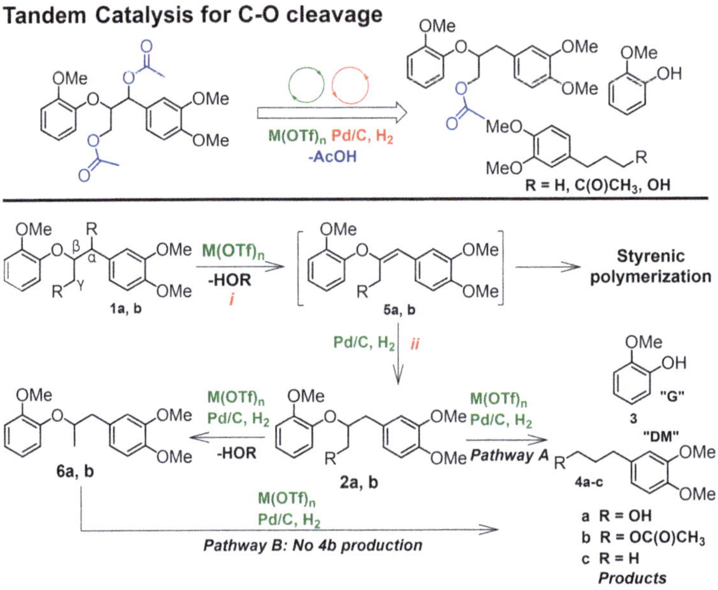

Figure 6.3 Step-wise Reaction Profile for Lignin C_β–OAr Bond Hydrogenolysis Using a Tandem $M(OTf)_n$ + Pd/C Catalytic System. *Source:* Lohr et al. [29] with Permission, © 2015 American Chemical Society.

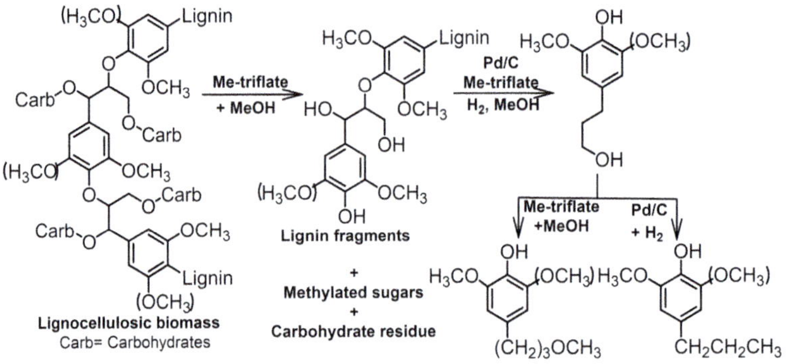

Figure 6.4 Tandem Catalytic System for the Reductive Fractionation of Woody Biomass into Sugars and Alkylmethoxyphenols over Pd/C and Metal Triflate in Methanol. *Source:* Huang et al. [31] with Permission, © 2017 The Royal Society of Chemistry.

feature of this LA/$NiAlO_x$ catalytic system was the formation of substituted cyclohexanols from the corresponding aryl ether (Figure 6.5).

Furthermore, besides the homogeneous Lewis acids species, in 2014, Yang and co-workers [33] developed an aqueous phase catalytic process to selectively depolymerize lignin and remove the oxygen (Figure 6.6). Combinations of noble metal catalysts in the presence of various solid acid zeolites were tested for HDO activity of the oligomeric technical lignins from ethanol and dilute alkali extraction of corn stover. Results showed that 35%–60% conversion of lignin with 65%–70% product

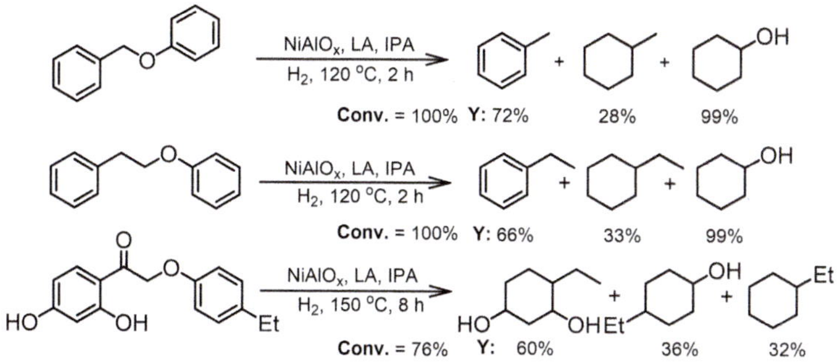

Figure 6.5 Hydrogenolysis of Lignin-derived Aromatic Ethers Catalyzed by a Stable Heterogeneous $NiAlO_x$ Catalyst in the Presence of a Lewis Acid (LA, $La(OTf)_3$). *Source:* Reaction Data Were Abstracted from Cui et al. [32].

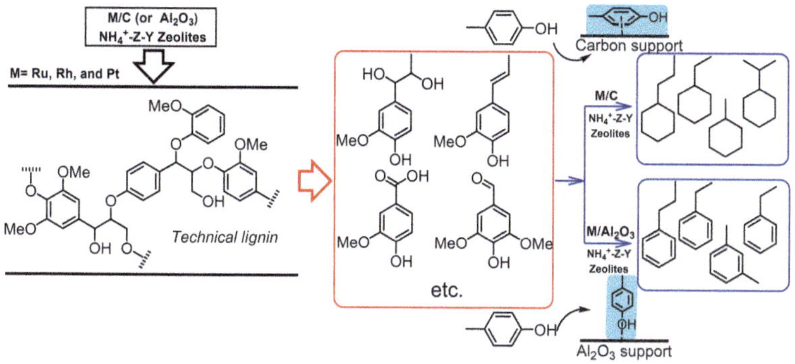

Figure 6.6 Noble-metal Catalyzed Hydrodeoxygenation of Biomass-derived Lignin to Aromatic Hydrocarbons. *Source:* Laskar et al. [33] with Permission, © 2014 The Royal Society of Chemistry.

selectivity for aromatic hydrocarbons (e.g. toluene) under various HDO conditions in the presence of noble metals (Ru, Rh, and Pt) over Al_2O_3 (or C) supports, and solid acid zeolites (e.g., NH_4^+ Z-Y) catalyst matrices were obtained. For the different performance of the supports between Al_2O_3 and active carbon, the author speculates the absorption model of the phenol intermediates over the catalysts plays a critical role.

Referring to the above catalytic biomass conversion, if the hydrogenation process was removed from the catalytic process, the lignin conversion process was called catalytic fast pyrolysis. The catalytic pyrolysis process is usually carried out above 400 °C, during which the lignin ether and C–C linkages can yield cleavage. The generated fragments or monomers can be further cleaved and produce a certain amount of liquid product, known as bio-oil products, which is a mixture of phenols, low-carbon alcohol, acid, ester, ether, and gas products of low-carbon alkane, olefines, CO, CH_4, and CO_2. To control the distribution of the products, in 2018, Triantafyllidis and co-worker [34] present a study of thermal (non-catalytic) and catalytic fast pyrolysis of lignin over the zeolite catalyst, which emphasizes the effect of process parameters and type of ZSM-5 zeolite catalyst on the selectivity of phenolics and aromatics in the bio-oil. It was shown that the higher temperature (i.e., from 400 to 600 °C) and the higher catalyst-to-lignin ratio favored the conversion of the initially formed alkoxy-phenols with C_2–C_3 alkyl side chains to smaller alkoxy-phenols and mono-aromatics (Figure 6.7). In addition, they compared the performance of three ZSM-5 zeolites with different crystal sizes but similar Si/Al ratio, i.e., a commercial microporous ZSM-5(40) zeolite (0.3–1μm), a nano-sized ZSM-5 (≤20 nm), and a mesoporous ZSM-5 (9 nm) prepared via the alkaline treatment of the ZSM-5(40) zeolite, it was shown that the well-crystallized ZSM-5(40) catalyst exhibited slightly higher selectivity toward mono-aromatics compared to the other two catalysts. The nano-sized ZSM-5 and especially the mesoporous ZSM-5 showed in addition remarkably higher selectivity toward alkyl-phenols. On the basis of the NMR data, the main composition profile of the lignin feedstock, in terms of phenylpropane units, is being "transferred" to the bio-oil, but the harsh condition with high temperature can cause the vast generation of carbon deposit or unconverted carbon polymer from lignin carbon constituents.

To get a higher lignin conversion, the cooperation between support acid and hydrogenation metal sites is very important, which may allow the reaction can be carried out in a mild condition. In 2015, Heeres and co-workers [35] screened the activity of supported Ru (C, Al_2O_3, TiO_2), Pd (C, Al_2O_3), and a Cu/ZrO_2 catalyst in catalytic hydrotreatment of Alcell lignin, to determine the best catalyst for high yields of biobased aromatics and alkylphenolics. Best results were obtained with Ru/TiO_2 and a lignin oil (78 wt % on lignin intake) with 9.1 wt % alkylphenolics, 2.5 wt % aromatics, and 3.5 wt % catechols on lignin intake were obtained in 4 h at 400 °C and an initial H_2 pressure of 100 bar. In 2016, Wang and co-workers [36] began using Nb-based catalysts with unique Brönsted/Lewis acid properties in the

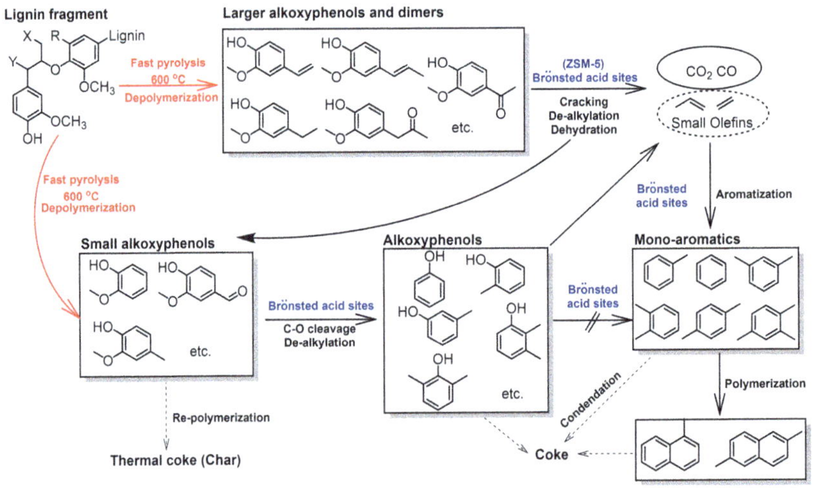

Figure 6.7 Catalytic Fast Pyrolysis of Lignin over ZSM-5 Zeolite for the Production of Alkyl-Phenols and Aromatics. *Source:* Lazaridis et al. [34] with Permission, © 2018 Lazaridis, Fotopoulos, Karakoulia, and Triantafyllidis.

lignin conversion. They firstly described the direct hydrodeoxygenation of raw woods into liquid alkanes with mass yields up to 28.1 wt% over a Pt/$NbOPO_4$ catalyst at 190 °C and 5 MPa H_2. Molecular mechanism investigation reveals that a synergistic effect between Pt and NbO_x species, and acidic sites promote this highly efficient hydrodeoxygenation of bulk lignocellulose (Figure 6.8a). After this work, in 2017, the Wang group [37] reported the selective production of arenes via hydrodeoxygenation of organosolv lignin over a porous Ru/Nb_2O_5 catalyst at 250 °C and 0.7 MPa H_2 (Figure 6.8b). Furthermore, in 2019, Wang and co-workers [38] reported the Ru/$NbOPO_4$ could achieve the catalytic cleavage of both inter-unit C–C and C–O bonds in one-pot lignin conversions at 310 °C at 0.5 MPa H_2, which provided a 20~50% higher monocyclic hydrocarbons yield than the theoretical yield based on the established nitrobenzene oxidation method. Further control experiments and in-situ inelastic neutron scattering characterization suggested the activity of Ru/$NbOPO_4$ in the cleavage of 5–5 biphenyl C–C bond and other Ar–C bonds (Figure 6.8c). A combined diffuse reflectance infrared Fourier Transform spectra (DRIFTS) and density functional theory (DFT) calculation analysis unambiguously confirmed that (i) NbO_x not only can strongly adsorb phenylcyclohexane via benzene ring but also selectively activate Caromatic–C bond between the benzene ring and cyclohexane; (ii) NbO_x can promote the protonation of phenylcyclohexane with the assistance of H_3PO_4; and (iii) NbO_x can significantly reduce the barrier of Caromatic–C bonds cleavage of the carbocation intermediate ($C_6H_5(H)$–$C_6H_{11}^+$) [39].

Wang and co-worker [40] in their further research showed the performances of Ru/$NbOPO_4$ catalysts in the hydrogenolysis of β-O-4 linkage and the product

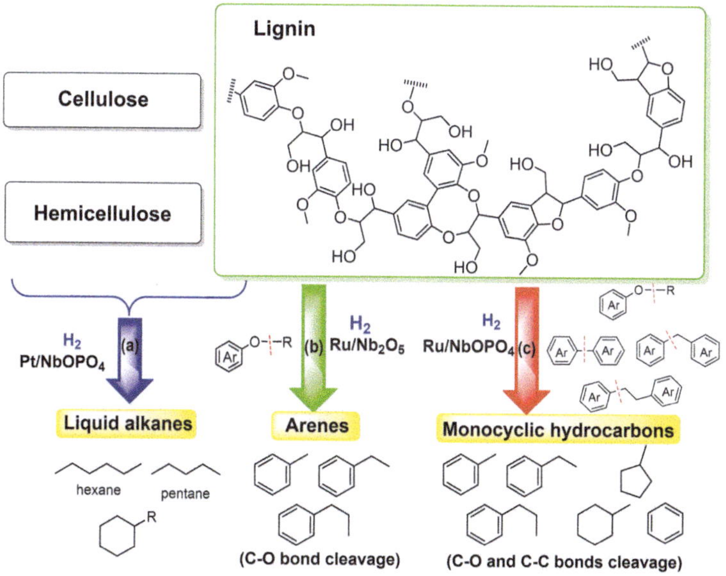

Figure 6.8 Direct Hydrodeoxygenation of Woody Biomass into Liquid Alkanes and Arenes over Nb-based Catalysts. *Source:* Based on Xia et al. [36], Shao et al. [37], Dong et al. [39].

distributions were strongly dependent on the size of Ru nanoparticles (NPs): Although dehydration/hydrolysis rate is similar over differently sized Ru/$NbOPO_4$, Ru NPs with smaller diameters (1.2 and 1.8 nm) have higher activity for the cleavage of $C_{aliphatic}$–O and $C_{aromatic}$–O bonds, whereas Ru NPs with larger diameters (2.5 and 5.3 nm) show a much lower activity for C–O cleavage, especially for the $C_{aromatic}$–O bond, resulting in Ru/$NbOPO_4$ with larger Ru NPs preferred to firstly catalyze the hydrolysis of the β-O-4 bond (Figure 6.9, Route 1) rather than promote the hydrogenative cleavage of C–O like the Ru/$NbOPO_4$ with small Ru NPs (Figure 6.9, Route 2). Their DFT calculations revealed that the presence of step and edge sites in smaller Ru particles can significantly lower the barriers for both $C_{aromatic}$–O and $C_{aliphatic}$–O bond cleavage.

Besides noble metal loading on the Nb-based support, in 2016, Parvulescu and co-workers [41] synthesized a series of Co@Nb_2O_5@Fe_3O_4 catalysts for lignin fragmentation (Figure 6.10). Further experiment optimization indicated that the 4 wt% Co loading catalyst allowed a 53% lignin conversion leading to a mixture containing over 96% in C_{20}–C_{28} and C_{29}–C_{37} fragments at 180 °C and 10 atm H_2. This catalytic behavior was explained on the capability of Nb_2O_5 to catalyze the acidic hydrolysis of the β-O-4 bonds and Co to break the C–C bonds via hydrogenolysis, which was different from catalysis of Ru/$NbOPO_4$ reported by Wang, [38] especially on the effect of Nb_2O_5/$NbOPO_4$ in β-O-4 ether bond cleavage.

Besides Lewis acid species of metal oxide and acid zeolite, inspired by the fact that the open metal sites (OMS) in some MOFs can behave as Lewis acids to activate C–O bonds, in 2016, Stavila and co-workers [42] demonstrated that Mg-IRMOF-74 catalysts with Ti or Ni doping could selectively cleave the phenyl ethers bonds of β-O-4, α-O-4, and 4-O-5 linkages to the corresponding hydrocarbons and phenols at 10 bar H_2 and 120 °C without added base (Figure 6.11). Their further DFT research showed that one of the substrate aromatic rings can directly adsorb over an Mg^{2+} ion. Further examination of the predicted binding energies ΔH°B for the reactants and products chemisorbed to the catalyst cluster model indicates that the cluster facilitates the reaction by stabilizing the products relative to the reactants. Furthermore, based on previous work [43, 44], the author points that the Mg^{2+} open metal sites (OMS) in Mg-IRMOF-74 are the strongest H_2 binding sites. The transition metal species inside the MOF pores activate H_2 more efficiently than the pure

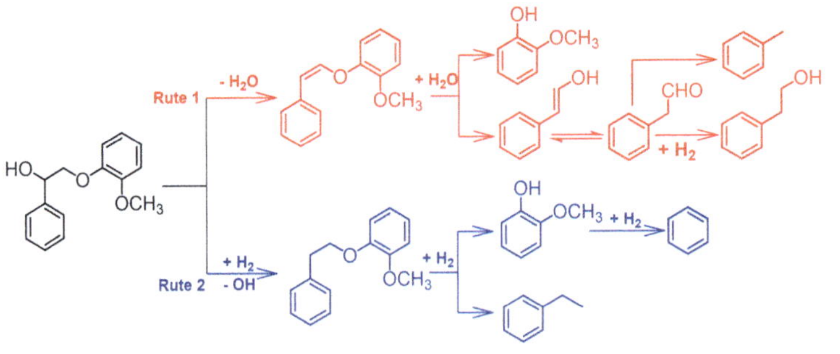

Figure 6.9 Size-dependent Catalytic Performance of Ru Nanoparticles of Ru/$NbOPO_4$ in the Hydrogenolysis of a β-O-4 Lignin Model Compound Linkage. *Source:* Dong et al. [40] with Permission, © 2018 The Royal Society of Chemistry.

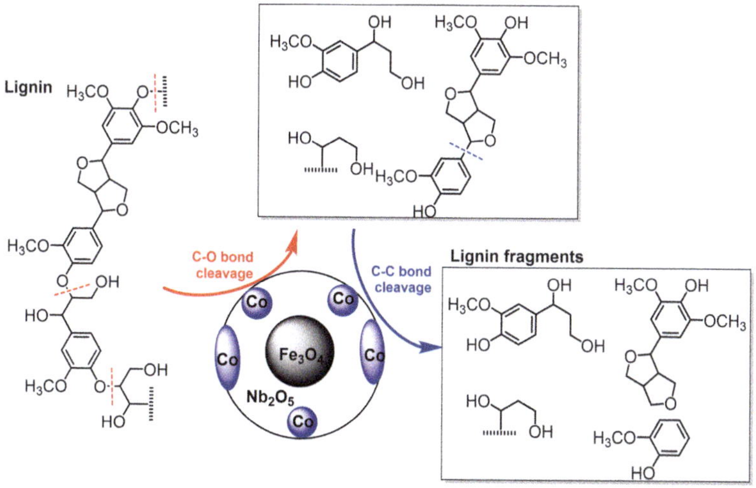

Figure 6.10 Lignin Fragmentation over Magnetically Recyclable Composite $Co@Nb_2O_5@Fe_3O_4$ Catalysts. *Source:* Opris et al. [41] with Permission, © 2016 Elsevier.

(a)

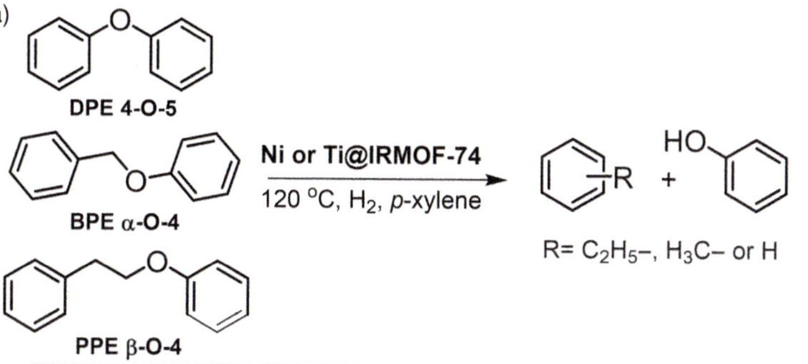

(b)

Binding Energies to IRMOF-74 ΔH^{o}_{B} (kJ/mol)		ΔG (kJ/mol) substrate + H_2 → PhOH + monomers Gas phase	On MOF
DPE	37.2	−80.3	−81.8
BPE	75.2	−106.6	−129.1
PPE	81.9	−103.5	−142.3

Figure 6.11 MOF-based Catalysts for Selective Hydrogenolysis of Carbon-oxygen Ether Bonds. *Source:* Data Abstracted from Stavila et al. [42].

MOFs, reducing the activation energy of the hydrogenolysis reaction and generating higher concentrations of reactive H-species.

6.2 Base/Organometallic Systems for the Direct Hydrogenative Cleavage of Ether Bonds

Besides the hydrogenative process promoted by the Brönsted/Lewis acid reagent or species, the catalytic base + organometallic systems can also achieve the cleavage of aryl-O-alkyl and aryl-O-aryl ether bonds. In 2011, Hartwig and co-workers [45] report that a soluble nickel carbene complex can catalyze the hydrogenolyses of aromatic C–O bonds in alkyl aryl and diaryl ethers that form exclusively arenes and phenols or alcohols under just 1 bar of H_2 at temperatures of 80 to 120 °C (Figure 6.12). The relative reactivity of ether substrates scale as Ar–OAr>>Ar–OMe>ArCH$_2$–OMe.

Agapie and co-workers [46] then gave research about the mechanism of aryl-O-alkyl bond cleavage over the nickel carbene catalyst. They found that the Ni(0) catalysts can catalyze the selective cleavage of the Ar–OR ether bond to arenes rather than the ArO–R ether bond to phenols with H_2. As shown in Figure 6.13, the isotopic labeling study supports a mechanism in which the aryl C–H is derived primarily from the ether substrate rather than H_2. The authors then proposed the mechanism for the conversion of aryl alkyl ethers to arenes (Figure 6.13, Route I): (1) aryl C–O bond is activated by nickel center to an Ar–Ni–OCH$_2$R intermediate; (2) Ar–Ni–OCH$_2$R transforms to Ar–H after the β–H elimination and reductive elimination; (3) the other intermediate with carbonyl group is further hydrogenated to alcohols. Although the mechanism above does not require H_2 for the cleavage of the aryl ether bond, further experiments support that H_2 is necessary for the formation of the active catalyst but not for the actual catalytic transformation. In this process, the direct oxidative addition of Ni(0) with aryl C–OR structure after the Ni(0) coordination to arene is the key step. Furthermore, as reasonable speculation, they suppose that route II could be the main route for the conversion of 4-O-5 linkage without CH$_2$ nearby [45].

In 2014, Surawatanawong [47] presented a DFT study of the mechanism of Ni N-heterocyclic carbene catalyst (Ni-SIPr) for C–O bond hydrogenolysis of diphenyl ether (PhOPh) to produce benzene and phenol as products (Figure 6.14). On the basis of DFT results, the Ni(SIPr)(η^6-PhOPh) complex is initially formed before rearranging to Ni(SIPr)(η^2-PhOPh), the active species for C–O bond activation. The

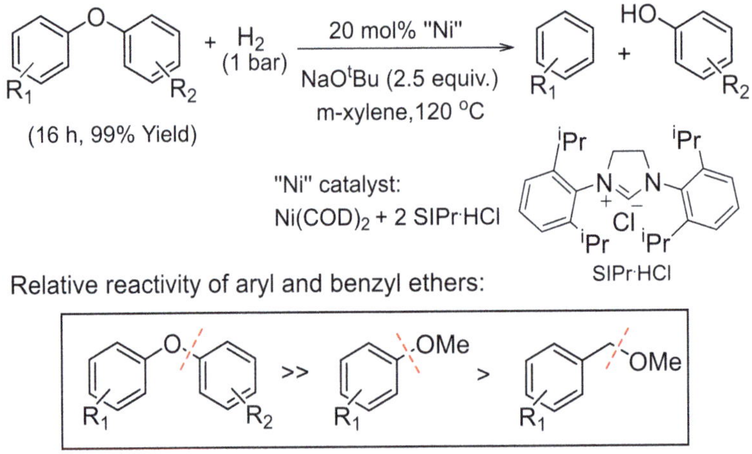

Figure 6.12 Hydrogenolysis of Aryl Ether Bonds over a Nickel Ni Carbene Catalyst with Base.

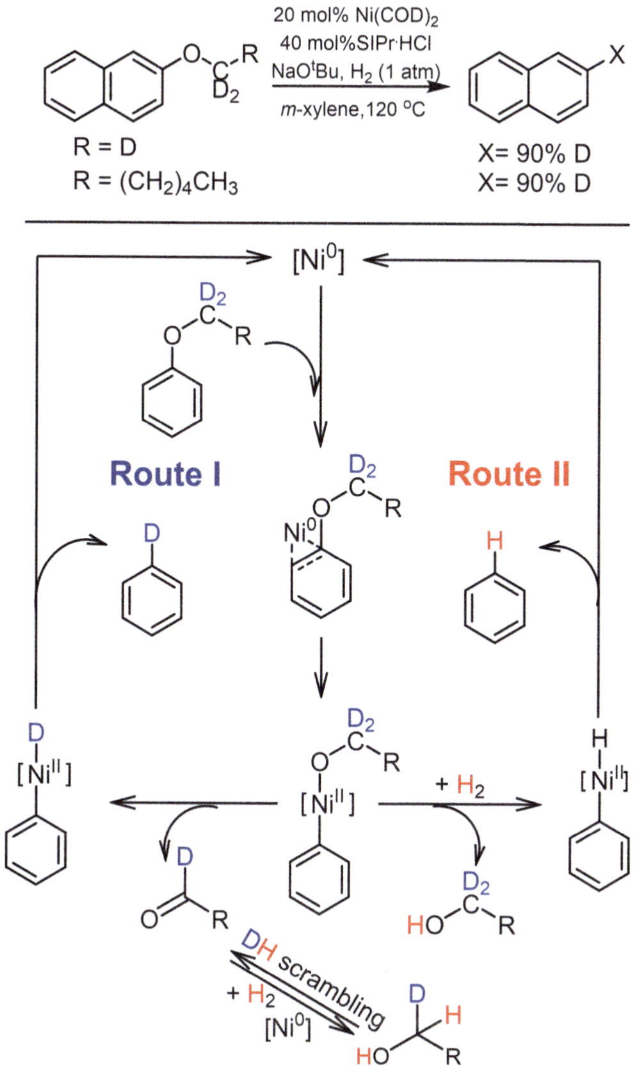

Figure 6.13 Mechanisms for the Conversion of Aryl Alkyl Ethers to Arenes. *Source:* Kelley et al. [46] with Permission, © 2012 American Chemical Society.

following catalytic reaction (Figure 6.14) has three steps: (i) oxidative addition of Ni(SIPr)(η^2-PhOPh) to form [Ni(SIPr)(OPh)(Ph)]^0, (ii) σ-complex-assisted metathesis, in which H_2 binds to the nickel to form [Ni(SIPr)(OPh)(Ph)(H2)]^0, and then benzene (or phenol) is eliminated, and (iii) reductive elimination of phenol (or benzene) and the binding of PhOPh to regenerate Ni(SIPr)(η^2-PhOPh) for the next cycle. During these steps, the oxidative addition step is the rate-determining step, and they also calculated that the diaryl ether containing an electron-withdrawing group can decrease the barriers for the oxidative addition. Given the fact that the highest energy barrier in cycle A is the reductive elimination step (+19.4 kcal·mol^{-1})

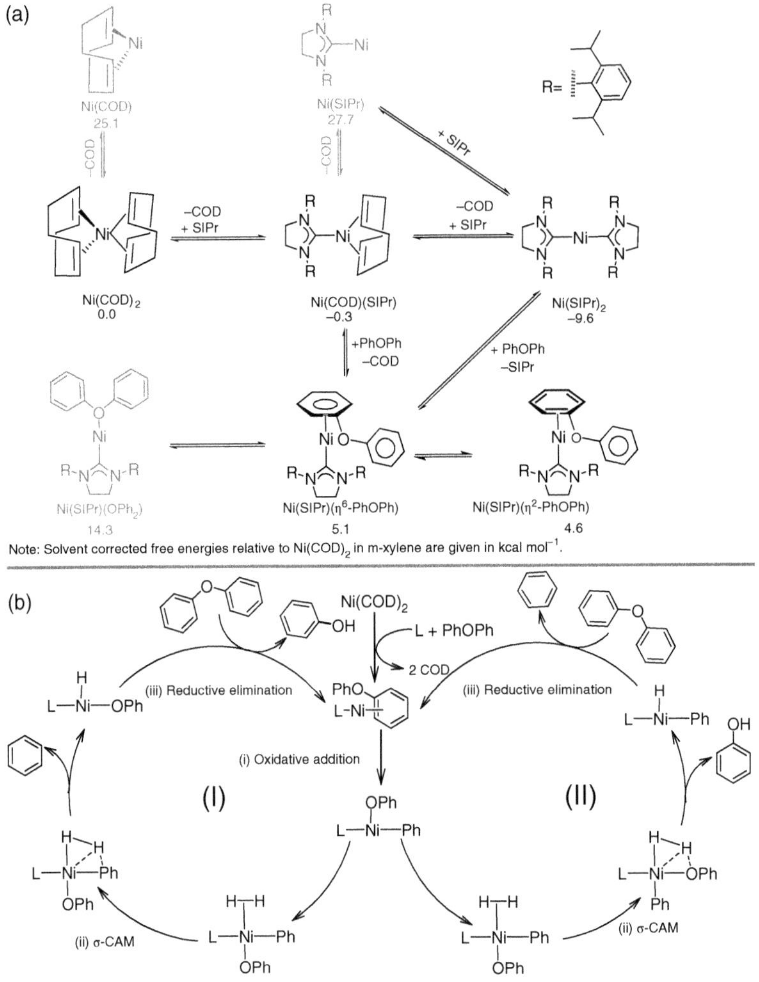

Figure 6.14 Reaction Mechanism for C–O Bond Hydrogenolysis of DiphenylEther by the Nickel N-Heterocyclic Carbene Complex (L = SIPr). *Source:* Sawatlon et al. [47] with Permission, © 2014 The Royal Society of Chemistry.

and the highest energy barrier in cycle B is the σ-CAM step (+19.3 kcal·mol^{-1}), comparable energy barriers suggested both pathways are possible. Moreover, they found the hydrogenation of benzene via Ni(SIPr)(η^2-C_6H_6) requires a high energy barrier (+39 kcal mol^{-1}), which explained the fact that the hydrogenation products, e.g. cyclohexane and cyclohexadiene, were not observed in the experiment.

Then, Surawatanawong and co-workers [48] further studied the reaction mechanism of the C–O bond hydrogenolysis of methyl phenyl ether (PhOMe) by using density functional theory (DFT). In the presence of H_2, the catalytic cycle involves the following: (i) aromatic C–O bond oxidative addition of Ni(SIPr)(η^2-PhOMe) to form Ni(SIPr)-(OMe)(Ph), (ii) β-H transfer from the methoxy to phenyl group in Ni(SIPr)(OMe)(Ph) via σ-complex-assisted metathesis (σ-CAM), which eliminates benzene and forms Ni(SIPr)(η^2-CH_2O), (iii) H_2 binding to form Ni(SIPr)(H_2)(η^2-CH_2O) prior to H-transfer from H_2 to the formaldehyde carbon via σ-CAM to generate Ni(SIPr)(H)(OMe), and (iv) reductive elimination to form methanol and the binding of methyl phenyl ether to regenerate Ni(SIPr)(η^2-PhOMe). Their study showed that the C—O bond oxidative addition is the rate-determining step and that the aromatic C—O bond cleavage to form Ni(SIPr)(OMe)(Ph) is more favorable than the aliphatic C—O bond cleavage to form Ni(SIPr)(OPh)(Me), consistent with the arene and alcohol products obtained from the experiment and previous studies. [45, 46] Notably, the β–H transfer from the methoxyl to phenyl group on Ni-SIPr is not a stepwise β-H elimination as generally perceived, but rather a concerted process that occurs via σ-CAM (Figure 6.15). This leads to benzene elimination before H_2 binding, in accordance with the results of the isotope labeling experiment of C—O bond hydrogenolysis of 2-methoxynaphthalene. In the absence of H_2, Ni(SIPr)(η^2-CH_2O) tends to undergo C—H bond activation and α-H elimination to release H_2 and generate a nickel carbonyl complex, the catalytically inactive species (Figure 6.15). This was reflected by experimental results which demonstrated low conversion of 2-methoxynaphthalene in the absence of H_2. Thus, H_2 is crucial to the catalytic reaction through its role in suppressing the formation of the inactive nickel carbonyl species.

In addition, besides the H_2, the excess base NaO^{t}Bu plays a crucial role in the catalytic cleavage of aromatic C–O bonds in alkyl aryl ethers and diaryl ethers conducted by the Ni-NHC (N-heterocyclic carbene). In 2016, Chung and co-workers [49] give detailed research on this phenomenon with DFT calculation. DFT calculations of several possible pathways suggest a new mechanistic pathway that involves

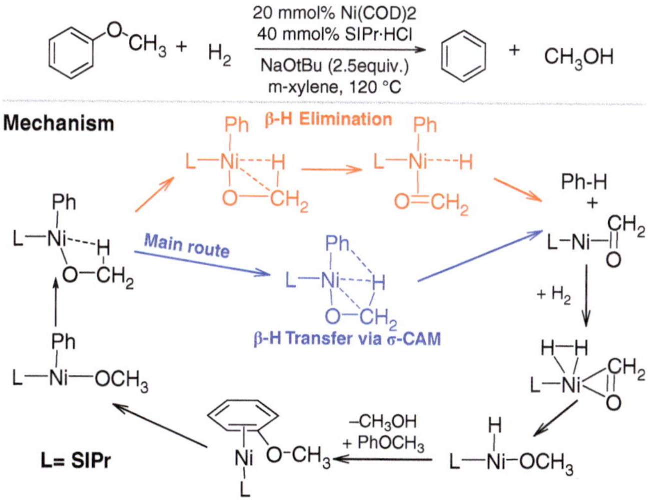

Figure 6.15 Catalytic C–OAr Bond Hydrogenolysis of Methyl Phenyl Ether with a Nickel N-heterocyclic Carbene Catalyst. *Source:* Wititsuwannakul et al. [48] with Permission, © 2016 American Chemical Society.

the coordination of the excess base (${}^{t}BuO^-$) to facilitate the rate-determining C—O activation step, dissociation of the ArO^- ligand, H_2 activation through a Ni—O^{t}Bu bond to give HO^{t}Bu, and finally reductive elimination to afford the arene product Figure 6.16). Another new ion-pair (S_NAr-like) pathway for the base-assisted C—O bond activation could compete with the above base-assisted oxidative addition pathway for some diaryl ethers. Their results demonstrate that the bulkiness and strong donating ability of the NHC ligand and the presence of excess base are the keys to a Ni(0)/Ni(II) catalytic cycle.

Different from the above catalytic systems carried out under the H_2 atmosphere, in 2015, Tobisu and co-workers [50] reported that a nickel catalyst with the N-heterocyclic carbene (NHC)-based ligand, containing 2-adamantyl groups, can cleave the C–O bonds of others in the absence of an external reductant (Figure 6.17). The hydrogen atom required in this reductive cleavage reaction is provided by the alkoxy group of the substrate, which serves as an internal reductant. Their further research shows that the absence of an external reductant enables the unique chemoselectivity, i.e., the selective reduction of an alkoxy group over alkenes and ketones.

For the catalytic hydrogenolysis of C–O bonds in aryl ethers, besides the generation of arenes as the product, in the perspective work of Tobisu and Chatani [51], they pointed out that nucleophile can be introduced in the ether bond catalytic cleavage to generate new R-Nu bonds (Figure 6.18). For an example of the published works, Wenkert [52] first documented the potential activity of nickel complexes

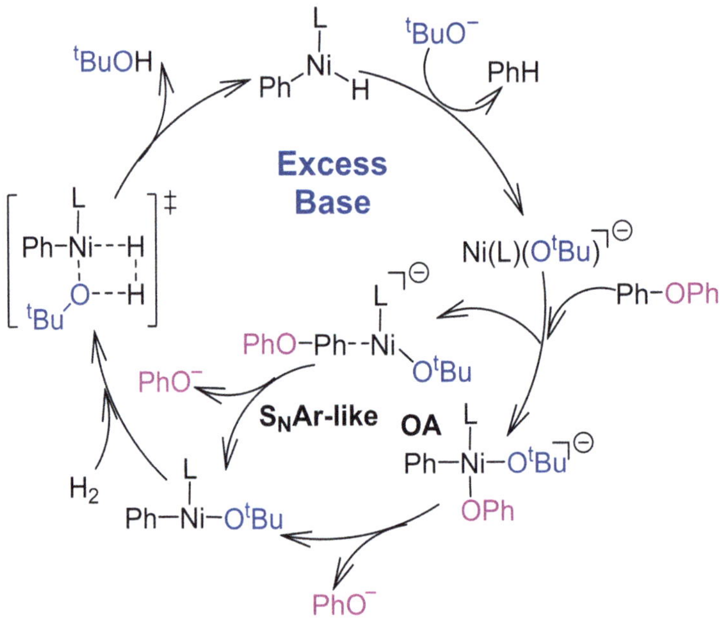

Figure 6.16 The Role of the Excess Base on the Mechanism of Ni-NHC Catalyzed Hydrogenolysis of Aryl Ethers. *Source:* Xu et al. [49] with Permission, © 2016 American Chemical Society.

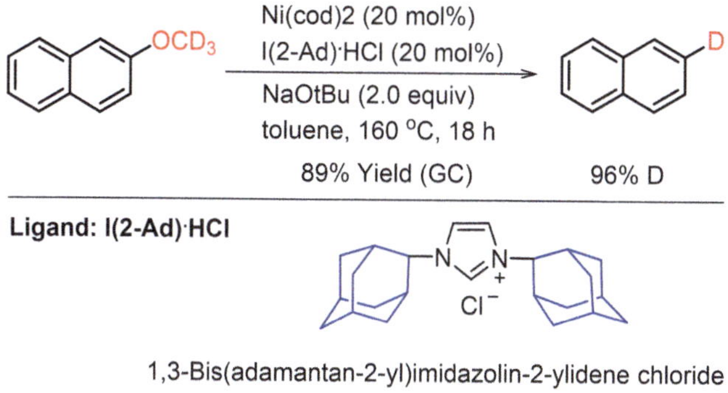

Figure 6.17 Nickel-catalyzed Reductive Cleavage of Aryl Alkyl Ethers to Arenes in Absence of H_2. *Source:* Reaction Data Abstracted from Tobisu et al. [50].

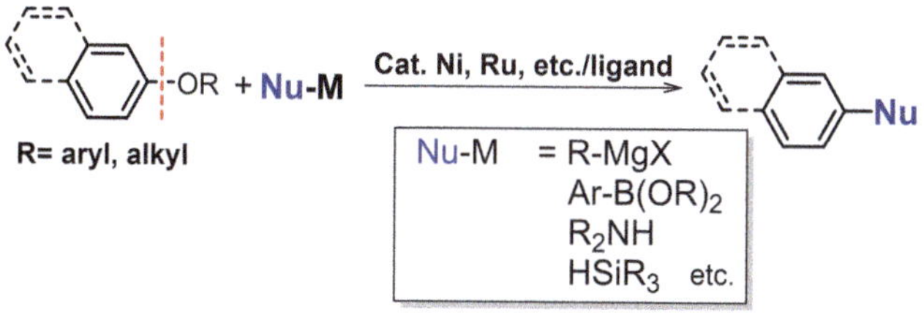

Figure 6.18 Metal-catalyzed Transformations of Aryl Ethers with Various Nucleophiles.

toward the cleavage of C(aryl)–O bonds and following cross-coupling reaction with Grignard reagents in 1979. Then, in 2004, Dankwardt [53, 54] found that the implementation of a bulky monodentate phosphine ligand such as PCy_3 or $PPhCy_2$ could significantly expand the scope of the Wenkert reaction. Then Wang [55] reported a Ni-catalyzed cross-coupling of aryl/alkenyl ethers with aryl Grignard reagents, and Uchiyama [56] put aryl ether in the Negishi coupling to generate the C(aryl)–C bonds. Further reports indicate the nucleophile can be organoboron [57, 58], and amine [59] reagents. In addition, the hydrosilane [60, 61] can be used as a suitable reductant for the Ni-catalyzed reductive cleavage of aryl ethers, a transformation synthetically equivalent to hydrogenolysis. While the activation mechanism of ether bond with Ni catalyst and hydrosilane is different, a more detailed mechanism discussion will be presented in Chapter 6.4.

6.3 Other Heterogeneous Catalytic Systems for the Direct Hydrogenative Cleavage of C–OAr Ether Bonds

Besides the homogeneous Ni-based catalytic systems used in the direct hydrogenative cleavage of C–O ether bonds, robust heterogeneous Ni catalysts have also drawn much attention to lignin ether bond cleavage. After their previous work [45], in 2012, Hartwig and co-workers [62] found that the in-situ generated Ni catalyst, from the well defined soluble nickel precursor $Ni(COD)_2$ or $Ni(CH_2TMS)_2(TMEDA)$ in the presence of ^tBuONa, can selectively cleave C_{Ar}–OAr bonds in aryl ether models of lignin without hydrogenation of aromatic rings (Figure 6.19).

In the same year, Lercher and co-workers [63, 64] gave detailed kinetics research about the direct cleavage of ether bonds of lignin-derived aromatic ethers and the hydrogenation of the oxygen-containing intermediates over the Ni/SiO_2 catalyst at 120 °C in the aqueous phase (Figure 6.20). Results show that the C–O bonds of α-O-4 and β-O-4 linkages are cleaved by hydrogenolysis on Ni, while the C–O bond of

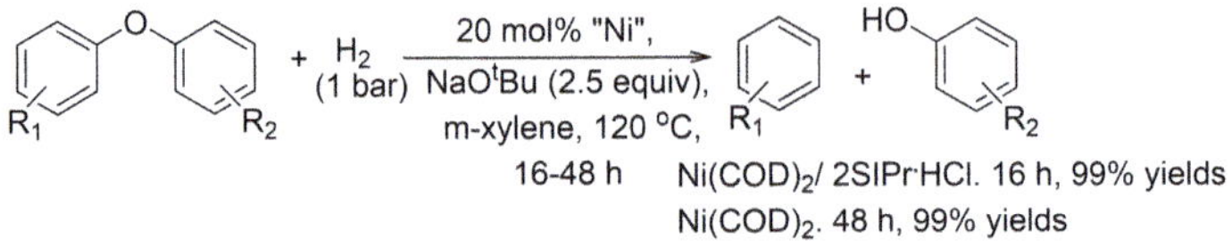

Figure 6.19 Hydrogenolysis of Diaryl Ethers over a Heterogeneous Nickel Catalyst. *Source:* Reaction Data Abstracted from Sergeev et al. [62].

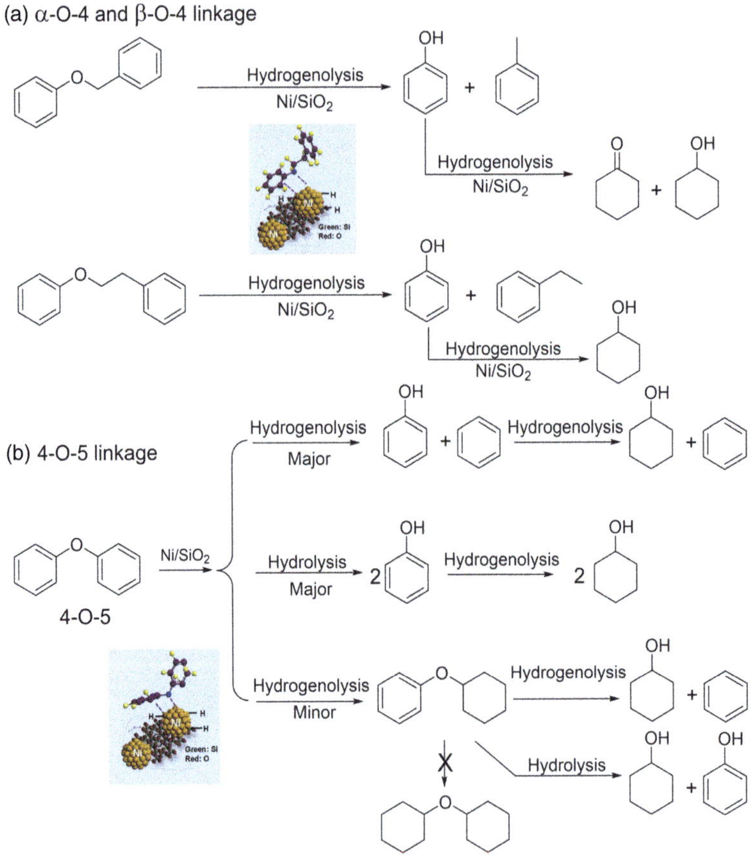

Figure 6.20 Ni-catalyzed Cleavage of Aryl Ethers in the Aqueous Phase. *Source:* He et al. [63] with Permission, © 2012 American Chemical Society.

the 4-O-5 linkage is cleaved via parallel hydrogenolysis and hydrolysis. They further studied the reductive deconstruction of organosolv lignin with zeolite-supported nickel nanoparticles (Ni/H-BEA) [65]. The organosolv lignin, consisting of seven to eight monolignol subunits and has an average molecular weight of ca. 1200 g mol^{-1}, was converted to substituted alicyclic and aromatic hydrocarbons with 5 to 14 carbon atoms. Full conversion with a total yield of 70 ± 5 wt% hydrocarbons was achieved at 593 K and 20 bar H_2. Besides the NMR spectra method, in situ IR spectroscopy was used to follow the changes in lignin constituents during the reaction. The transformation of organosolv lignin starts with the hydrogenolysis of aryl alkyl ether bonds, followed by the hydrogenation of the aromatic rings to cyclic alcohols. The cyclic alcohols further lose the oxygen is via dehydration on Brønsted acid sites to cyclic alkenes that are further hydrogenated.

In addition, loading Ni species onto some supporters or catalysts with hydrogenative ability can promote the catalytic transformation of lignin ether bonds. In 2014, Esposito and co-workers [66, 67] reported that the fixed bed reactor with Ni particles loading on the titanium nitride (TiN-Ni) catalyst can achieve the efficient conversion of different aryl ethers under relatively mild conditions within minutes. Although the generated phenols from lignin models tend to be hydrogenated to aliphatic alcohol and lose the aromatic structure, the TiN-Ni catalyst can exhibit high activity for the hydrogenolysis of raft lignin, one of the most available types of lignin yet the most difficult to upgrade via catalytic processes due to the presence of sulfur in its structure. The depolymerization of raft lignin yielded the formation of substituted phenols (3.2 wt %) together with many aromatic fragments of smaller molecular weight (up to ~60 wt %). Compared with other known catalysts used in lignin depolymerization, such as Raney Ni and Pd/C, the TiN-Ni catalyst appeared to be superior stability and a lower tendency to deactivation. In addition, the better catalytic performance of TiN-Ni may be due to the better dispersion of Ni in the TiN phase as well as the different oxidation state of Ti(III) in comparison to the titanium(IV) dioxide, which is known to alter the behavior of the material toward hydrogen uptake.

Besides the nitride supporter, in 2012, Li and co-workers [68] reported that Ni-W_2C can efficiently catalyze the transformation of carbohydrate fraction in the woody biomass, i.e., cellulose and hemicellulose were converted to ethylene glycol and other diols with a total yield of up to 75.6% (based on the amount of cellulose & hemicellulose), while the lignin component was converted selectively into monophenols with a yield of 46.5% (based on lignin) at 235 °C and 6 MPa H_2 (Figure 6.21). The proposed mechanism shows that Ni-W_2C can catalyze the cleavage of C–C bonds in glucose and xylose via the retro-aldol process, while more information about the cleavage mechanism of ether bonds in lignin is not involved in this work.

Then, Song and co-workers [69] found that Ni/AC catalyst prepared by the carbothermic reduction could efficiently catalyze the cleavage of C_β–OAr bond in birch lignin with a high yield of aromatic rings. The reaction was carried out at 200 °C in methanol and no extra H_2 was needed, and the best selectivity of 97% toward monomeric phenols is achieved at 50% conversion of birch wood lignin (Figure 6.22). Analyses with MALDI-TOF and NMR show that birch lignin is first fragmented into smaller lignin species consisting of several benzenes rings with a molecular weight

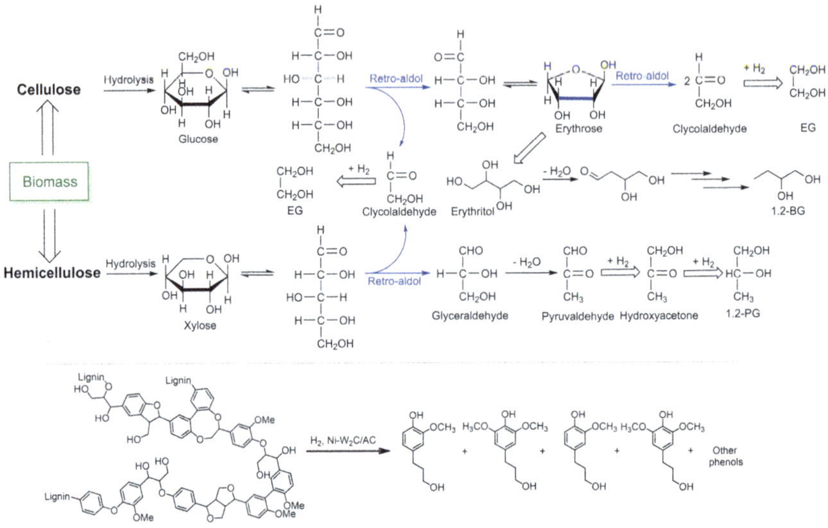

Figure 6.21 One-pot Catalytic Hydrocracking of Raw Woody Biomass into Chemicals over the Ni/W$_2$C Catalyst. *Source:* Li et al. [68] with Permission, © 2012 The Royal Society of Chemistry.

of m/z ca. 1100 to ca. 1600 via alcoholysis reaction. The second step involves the hydrogenolysis of the fragments into phenols, during which the alcohols provide active hydrogen species. In addition, the Ni/AC catalysts were recycled by magnetic separation and could be reused four times without losing activity. Furthermore, the deuterium distribution experiments (inserted Table in Figure 6.22) suggest that the unique cooperation between solvent molecules and Ni/AC catalyst centers, such as hydrogenation and dehydration sites, can realize the direct hydrogenative cleavage of C_β–OAr ether bond in β-O-4 linkage with a C_α–OH structure, which further transformed to the phenylpropane structure via the Route 3 (Figure 6.22).

It should be pointed that the carbothermic reduction with carbon element modifying the Ni particle makes the obtained Ni/AC catalyst quite different from other NiM/supported catalysts and noble metal-supported catalysts prepared from H_2 reduction (Figure 6.23a and 6-23b). Even under the extra H_2 condition, this Ni/AC can selectively catalyze the cleavage of ether bonds in the lignin β–O-4 model with both high conversion and high yield of aromatic rings (Figure 6.23c) [70]. In addition, in 2012, Song and co-worker [71] reported that the heterogeneous Ni/AC catalyst could efficiently catalyze lignosulfonate (LS) hydrogenolysis to phenols without arenes hydrogenation, a 91% LS conversion was obtained within 6 h (Figure 6.23d). The activated hydrogen over Ni(0) sites selectively broke C_{aryl}–O–C_{alkyl} bonds to phenols and methylene structure (-CH_2-), and cleaved hydroxyl groups at side chains to –CH_3 or –CH_2-, producing C_1–C_3 alkane-substituted guaiacols as major products. Furthermore, carbon-sulfur bonds could also be reductively cleaved on Ni(0) sites to –CH_2- and H_2S.

Then, Abu-Omar and co-workers [72] further investigated and expanded the application of the Ni/C system to poplar and eucalyptus wood by adopting the reported reaction conditions of Song's (Figure 6.24). Their results show that the propenyl

(a)

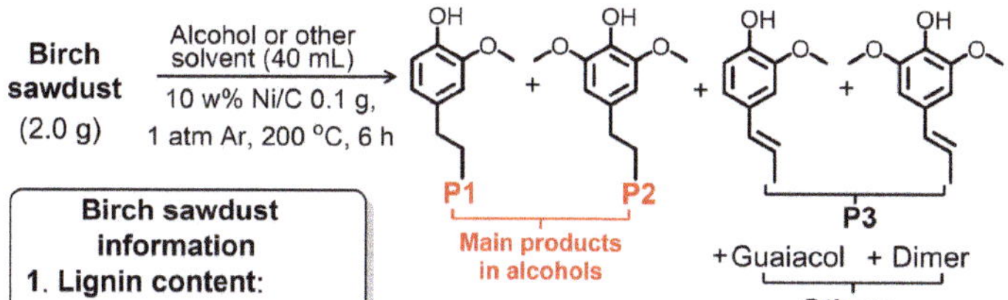

Birch sawdust information
1. **Lignin content:** 19 w% (TAPPI Method)
2. **Ether linkages in birch lignin:** β-O-4 60%; α-O-4 6-8%; 4-O-5 6.5%; Glyceraldehyde-2-aryl ether 2%.
3. **Theoretical yield of phenol monomers from C-OAr bonds cleavage: 74.5~76.5%**

Conversion Results:

Solvent	Lignin Conv. (%)	Products Sel. (%) P1+P2	P3	Others
CH_3OH	54	89	1	10
C_2H_5OH	48	95	3	2
EG	50	97	1	2
Dioxane	15	6	93	<1
Cyclohexane	<1	-	-	-
CH_3OH	62 (220 °C)	80	18	2

(b)

Deuterium distribution for the propyl group in 4-propylsyringol

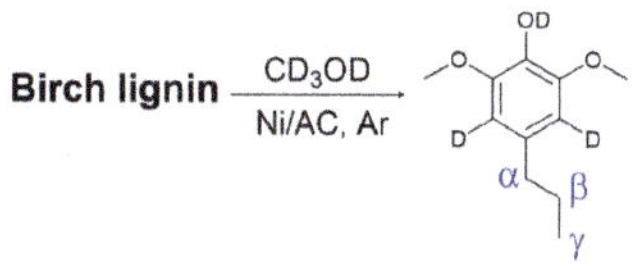

Entry	Deuterium Content α	β	γ
NMR Result	1.00	0.54	1.60
Route 1	1.00	2.00	2.00
Route 2	1.00	0.25	2.00
Route 3	1.00	0.50	2.00

Transfer hydrogenation routes for the cleavage of C_β-OPh bond

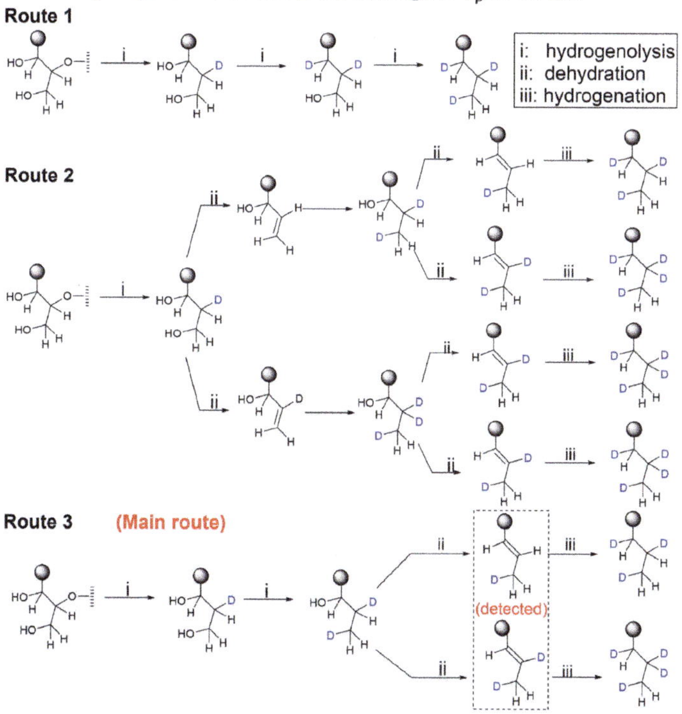

Figure 6.22 Lignin Conversion in Methanol with Ni/AC Catalyst. *Source:* Song et al. [69] with Permission, © 2013 The Royal Society of Chemistry.

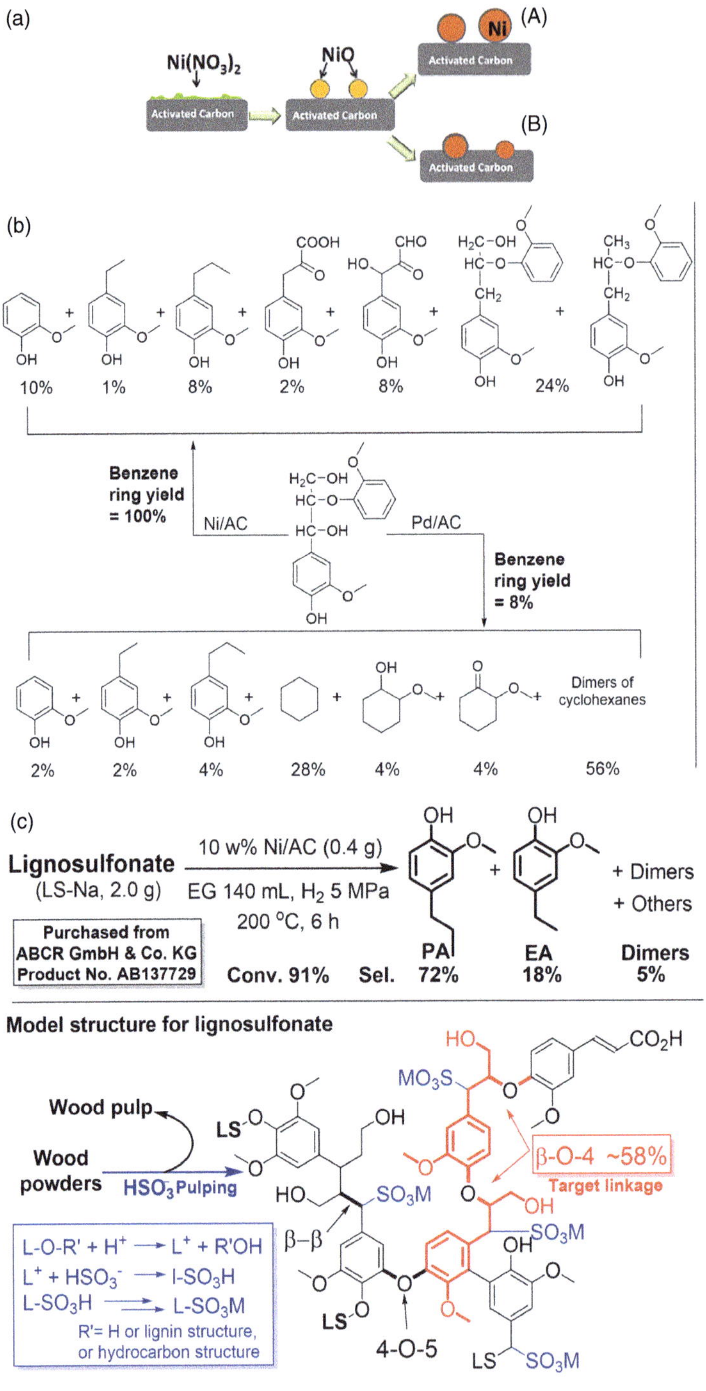

Figure 6.23 Representation of the Ni/AC Catalyst Reduced by H_2 (a) and Activated Carbon (b); The Conversion of Lignin β-O-4 Model Compound over Ni/AC and Pd/AC (c); Catalytic Lignosulfonate Hydrogenation to Phenols over Ni/AC (d). *Source:* Reaction Data Abstracted from Song et al. [70, 71].

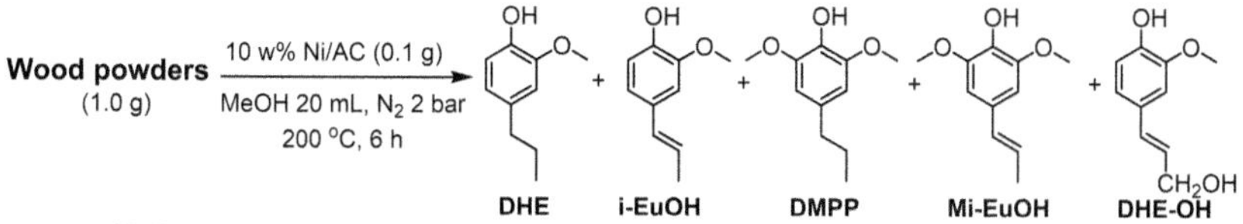

			DHE	i-EuOH	DMPP	Mi-EuOH	DHE-OH
Birch	Conv. 32%	Yield:	10%	1%	18%	3%	0%
Poplar	Conv. 26%	Yield:	1%	8%	2%	15%	0%
Eucalyptus	Conv. 28%	Yield:	6%	3%	8%	11%	0%

Figure 6.24 Depolymerization of Various Lignin over the Ni/C Catalyst in Methanol. *Source:* Reaction Data Abstracted from Klein et al. [72].

analogs of 2-methoxy-4-propylphenol (DHE) and 2,6-dimethoxy-4-propyl-phenol (DMPP), propenyl derivatives isoeugenol (i-EuOH) and methoxyisoeugenol (Mi-EuOH), are the main products from all wood substrates using 5 wt% Ni/C and that total yields range from <10% to >30% depending on the biomass substrate. In the presence of higher Ni/C (10 wt%), DHE and DMPP become the major products from birch wood, which is presumably due to the presence of more H_2 produced from methanol reforming over Ni/C. For the discrepancies of birch lignin conversion between two research groups, Abu-Omar suggested it may be due to the biomass composition, which is known to vary across regions and growing seasons and place of origin. These results suggest that catalytic systems for selective lignin depolymerization must carefully consider the nature of the substrate to achieve high yields of individual, upgradeable products. Furthermore, based on the preparation introduction of Ni/AC reported by Song, Ni/AC was sensitive to the air, which needs N_2 protection during the transfer. The slight surface oxidation of Ni could restrain or affect the activity of Ni/AC catalysts in lignin conversion with alcohol.

Besides the catalytic cleavage of β–O-4 and α-O-4 linkages, in 2016, Hartwig and co-workers [73] then reported that another Ni/C catalyst, prepared from of decomposition of $Ni(COD)_2$ on active carbon, can efficiently achieve the chemo- and regioselective hydrogenolysis of the C–O bonds in di-ortho-substituted diaryl ethers with the assistance of the base NaO^tBu or potassium bis(trimethylsilyl)amide (KHMDS), and no aromatic rings was further hydrogenated (Figure 6.25).

Besides the supported Ni catalyst loading on the TiN, W_2C, and active carbon for lignin hydrogenolysis, in 2014, the Li group [74] reported the complete catalytic ethanolysis of raft lignin over an α-MoC_{1-x}/AC catalyst in pure ethanol at 280 °C to give high-value chemicals of low molecular weight with a maximum overall yield of the 25 most abundant liquid products (LP25) of 1.64 g per gram of lignin. The LP25 products consisted of C_6–C_{10} esters, alcohols, arenes, phenols, and benzyl alcohols. In 2017, Rinaldi and co-workers [75] provided a detailed hydrodeoxygenation study of the pyrolysis bio-oil to alkylphenols over the MoC_x/C catalyst. Then, in 2016, the Jiang group [76] prepared a mesoporous carbon-supported Ni–Mo_2C catalyst by one-pot fast pyrolysis of Ni-Mo preloaded sawdust and demonstrated its catalytic performance for hydrogenation of lignin. Under the optimized reaction condition (250 °C, 2 h, and 2.0 MPa initial H_2 pressure), the as-prepared Ni-Mo_2C/C catalyst provided a 61.3 wt% of lignin conversion to liquid products in the presence of H_2 and isopropanol solvent, and 80 wt% of which are phenols, guaiacols, and

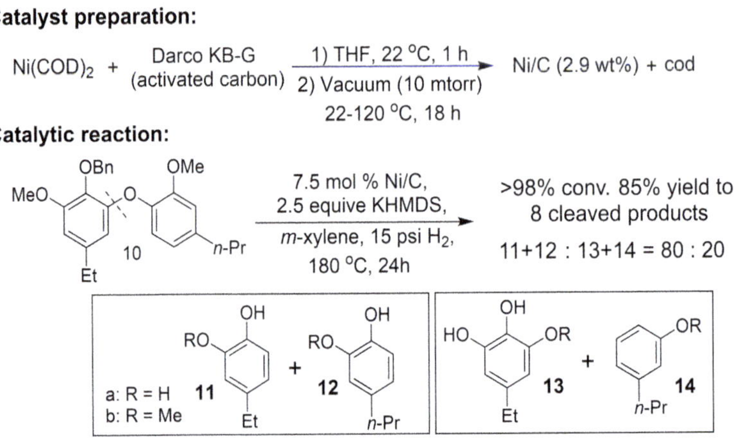

Figure 6.25 Hydrogenolysis of Diaryl Ether C–O Bonds by a Heterogeneous Ni/C Catalyst. *Source:* Reaction Data Abstracted from Gao et al. [73].

trimethoxybenzenes. The author suggested that the high catalytic performance of Ni–Mo_2C/C toward lignin hydrogenation may be attributed to the synergistic effect of the graphitized biochar matrix and the Ni–Mo_2C nanoparticles which facilitates electron transfer. At first, evenly distributed Ni NPs in the catalytic system can adsorb H_2 or react with isopropyl alcohol to form active hydrogen. Meanwhile, the O–H bond of the isopropanol molecule on the catalyst surface is broken by the metallic nickel to form isopropyl radicals. Coordination bonds can form between the oxygen atoms of isopropyl radicals and the lone electron pairs of the unoccupied d orbitals of the Mo ions to form the fully isopropylyzed active Mo species. Subsequently, the active hydrogen and active Mo species attack the C–O or C–C bonds of lignin segments, which causes the formation of benzodiazepine molecules. And the active hydrogen can incorporate with the oxygen atom which was preliminarily activated by the active Mo species. Finally, a series of small molecule compounds are produced by the C–O bonds cleavage (Figure 6.26).

Besides the single Ni supported catalysts presented above, other noble metals, such as Ru, [77, 78] Rh [79], Pd, and Au [80] supported catalysts were known

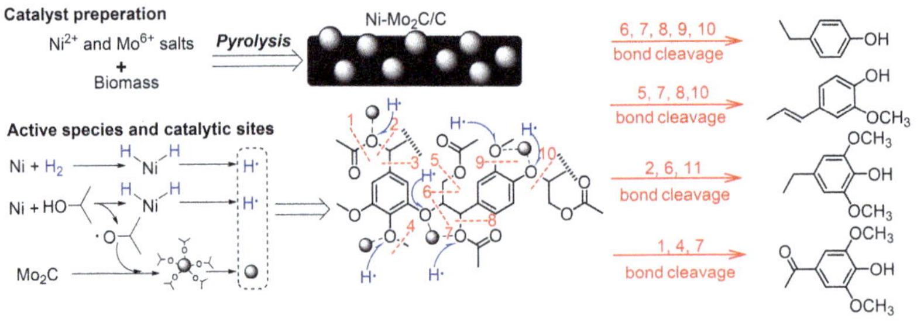

Figure 6.26 Hydrogenation of Lignin to Aromatic Chemicals over a Ni-Mo_2C/C Catalyst. *Source:* Wang et al. [76] with Permission, © 2016 The Royal Society of Chemistry.

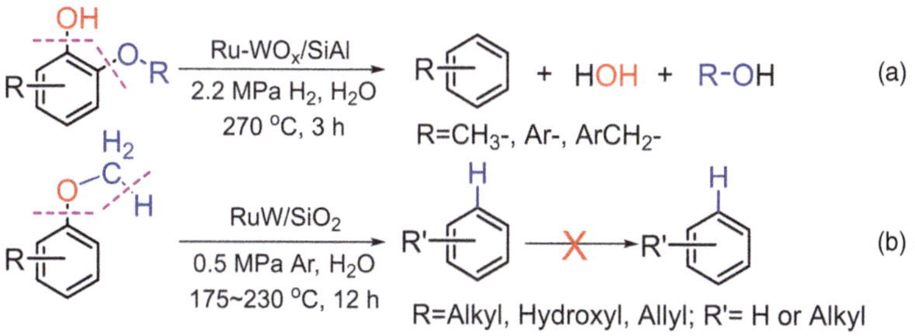

Figure 6.27 Catalytic Lignin Ether Bond Cleavage over the Supported Ru-WO_x Catalyst and RuW Alloy Catalyst. *Source:* Based on Huang et al. [81] and Meng et al. [82].

heterogeneous catalysts for lignin, lignin models, or lignin monomers hydrogenolysis (List in Chapter 4.4), focusing the cleavage of lignin ether bonds.

In 2015, Fu and co-workers [81] reported a Ru–WO_x bifunctional catalyst that showed high catalytic activities for the hydrogenolysis of a wide range of phenols and phenyl ethers, including dimeric lignin model compounds and the primitive phenols separated from pyrolysis lignin, to form arenes selectively in water (Figure 6.27a). Preliminary mechanistic studies supported that the reactions occurred via a direct cleavage of the C(sp^2)–O bonds and the concerted effects of the hydrogenating Ru sites and the Lewis acidic W sites are the key to such an unusual reactivity. In 2017, Han and co-workers [82] further found that the RuW alloy supported on the SiO_2 could efficiently catalyze self-supported hydrogenolysis of aromatic ethers to produce arenes using the hydrogen source within the reactants (Figure 6.27b). In 2018, the Han group [77] found that Ru/C could efficiently catalyze the cleavage of the aromatic ether bonds in various lignin-derived 4-O-5 compounds via a transfer hydrogenolysis route using isopropanol as the hydrogen resource. Various lignin-derived compounds could be efficiently cleaved over commercial Ru/C to generate the corresponding aliphatic alkanes, aliphatic alcohols, and aromatic derivatives under milder conditions. A mechanism study indicated that the reaction occurred through the direct cleavage of aromatic ether bonds or the formation of the reaction intermediate cyclohexyl phenyl ether. While, given the fact that most processes focussing on the native lignin transformation with a more complex structure usually do not begin with the direct cleavage of C–OAr bonds, such as the Pd/C mediating process, these processes will be discussed in the following corresponding chapters. While, in the following part, we will present some alloy works, which involve the direct cleavage of lignin C–OAr bonds in lignin or its model conversion.

In 2014, Yan and co-weokers [83–86] synthesized NiRu, NiRh, NiAu, NiAg, and NiPd catalysts and evaluated their catalytic ability in the hydrogenolysis of lignin C–O bonds, and results proved that the alloys were superior over single-component catalysts. The optimized NiRu catalyst contains 85% Ni and 15% Ru, composed of Ni surface-enriched, Ru–Ni atomically mixed, ultrasmall nanoparticles, which were checked by XPS, TEM, and EXAFS. The $Ni_{85}Ru_{15}$ catalyst showed high activity in β-O-4 type C–O bond hydrogenolysis and significantly higher activity over Ni and Ru

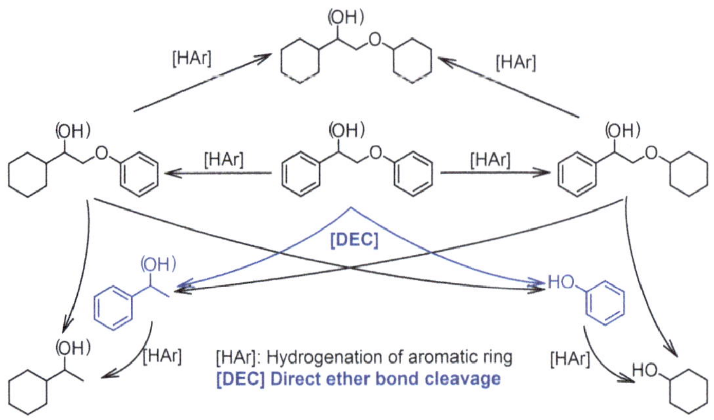

Figure 6.28 Reaction Pathways of β-O-4 Model Transformation into Monomeric and Dimeric Products over a $Ni_{85}Ru_{15}$ Catalyst. *Source:* Zhang et al. [83] with Permission, © 2014, American Chemical Society.

catalysts in the direct conversion of organosolv lignin into monomeric aromatic chemicals. Further investigation indicates that the synergistic effect of NiRu can be attributed to three factors: (1) increased fraction of surface atoms (compared with Ni), (2) enhanced H_2 and substrate activation (compared with Ni), and (3) inhibited benzene ring hydrogenation (compared with Ru). The same results can be found in NiRh and NiPd catalysts than their single-component counterparts in hydrogenative cleavage of lignin ether bonds. However, the mechanism research about the lignin model compounds conversion shows the detailed process is complex, which contains the direct cleavage of target ether bond as well as the first transformation of the neighboring hydroxyl group and aromatic rings in β-O-4 linkage (Figure 6.28).

Then, various alloy catalysts were reported in the applications of lignin and its models' hydrogenolysis research, such as the bimetallic Ru-Ni, [87] Ru-Fe [88], Rh-Ni [87], Pd-Ni [89], Pd-Fe [90], Pd-Ag [91], and W-based M–W/AC (M=Ru, Pt, and Pd) [92]. Besides the alloy catalysts with noble metals, in 2017, Zhang and co-workers [93] achieved the depolymerization of organosolv lignin and native birch sawdust via a non-precious Ni_1-Fe_1 alloy catalyst. In addition, besides the MoC_x and W_2C-based catalysts shown above, more metal-nonmetal compounds (not the metal oxides) are used in lignin hydrogenolysis, focusing on the direct cleavage of lignin ether bonds, such as the catalytic systems with boride (NiBi, etc), carbide, nitride (MoN_x, etc), phosphide (FeP_x, CoP_x, NiP_x, MoP_x, $FeMoP_x$, etc), sulfide ($NiMoS_x$, $CoMoS_x$, etc) catalysts (listed in Chapter 4.4).

6.4 Direct Reductive Cleavage of Ether Bond with Hydride Reagents

Besides the hydrogen as the cleavage agent to cleave the lignin ether bonds, some hydride reagents like $NaBH_4$, $LiAlH_4$, and silane (R_3SiH) can also achieve the cleavage of lignin ether bond with the assistance of various catalysts.

In 1964–1965, during examining the chemistry of boron, Freeguar and co-worker reported that in conjunction with a halogen such as iodine, boranes [94], and metal borohydrides [95] will destroy the hydroxyl groups of alcohols, which are converted into organic halides. They further found that the diborane on its own does not normally react with ethers near room temperature, except possibly to form a very loose addition compound, but a comparable reaction occurs with ethers in the presence of free halogen. [96] An oxygen-carbon link is ruptured, the oxygen attaching itself to the boron while the carbon links up with halogen. With anisole as the substrate example, the reaction with I_2/B_2H_6 thus gives methyl iodide and triphenyl borate, from which phenol is immediately obtained on subsequent hydrolysis in the cold (Figure 6.29). Since one B_2H_6 molecule can cleave six molecules of the ether, the relative amount of diborane required is small. In addition, although the reaction needs a dry ether and carries out at room temperature or even below, the reaction is moreover very widely applicable. The ether may be symmetric, unsymmetric, aliphatic, aromatic, or cyclic. With cyclic ethers, a boron ester is first produced which on subsequent hydrolysis yields the ω-iodo derivative of an alcohol as the ultimate product, such as the transformation of tetrahydrofuran to 4-iodobutan-1-ol (Figure 6.29).

In general, the ether bond cleavage reaction with halogen and borohydrides is efficient and rapid, which proceeds quietly to completion within minutes at room temperature. Higher boranes are inclined to react more slowly than diborane. Metal borohydrides are comparable in reactivity to diborane, although sodium borohydride is less reactive than lithium salt. Bromine and interhalogen compounds such as iodine chloride react more vigorously than iodine but are in general less convenient. In 2006, Haenel et al. [97] reported in the context of hydrogenation studies with $NaBH_4/I_2$ catalysts for coal liquefaction catalytic activity towards aromatic compounds, increasing the original coal's aliphatic carbon content from 11% up to 60%. Then, in 2009, Ragauskas and co-workers [98] investigated the performance of the $NaBH_4/I_2$ system on the hydrogenolysis of ethanol organosolv lignin (EOL). On the basis of the fact that the hydrogenolysis cleavage of an interlinking aryl-O-aryl and aryl-O-aliphatic linkage increases the hydroxyl group content of the final product, and the hydrogenation of the aromatic ring decreases the phenol concentration and elevates the concentration of aliphatic hydroxyl groups. They used the selective phosphitylation of the hydroxyl groups on the lignin polymer with

Aromatic ester:

$$6PhOMe + 3I_2 + B_2H_6 \longrightarrow 2B(OPh)_3 + 6MeI + 3H_2$$

$$3PhOMe + 2I_2 + NaBH_4 \longrightarrow B(OPh)_3 + 3MeI + NaI + 2H_2$$

$$B(OPh)_3 + 3H_2O \longrightarrow 3PhOH + H_3BO_3$$

Cyclic ethers:

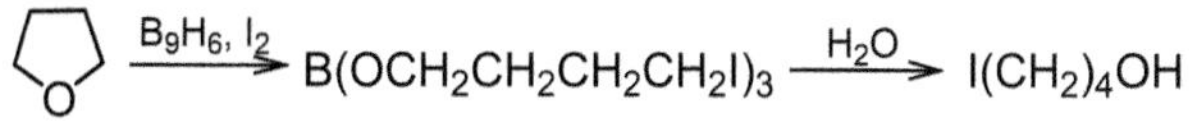

Figure 6.29 Low-temperature Cleavage of Ethers with Boranes/Metal Borohydrides and Halogen.

2-chloro-4, 4, 5, 5,-tetramethyl-1, 3, 2,-dioxaphospholane (TMDP), followed by quantitative ^{31}P-NMR measurements, to provide a reliable means of monitoring lignin changes (^{31}P-NMR, aliphatic hydroxyl 149.0–145.6 ppm; phenolic hydroxyl 144.4–137.6 ppm; carboxylic hydroxyl 136.0–133.8 ppm). When $NaBH_4/I_2$ was employed, besides the obvious increase of carboxylic acid hydroxyl, the observed increase in the overall hydroxyl content on the treated EOL proposed that hydrogenolysis cleavage of the aryl-O-aryl and aryl-O-aliphatic linkages occurred.

Besides using the halogen, in 2013, Nair [99] achieved the deprotection of allyl/benzyl aryl ethers with the mild and inexpensive reagent combination of $NiCl_2{\cdot}6H_2O/NaBH_4$ in methanol at 0 °C, which afforded the products in high yields > 80%, within a reaction time of 5–10 min. A plausible mechanism for the reaction is illustrated in Figure 6.30. It is expected that nickel forms a chelated complex with the p-electrons of the aryl ring, which weakens the aryl-oxygen bond significantly. This facilitates the donation of electrons from the oxygen atom to the electron-deficient boron resulting in the attack of hydride on the benzylic/allylic carbon, and consequently the cleavage of the ether linkage. The high bonding affinity between oxygen and boron aided by the polar protic medium which stabilizes the partial charges of the transition state could be attributed to the faster reaction kinetics and chemoselectivity. The evidence for complexation is apparent from the inert nature of the alkyl ethers which did not cleave even with an excess amount of the reagents. In addition, compared with other metal salts like $FeCl_3$, $CoCl_2$, CuCl, and $PdCl_2$, the cooperation between $NiCl_2$ and $NaBH_4$ is more efficient.

When the catalytic reduction with $NaBH_4$ over the active metal center like Pd, the phenol monomers generated from the direct cleavage of the ether bond can be further hydrogenated to cyclohexanone or cyclohexanol. In 2018, Li and co-workers [100] reported a Pd-catalyzed method for the direct cross-coupling of diaryl ether 4-O-5 linkage models with amines, in which dual $C_{(Ar)}$–O bond cleavages form valuable nitrogen-containing derivatives (Figure 6.31). Based on the experimental results, the cross-coupling reaction is proposed is begin with the first reaction between $NaBH_4$ and $Pd(OH)_2/C$ catalyst to form the H–Pd(II)–H species. Subsequently, hydrogenolysis of diphenyl ether generates phenol A by C(Ar)–O bond cleavage, which is further reduced by H–Pd(II)–H species to form cyclohexanone B. Intermediate B reacts with the amine to form iminium intermediate C, which is reduced to form compound D (Path a), and D is subsequently oxidized to generate intermediate E or H. Intermediate E, upon losing H^+, tautomerizes to form enamine F. Finally, enamine F is oxidatively aromatized to form the product pyrrole

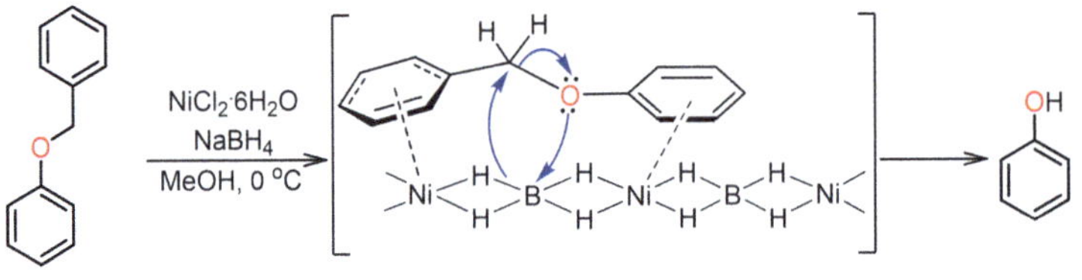

Figure 6.30 $NiCl_2/NaBH_4$ in Methanol Achieves the Dealkylation or Debenzylation of Aryl Ethers.

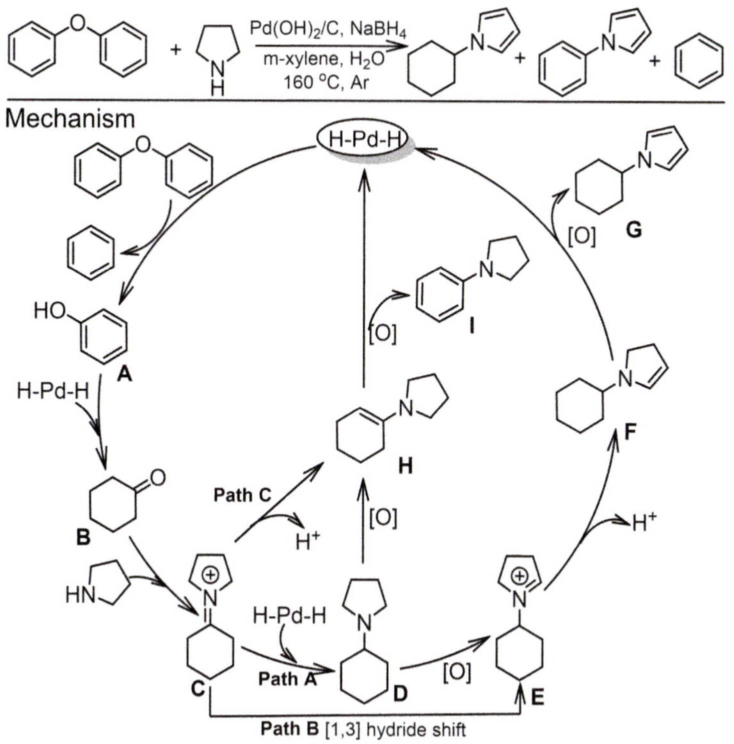

Figure 6.31 Palladium-Catalyzed Formal Cross-Coupling of Diaryl Ethers with Amines. *Source:* Zeng et al. [100] with Permission, © 2018 Wiley-VCH.

G, as well as regenerating the H–Pd(II)–H species. Another intermediate H undergoes oxidative aromatization to form product I and regenerates the H–Pd(II)–H species. It is also possible that intermediate C can undergo a [1, 3] hydride shift to generate intermediate E (Path B). Furthermore, a small portion of intermediate C can tautomerize to form enamine H (Path C).

Besides the reduction system with borohydride and $NaBH_4$, in 2013, Wang and co-workers [101] reported a selective iron-based catalytic system for the reductive cleavage of aryl C-O bonds that employed iron(III) acetylacetonate ($Fe(acac)_3$) as the catalyst precursor and $LiAlH_4$ as the reducing agent (Figure 6.32). The present catalyst system was further shown to be effective for the selective cleavage of the

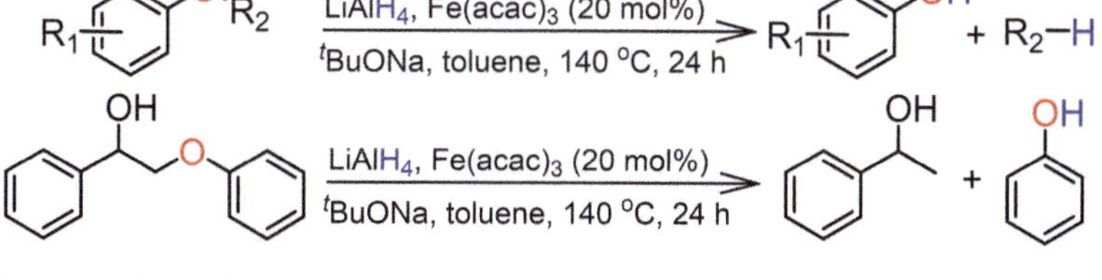

Figure 6.32 Selective Reductive Cleavage of Inert aryl C–O Bonds by an Iron Catalyst,

β-O-4 linkage of lignin model compounds under a hydrogen atmosphere in the absence of $LiAlH_4$ (Figure 6.32).

Besides the B–H and Al–H systems shown above, the Si–H is also employed to cleave the ether linkages. In 2003, Ohta and co-workers [102] reported an Rh–PPh_3-catalyzed method for reductive cleavage of the C–O bond of acetals and orthoesters with silane (Figure 6.33). This rhodium–silane reduction system is effective for the reductive cleavage of sp^3-carbon sp^3-oxygen bonds.

In addition, hydrosilane can be used as a suitable reductant for the Ni-catalyzed reductive cleavage of aryl ethers, a transformation synthetically equivalent to hydrogenolysis. In 2010, Martin and co-worker [61] developed an efficient Ni-catalyzed protocol for the reductive cleavage of inert Ar–OR bonds with TMDSO (1, 1, 3, 3-tetramethyldisiloxane) (Figure 6.34). Shortly after, in 2011, the Tobisu group [60] described a related Ni-catalyzed procedure of aryl methyl ethers and aryl pivaloates using $(MeO)_2MeSiH$ as the reducing agent (Figure 6.34). It is noteworthy that almost no products from Ar–OMe bond cleavage were formed without the addition of reducing agents, which shows that the incorporated hydride is derived from the reducing agent, rather than from a methoxy group, via β-hydride elimination [103]. This is confirmed by a labeling experiment, in which a deuterated product was obtained exclusively by using $DSiEt_3$ as a reducing agent (Figure 6.34), which is also confirmed in Martin's research [61]. In addition, the superior reactivity of alkoxysilanes to trialkylsilanes in this catalysis can be attributed to their higher Lewis acidity, [104] which should facilitate a key σ-bond metathesis between nickel–methoxide A and hydrosilane by strengthening the silicon-oxygen interaction in B (Figure 6.34) [105].

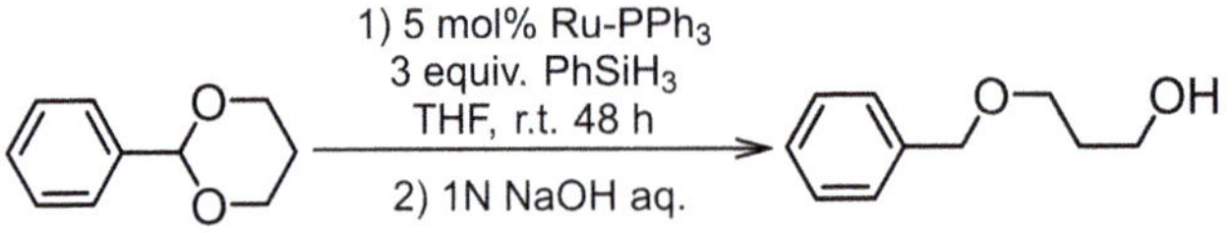

Figure 6.33 Rh-catalyzed Reductive Cleavage of the C–O Bond of Acetals and with Silane.

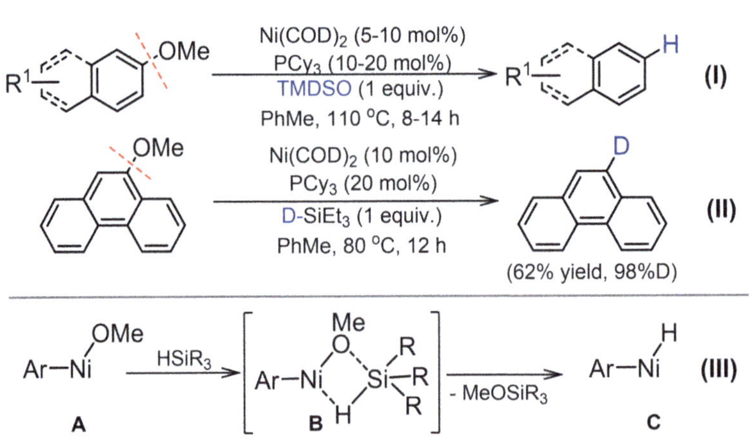

Figure 6.34 Ni-catalyzed Reductive Cleavage of Aryl ether with Hydrosilane. *Source:* Based on Li et al. [105].

Then, in 2013, the Martin group [106] presented detailed research about the mechanism with X-ray crystallography, isotopic-labeling experiments, kinetic isotope effects, and computational studies. Different from the classical Ni(0)/Ni(II) for C–OMe bond cleavage via the first oxidation addition of Ni(0) precatalyst into the C–OMe bond (Figure 6.35, Left), they confirmed that the mechanism is initiated by Ni(I)–$SiEt_3$ that are generated from $Ni(COD)_2$, PCy_3, and Et_3SiH via a comproportionation event. Then, the initially generated $L_2Ni(I)–SiEt_3$ coordinates to 2-methoxynaphthalene in a η^2-fashion, and after a migratory insertion event, a benzyl nickel species VII is generated (Figure 6.35, Right). Elimination of $MeOSiR_3$ with the concomitant migration of the Ni atom results in IX that undergoes a reversible σ-bond metathesis, delivering the final product naphthalene while generating back the Ni(I)–$SiEt_3$ species VI.

Then, in the same year 2013, Grubbs and co-workers [107] reported a transition-metal-free protocol, with a powerful reductive couple of triethylsilane and common bases KO^{t}Bu, for the efficient reductive cleavage of diaryl and aryl alkyl ethers (Figure 6.36). After carefully optimizing the reaction conditions, the competitive aromatic C–H silylation reaction accompanied with the cleavage of the C–OAr bond

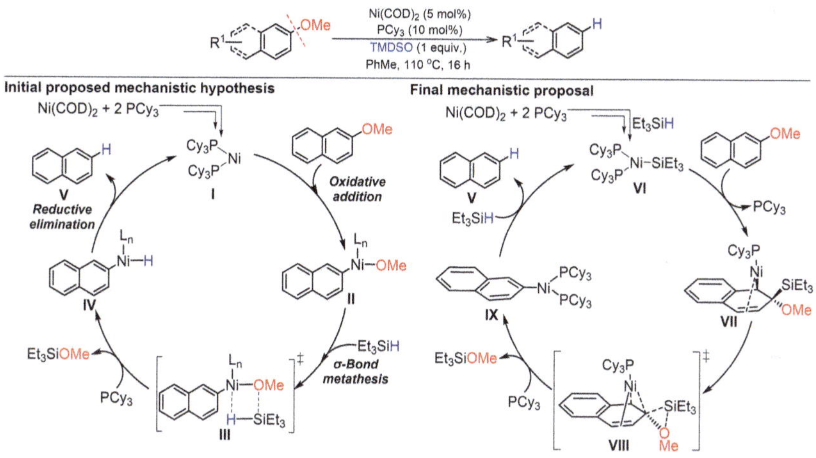

Figure 6.35 Ni-catalyzed Protocol for the Reductive Cleavage of Inert Ar–OR Bonds with Silane. *Source:* Cornella et al. [106] with Permission, © 2013 American Chemical Society.

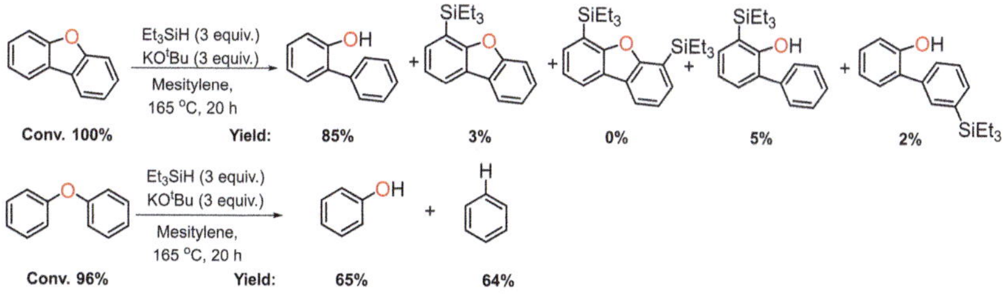

Figure 6.36 Reductive Cleavage of Aryl ethers with Silane and Base. *Source:* Reaction Data Abstracted from Fedorov et al. [107].

can be restrained, and the selective reductive cleavage of lignin 4-O-5 ether bonds leads to the generation of corresponding phenolic and aromatic constituents.

Besides the base promoting method, then, in 2014, Cantat and co-workers [108] further investigate the application of Et_3SiH, $Me_3Si(OSiMeH)_nOSiMe_3$ (PMHS), and $Me_2SiHOSiHMe_2$ (TMDS) silanes as reductants for the reductive cleavage of lignin α-O-4 and β-O-4 linkages with the Lewis acid. Under the mild condition, reaction with catalyst $B(C_6F_5)_3$ efficiently afforded products of primary alcohols and phenols (Figure 6.37a). Based on the control experiments, DFT calculation, and isotopic

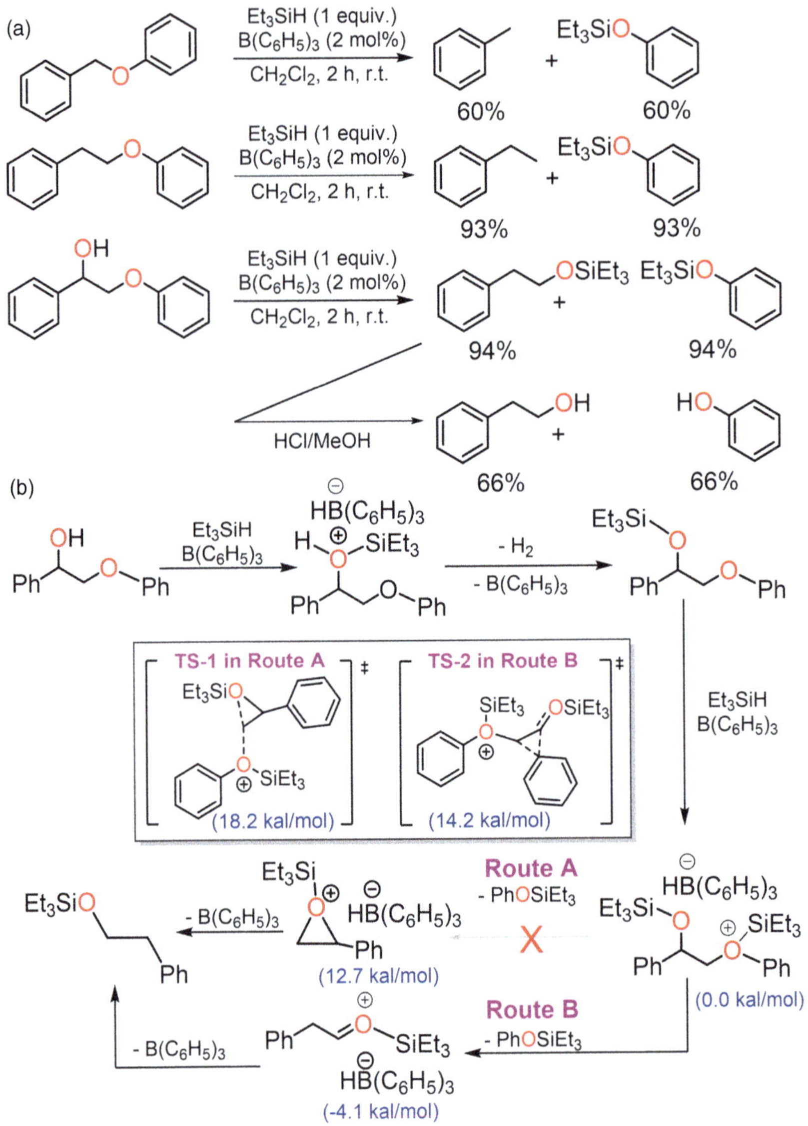

Figure 6.37 Reductive Cleavage of Lignin Model Compounds to Phenols and Primary Alcohols with Silane and $B(C_6F_5)_3$. Note: ΔG Values Are Listed in the Parentheses. *Source:* Feghali and Cantat [108] with Permission, © 2014 The Royal Society of Chemistry.

labeling studies, the author proposed a mechanism with a more favorite intermediate TS-2 than TS-1 (Figure 6.37b) to explain the phenomenon of semipinacol rearrangement, which means the positions change between C_α and C_β to produce primary alcohols when the β-O-4 substrate has the C_α–OH.

In the same year, the Brook group [109] reported that hydrosilanes could be used in the conversion of lignin aryl ethers to hydrolyzable silyl ethers in the presence of $B(C_6F_5)_3$. The relative reactivity of functional groups on model lignin compounds was found to be phenol > primary alcohol > methoxybenzene > alkyl silyl ethers. The process could lead to the cleavage of β-O-4, α-O-4, and methoxybenzene groups with concomitant silylation of phenolic and secondary alcohol groups, but no reaction was obtained with 4-O-5 linkage (Figure 6.38). In addition, over 95% of hardwood lignin could be efficiently reduced/degraded into organosoluble products by the monofunctional hydrosilane $HMe_2SiOSiMe_3$ and $B(C_6F_5)_3$ at 50 °C. The molecular weight of the silylated products was consistent with oligomeric structures comprised of 3–8 linked aryl groups.

Besides using as the nucleophile, in 2017, the Samec group [110] reported that the polymethylhydrosiloxane (PMHS) and water could work as the hydrogen sources in the hydrogenolysis of C–O bonds of lignin dibenzodioxocin motif with Pd/C catalyst (Figure 6.39). The efficiency of the reaction is highly dependent on both water concentration and the addition of a base. The mechanistic studies showed that the benzylic C–O bond was cleaved faster than the terminal C–O bond, which only cleaves upon the presence of the neighboring phenol. They proposed a hydrogen bond formation between an oxygen atom of an ether group and a proton of a neighboring phenol under the employed mild reaction conditions, which facilitated cleavage of the C–O bond.

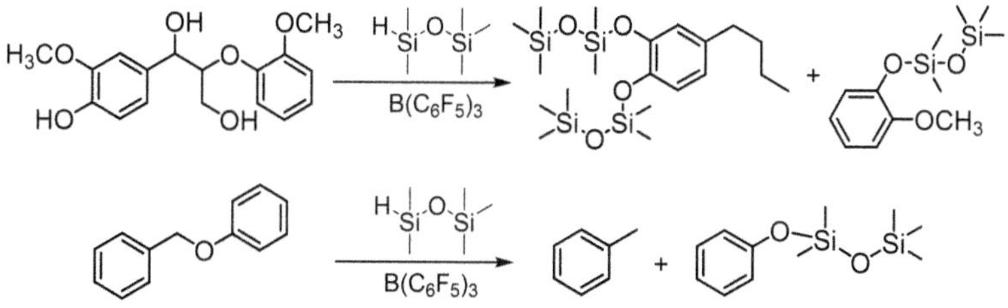

Figure 6.38 Reductive Degradation of Lignin and Model Compounds by Hydrosilane and $B(C_6F_5)_3$.

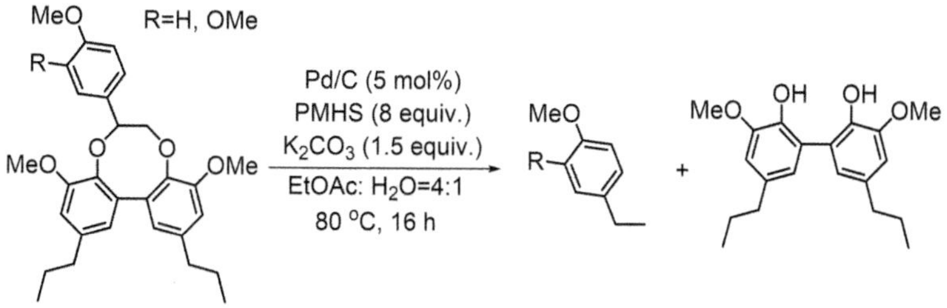

Figure 6.39 Pd-catalyzed Hydrogenolysis of Dibenzodioxocin Lignin Model Compounds with Silanes and Water.

6.5 Direct Reductive Cleavage of Lignin Ether Bond with e^-

Besides the methods with the H_2 and hydride as the reductive cleavage reagents, the electron with high reducibility can also cleave the stable lignin ether bonds. In 2017, Rothenberg and co-workers [111] showed a reductive depolymerization study of various lignins with solvated electrons at room temperature by adding sodium metal to the liquid ammonia without any external hydrogen donor, which obtained 10-40 % yields of bio-oil and higher yields of alkylphenols. However, the use of solvated electrons from sodium metal is not convenient. Compared with solvated electrons, electrocatalytic hydrogenolysis or reduction with an electron at the cathode should be a more accepted approach [112, 113].

In 1983, Stromskaya and co-workers [114] study the electrochemical reduction of lignin in liquid ammonia. Given the inconvenience of the liquid NH_3, the following electrocatalytic methods usually use a greener electrolyte. In 1997, Lessard and co-workers [115] investigated the electrocatalytic hydrogenolysis (ECH) of lignin β-O-4 and α-O-4 models without C_α–OH under galvanostatic control at Raney nickel electrodes in aqueous ethanol. The efficiency of hydrogenolysis of the C–O bond can be optimized to 100% by choosing the proper substrate concentration, current density, and temperature. Under the optimum conditions, further hydrogenation of the products formed (phenol to cyclohexanol, 2-methoxy phenol to phenol) occurred to a small extent (6 to 8%).

In 2000, Cyr and co-workers [116] carried out the electrocatalytic hydrogenation (ECH) study of β-O-4 lignin model compounds at constant current with Raney nickel and palladium-based cathodes in aqueous sodium hydroxide solution at temperatures ranging from 25 to 75°C. The ECH of 1-(4-hydroxy-3-methoxyphenyl)-2-(2-methoxyphenoxy)-1-ethanol at Raney nickel gives guaiacol, acetovanillone, α-methylvanillyl alcohol, and products derived from their further hydrogenation. At palladium, the hydrogenolysis of the same β-O-4 lignin model required only half of the theoretical amount of electricity due to the generation of chemisorbed hydrogen by dehydrogenation at C_α–OH followed by hydrogenolysis of the ether bond in the resulting ketone.

In 2014, Huang and co-workers [117] reported that the electrochemical reductive cleavage of diaryl ether C–O bonds, 4-O-5 lignin linkage, could be promoted by $NaBH_4$. Operating in an undivided cell with two electrodes system and an air-tolerable reaction condition, the reaction occurred highly selectively with a minimum further reduction on the aryl rings. According to their results, a one-electron mechanism (Figure 6.40) might be considered as the major path for the electrochemical reductive cleavage of diaryl ether bond. The diaryl ether model first transforms to an anion species with the electron from the cathode, which further yields diaryl ether bond cleavage to a phenol anion and a phenyl radical. The phenyl radical prefers to get a hydrogen atom from solvent or $NaBH_4$ to generate benzene rather than occurs further reduction of the phenyl radical to the anion. For the promotion effect of $NaBH_4$, besides preventing the oxidation of phenol products by the anode, $NaBH_4$ contributed about 20% of the protons on the product of benzene. In addition,

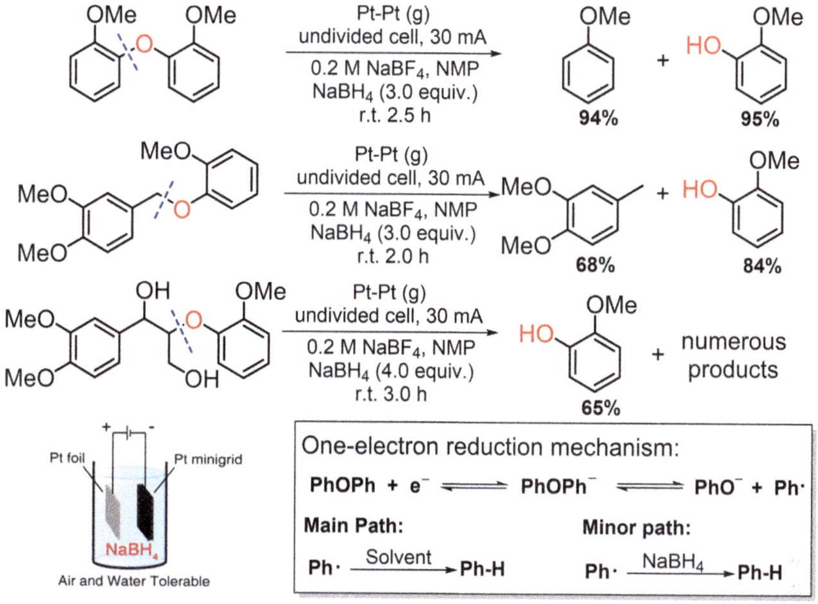

Figure 6.40 Electrochemical Cleavage of Aryl Ethers Promoted by Sodium Borohydride. *Source:* Reaction Data Abstracted from Wu and Huang [117].

this reaction system was also workable on the cleavage of alkyl C–OAr bonds in α-O-4 and β-O-4 model compounds of lignin.

Besides the electron from electrocatalysis at the cathode [118], the photo-induced electron from activated homogeneous photo- photosensitizer/photocatalysts or at conduction band of the heterogeneous semiconductors could also be used in aryl ether bond reductive cleavage. While the existence of neighboring groups such as the various hydroxyl groups and these catalytic systems with both indiscrete oxidation and reduction centers will make the mechanism different from the direct ether bond cleavage. As shown in Figure 6.40, the existence of hydroxy groups in the β-O-4 linkage can make the conversion process more complex than the direct ether bond cleavage, which is different from the electrocatalytic cleavage of lignin 4-O-5, α-O-4 ether models [119] and ArO–Me bonds cleavage to the phenolic monomers observed after lignin pyrolysis [120, 121]. The corresponding researches for the electrocatalytic ether bonds reductive cleavage and lignin C–C bonds oxidized cleavage will be introduced in the following chapters.

6.6 Chapter Summary

For lignin depolymerization, the direct cleavage of C_β–OAr, C_α–OAr or ArO–Ar ether bonds, and C–Ar bonds provide the simplest mode to tailor the β-O-4 linkage and other lignin linkages. In this connection, some stoichiometric reaction and their derived catalytic systems have been founded for the direct cleavage of lignin linkages, but these have focused primarily on the C–O bonds rather than the cleavage of the C–C bonds. In this chapter, we first review the stoichiometric reactions and catalytic systems for the direct cleavage of ether bonds with a strong nucleophile and electrophilic reagent. Then, as the modification and update of the above methods, Brönsted/Lewis acid + metal systems, base/organometallic systems, and other heterogeneous catalysts used in the direct hydrogenative cleavage of C–OAr ether bonds are systemically introduced. In addition, for the direct reductive cleavage of ether bond, this chapter also gives a brief review of the lignin conversion methods with hydride reagents, and e^- species from chemical conversion or surface of the cathode or semiconductors. The keywords for this chapter could be: Lignin depolymerization, Ether bond cleavage, C–C bond cleavage, Hydrogenative cleavage, Reductive cleavage

References

1 Maercker, A. (1987). Ether cleavage with organo-alkali-metal compounds and alkali-metals. *Angew. Chem. Int. Ed.* 26: 972–989.

2 Greene, T.W., Wuts, P.G., and Wiley, J. (1999). *Protective Groups in Organic Synthesis*, vol. 168. New York: Wiley.

3 Rolando, C., Monties, B., and Lapierre, C. (1992). Thioacidolysis. In: *Methods in Lignin Chemistry* (ed. S.Y. Lin and C.W. Dence), 334–349. Berlin, Heidelberg: Springer Berlin Heidelberg.

4 Mimero, P., Saluzzo, C., and Amouroux, R. (1994). Regiocontrolled ring-opening of 2-methyltetrahydrofuran with acid-chlorides and iodides. *Tetrahedron Lett.* 35: 1553–1556.

5 Ganem, B. and Small, V.R. (1974). Ferric chloride in acetic anhydride. A mild and versatile reagent for the cleavage of ethers. *J. Org. Chem.* 39: 3728–3730.

6 Oriyama, T., Kimura, M., Oda, M., and Koga, G. (1993). Tin(II) bromide-acetyl bromide-A mild and efficient reagent for direct conversion of alcohol benzyl ethers into acetates. *Synlett* 437–440.

7 Ahmad, S. and Iqbal, J. (1987). Cobalt(II) chloride catalyzed acylative cleavage of ethers. *Chem. Lett.* 953–956.

8 Iqbal, J., Khan, M.A., and Srivastava, R.R. (1988). Cobalt catalyzed regioselective cleavage of oxiranes with acylchlorides. *Tetrahedron Lett.* 29: 4985–4986.

9 Iqbal, J. and Srivastava, R.R. (1991). Cobalt(II) chloride catalyzed cleavage of ethers with acyl halides-Scope and mechanism. *Tetrahedron* 47: 3155–3170.

10 Green, L., Hemeon, I., and Singer, R.D. (2000). 1-Ethyl-3-methylimidazolium halogenoaluminate ionic liquids as reaction media for the acylative cleavage of ethers. *Tetrahedron Lett.* 41: 1343–1346.

11 Qian, C.T., Qiu, A.E., Huang, Y.Z., and Chen, W.J. (1991). Study on organolanthanide complexes. 37. Reaction of dicyclopentadienylyttrium chloride with acyl Chlorides in tetrahydrofuran-acylative cleavage of the Cp-Y-pi-bond and tetrahydrofuran ring. *J. Organomet. Chem.* 412: 53–59.

12 Qian, C.T., Qiu, A.N., Zhu, D.M., and Yang, X.W. (1994). Regioselective acylative cleavage of cyclic ethers catalyzed by rare-earth compounds. *J. Mol. Catal.* 87: 357–359.
13 Tian, J., Yue, H., Yang, P., and Sang, D. (2019). One-pot cleavage of aryl alkyl ethers by aluminum and iodine in acetonitrile. *ChemistrySelect* 4: 38–41.
14 Luzzio, F.A. and Bobb, R.A. (1999). Organomercury/aluminum-mediated acylative cleavage of cyclic ethers. *Tetrahedron* 55: 1851–1858.
15 Bhar, S. and Ranu, B.C. (1995). Zinc-promoted selective cleavage of ethers in presence of acyl chloride. *J. Org. Chem.* 60: 745–747.
16 Suzuki, Y., Matsushima, M., and Kodomari, M. (1998). Graphite-catalyzed acylative cleavage of ethers with acyl halides. *Chem. Lett.* 319–320.
17 Alper, H. and Huang, C.C. (1973). Group VI metal-carbonyl catalyzed reaction of ethers and acid halides. *J. Org. Chem.* 38: 64–71.
18 Tsonis, C.P. (1988). Cleavage of ethers with benzoyl chloride catalyzed by arenetricarbon-metal complexes. *J. Mol. Catal.* 45: 145–150.
19 Guo, Q.X., Miyaji, T., Gao, G.H. et al. (2001). Catalytic C—O bond cleavage of ethers using group 5 or 6 metal halide/acid chloride systems. *Chem. Commun.* 1018–1019.
20 Ganem, B., Vernon, R., and Small, J. (1974). Ferric chloride in acetic anhydride. A mild and versatile reagent for the cleavage of ethers. *J. Org. Chem.* 39: 3728–3730.
21 Trost, B.M. (1991). The atom economy-A search for synthetic efficiency. *Science* 254: 1471–1477.
22 Trost, B.M. (1995). Atom economy -A challenge for organic-synthesis: Homogeneous catalysis leads the way. *Angew. Chem. Int. Ed.* 34: 259–281.
23 Lohr, T.L., Li, Z., and Marks, T.J. (2016). Thermodynamic strategies for C—O Bond formation and cleavage via tandem catalysis. *Acc. Chem. Res.* 49: 824–834.
24 Atesin, A.C., Ray, N.A., Stair, P.C., and Marks, T.J. (2012). Etheric C—O bond hydrogenolysis using a tandem lanthanide triflate/supported palladium nanoparticle catalyst system. *J. Am. Chem. Soc.* 134: 14682–14685.
25 Assary, R.S., Atesin, A.C., Li, Z. et al. (2013). Reaction pathways and energetics of etheric C—O bond cleavage catalyzed by lanthanide triflates. *ACS Catal.* 3: 1908–1914.
26 Li, Z., Assary, R.S., Atesin, A.C. et al. (2013). Rapid ether and alcohol C–O bond hydrogenolysis catalyzed by tandem high-valent metal triflate + supported Pd catalysts. *J. Am. Chem. Soc.* 136: 104–107.
27 Dzudza, A. and Marks, T.J. (2010). Efficient intramolecular hydroalkoxylation of unactivated alkenols mediated by recyclable lanthanide triflate ionic liquids: Scope and mechanism. *Chem. Eur. J.* 16: 3403–3422.
28 Dzudza, A. and Marks, T.J. (2009). Efficient intramolecular hydroalkoxylation/cyclization of unactivated alkenols mediated by lanthanide triflate ionic liquids. *Org. Lett.* 11: 1523–1526.
29 Lohr, T.L., Li, Z., and Marks, T.J. (2015). Selective Ether/Ester C—O Cleavage of an Acetylated Lignin Model via Tandem Catalysis. *ACS Catal.* 5: 7004–7007.
30 Huang, X., Zhu, J., Koranyi, T.I. et al. (2016). Effective release of lignin fragments from lignocellulose by Lewis acid metal triflates in the lignin-first approach. *ChemSusChem* 9: 3262–3267.

31 Huang, X.M., Gonzalez, O.M.M., Zhu, J.D. et al. (2017). Reductive fractionation of woody biomass into lignin monomers and cellulose by tandem metal triflate and Pd/C catalysis. *Green Chem.* 19: 175–187.
32 Cui, X.J., Yuan, H.K., Junge, K. et al. (2017). A stable and practical nickel catalyst for the hydrogenolysis of C—O bonds. *Green Chem.* 19: 305–310.
33 Laskar, D.D., Tucker, M.P., Chen, X. et al. (2014). Noble-metal catalyzed hydrodeoxygenation of biomass-derived lignin to aromatic hydrocarbons. *Green Chem.* 16: 897–910.
34 Lazaridis, P.A., Fotopoulos, A.P., Karakoulia, S.A., and Triantafyllidis, K.S. (2018). Catalytic fast pyrolysis of kraft lignin with conventional, mesoporous and nanosized ZSM-5 zeolite folr the production of alkyl-phenols and aromatics. *Front. Chem.* 6: 295.
35 Kloekhorst, A. and Heeres, H.J. (2015). Catalytic hydrotreatment of Alcell ignin using supported Ru, Pd, and Cu catalysts. *ACS Sustainable Chem. Eng.* 3: 1905–1914.
36 Xia, Q.N., Chen, Z.J., Shao, Y. et al. (2016). Direct hydrodeoxygenation of raw woody biomass into liquid alkanes. *Nat. Commun.* 7: 11162.
37 Shao, Y., Xia, Q., Dong, L. et al. (2017). Selective production of arenes via direct lignin upgrading over a niobium-based catalyst. *Nat. Commun.* 8: 16104.
38 Dong, L., Lin, L.F., Han, X. et al. (2019). Breaking the limit of lignin monomer production via cleavage of interunit carbon-carbon linkages. *Chem* 5: 1521–1536.
39 Dong, L., Xia, J., Guo, Y. et al. (2021). Mechanisms of $C_{aromatic}$—C bonds cleavage in lignin over NbO_x-supported Ru catalyst. *J. Catal.* 394: 94–103.
40 Dong, L., Yin, L.-L., Xia, Q. et al. (2018). Size-dependent catalytic performance of ruthenium nanoparticles in the hydrogenolysis of a β-O-4 lignin model compound. *Catal. Sci. Technol.* 8: 735–745.
41 Opris, C., Cojocaru, B., Gheorghe, N. et al. (2016). Lignin fragmentation over magnetically recyclable composite $Co@Nb_2O_5@Fe_3O_4$ catalysts. *J. Catal.* 339: 209–227.
42 Stavila, V., Parthasarathi, R., Davis, R.W. et al. (2016). MOF-based catalysts for selective hydrogenolysis of carbon-oxygen ether bonds. *ACS Catal.* 6: 55–59.
43 Sumida, K., Brown, C.M., Herm, Z.R. et al. (2011). Hydrogen storage properties and neutron scattering studies of Mg-2(dobdc)-a metal-organic framework with open Mg^{2+} adsorption sites. *Chem. Commun.* 47: 1157–1159.
44 Lee, K., Howe, J.D., Lin, L.C. et al. (2015). Small-molecule adsorption in open-site metal–organic frameworks: A systematic density functional theory study for rational design. *Chem. Mater.* 27: 668–678.
45 Sergeev, A.G. and Hartwig, J.F. (2011). Selective, Nickel-catalyzed hydrogenolysis of aryl ethers. *Science* 332: 439–443.
46 Kelley, P., Lin, S., Edouard, G. et al. (2012). Nickel-mediated hydrogenolysis of C—O bonds of aryl ethers: What is the source of the hydrogen? *J. Am. Chem. Soc.* 134: 5480–5483.
47 Sawatlon, B., Wititsuwannakul, T., Tantirungrotechai, Y., and Surawatanawong, P. (2014). Mechanism of Ni N-heterocyclic carbene catalyst for C—O bond

hydrogenolysis of diphenyl ether: a density functional study. *Dalton Trans.* 43: 18123–18133.

48 Wititsuwannakul, T., Tantirungrotechai, Y., and Surawatanawong, P. (2016). Density functional study of nickel N-heterocyclic carbene catalyzed C—O bond hydrogenolysis of methyl phenyl ether: The concerted β-H transfer mechanism. *ACS Catal.* 6: 1477–1486.

49 Xu, L.P., Chung, L.W., and Wu, Y.D. (2016). Mechanism of Ni-NHC catalyzed hydrogenolysis of aryl ethers: Roles of the excess base. *ACS Catal.* 6: 483–493.

50 Tobisu, M., Morioka, T., Ohtsuki, A., and Chatani, N. (2015). Nickel-catalyzed reductive cleavage of aryl alkyl ethers to arenes in absence of external reductant. *Chem. Sci.* 6: 3410–3414.

51 Tobisu, M. and Chatani, N. (2011). Catalytic hydrogenolysis of C—O bonds in aryl ethers. *ChemCatChem* 3: 1410–1411.

52 Wenkert, E., Michelotti, E.L., and Swindell, C.S. (1979). Nickel-induced conversion of carbon-oxygen into carbon-carbon bonds - One-step transformations of enol ethers into olefins and aryl ethers into biaryls. *J. Am. Chem. Soc.* 101: 2246–2247.

53 Dankwardt, J.W. (2004). Nickel-catalyzed cross-coupling of aryl Grignard reagents with aromatic alkyl ethers: An efficient synthesis of unsymmetrical biaryls. *Angew. Chem. Int. Ed.* 43: 2428–2432.

54 Guan, B.T., Xiang, S.K., Wu, T. et al. (2008). Methylation of arenes via Ni-catalyzed aryl C—O/F activation. *Chem. Commun.* 1437–1439.

55 Xie, L.G. and Wang, Z.X. (2011). Cross-coupling of aryl/alkenyl ethers with aryl Grignard reagents through nickel-catalyzed C—O activation. *Chem. Eur. J.* 17: 4972–4975.

56 Wang, C., Ozaki, T., Takita, R., and Uchiyama, M. (2012). Aryl ether as a Negishi coupling prtner: An approach for constructing C—C bonds under mild conditions. *Chem. Eur. J.* 18: 3482–3485.

57 Tobisu, M., Shimasaki, T., and Chatani, N. (2008). Nickel-catalyzed cross-coupling of aryl methyl ethers with aryl boronic esters. *Angew. Chem. Int. Ed.* 47: 4866–4869.

58 Kakiuchi, F., Usui, M., Ueno, S. et al. (2004). Ruthenium-catalyzed functionalization of aryl carbon-oxygen bonds in aromatic ethers with organoboron compounds. *J. Am. Chem. Soc.* 126: 2706–2707.

59 Tobisu, M., Shimasaki, T., and Chatani, N. (2009). Ni^0-catalyzed direct amination of anisoles involving the cleavage of carbon-oxygen bonds. *Chem. Lett.* 38: 710–711.

60 Tobisu, M., Yamakawa, K., Shimasaki, T., and Chatani, N. (2011). Nickel-catalyzed reductive cleavage of aryl-oxygen bonds in alkoxy- and pivaloxyarenes using hydrosilanes as a mild reducing agent. *Chem. Commun.* 47: 2946–2948.

61 Alvarez-Bercedo, P. and Martin, R. (2010). Ni-catalyzed reduction of inert C—O bonds: A new strategy for using aryl ethers as easily removable directing groups. *J. Am. Chem. Soc.* 132: 17352–17353.

62 Sergeev, A.G., Webb, J.D., and Hartwig, J.F. (2012). A heterogeneous nickel catalyst for the hydrogenolysis of aryl ethers without arene hydrogenation. *J. Am. Chem. Soc.* 134: 20226–20229.

63 He, J., Zhao, C., and Lercher, J.A. (2012). Ni-catalyzed cleavage of aryl ethers in the aqueous phase. *J. Am. Chem. Soc.* 134: 20768–20775.

64 He, J.Y., Zhao, C., Mei, D.H., and Lercher, J.A. (2014). Mechanisms of selective cleavage of C—O bonds in di-aryl ethers in aqueous phase. *J. Catal.* 309: 280–290.

65 Kasakov, S., Shi, H., Camaioni, D.M. et al. (2015). Reductive deconstruction of organosolv lignin catalyzed by zeolite supported nickel nanoparticles. *Green Chem.* 17: 5079–5090.

66 Molinari, V., Clavel, G., Graglia, M. et al. (2016). Mild continuous hydrogenolysis of kraft lignin over titanium nitride-nickel catalyst. *ACS Catal.* 6: 1663–1670.

67 Molinari, V., Giordano, C., Antonietti, M., and Esposito, D. (2014). Titanium nitride-nickel nanocomposite as heterogeneous catalyst for the hydrogenolysis of aryl ethers. *J. Am. Chem. Soc.* 136: 1758–1761.

68 Li, C.Z., Zheng, M.Y., Wang, A.Q., and Zhang, T. (2012). One-pot catalytic hydrocracking of raw woody biomass into chemicals over supported carbide catalysts: simultaneous conversion of cellulose, hemicellulose and lignin. *Energy Environ. Sci.* 5: 6383–6390.

69 Song, Q., Wang, F., Cai, J. et al. (2013). Lignin depolymerization (LDP) in alcohol over nickel-based catalysts via a fragmentation–hydrogenolysis process. *Energy Environ. Sci.* 6: 994–1007.

70 Song, Q., Cai, J., Zhang, J. et al. (2013). Hydrogenation and cleavage of the C—O bonds in the lignin model compound phenethyl phenyl ether over a nickel-based catalyst. *Chin. J. Catal.* 34: 651–658.

71 Song, Q., Wang, F., and Xu, J. (2012). Hydrogenolysis of lignosulfonate into phenols over heterogeneous nickel catalysts. *Chem. Commun.* 48: 7019–7021.

72 Klein, I., Saha, B., and Abu-Omar, M.M. (2015). Lignin depolymerization over Ni/C catalyst in methanol, a continuation: effect of substrate and catalyst loading. *Catal. Sci. Technol.* 5: 3242–3245.

73 Gao, F., Webb, J.D., and Hartwig, J.F. (2016). Chemo- and regioselective hydrogenolysis of diaryl ether C—O Bonds by a robust heterogeneous Ni/C catalyst: Applications to the cleavage of complex lignin-related fragments. *Angew. Chem. Int. Ed.* 55: 1474–1478.

74 Ma, R., Hao, W., Ma, X. et al. (2014). Catalytic ethanolysis of kraft lignin into high-value small-molecular chemicals over a nanostructured α-molybdenum carbide catalyst. *Angew. Chem. Int. Ed.* 53: 7310–7315.

75 Cao, Z.W., Engelhardt, J., Dierks, M. et al. (2017). Catalysis meets nonthermal separation for the production of (alkyl)phenols and hydrocarbons from pyrolysis oil. *Angew. Chem. Int. Ed.* 56: 2334–2339.

76 Wang, Y.-Y., Ling, L.-L., and Jiang, H. (2016). Selective hydrogenation of lignin to produce chemical commodities by using a biochar supported Ni–Mo_2C catalyst obtained from biomass. *Green Chem.* 18: 4032–4041.

77 Wu, H., Song, J., Xie, C. et al. (2018). Efficient and mild transfer hydrogenolytic cleavage of aromatic ether bonds in lignin-derived compounds over Ru/C. *ACS Sustainable Chem. Eng.* 6: 2872–2877.

78 Chen, M.Y., Huang, Y.B., Pang, H. et al. (2015). Hydrodeoxygenation of lignin-derived phenols into alkanes over carbon nanotube supported Ru catalysts in biphasic systems. *Green Chem.* 17: 1710–1717.

79 Chatterjee, M., Ishizaka, T., Suzuki, A., and Kawanami, H. (2013). An efficient cleavage of the aryl ether C—O bond in supercritical carbon dioxide-water. *Chem. Commun.* 49: 4567–4569.

80 Mao, J.B., Zhou, J.X., Xia, Z. et al. (2017). Anatase TiO_2 activated by gold nanoparticles for selective hydrodeoxygenation of guaiacol to phenolics. *ACS Catal.* 7: 695–705.

81 Huang, Y.B., Yan, L., Chen, M.Y. et al. (2015). Selective hydrogenolysis of phenols and phenyl ethers to arenes through direct C—O cleavage over ruthenium-tungsten bifunctional catalysts. *Green Chem.* 17: 3010–3017.

82 Meng, Q.L., Yan, J., Liu, H.Z. et al. (2019). Self-supported hydrogenolysis of aromatic ethers to arenes. *Sci. Adv.* 5.

83 Zhang, J., Teo, J., Chen, X. et al. (2014). A series of NiM (M = Ru, Rh, and Pd) bimetallic catalysts for effective lignin hydrogenolysis in water. *ACS Catal.* 4: 1574–1583.

84 Zhang, J., Asakura, H., van Rijn, J. et al. (2014). Highly efficient, NiAu-catalyzed hydrogenolysis of lignin into phenolic chemicals. *Green Chem.* 16: 2432–2437.

85 Konnerth, H., Zhang, J., Ma, D. et al. (2015). Base promoted hydrogenolysis of lignin model compounds and organosolv lignin over metal catalysts in water. *Chem. Eng. Sci.* 123: 155–163.

86 De, S., Zhang, J., Luque, R., and Yan, N. (2016). Ni-based bimetallic heterogeneous catalysts for energy and environmental applications. *Energy Environ. Sci.* 9: 3314–3347.

87 Bulut, S., Siankevich, S., van Muyden, A.P. et al. (2018). Efficient cleavage of aryl ether C—O linkages by Rh-Ni and Ru-Ni nanoscale catalysts operating in water. *Chem. Sci.* 9: 5530–5535.

88 Zhang, L., Wang, Y., Yang, Y. et al. (2021). Selective hydrogenolysis of aryl ether bond over Ru-Fe bimetallic catalyst. *Catal. Today* 365: 199–205.

89 Zhang, J.W., Cai, Y., Lu, G.P., and Cai, C. (2016). Facile and selective hydrogenolysis of β-O-4 linkages in lignin catalyzed by Pd-Ni bimetallic nanoparticles supported on ZrO_2. *Green Chem.* 18: 6229–6235.

90 Kim, J.K., Lee, J.K., Kang, K.H. et al. (2015). Selective cleavage of C—O bond in benzyl phenyl ether to aromatics over Pd-Fe bimetallic catalyst supported on ordered mesoporous carbon. *Appl. Catal., A* 498: 142–149.

91 Singh, A.K., Jang, S., Kim, J.Y. et al. (2015). One-pot defunctionalization of lignin-derived compounds by dual-functional $Pd_{50}Ag_{50}/Fe_3O_4$/N-rGO Catalyst. *ACS Catal.* 5: 6964–6972.

92 Ji, J., Guo, H., Li, C. et al. (2018). Tungsten-based bimetallic catalysts for selective cleavage of lignin C—O bonds. *ChemCatChem* 10: 415–421.

93 Zhai, Y.X., Li, C., Xu, G.Y. et al. (2017). Depolymerization of lignin via a non-precious Ni-Fe alloy catalyst supported on activated carbon. *Green Chem.* 19: 1895–1903.

94 Freeguard, G.F. and Long, L.H. (1964). A novel method for preparing organic iodides. *Chem Ind-London* 1582–1583.

95 Freeguard, G.F. and Long, L.H. (1965). Improved preparation of diborane. *Chem Ind-London* 471.

96 Long, L.H. and Freeguar., G. (1965). Low-temperature cleavage of ethers. *Nature* 207: 403–403.
97 Haenel, M.W., Narangerel, J., Richter, U.B., and Rufinska, A. (2006). The first liquefaction of high-rank bituminous coals by preceding hydrogenation with homogeneous borane or iodine catalysts. *Angew. Chem. Int. Ed.* 45: 1061–1066.
98 Nagy, M., David, K., Britovsek, G.J.P., and Ragauskas, A.J. (2009). Catalytic hydrogenolysis of ethanol organosolv lignin. *Holzforschung* 63: 513–520.
99 Chouhan, M., Kumar, K., Sharma, R. et al. (2013). $NiCl_2 \cdot 6H_2O/NaBH_4$ in methanol: a mild and efficient strategy for chemoselective deallylation/debenzylation of aryl ethers. *Tetrahedron Lett.* 54: 4540–4543.
100 Zeng, H., Cao, D., Qiu, Z., and Li, C.J. (2018). Palladium-catalyzed formal cross-coupling of diaryl ethers with amines: Slicing the 4-O-5 linkage in lignin models. *Angew. Chem. Int. Ed.* 57: 3752–3757.
101 Ren, Y.L., Yan, M.J., Wang, J.J. et al. (2013). Selective reductive cleavage of inert aryl C—O bonds by an iron catalyst. *Angew. Chem. Int. Ed.* 52: 12674–12678.
102 Ohta, T., Michibata, T., Yamada, K. et al. (2003). Reductive cleavage of the C–O bond of acetals and orthoesters: reduction by silane in the presence of a Rh-PPh_3 complex. *Chem. Commun.* 1192–1193.
103 van der Boom, M.E., Liou, S.Y., Ben-David, Y. et al. (1998). Alkyl- and aryl-oxygen bond activation in solution by rhodium(I), palladium(II), and nickel(II). Transition-metal-based selectivity. *J. Am. Chem. Soc.* 120: 6531–6541.
104 Orito, Y. and Nakajima, M. (2006). Lewis base catalyzed asymmetric reactions involving hypervalent silicate intermediates. *Synthesis-Stuttgart* 1391–1401.
105 Li, Z., Zhang, S.L., Fu, Y. et al. (2009). Mechanism of Ni-catalyzed selective C–O bond activation in cross-coupling of aryl esters. *J. Am. Chem. Soc.* 131: 8815–8823.
106 Cornella, J., Gomez-Bengoa, E., and Martin, R. (2013). Combined experimental and theoretical study on the reductive cleavage of inert C—O bonds with silanes: Ruling out a classical Ni(0)/Ni(II) catalytic couple and evidence for Ni(I) intermediates. *J. Am. Chem. Soc.* 135: 1997–2009.
107 Fedorov, A., Toutov, A.A., Swisher, N.A., and Grubbs, R.H. (2013). Lewis-base silane activation: from reductive cleavage of aryl ethers to selective ortho-silylation. *Chem. Sci.* 4: 1640–1645.
108 Feghali, E. and Cantat, T. (2014). Unprecedented organocatalytic reduction of lignin model compounds to phenols and primary alcohols using hydrosilanes. *Chem. Commun.* 50: 862–865.
109 Zhang, J., Chen, Y., and Brook, M.A. (2014). Reductive degradation of lignin and model compounds by hydrosilanes. *ACS Sustainable Chem. Eng.* 2: 1983–1991.
110 Subbotina, E., Galkin, M.V., and Samec, J.S.M. (2017). Pd/C-catalyzed hydrogenolysis of dibenzodioxocin lignin model compounds using silanes and water as hydrogen source. *ACS Sustainable Chem. Eng.* 5: 3726–3731.
111 Prinsen, P., Narani, A., and Rothenberg, G. (2017). Lignin depolymerisation and lignocellulose fractionation by solvated electrons in liquid ammonia. *ChemSusChem* 10: 1022–1032.
112 Lam, C.H., Lowe, C.B., Li, Z. et al. (2015). Electrocatalytic upgrading of model lignin monomers with earth abundant metal electrodes. *Green Chem.* 17: 601–609.

113 Garedew, M., Lin, F., Song, B. et al. (2020). Greener Routes to biomass waste valorization: Lignin transformation through electrocatalysis for renewable chemicals and fuels production. *ChemSusChem* 13: 4214–4237.

114 Stromskaya, G.I. and Chupka, E.I. (1983). Electrochemical reduction of lignin in liquid ammonia. 1. Change in the chromophore and molecularweight composition during electrochemical reduction in liquid ammonia. *Koksnes Kimija* 2: 60–63.

115 Mahdavi, B., Lafrance, A., Martel, A. et al. (1997). Electrocatalytic hydrogenolysis of lignin model dimers at Raney nickel electrodes. *J. Appl. Electrochem.* 27: 605–611.

116 Cyr, A., Chiltz, F., Jeanson, P. et al. (2000). Electrocatalytic hydrogenation of lignin models at Raney nickel and palladium-based electrodes. *Can. J. Chem.* 78: 307–315.

117 Wu, W.B. and Huang, J.M. (2014). Electrochemical cleavage of aryl ethers promoted by sodium borohydride. *J. Org. Chem.* 79: 10189–10195.

118 Zhou, Y.T., Klinger, G.E., Hegg, E.L. et al. (2020). Multiple mechanisms mapped in aryl alkyl ether cleavage via aqueous electrocatalytic hydrogenation over skeletal nickel. *J. Am. Chem. Soc.* 142: 4037–4050.

119 Garedew, M., Young-Farhat, D., Bhatia, S. et al. (2020). Electrocatalytic cleavage of lignin model dimers using ruthenium supported on activated carbon cloth. *Sustain. Energy Fuels* 4: 1340–1350.

120 Lam, C.H., Das, S., Erickson, N.C. et al. (2017). Towards sustainable hydrocarbon fuels with biomass fast pyrolysis oil and electrocatalytic upgrading. *Sustain. Energy Fuels* 1: 258–266.

121 Garedew, M., Young-Farhat, D., Jackson, J.E., and Saffron, C.M. (2019). Electrocatalytic upgrading of phenolic compounds observed after lignin pyrolysis. *ACS Sustainable Chem. Eng.* 7: 8375–8386.

7

Lignin C–C/C–O Bonds Cleavage via First Phenolic Hydroxyl Group Dehydrogenation or First Aromatic Rings Activation

7.1 Lignin C_{Ar}–C_{α}/C_{α}–C_{β} Bonds Cleavage after the First Phenolic Hydroxyl Group Dehydrogenation to the Phenolic Radical

As discussed in Chapter 2, phenolic hydroxyl group dehydrogenation in sinapyl alcohol (S-unit), coniferyl alcohol (G-unit), and *p*-coumaryl alcohol (H-unit) generate phenoxyl radicals and their resonant carbon radicals (Figure 7.1a), which further combine and polymerize to the 3D lignin structure with C–C/C–O linkage generation. For lignin depolymerization to aromatic monomers, the critical issue is to efficiently cleave the C–C/C–O linkage bonds. For establishing the lignin depolymerization methods/strategies, referring to the proverb that *the trouble should end it*, introducing/generating active radicals in lignin polymer structure should be a promising thought. Given the fact that the lignin phenolic group can be easily transformed to the O-centered radical which can easily resonate into various C-centered radicals and carbon-centered organic radicals which are unstable and high-active, the C-centered radicals can further work as the substrate sites to react with the ground triplet state $^3\Sigma_g^-O_2$ and other active O* ($^1\Delta_gO_2$, M-OO$^\bullet$, etc) for C–C bonds cleavage [1, 2], or the carbon radical after further oxidation and structure reforming can decompose to smaller species [3–5], the first transformation of lignin structure to the active phenoxyl radical intermediate could be a potential strategy for lignin decomposition (Figure 7.1b). In this subsection, the thermal systems (7.1.1), electro/photo-electro systems (7.1.2), and biodegradation systems (7.1.3) for the cleavage of lignin C_{Ar}–C_{α}/C_{α}–C_{β} bonds after the first phenolic hydroxyl group dehydrogenation to the phenolic radical will be summarized.

7.1.1 Thermal Systems via the Phenolic Radical

Besides some peroxidase [6] that can achieve the first hydrogen atom abstraction to afford the phenoxyl radical, the M(salen)/M(Schiff)-based homogeneous catalysts, especially M = Co [7–15], Mn [16], Cu [17], and V catalysts [18], can achieve the first

Lignin Conversion Catalysis: Transformation to Aromatic Chemicals, First Edition.
Chaofeng Zhang and Feng Wang.

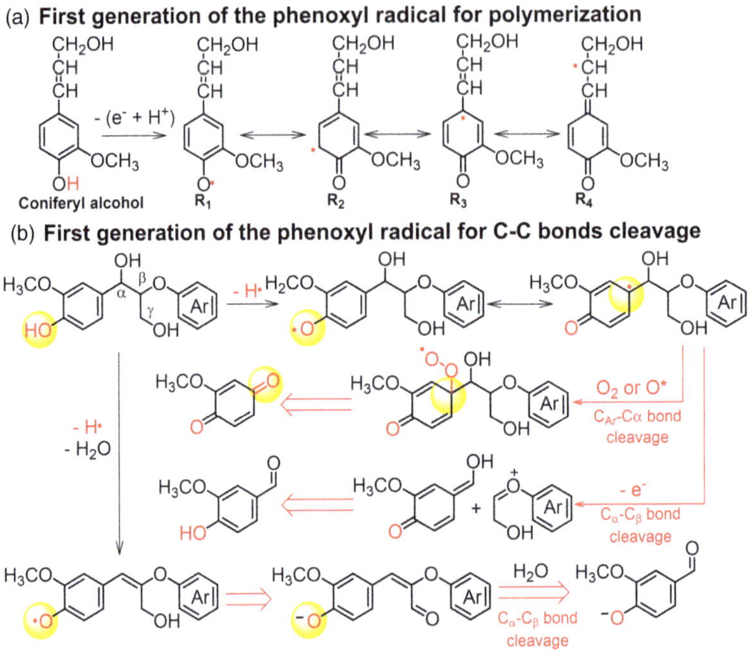

Figure 7.1 Phenoxyl Radical Induce the Cleavage of Lignin C–C Bonds.

dehydrogenation of the phenolic hydroxyl group at lignin model and lignin to phenolic radicals, which further induce the cleavage of lignin C–C bond.

Among these catalysts, cobalt salen ([Co(salen)]) complexes are used most frequently in lignin oxidative degradation research. In 1995, Bozel and co-workers [8] reported that the para-substituted phenolic, serving as models for lignin, could be oxidized to the corresponding benzoquinone with O_2 in the presence of catalytic amounts of Co-Schiff base complexes (Figure 7.2a). They observed the reactivity between the Co/O_2 complexes and para-substituted phenolics was strongly affected by the coordination about the cobalt. 4-Coordinate Co complexes (4C-Co(A)) binding oxygen poorly at room temperature in MeOH solution; the amount of superoxo complex (Salen)Co-OO$^{\bullet}$ present was normally low. In contrast, 5-coordinate complexes, such as 5C-Co(A) and 5C-Co(B), bind oxygen strongly. In addition, the presence of a donor ligand in the axial position of the complex can further stabilize the Co–O_2 bond for the following oxidation steps [19]. Treatment of syringyl alcohol with O_2 in the presence of 10 mol% of 5-coordinate catalysts 5C-Co(A) or 5C-Co(B) gave 2,6-dimethoxybenzoquinone with Ar–C bond cleavage in 71 and 88% yield, respectively. 4-Coordinate 4C-Co(A) only provides a 28% quinone yield, but a 50% aldehyde yield without Ar–C bond cleavage. When the substrate is changed from a species bearing two methoxy groups to one bearing a single methoxy group, little or no oxidation to quinone is observed in MeOH using either 4- or 5-coordinate Co catalysts. Bozel and co-workers further ruled out the possible aromatic radicals coupling reaction and confirmed that the failure of these substrates to undergo oxidation is not a function of the presence or absence of a substituent para to the hydroxyl

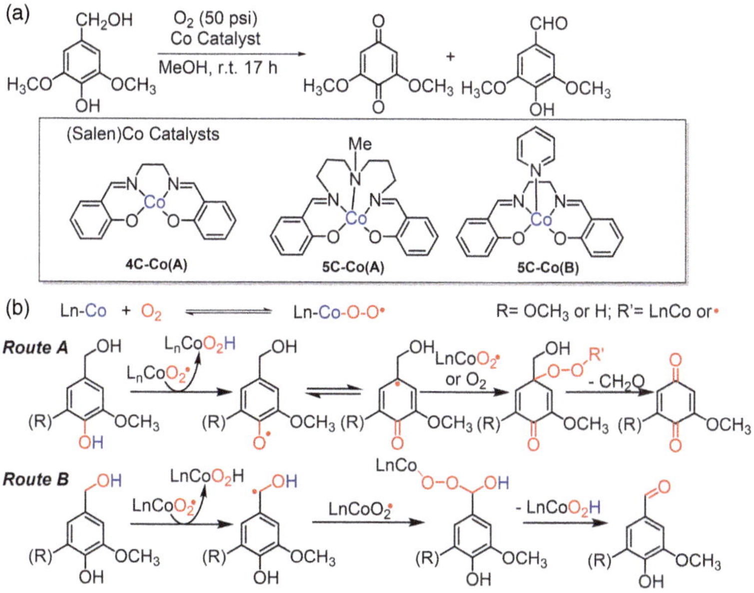

Figure 7.2 Catalytic Oxidation of Syringyl Alcohols to the Corresponding Benzoquinone with Co/O_2 System.

group. In general, electron-donating substituents on the ring promote the formation of phenoxy radicals while electron-withdrawing groups slow the reaction. In addition, electron-donating groups on the ring can lower the bond energy of the ArO–H bond. Therefore, the electronic nature of the aromatic ring and BDE of the ArO–H bond may have a significant bearing on the ability of the Co-catalyst system to convert the starting phenol into a phenoxy radical. As mentioned in their study, the irreversible oxidation of the catalyst is a primary deactivation pathway in many processes involving Co-Schiff base complexes [20]. The slow reaction rate of hydrogen abstraction from the phenolic hydroxyl group to a phenolic radical with Co-catalyst means more part of Co-Schiff catalyst can occur oxidation deactivation in MeOH, which further weakens the reaction performance. To overcome this difficulty, they further changed the solvent from MeOH to CH_2Cl_2 to weaken the hydrogen-bonding ability between the solvent and phenolic substrate, and added co-catalyst $CuCl_2$ or reagent like Frémy's salts to promote the generation of phenoxy radicals, results indicating that these methods could increase the reactivity. In the proposed mechanism (Figure 7.2b), the reaction of the Co complex with oxygen gives a superoxo (Salen)Co–OO$^{\bullet}$ adduct. This complex abstracts the phenolic hydrogen giving a phenoxy radical (Route A), whose mesomeric radical form with a C_1 radical is further trapped by a second molecule of the Co superoxo complex or O_2. Elimination of a molecule of formaldehyde from the intermediate generarated above generates quinone and, when $R' = L_nCo$, a Co-hydroxy species, known to be catalytically active. Alternatively, L_nCoOOH, formed in the first step of the reaction, breaks down to regenerate the starting catalyst. If the substrate is a nonphenolic compound such as veratryl alcohol, the abstraction of a hydrogen happens on the substitutional group to generate a benzyl radical which reacts with a second molecule of Co superoxo complex to form the final product veratryl aldehyde (Route B).

In 1996, Bozell and co-worker [9] further studied the oxidation difference among syringy, guaiacyl, and *p*-hydroxyl compounds with Co-Schiff base catalysts, which involved the molecular orbital calculations on phenolic substrates. The energies of the highest occupied molecular orbital (HOMO), the charge on the phenolic oxygen, and frontier molecular orbital coefficients on the phenolic hydrogen are consistently greater for the syringyl compounds. In addition, syringyl compounds have lower BDEs than the similar guaiacyl and *p*-hydroxyl compounds for hydrogen abstraction. Based on their results, it appears that the oxidation reaction is controlled by the dehydrogenation step, rather than a later attack of the superoxo Co(salen) compound on the free radical.

For the spectral information of the active Co-centered species, in 1996-1997, Canevali and co-workers [21, 22] presented a detailed oxidation study of propenoidic phenols catalyzed by Cobalt(II)-Salen catalysts. The reaction of methyl *E*-4-hydroxy-3-methoxycinnamate (*E*-methyl ferulate) selectively gave the corresponding 4-hydroxy-3-methoxybenzoic acid and 4-hydroxy-3-methoxy benzaldehyde in chloroform. Their further EPR and electronic spectra investigations suggest that organometallic radical acts as the intermediate active species. When chloroform solutions of Co(salen) and E-methyl ferulate were exposed to oxygen and left for 15 min at 25°C under 10 bar pressure, the EPR spectrum showed a new radical species which has eight resonance lines centered at g = 2.0028, due to the hyperfine interaction with ^{59}Co nucleus (I = 7/2) of 10 Gauss. The value of the Co hyperfine interaction indicates that the unpaired electron is mainly located on the ligand and that another nucleus, probably ^{1}H (I = 1/2), interacts with it. Their EPR investigation of the reaction mixtures, performed on samples taken at different reaction times, demonstrated that the most probable mechanism involves reactions (i)–(iii) [22].

$$[\mathrm{Co^{II}(salen)}] + \mathrm{ArOH} + \mathrm{O_2} \rightleftharpoons [\mathrm{Co^{III}(salen)(ArOH)(O_2)}]^- \tag{i}$$

$$[\mathrm{Co^{III}(salen)(ArOH)(O_2)}]^- + \mathrm{ROH} \rightleftharpoons [\mathrm{Co^{III}(salen)(ArOH)(ArO^\bullet)}] + \mathrm{HO_2^-} \tag{ii}$$

$$[\mathrm{Co^{III}(salen)(ArOH)(ArO^\bullet)}] + \mathrm{HO_2^-} \rightleftharpoons [\mathrm{Co^{III}(salen)(ArO^-)(ArO^\bullet)}] + \mathrm{H_2O_2} \tag{iii}$$

The superoxocobalt radical, [Co^{II}I(salen)(ArOH)(O2)]$^-$, and the phenoxy cobalt radical, [Co^{II}I(salen)(ArO$^-$)(ArO$^\bullet$)], are the EPR-active species; ArO$^\bullet$ is suggested easily to dissociate from [Co^{II}I(salen)(ArO$^-$)(ArO$^\bullet$)] and, reacting with O_2 at the β-carbon of the allyl substituent, probably gives a dioxethane, which can decompose to aldehyde and acid. In addition, the phenol electron-donor properties promote the formation of the superoxocobalt derivative and consequently favor the formation of the phenoxy cobalt radical; too high stability of this radical does not favor its evolution into oxidation products. In order to confirm their previously hypothesized mechanism, in 2002, Canevali and co-workers [23] presented a further oxidative degradation study of lignin related model compounds, including apocynol, various arylglycerol β-aryl ethers (β-O-4), and phenylcoumarans (α-O-4 and β-5) with a phenolic hydroxyl group, catalyzed by [N,N′-bis(salicylidene)ethane-1,2-diaminato]cobalt(II) (Figure 7.3). The continuous wave (CW) electron paramagnetic resonance (EPR) investigation using the X-band, performed at the reaction temperature (298 K) and frozen solution condition (200-10 K), confirmed that two radicals ([Co^{II}I(salen)(ArOH)(RO•)] and [Co^{II}I(salen)(ArO^-)(ArO•)]) are involved in the oxidation mechanism of all the phenol compounds. Then, in 2008, Canevali and co-workers [24] applied the [Co(salen)] oxidation method in the treatments of unbleached thermomechanical pulp (TMP) fibers with molecular O_2. Besides the treatment in the organic solvent, they modified the ligand of the [Co(salen)] with sulpho groups, which can let the treatment carry out in the water.

Besides the mechanism shown above with $L_nCoO_2^•$ as the original species to achieve the first dehydrogenation of phenol group or C_α–H, in 2005, Weckhuysen and co-workers [25] elucidated a somewhat different mechanism of Co(salen)-catalyzed [salen = N, N′-bis(salicylidene)ethylenediamine] oxidization of veratryl alcohol (3,4-dimethoxybenzyl alcohol) without phenolic hydroxyl group by O_2 in alkaline aqueous solution with in situ ATR-IR, Raman, and UV/vis spectroscopy (Figure 7.4). In their proposed mechanism, the reaction starts with the formation of a bis-μ-hydroxo[Co(salen)]$_2$ species (**A**), which explains the dramatic effect of pH on the reaction rate. Alcohol substrate coordination to this species leads to the formation of a cobalt-bound veratryl alkoxo intermediate (**B**). Intermediate (**B**) then bonds with an O_2, which generates a μ-peroxo bridge between two such Co(salen) substrate units (**C**). Finally, transfer of a hydrogen atom from the substrate to the peroxo bridge results in detachment of the product aldehyde and regeneration of the initial bis-μ-hydroxo[(Co(salen)]$_2$ species. In the overall cycle, two substrate molecules are oxidized to aldehydes and molecular oxygen is reduced to water. The rate-limiting step is the detachment of the product molecule, which is aided by the methoxy substituents in the aromatic ring of the benzylic alcohol.

To accelerate the oxidation activity of Co systems for biomass conversion, in 2011, Weckhuysen and co-workers [26, 27] carried out an in situ spectroscopic investigation of the cobalt-catalyzed oxidation of lignin model compounds in ionic liquids, in

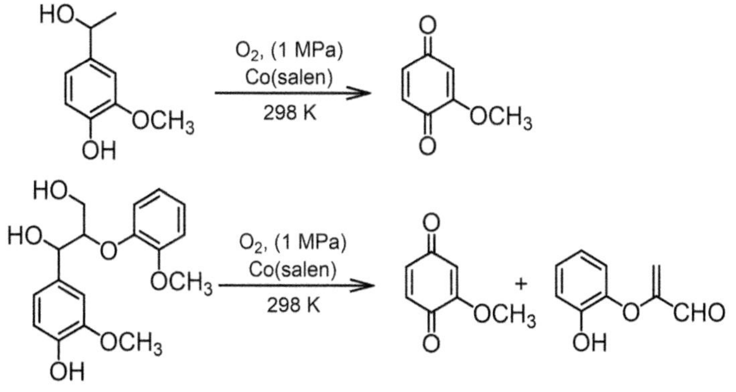

Figure 7.3 Oxidative Degradation of Monomeric and Dimeric Phenylpropanoids with Co(salen).

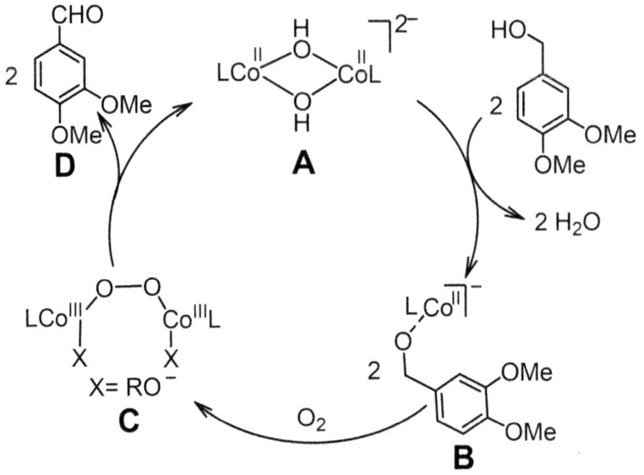

Figure 7.4 The Oxidation Mechanism of Veratryl Alcohol with Co(salen) Media. *Source:* Kervinen et al. [25] with Permission, © 2005 Wiley-VCH.

which high O_2 and lignin solubility make it possible that the oxidation proceeds under mild conditions. They reported that the veratryl alcohol could be rapidly oxidized to veratraldehyde over Co catalyst in ionic liquid (IL, 1-ethyl-3-methylimidazolium diethylphosphate, [Emim][DEP]) in the presence of NaOH at 25 °C with a maximum TOF (turnover frequency) of 1300 h^{-1}. In their proposed mechanism (Figure 7.5), the interaction of the ionic liquid with the cobalt catalyst provided a $Co^{II}(DEP)_2$ (species **B**), which further transformed to a catalytic species Co^{II}/ [EMIM] [DEP]/OH (species **C**) with NaOH. The hydroxide in activated cobalt species **C** promoted the deprotonation of the alcohol substrate, resulting in the formation of water and an alkoxy-cobalt intermediate (species **C**). Then species **D** reacted with an O_2 and generated a superoxocobalt complex (species **E**, Co^{III}), which further transformed to a cobaltperoxo complex (species **F**, Co^{II}-OOH) and released the aldehyde product. Finally, the cobalt-peroxo species reacted with water to form H_2O_2, and the regeneration of Co^{II}/ [EMIM][DEP]/OH closed the catalytic cycle. In addition, hydrogen peroxide that forms as a reaction by-product underwent rapid disproportionation to yield water and O_2. Furthermore, the properties of the ionic liquid greatly influence the catalytic activity both by stabilizing reactive intermediates and by favoring the coordination of the substrate to the cobalt over the direct oxidation of the cobalt without substrate.

Besides the various homogeneous Co catalysts in lignin models oxidation, to overcome the recollection problem of homogeneous catalyst in lignin oxidation, in 2009 Luque and co-workers [11] immobilized Co(salen) on SBA-15, and used it for phenolic dimer apocynol oxidation. They presented that the heterogeneous Co(salen)/SBA-15 catalyzed reaction under microwave irradiation at 150 °C could give a significantly superior TON (turnover number) and much higher yield of 2-methoxyphenol as compared to the homogeneous Co(salen) catalyst in the thermal reaction. Then, Zhou and co-workers [17, 28] synthesize two Y zeolite catalysts with encapsulated salen metal complex by "impregnation" and flexible ligand

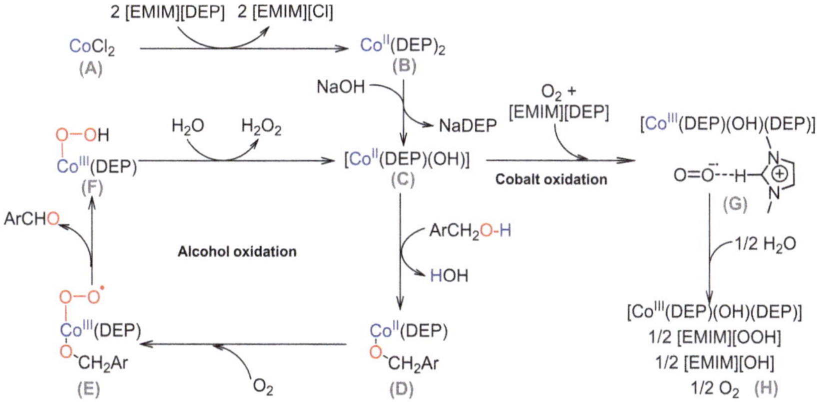

Figure 7.5 Catalytic Oxidation of Veratryl Alcohol via a Co/ [EMIM][DEP]/OH System. *Source:* Zakzeski et al. [27], © 2011 The Royal Society of Chemistry.

ship-in-a-bottle methods. They used them in the selective oxidation of kraft lignin and TCF pulp bleaching with peracetic acid as the oxidant.

In addition, besides the effects of the reaction medium, the coordination state of the Co complex, the activity and stability of Co-Schiff base catalysts are determined by the reaction pH in the oxidation of the phenolic compounds. In 2003, Sippola and co-workers [29] presented a study about the oxidation activity and stability of homogeneous cobalt-sulphosalen catalyst with a phenolic and a non-phenolic lignin model compound in an aqueous alkaline medium. Their studies showed that Co-salen was the promising delignification catalyst at pH 8–10. The catalytic activity decreases rapidly at pH higher than 10, mainly because of the hydrolysis of the imine structure.

Therefore, besides the direct dehydrogenation of phenol group with ArO–H bond homolysis cleavage, adding an appropriate base may promote the ArO–H bond activation. In 2012, Bozell and co-workers [13] carried out a catalytic study of para-substituted phenols oxidation to benzoquinones over the Co–Schiff base complexes/O_2 system. They found the yield of *p*-quinones from vanillyl alcohols are significantly improved by adding a sterically hindered base such as N,N-diisopropylethylamine (DIPEA), diisopropylamine (DIPA), or triethylamine (TEA), which does not coordinate with the catalyst. The proposed mechanism involves deprotonation of the substrate by the sterically hindered base to form a more readily oxidized phenolate anion, which is oxidized by the superoxocobalt radical. The generated phenoxy radical undergoes conversion to benzoquinones, formaldehyde, and Co(ligand)–OH as shown in (Figure 7.6). In their following work [15], new Co-Schiff base complexes that incorporate a sterically hindered ligand and an intramolecular bulky piperazine base in close proximity to the Co center are synthesized. Their results indicate that the increased steric interference is suggested as the source of increased reactivity in guaiacyl alcohol oxidation.

Besides the base-promoted methods targeting at the cleavage of lignin C_β–OAr bond which will be discussed in Chapter 8.1, some base-promoted alkaline oxidation of lignin can achieve the lignin C–C bonds, especially for the C_α–C_β in β-O-4 linkage after the lignin fragmentation from the first C_β–OAr bond cleavage,

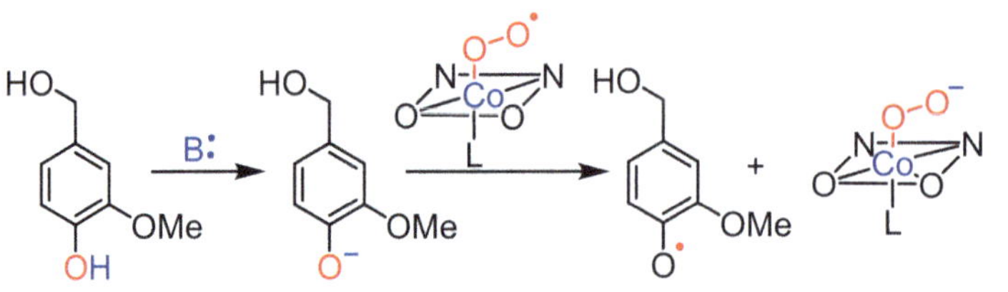

Figure 7.6 The Catalytic Oxidation Mechanism of Guaiacyl Models with a Hindered Base.

which provides aromatic aldehydes/acids as the final product. Different mechanisms of alkaline catalyzed lignin oxidation to produce aldehyde have been proposed in the literature. The most acceptable one is the radical chain mechanism proposed by Tarabanko and co-workers (Figure 7.7) [4, 30]. In 2000, Tarabanko and co-workers [4] studied the alkaline oxidation of lignin to produce vanillina with NaOH. According to the proposed mechanism, the reaction begins with dehydration of the lignin structure units (**I**), followed by the detachment of one electron from the phenoxyl anion **II** to yield the phenoxyl radical **III**. This radical then undergoes either disproportionation or proton detachment, followed by oxidation to form quinonemethide **IV**. The nucleophilic addition of the hydroxide ion to the C_γ of **IV** generates coniferaldehyde **V**. Finally, the retro-aldol cleavage of C_α–C_β bond in **V** gives vanillin **VI** with the assistance of H_2O [30]

The addition of the redox metal that can activate the O_2 to the active O* in the alkaline reaction system can change the C–C bond cleavage mechanism. In 1995, Wu and co-workers [31, 32] obtained 14.6 wt % aromatic aldehydes (including 4.7 wt % vanillin) from steam-explosion hardwood lignin with $CuSO_4$ and $FeCl_3$ catalysts in a NaOH solution. During the oxidation process, as an electron acceptor, Cu^{2+} accelerates the formation of the phenoxy radical and thus speeds up the radical oxidation reaction. As for the role of Fe^{3+} in the oxidative degradation of lignin, it acts as an oxygen carrier to form an intermediate O_2–Fe^{3+}-lignin complex in the oxidation process. As illustrated in Figure 7.8, the O_2–Fe^{3+}-lignin intermediate attacks the phenolate anion in the lignin structure to form the oxidation intermediates, which are further oxidized to acids and aldehydes via the ionic reaction route.

In addition, besides the homogeneous base species and redox metals, the affordability and stability properties of perovskite-type oxides (with the chemical formula ABO_3) have been also used as promising catalysts for wet air oxidation of lignin. These perovskites include $LaCoO_3$, $LaMnO_3$, and $LaFeO_3$ [33–37]. They possess high activity and stability in the lignin oxidation reaction, and the catalytic activity can be increased by the partial substitution of A- or B-site cations [38–40]. As an example,

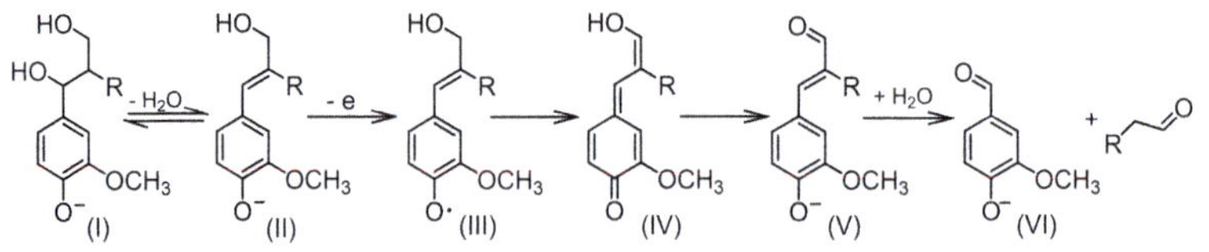

Figure 7.7 Reaction Mechanism Proposed for Alkaline Oxidation of Lignin to Vanillina.

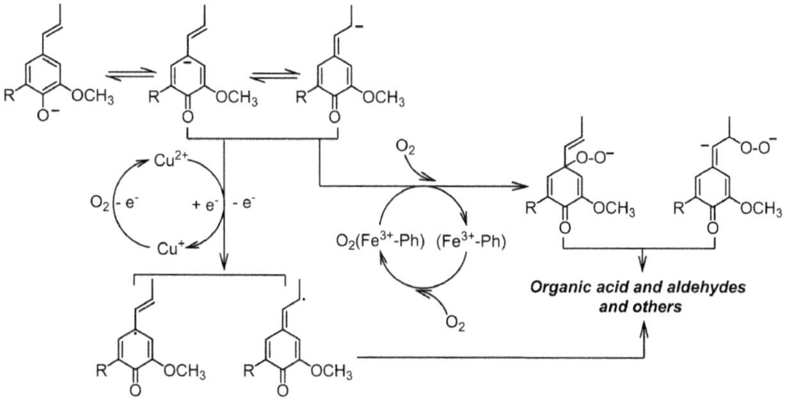

Figure 7.8 Lignin Oxidation Mechanism with Cu^{2+} and Fe^{3+} in NaOH. *Source:* Wu and Heitz [32] with Permission, © 1995 Taylor & Francis.

Wang and co-workers [38] reported that $LaFe_xMn_{1-x}O_3$ and $La_{0.9}Sr_{0.1}MnO_3$ hollow nanospheres synthesized via a urea-assisted solvothermal-calcination method showed much better catalytic performance in catalytic wet air oxidation of lignin than perovskite prepared by the traditional sol-gel method. Lin and co-workers also disclosed that Cu-doped Fe- and Co-based perovskite-type oxides $LaFe_{1-x}Cu_xO_3$ [39] and $LaCo_{1-x}Cu_xO_3$ [36] exhibited enhanced catalytic activity in comparison to the copper-free counterparts. These substitutions were suggested to increase the concentration of oxygen vacancies of perovskites, which promoted the amount of adsorbed oxygen surface active site $[Fe(surf)^{3+}O_2^-]$ or $[Co(surf)^{3+}O_2^-]$ species and improved the yield of activated species $[Cu(surf)^{2+}O_2^-]$. The proposed oxidation mechanism [36, 39] over the Cu-doped perovskite is illustrated in Figure 7.9.

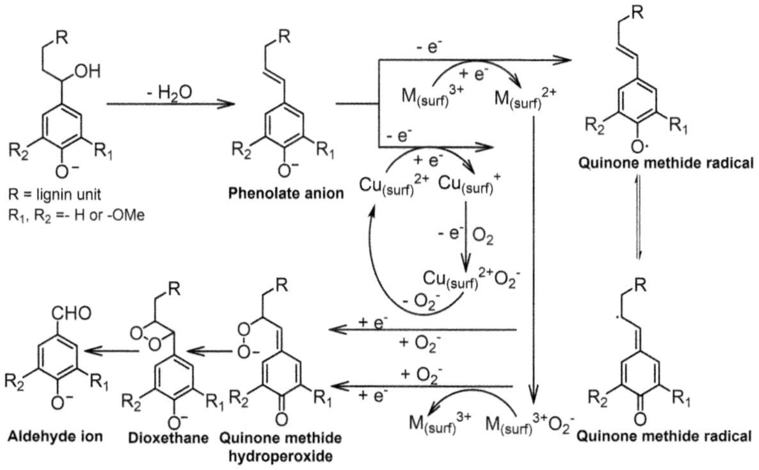

Figure 7.9 Proposed Mechanism of Lignin Oxidation in the Presence of Cu-Substituted Perovskite Catalysts (M = Fe or Co). *Source:* Deng et al. [36] with Permission, © 2010 American Chemical Society.

7.1.2 Electro/Photo-Electro Systems via the Phenolic Radical

For the cleavage of lignin C–C bond after the first phenolic hydroxyl group dehydrogenation, besides the thermal process shown above, the electrocatalytic oxidation of lignin phenolic group at cathode by reactive oxygen species would be another green and economic choice. As discussed in Figure 7.6, compared with the direct dehydrogenation of phenol group with ArO–H bond homolysis cleavage, adding an appropriate base in the electrolyte to first transform the phenolic ArO–H to phenoxyl anion (ArO^-) can promote the ArO–H bond activation to a phenoxyl radical.

In 1991, Utley and co-workers [5] studied the oxidative cleavage of several phenolic ethers, models for lignins at nickel anodes in an alkaline electrolyte. At nickel anodes, although the aldehydes were formed together with larger amounts of the corresponding carboxylic acids, polymerization was predominant. Combinations of 4-hydroxyl, α-keto, β-O-aryl, and β-hydroxymethyl functionality were shown to be crucial for the oxidation at nickel. The carboxylic acid formation probably involves a route with initial hydrogen atom abstraction (HAA) at the surface. The suggested routes are summarized in Figure 7.10. In the alkaline condition, the ArO^- anion generated from phenol **1** can be oxidized to the phenolic radical (or its resonant structure **2**), which further induced the cleavage of the C_α–C_β bond to provide the corresponding aldehyde after the SET oxidation (**Route I**). Otherwise, the Ni(III) species over the anode could induce the oxidation of the C_α–OH to $C_\alpha = O$ after the hydrogen atom abstraction at C_α–H(**Route II,** intermediate **6→7→8**), the generated

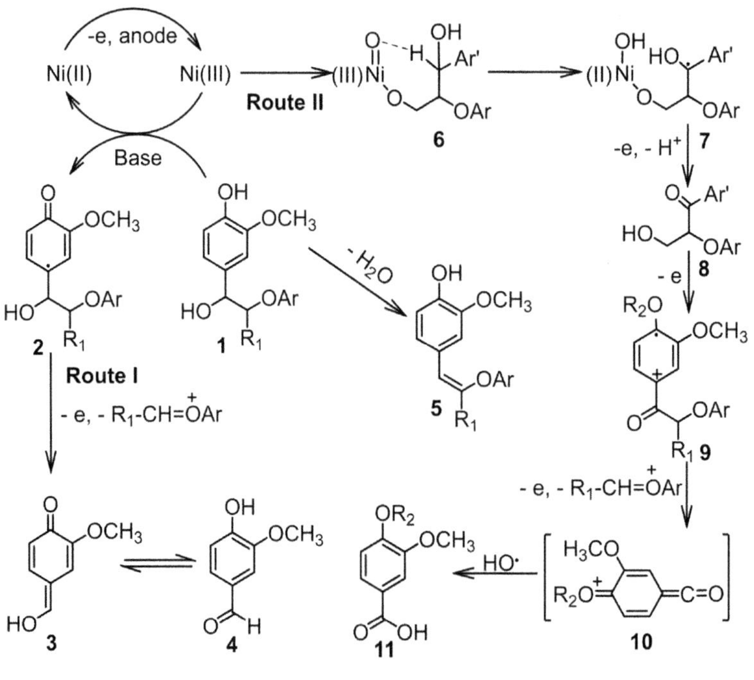

Figure 7.10 Mechanism of Electrooxidation Cleavage of Lignin Model Dimers. *Source:* Pardini et al. [5] with Permission, © 1991 American Chemical Society.

ketone intermediate **8** could further occur C_α–C_β bonds cleavage with releasing the $RC_\beta H = O^+Ar$ fragment after the SET oxidation of the aromatic ring. The left fragment **10** could be further oxidized to the aromatic acid.

In 2011, Utley and co-workers [41] explored the electrochemical conversion of a spruce lignosulfonate into vanillin at nickel anodes. A flow reactor (FM01), in a rig that permitted considerable variation of electrolysis conditions, allowed up to 150 g to be electrolyzed at up to 12 A at a variety of electrode configurations. Samples taken during electrolysis gave detailed reaction profiles. The electrolyser operated at 145 °C/500 kPa/3 M NaOH and yields of vanillin were similar to those obtained industrially using chemical oxidants (about 5–7% w/w). Vanillin production was favoured by low current density and low initial concentration of lignosulfonate. Vanillin, alone, was consumed in a 2.7 F process under the above conditions. Given the fact that yields in chemical and electrochemical conversions of lignins into vanillin do not exceed 10%, the authors explain this apparent limit as the equilibrium between the formation of vanillin, its concomitant oxidative destruction, and further condensation of lignins. In addition, besides the lignin conversion discussion, the authors also presented a detailed introduction about the structure of the flow cell reactor. The people who are interested in the equipment can check the original setting drawing.

In 2014, Chen and co-workers [42] presented a study about electrochemical depolymerization of lignin into renewable aromatic compounds in a non-diaphragm electrolytic cell. The lignin fragments with a phenol hydroxyl anion structure in the alkaline electrolyte were directly electro-oxidized on the anode and chemically oxidized by the electro-generated H_2O_2 formed on the cathode simultaneously (Figure 7.11a). 59.2% of lignin was depolymerized into more than 20 kinds of low-molecular-weight aromatic compounds, containing phenolic hydroxyl, aldehyde,

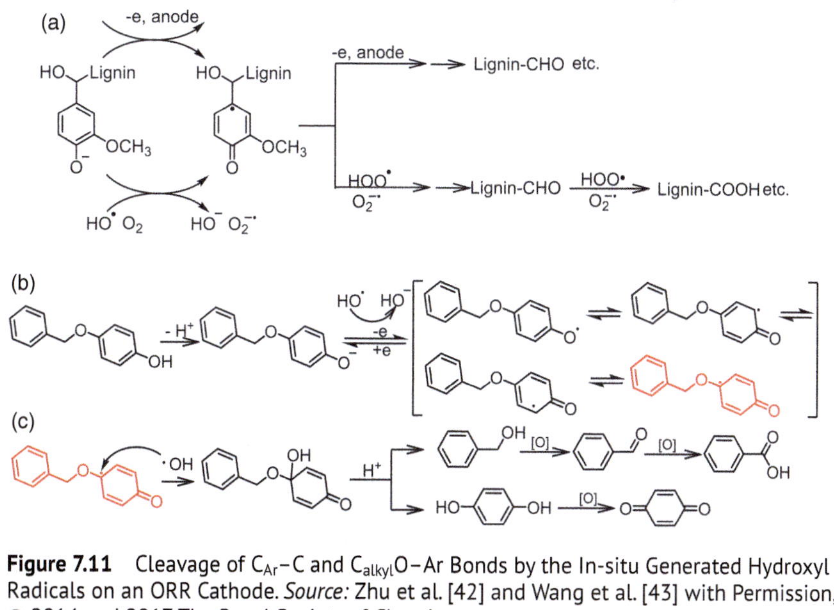

Figure 7.11 Cleavage of C_{Ar}–C and C_{alkyl}O–Ar Bonds by the In-situ Generated Hydroxyl Radicals on an ORR Cathode. *Source:* Zhu et al. [42] and Wang et al. [43] with Permission, © 2014 and 2017 The Royal Society of Chemistry.

carbonyl, and carboxyl groups, after 1 hour-electrolysis at 80 °C under a current density of 8 mA cm^{-2} with extra O_2 supplement. Then, in 2017, Chen and co-workers [43] further published a study on the cleavage of $C_{alkyl}O–Ar$ bonds by in situ generated hydroxyl radicals on an ORR cathode (Figure 7.11b and c). When 4-benzyloxy phenol (PBP) were employed as lignin model compounds, PBP exists the formation of phenol hydroxyl anion due to alkaline medium, and then being oxidized by $HO^•$ radical to form phenoxy radical. After that, the phenoxy radical transforms to their resonant quinone structure, causing the C atom on the opposite of phenolic hydroxyl group to become an activated site for electrophilic attack. Finally, one $HO^•$ radical further attacks the active site as electrophilic reagents and provides an active hemiketal structure, which leads to the cleavage of the $C_{alkyl}O–Ar$ ether bond and degradation of the compound.

In 2020, Waldvogel and co-workers [44] reported an electrochemical method to depolymerize technical Kraft lignin to the aromatic chemical vanillin. The electro-degradation of Kraft lignin was performed at 160 °C in a simple undivided high-temperature electrolysis cell with active nickel (based) electrodes. Using the optimized condition, the vanillin could be obtained in a good yield of 4.2 wt %, which is equal to 67% of the common nitrobenzene oxidation yield. Additionally, the established high-temperature electrolysis indicated a reliable process and could be easily adapted to a variety of kraft lignins with different sources and provided a 2.2~4.4% vanillin yield. For the generation of vanillin via this electrochemical process, as shown in Figure 7.12, the lignin fragment with phenolic hydroxyl groups can be dissolved in sodium hydroxide solution to a sodium phenolate **A**, which can be further oxidized to the carbon radical **B**. The further oxidation of intermediate **B** on the anode leads to the cleavage of the $C_\alpha–C_\beta$ bond in the β-O-4 linkage. The generated intermediate **C** transforms to its resonant vanillin structure [41].

Besides, the direct electro-oxidation of lignin structure to the corresponding oxygen-contained products, like aldehyde, acid, or quinones, some dissociated structure from lignin degradation can be further transformed to some active species over some electrode catalysts, which can further combine with the lignin depolymerizing fragments to some products. In 2014, Li and a co-workers [45] reported a BHT (2,6-bis(1,1-dimethylethyl)-4-methylphenol) synthesis method via the electrochemical degradation of lignin on a Pb/PbO_2 anode in alkaline solution, with a copper auxiliary cathode and a reference saturated calomel electrode. The optimum operating conditions for the production of BHT (5.61 $g·L^{-1}$) from lignin

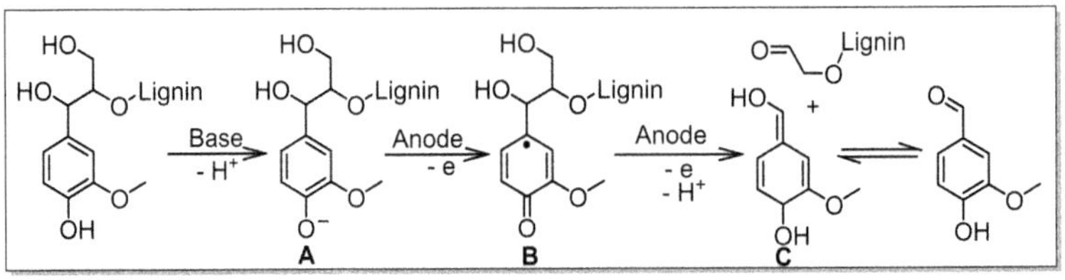

Figure 7.12 High-temperature Electrolysis of Kraft Lignin for Selective Vanillin Formation. *Source:* Zirbes et al. [44] with Permission, © 2020 American Chemical Society.

degradation were found to be a current density of 25 mA cm^{-2} and electrolysis time of 3 h. A reaction mechanism was then proposed for the generation of BHT from lignin (Figure 7.13). Initially, lignin from non-wood biomass(straw) contains phenolic hydroxyl groups on its molecular structure, which can be dissolved in sodium hydroxide solution with the formation of sodium phenolate. The selective oxidation of the carbon-carbon (C–C) bond in aryl-substituted aromatic ring is initiated by superoxide anion radical ($O_2^{\bullet-}$) on the Pb/PbO_2 anode surface, which yields lignin superoxide anion radical and results in cleavage of the side chain from the benzene ring. And then the intermediate of the side-chain product is reduced on the cathode surface with H^+ and gives the electrophilic reagent of the isobutylene carbenium ion. On the other hand, the 4-methyl-phenol monomer degraded from lignin is subsequently electrophilically attacked by the isobutylene carbenium ion to give 2,6-bis(1,1-dimethylethyl)-4-methylphenol.

Besides the pure electro-methods shown above, in 2021, Leem and co-workers [46] first introduced the dye-sensitized photoelectrochemical cell (DSPEC) system, which generally targets the water-splitting reaction, with a hydrogen atom transfer (HAT) mediator (HAT-DSPEC) into the lignin conversion. Leem reported the solar-driven selective cleavage of the C(aryl)–C(alkyl) σ-bond in lignin at ambient temperature using a HAT-DSPEC under redox-neutral conditions (Figure 7.14). The photocatalyst (bis-2,2′-bipyridine)(2,2′-bipyridine-4,4′-dicarboxylic acid)Ru(II) (RuC) adsorbed onto a TiO_2 nanorod array with the length of ~1.6 μm and a rod diameter of 100 nm atop fluorine-doped tin oxide (FTO|TiO_2 NRAs|RuC) film was prepared and investigated with an HAT mediator, 4-acetamido 2,2,6,6-tetramethyl-piperidine-1-oxyl (ACT), in solution. Photophysical and electrochemical studies of RuC and ACT with a β-O-4 lignin model compound (LMC), reveal that the metal-to-ligand charge transfer (MLCT) excited states from the RuC are efficiently quenched in the presence of ACT with LMC. The HAT-DSPEC photoanode, containing the surface-bound photocatalyst RuC at the photoanode with ACT and LMC in solution, sustained an excellent photocurrent density, significantly outperforming that with the photocatalyst RuC alone. The mechanism of the HAT-DSPEC

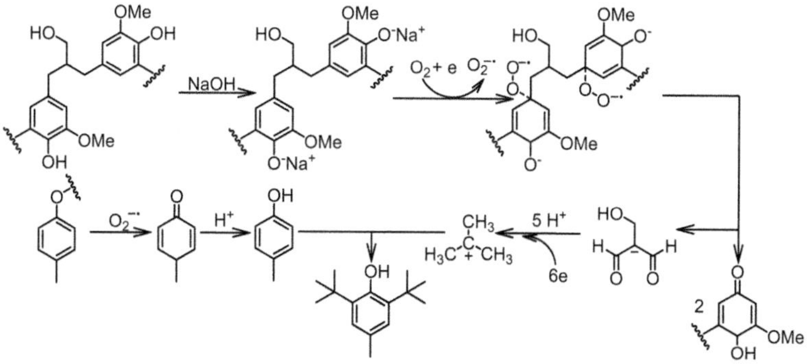

Figure 7.13 Degradation of Lignin to BHT by Electrochemical Catalysis on Pb/PbO_2 Anode in Alkaline Solution. *Source:* Zhang et al. [45] with Permission, © 2013 Society of Chemical Industry.

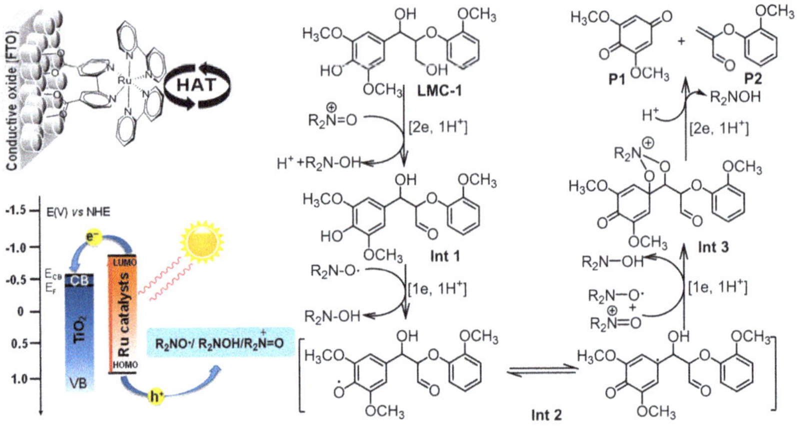

Figure 7.14 Photocatalytic Chemoselective C–C Bond Cleavage at Room Temperature in a Dye-sensitized Photoelectrochemical Cell. *Source:* Li et al. [46] with Permission, © 2021 American Chemical Society.

process with oxidative C_{aryl}–C_{α} bond cleavage contains a series of photodynamic events (Figure 7.14): After photoexcitation of the surface-bound photocatalyst RuC on the FTO|TiO_2 NRAs (step 1), the excited-state RuC* is formed. Electron and hole generation (TiO_2 NRAs (e^-) and RuC(III), respectively) occur following electron injection into the conduction band of TiO_2 NRAs (step 2). Upon the hole transfer from Ru(III) to ACT (R_2N–$O^{\bullet}$) mediator in solution, the ground-state RuC(II) regenerates (step 3). Oxidation of R_2N–$O^{\bullet}$ after hole transfer generates the oxoammonium $R_2N^+ = O$ (ACT^+), which is a strong oxidizing agent and the active species for alcohol oxidation of LMC (step 4). The photochemical formation of ACT^+, and the buildup of a pool of ACT^+ near the photoanode surface, then initiates a series of steps that ultimately lead to the formation of quinone **P1** and aromatic fragment **P2**. As proposed by the authors, one equivalent of the oxoammonium likely oxidizes the γ-hydroxyl group in **LMC-1** (the less sterically hindered primary alcohol), resulting in the formation of **Int1** (step 5). The phenoxy radical (in **Int2**) in the para position might form following hydrogen atom abstraction by the R_2N–$O^{\bullet}$, and the resulting semiquinone intermediate is imperative to observe the products of C–C bond cleavage (step 6). The consumption of an additional 1 equivalent of ACT^+ and 1 equivalent of ACT possibly proceeds through the pentacyclic intermediate **Int3** (step 7), which provides the final products **P1** and **P2** with C_{α}–C_{β} bond cleavage.

In regard to the application of $R_2N^+ = O$ (Oxoammonium) in lignin depolymerization, Samec and co-workers [47] presented a method in 2021 to produce additional valuable monomers from the high-molecular-weight lignin fraction through oxidative C–C bond cleavage with (2,2,6,6-tetramethylpiperidin-1-yl)oxyl (TEMPO)-based oxidants (oxoammonium salts). After the reaction condition optimization, the commercially available Bobbitt's salt is used as the final oxidant. Based on the results of the experiment about the conversion of β-β models and literature reports, they propose the following reaction mechanism (Figure 7.15a): the reaction begins with the oxidation of the phenolic moiety of **1** to form

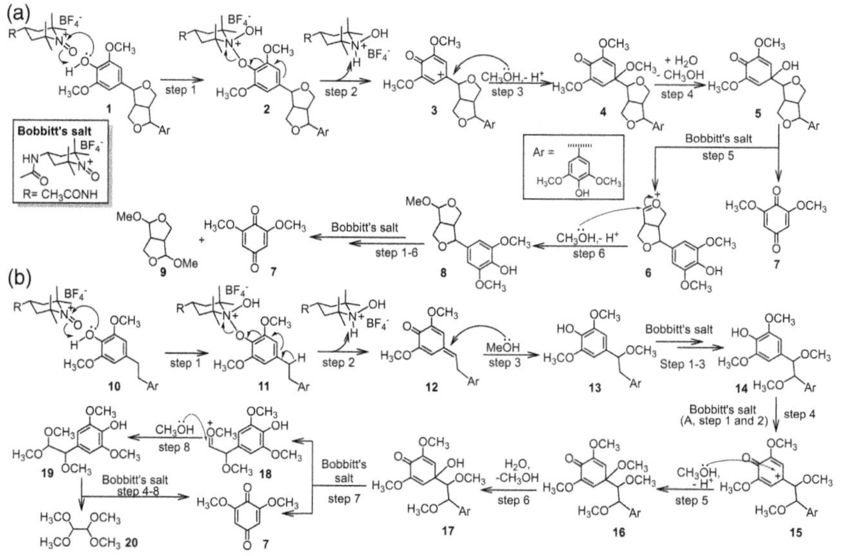

Figure 7.15 Mechanism of C–C Bonds Cleavage in the Catalytic Oxidation the β-β (a) and d β-1′ (b) Model Compounds with Oxoammonium. *Source:* Subbotina et al. [47] with Permission, © 2021 Springer Nature.

dearomatized carbocation **3** with Bobbitt's salt, which is probably stabilized by a neighboring oxygen atom from an ether group. Based on isotopic labeling experiments, the nucleophilic attack by methanol followed by an ipso-substitution by water leads to **5**. Oxidation of **5** with oxoammonium salt furnishes product **7** (2,6-dimethoxybenzoquinone, DMBQ) and a fragment **7**. The conversion of fragment **6** just repeats steps 1-6 and provides another product **7**, along with a bicyclic acetal **9**. In the oxidation of the β-1′ linkage model without a C_α–O structure to stabilize the dearomatized carbocation (Figure 7.15b), the first oxidation of the phenolic group with oxoammonium salts can induce the generation of quinone methide intermediate **12**, which can be further attacked by the MeOH to generate the phenolic intermediate **13** with a C_α–OMe ester structure. The next oxidation process of β-1′ linkage with C_α–C_{Ar} bonds cleavage is similar to the β-β conversion after the generation of C_β–OMe in the intermediate **14**. As discussed above, the overall transformation of 1 mol β-β compound thus requires 4 mol of oxidant and two times higher for the β-1′ compound. For the regeneration of the oxoammonium oxidant, they tested the electrochemical oxidation method. They pointed that separating the oxidant from $NaBF_4$ was not required, and $NaBF_4$ can be collected together with the reduced form of Bobbitt's salt from the solution of the crude reaction mixture in DCM via extraction with water. The collected mixture can be subjected directly to electrochemical regeneration. Likewise, following the completion of the electrochemical reaction and removal of water, the mixture of the oxidant and electrolyte can be directly used in the next reaction cycle. Besides the lignin models, when a mixture of dimers and oligomers obtained after pulping of birch wood is used as the substrate, this oxidation reaction proceeds with high selectivity to give 2,6-dimethoxybenzoquinone (DMBQ) from high-molecular-weight lignin in an 18 wt% yield, thus increasing the yield of monomers by 32%.

7.1.3 Biodegradation Systems via the Phenolic Radical

However, for the lignin biodegradation, the corresponding oxidation process is much more complex. In 1989, Higuchi and co-workers [48] investigated the degradation of syringylglycerol-β-guaiacyl ether (**A**) by laccase of *C. versicolor* (Figure 7.16). The substrate was mainly converted to the α-carbonyl dimer (**B**), 2,6-dimethoxyhydroquinone (**I**), and glyceraldehyde-2-guaiacyl ether (**H**) by alkyl-phenyl bond cleavage, and, guaiacol (**F**) by ArO–C_β bond cleavage. They could not find syringaldehyde and guaiacoxyethanol as direct C_α–C_β cleavage products of the substrate. Their subsequent investigation to identify the pathway to give guaiacol showed that the C_α-carbonyl dimer (**B**) used as substrate was cleaved between C_α and C_β to give syringic acid (**G**) and guaiacol (**F**) as shown in Figure 7.16. The result indicated that phenolic β-O-4 compound is degraded not only by alkyl-phenyl cleavage, which was proposed as a major laccase-mediated degradative reaction but also by C_α–C_β cleavage of the C_α–carbonyl dimer previously formed by C_α oxidation by laccase.

In addition, besides the biodegradation of the β-O-4 linkages, based on the structures of the degradation products and the isotopic experiments of the β-1model compounds, in 1990, Higuchi and co-workers [48] concluded that three types of reactions proceeded via the first generation of phenolic radicals from β-1 substrates catalyzed by laccase: (1) C_α–C_β cleavage between C_α and C_β carbons, (2) alkyl-phenyl cleavage between C_α and aryl groups, and (3) C_α oxidation (Figure 7.17 and Figure 7.18).

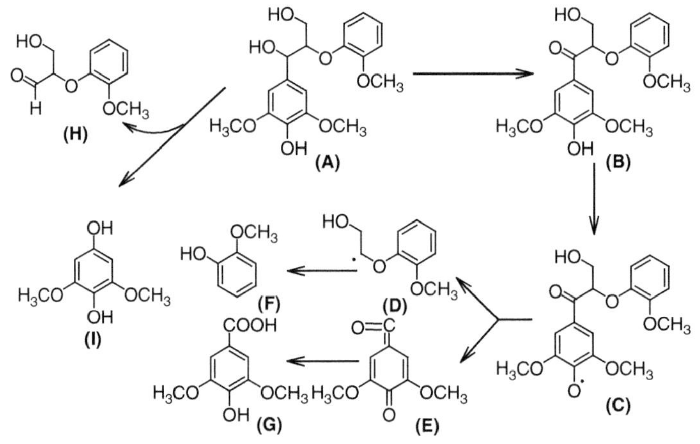

Figure 7.16 Proposed Mechanism for Side-chain Cleavage of a Phenolic β-O-4 Lignin Substructure Model by Laccase of *Coriolus versicolor*. *Source:* Higuchi [48] with Permission, © 1990 Springer-Verlag.

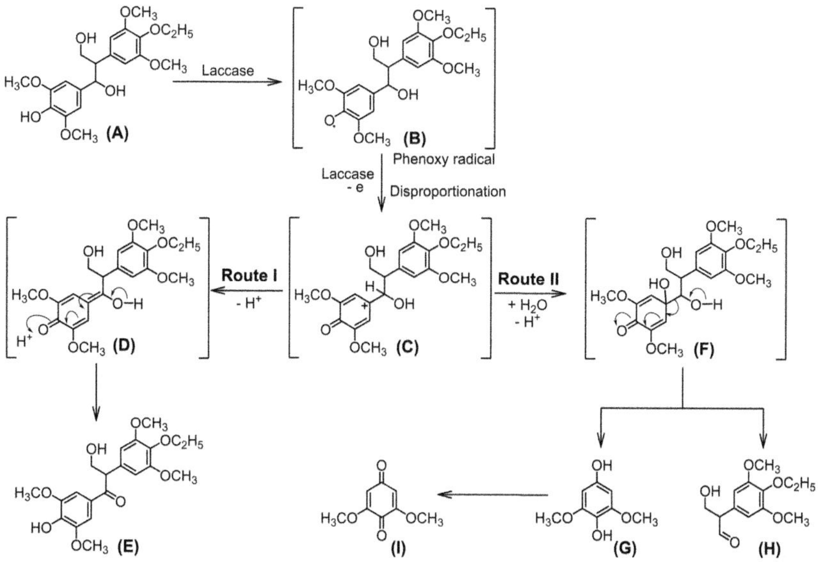

Figure 7.17 Proposed Mechanisms for Cα Oxidation (A) and Alkyl-phenyl Bond Cleavage (B) of Phenolic β-1 Model Compounds by Laccase of *C. versicolor*. *Source:* Higuchi [48] with Permission, © 1990 Springer-Verlag.

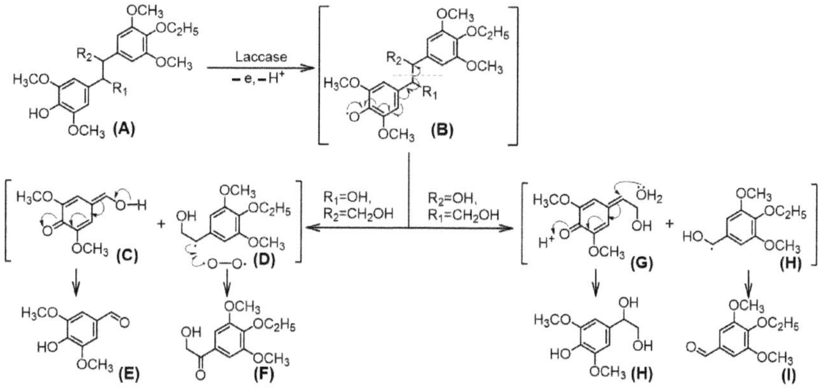

Figure 7.18 Proposed Mechanism for C_α–C_β Cleavage of Phenolic β-1 Model Compounds by Laccase of *C. versicolor*. *Source:* Higuchi [48] with Permission, © 1990 Springer-Verlag.

7.2 Lignin C_α–C_β bonds Cleavage via the First Single-Electron Transfer (SET) of the Aromatic Ring

In 1989, Higuchi and co-workers [49] further studied the side-chain cleavage of phenolic β-O-4 lignin substructure model compounds with lignin peroxidase and laccase, and the investigations suggested that the same chemical principle, phenoxy radical as intermediate, is involved in the degradation of phenolic lignin

Figure 7.19 Lignin C_α–C_β Bonds Cleavage via the First Generation of Aryl Cation Radical.

substructure model compounds by both enzymes. Laccase only catalyzes the oxidation of phenolic substrates to give phenoxy radicals; nonphenolic compounds are not oxidized by laccase, while lignin peroxidase catalyzes one-electron oxidation of both phenolic and nonphenolic β-O-4 model compounds to give the corresponding phenoxy radicals and aryl cation radicals. Besides the further transformation of the phenolic radical shown above, the aryl cation radicals can be attacked by nucleophiles such as water or the hydroxyl groups of the substrate to produce C-radical intermediates with side-chain C–C bond cleavage (Figure 7.19). These carbon-centered radical intermediates and phenoxy radicals are either attacked by O_2 to give various degradation products, or repolymerization by radical-radical couplings. In Chapter 7.2, the methods, including the biocatalytic and non-biocatalytic methods, for the lignin C–C bonds cleavage via the first single-electron transfer (SET) of the aromatic ring to an aryl cation radical will be discussed.

7.2.1 Biocatalytic Oxidation Systems and the Chemically Mimetic Systems

In nature, certain fungi, for example, the white-rot fungi [50], can degrade the lignin with complex inter-monomeric linkages to CO_2. For this phenomenon, intensive research has indicated that the ligninase peroxidase (LiP)-catalyzed mechanism usually involved an enzymatic single-electron transfer (SET) oxidation of lignin yielding aromatic radical cations, which can further induce the cleavage of phenylpropyl C_α–C_β chains (Figure 7.20) [51–54]. More studies indicated the oxy-ferryl center ($Fe^{IV} = O$) with high redox potential [53], generated from the first reaction of haem-containing ligninase with hydrogen peroxide, plays a critical role in the

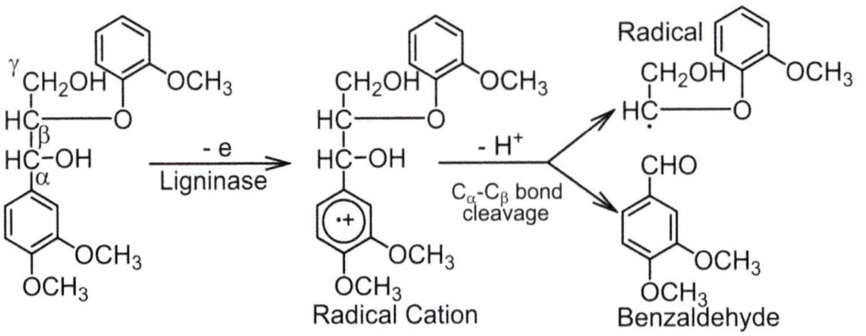

Figure 7.20 Proposed Radical Depolymerization Route of a Model Compound of β-aryl Ether Type.

one-electron oxidation of lignin rings to aromatic radical cations [52]. In addition, ligninase is distinct from the metalloenzymes laccase and peroxidase [48, 55–58]. Laccase is a copper-containing phenol oxidase, which oxidizes the substrates containing free phenolic hydroxyl groups, whereas the ligninase can direct oxidize the aromatic rings. The generated aldehyde part and another lignin carbon radical fragment from lignin radical cations decomposition can be further oxidized by the O_2, which was studied by ^{18}O labeling experiments [59].

At the beginning of the ligninase peroxidase study, the substrates were usually the alcohol monomers [60]. In 1986, Fee and co-workers [61] catalysis of the H_2O_2-dependent oxidation of 3,4-dimethoxybenzyl (veratryl) alcohol by the hemoprotein ligninase isolated from wood-decaying fungus, *Phanerochaete chrysosporium Burds*. Transient-state kinetic studies reveal that, upon the reaction of ligninase with H_2O_2, spectral changes occur in the Soret region, which, by analogy to previous studies of horseradish peroxidase, are consistent with the formation of Compounds I and II.

$$PFe^{3+}O_{(\text{Compound I})} + H_2O_2 \rightarrow PFe^{3+}OH_{(\text{Compound II})} + HO_2$$

The active form of the enzyme appears to react rapidly with H_2O_2; they observed a positive correlation between the turnover number of the enzyme preparation and the extent of a rapid reaction between H_2O_2 and ligninase to form Compound I. Although the free radical cations derived from veratryl alcohol do not appear to be released from the enzyme during catalysis, the authors believed that substrate was converted to cation radicals [54]. In 1987, Schoemaker and co-workers [62] studied the oxidation of veratryl alcohol by the lignin peroxidase of *Phanerochaete chrysosporium*. Five products were identified: veratraldehyde, two quinones, and two aromatic ring cleavage lactones. A similar product pattern was obtained with the l-electron oxidant Ce(IV). Under anaerobic reaction conditions or in the presence of Mn(II) only traces of quinones or lactones were detected besides veratraldehyde. This indicates the involvement of activated oxygen species in the enzyme reaction. Accoding to the generated products, the author proposed a mechanism with first single-electron transfer of the aromatic ring (Figure 7.21).

It is well known that besides the Fe-centered lignin peroxidase (LiP), the manganese-dependent peroxidase (MnP) [63–65] produced from white-rot basidiomycetes can also oxidatively catalyze the delignification of woody tissues at the mild condition [66–70]. To further improve the lignin oxidation efficiency and selectivity with the easily available catalysts, some metal complexes mimicking LiP and MnP have been used in lignin oxidation [71–76]. Synthetic metalloporphyrins (Figure 7.22, complexes 1-3) are typical biomimetic catalysts for lignin oxidation that can yield highly oxidized metallo-oxo species similar to those generated from LiP and MnP [77]. These catalysts have high activity for lignin oxidation because of the structural similarity between the active site of the enzyme and metalloporphyrin complexes. Metallophthalocyanines [78–80] (Figure 7.22, complexes 4) are another type of biomimetic peroxidases for lignin oxidation. This kind of catalyst

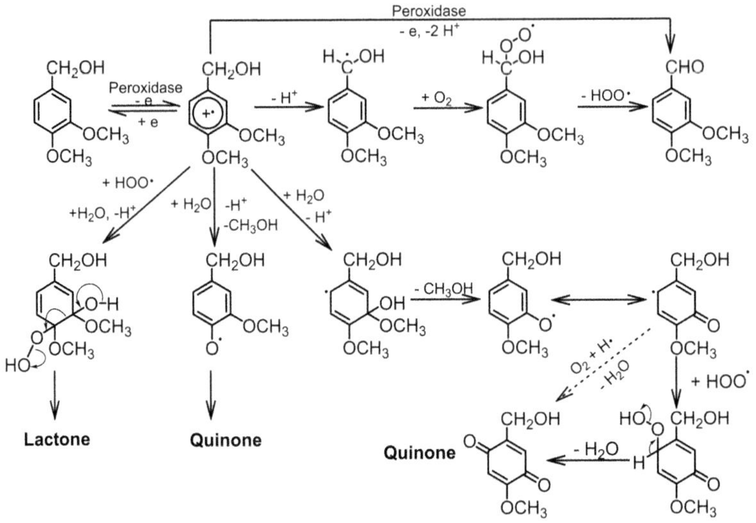

Figure 7.21 Possible Reactions of the Veratryl Alcohol Cation Radical and a Hypothetical Mechanism for the Formation of Quinone Products. *Source:* Haemmerli et al. [62] with Permission, © 1987 Federation of European Biochemical Societies.

has the advantage that the phthalocyanines are readily available and inexpensive, and the iron(III) and manganese(III) complexes are able to catalyze the oxidation of various lignin and model compounds. However, their stability is highly dependent upon pH and the oxidant property [79].

Oxidation of the lignin model compounds by metalloporphyrins probably takes place via a single electron transfer(SET) mechanism [70, 71, 73]. In 1987, Shimada and co-workers [71] investigated the application of Iron (III) porphyrin chloride in biomimetic oxidation of β-1 lignin model compounds; the general reaction mechanism for both C—C and C—H bond cleavage with iron(III) porphyrin catalyst in the presence of active oxygen donor is illustrated in Figure 7.23. First, the two electron-deficient oxo-iron(IV) porphyrincation radical species **X** which is formed from iron(III) porphyrin catalyst in the presence of active oxygen donor removes one electron from aromatic ring **B** of model compound **1**, forming cation radical **Ya**. Intermediate **Ya** undergoes heterolytic C_α–C_β bond cleavage yielding *p*-anisaldehyde (**2**), and the other hydroxymethyl radical intermediate **Z**. The radical **Z** is then attacked by O_2 to the oxygenated chemical **3** and **4** via different routes shown in Figure 7.23 (**Route a** and **Route b**). Alternatively, **Z** undergoes the second electron transfer to form the benzyl cation **Zb** and the subsequent hydroxylation to form 3 (**Route c**) with H_2O or the oxo-iron porphyrin complex (**X**) whose oxygen atom is exchanged with water (**Route d**). When another type of cation radical intermediate (**Yb**) is formed, besides the same homolytic C_α–C_β bond cleavage mechanism of like **Ya**, as another possibility, the C_α–H bond cleavage in **Yb** offers β-ketol without C_α–C_β bond cleavage. In addition, the mechanism beginning from the SET from the ring A can be also used in the β-O-4 linkage conversion.

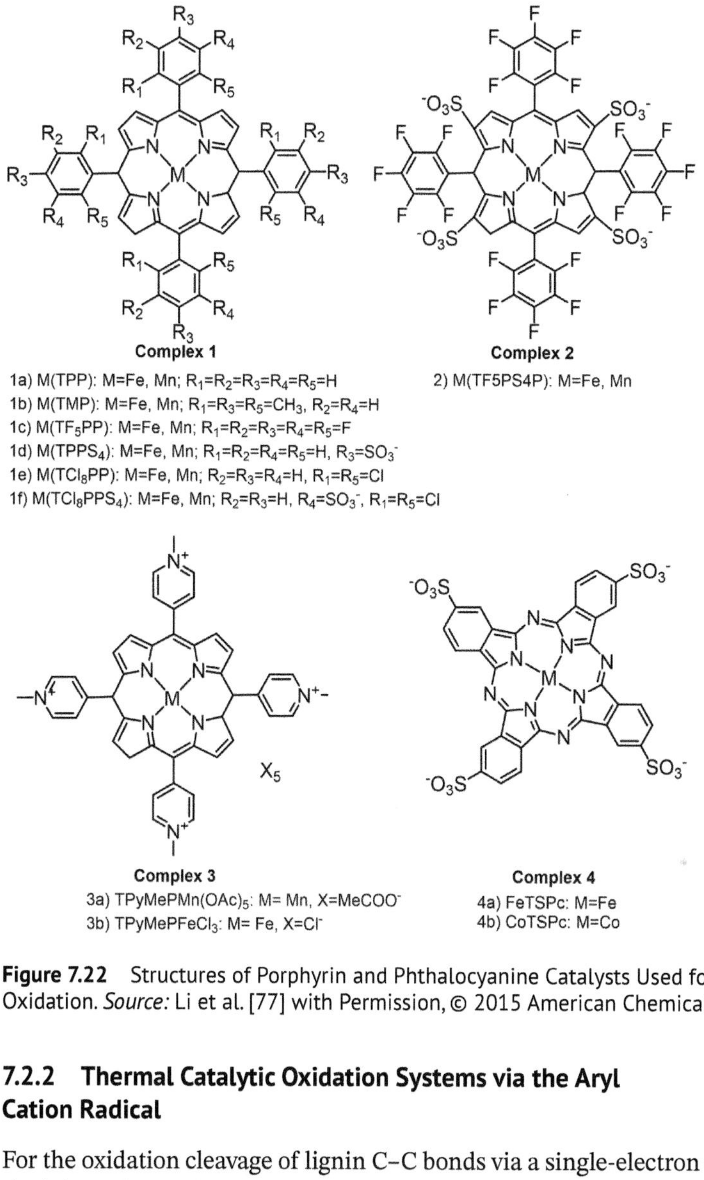

Figure 7.22 Structures of Porphyrin and Phthalocyanine Catalysts Used for Lignin Oxidation. *Source:* Li et al. [77] with Permission, © 2015 American Chemical Society.

7.2.2 Thermal Catalytic Oxidation Systems via the Aryl Cation Radical

For the oxidation cleavage of lignin C–C bonds via a single-electron transfer (oxidation) from the aromatic rings, besides the LiP and metal complexes mimicking LiP and MnP, some metal species in high valence can also induce the SET oxidation [81, 82], such as the metal salts and compounds of Fe, Cu, Co, Mn, V, Mo, Ag, and Ce [52, 82–88], which indicates that non-heme, non-oxy-ferryl type catalysts may also function in the SET mechanism for the lignin linkage transformation. To make the SET oxidation with high valence metal a catalytic cycle, an extra oxidant is necessary for the system to achieve the recovery of the high valent metal species from a reductive state. As proposed in Figure 7.24, the catalytic oxidation of lignin linkages to aldehydes and carboxylic acids, catalyzed by a number of different transition

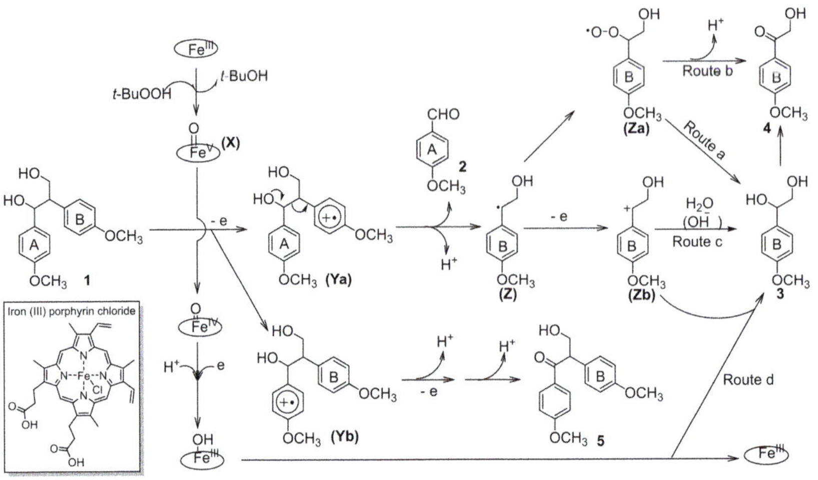

Figure 7.23 Single-electron Transfer Mechanism for Oxidation of the β-1 Lignin Model Compound Catalyzed by Iron(III) Porphyrin and ᵗBuOOH. *Source:* Shimada et al. [71] with Permission, © 1987 Walter de Gruyter.

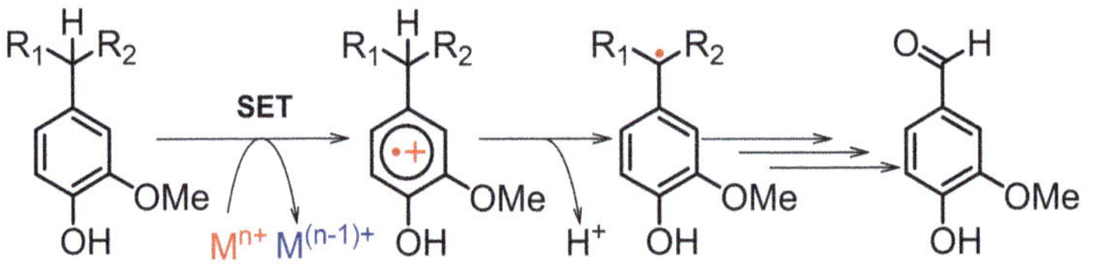

Figure 7.24 Oxidative Cleavage of Lignin to Aromatic Aldehydes by Metal Salts.

metals, proceeds by an initial electron transfer resulting in a one-electron reduction of the metal catalyst and concomitant formation of a substrate radical cation, which induces the further C_α–C_β bonds cleavage. Therefore, the ease of electron-transfer oxidation of aromatic hydrocarbons to produce radical cations is directly related to the ionization potential of these compounds and dominates the following C_α–C_β bonds cleavage. Meanwhile, electron-donating substituents such as methoxyl as the aryl group of lignin can lower the oxidation potential [89].

In 1988, DiCosimo and co-workers [88] studied the performance of Co(II), Mn(II), or Co(II)/Mn(II) systems in lignin models transformation with peracetic acid or O_2 as the oxidant in acetic acid. Based on the analysis of the product, the C_α–C_β bond cleavage of β-O-4 and β-1 models results from an initial single-electron oxidation to produce an intermediate aromatic radical cation with Co(III) and peracetic acid. According to the above mechanism, the author believed that using O_2 to replace peracetic acid would be an economical way to generate the desired electron-transfer oxidants in situ because the alkylperoxy or peroxyacid intermediates formed during the reaction of O_2 with the C_α–C_β bond-cleavage products of lignin could reoxidize the catalyst. When carried out the reaction with O_2 as the oxidant, acetaldehyde was added as a co-oxidant for the generation of Mn(III)/Co(III) in acetic acid. They further studied the effects of different Co/Mn ratios on catalytic performance. The mixed catalyst Mn/Co (1:9) gave the best selectivities and conversions in the presence of added acetaldehyde, which is because of a synergistic effect when using Co(II) and Mn(II) for the autoxidation [90].

In 2000, Evtuguin and co-workers [91] reported that both phenolic and nonphenolic units were reactive with a heteropolyanion $[PMo7V5O_{4}0]^{8-}$ (known as HPA-5) catalyst. Phenolic β-aryl ether structures have a conversion, at least five-times higher than the corresponding nonphenolic structures. Further study of the oxidation of β-aryl ether dimers showed that the reaction mechanisms involved are different. The cleavage of phenolic β-O-4 bonds occurs predominantly through a heterolytic mechanism which includes the hydrolytic cleavage of the alkyl-phenyl bonds in the phenolic structural units oxidized by $[PMo7V5O_{4}0]^{8-}$ (Figure 7.25a, Route II) and the typical acid-catalyzed splitting of the β-O-4 linkages (Figure 7.25a, Route I). In contrast, the oxidative cleavage of nonphenolic β-aryl ether proceeds by one-electron oxidation of aromatic groups with VO^{2+} released from $[PMo7V5O_{4}0]^{8-}$ followed by homolytic cleavage of β-O-4 ether and C_α–C_β linkages, and the detailed mechanism is illustrated in Figure 7.25b with the different SET position of the aromatic rings.

In 2011, Sedai and co-workers [85] reported that CuCl/TEMPO (TEMPO = tetramethylpiperidine N-oxide) in pyridine solvent could catalyze the aerobic oxidation of the lignin model compound 1,2-diphenyl-2-methoxyethanol (β-O-4 model) to

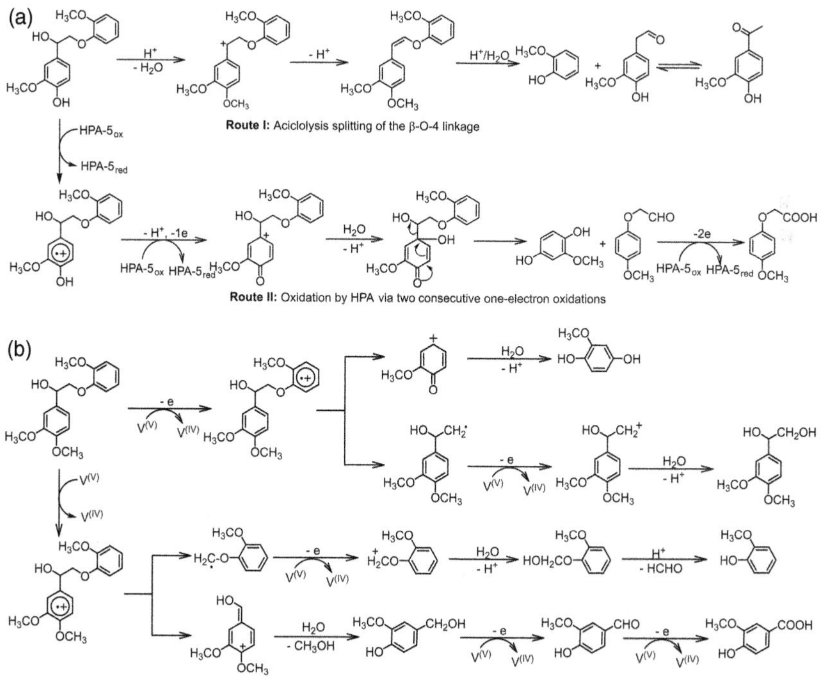

Figure 7.25 HPA-5 Catalyzed Oxidative Cleavage of Non-phenolic β-O-4 Bonds (A) and Phenolic β-O-4 Bonds (B); Note: V(V) and V(IV) are in the Composition of VO_2^+ and VO^{2+} Ions. *Source:* Evtuguin et al. [91] with Permission, © 2000 Elsevier Science B.V.

aldehydes with the cleavage of the C_α–C_β bond. In their following work, they found that the CuOTf/TEMPO in toluene with the assistance of 2,6-lutidine [86] was also efficient in the catalytic oxidative cleavage of the C_α–C_β bond in β-O-4 linkage. In 2015, Mottweiler and co-workers [87] reported that transition-metal-containing hydrotalcites (HTC) and $V(acac)_3/Cu(NO_3)_2{\cdot}3H_2O$ (acac = acetylacetonate) mixtures were efficient in the cleavage of C_α–C_β bond in β-O-4 lignin model with molecular oxygen as oxidant. Both catalytic systems displayed high activity and good selectivity and afforded veratric acid as the main product.

In addition, to increase the generation of metal species at a higher oxidation state to trigger the SET oxidation of the aromatic rings, introducing a stronger oxidant than O_2 should be a good choice [92, 93]. For example, the thermal or catalytic decomposition of persulphate with a metal ion, such as Fe, Cu, Ag, Co, can achieve the recovery of the metal ion for SET and generate the radical anion ($SO_4^{\cdot-}$), which can directly oxidize aromatic rings by electron transfer (Figure 7.26) [94, 95].

In 1999, Bjørsvik and co-workers [96] used persulphate as the oxidant for lignosulfonate oxidation with different metal salt catalysis. The results indicated that the radical anion ($SO_4^{\cdot-}$) is able to oxidize metal salt to higher oxidation states, such as Cu(II) to Cu(III), which can be responsible for electron-transfer oxidation. However, the conversions were low. The main difficulty, encountered by this approach, is concerned with the reaction temperature because the overall reaction is a combination of hydrolysis and oxidation of lignosulfonates (LS). Hydrolysis of lignosulfonate to smaller fragments requires a high temperature (160-200 °C), which is not suitable for the oxidation by persulphate at a lower temperature (100-110 °C). The lower conversion can be due to the low hydrolysis of LS.

To increase the oxidation efficiency and aromatic monomers yield, adding some mild co-catalyst to promote the SET process and match the reaction temperature of different steps in lignin conversion can be a worth-trying choice. Previous studies [97–100] indicated that nitrobenzene could be used as the oxidant to achieve the vanillin and syringaldehyde generation from lignin with an approximately 15 and 35% yield, respectively. The high yield of vanillin in the oxidation by a large excess of nitrobenzene can be related always to oxidation by electron-transfer processes, which avoid the degradation of vanillin by free-radical processes with the O_2.

$$S_2O_8^{2-} \longrightarrow 2\,SO_4^{\cdot-}$$

$$S_2O_8^{2-} + M^{n+} \longrightarrow SO_4^{2-} + SO_4^{\cdot-} + M^{(n+1)+}$$

$M = Fe^{2+}, Cu^{+}, Ag^{+}, Co^{2+}$; $Cu^{2+} + SO_4^{\cdot-} \longrightarrow Cu^{3+} + SO_4^{2-}$

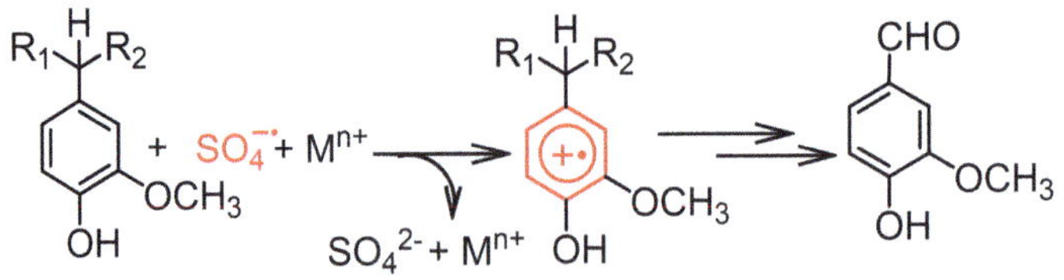

Figure 7.26 Oxidation of Lignosulfonate with Persulphate and Various Metal Salts.

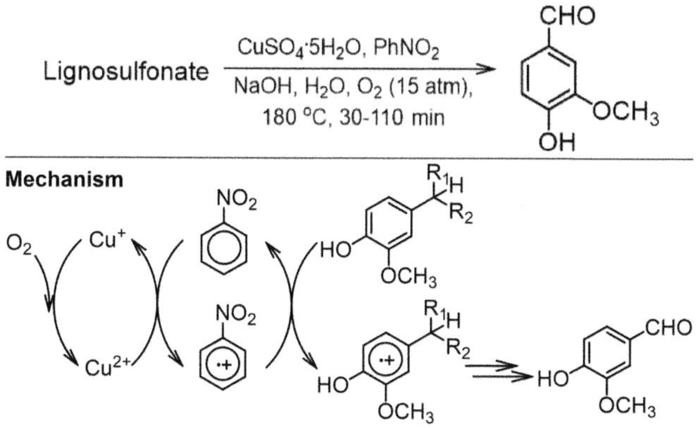

Figure 7.27 Oxidation of Lignosulfonate with Nitrobenzene and Metal Salts.

However, the amount of nitrobenzene, required by this procedure is too high for industrial application. Researchers then tried to make the overall process catalytic in nitrobenzene by combining the oxidation of LS by nitrobenzene and oxygen catalyzed by Cu(II) or other metal salts to obtain the catalytic cycle. The famous and industrial process is the synthesis of vanillin by oxidation of lignosulfonates (LS) or other technical lignin, involving the catalysis NaOH and Cu(II) with nitrobenzene as the co-catalyst. In 1999, Bjørsvik [96] presented a study of vanillin synthesis from lignosulfonates oxidation with Cu/nitrobenzene system, which provided a 5.92 % yield of vanillin (Figure 7.27). To increase the vanillin yield, the author changed the nitrobenzene with a more powerful oxidant 1,3,5-trinitrobenzene, while the yield was turned to 70% of the Cu/$PhNO_2$ system. In addition, besides the system with Cu(II) salt, Ce(III)/Ce(IV) and Co(II)/Co(III) as the well known catalytic cycle for the electron-transfer from aromatic substrates have been also used in the lignin aerobic oxidation with the assistance of NaOH. The results indicated that the efficiency of the catalysts is Co(II) > Cu(II) > Ce(IV) > Ce(IV)/Cu(II). It means that Co(II) can substitute for Cu(II) as the catalyst for the Cu-based industrial process, while Ce(IV) shows a lower efficiency in the oxidation of LS, which is likely related to the higher difficulty to reoxidize Ce(III) to Ce(IV) and low activity of Ce oxide than the CuO_x, which was generated from the reaction of Ce salts with NaOH.

7.2.3 Electrocatalytic Systems via the Aryl Cation Radical

For the in-situ regeneration of the desired metal species in high valence as the electron-transfer oxidants, besides the thermal methods by introducing another oxidant, using the oxidative ability of the anode would be a greener approach which can make the reaction occur at a mild condition. In 1991, Utley and co-workers [5] investigated the mechanisms for oxidative cleavage of several phenolic and non-phenolic lignin models by a detailed comparison of the result of anodic oxidation at nickel anodes in alkaline electrolyte with that of oxidation in acetonitrile in the presence of a triarylamine (Ar_3N) redox catalyst. During the anodic oxidation with Ar_3N, the non-phenolic lignin models oxidation was initiated by SET which was induced by the first generation of the $Ar_3N^{+\bullet}$ at the anode, and in this case, the major product is aromatic aldehyde derivatives with C_α–C_β bonds cleavage (Figure 7.28). In addition, the mechanism of direct anodic oxidation of the β-O-4 lignin dimer with a phenolic group was complex, which has been discussed in Chapter 7.1.2. In the alkaline condition, the ArO^- anion can be oxidized to phenolic radical which further induced the cleavage of the C_{Ar}–C_α bond. Otherwise, the Ni(III) species over the anode could induce the oxidation of the C_α–OH to C_α = O, the generated ketone intermediate could further occur C_α–C_β bonds cleavage after the SET oxidation of the aromatic ring (Figure 7.10).

In 1992, Viertler and co-workers [101] studied the anodic oxidation of some lignin model compounds in methanol to ascertain factors influencing cleavage of C_α–C_β bonds of substitution. Cyclic voltammetric oxidation peak potentials were influenced by substrate structure but not by pH. Reparative-scale electrolysis gave 3.4-dimethoxy-benzaldehyde as the major product from β-O-4 compounds with the C_α–OH, corresponding to cleavage of the C_α–C_β bond. In contrast, β-O-4 compounds with C_α = O were not cleaved with significant efficiency. The results are convincingly rationalized in terms of rapid cleavage of the C_α–C_β bonds in the first-formed radical cations.

To gain information about how alkoxy substitution in arene rings of β-O-4 structural units within lignin governs the efficiencies/rates of radical cation C_α–C_β bond cleavage reactions, in 2013, Mariano and co-workers [102] explored the single electron transfer photochemical and lignin peroxidase-catalyzed oxidation reactions of dimeric/tetrameric model compounds. Their results show that the radical cations derived from less alkoxy-substituted dimeric β-O-4 models undergo more rapid C_α–C_β bond cleavage than those of more alkoxy-substituted analogs. Their further DFT calculations results demonstrate that C_α–C_β bond dissociation energies of β-O-4 radical cations decrease as the degree of alkoxy substitution decreases. In addition, they further studied the SET oxidation mechanism of tetrameric compounds consisting of two β-O-4 units, containing different degrees of alkoxy substitution. Regioselective radical cation C–C bond cleavage was observed to occur in one case at the C_α–C_β bond in the less alkoxy-substituted β-O-4 moiety. These observations show that lignins containing greater proportions of less rather than more alkoxylated rings as part of β-O-4 units would be more efficiently cleaved by SET mechanisms.

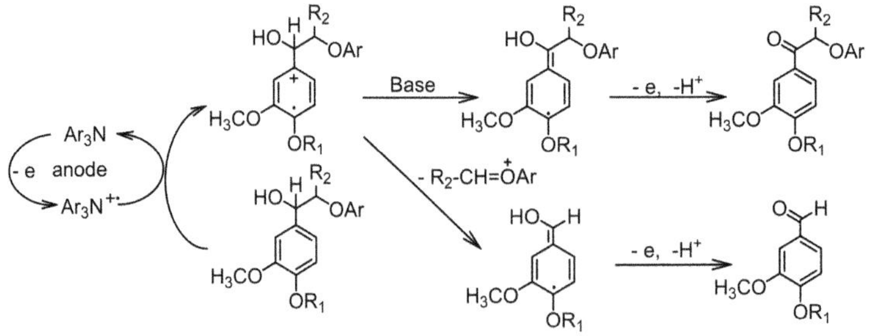

Figure 7.28 Triarylamine Mediated the Electrochemical Oxidation of Nonphenolic Lignin Dimers. *Source:* Pardini [5] with Permission, © 1991 American Chemical Society.

7.2.4 Photocatalytic Methods via the Aryl Cation Radical

For the cleavage of lignin C_α–C_β bonds beginning with the first SET oxidation of adjacent aromatic ring, besides the enzymatic process, thermal catalytic [85–87], and electrocatalytic methods shown above, various photocatalytic methods have also been founded for the lignin C–C bonds cleavage [103, 104].

In 2010, Mariano and co-workers [68] studied the generation of cation radicals from lignin β-1 and β-O-4 dimeric model compounds by using SET-sensitized photochemical, Ce(IV), and lignin peroxidase with H_2O_2, and the nature and kinetics of their C–C bond cleavage reactions were determined. The results show that significant differences exist between the rates of cation radical C_α–C_β bond cleavage reactions of 1,2-diaryl-(β-1) and 1-aryl-2-aryloxy-(β-O-4) propan-1,3-diol structural units. Specifically, under all conditions C_α–C_β bond cleavage reactions of cation radicals of the β-1 models take place more rapidly than those of the β-O-4 counterparts. The results of DFT calculations on cation radicals of the model compound show that the C_α–C_β bond dissociation energies of the β-1 lignin model compounds are significantly lower than those of the β-O-4 models, providing clear evidence for the source of the rate differences.

In 2011, Mariano and co-workers [70] further prepared new types of β-O-4 and β-1 tetrameric lignin model compounds and explored the photocatalytic reaction of carbon-carbon bond fragmentation reactions of their cation radicals, formed by photochemical (9,10-dicyanoanthracene) and enzymatic (lignin peroxidase) SET-promoted methods (Figure 7.29). The results show that cation radical intermediates generated from the tetrameric model compounds undergo highly regioselective C–C bond cleavage in their β-1 subunits. The products of these processes suggest that cation radicals of lignins formed by SET to excited states of sensitizers or heme-iron centers in enzymes degrade selectively through bond cleavage reactions in β-1 vs β-O-4 moieties. In addition, the findings made in the enzymatic studies demonstrate that the sterically large tetrameric lignin model compounds undergo lignin peroxidase-catalyzed cleavage via a mechanism involving the preliminary formation of an enzyme-substrate complex.

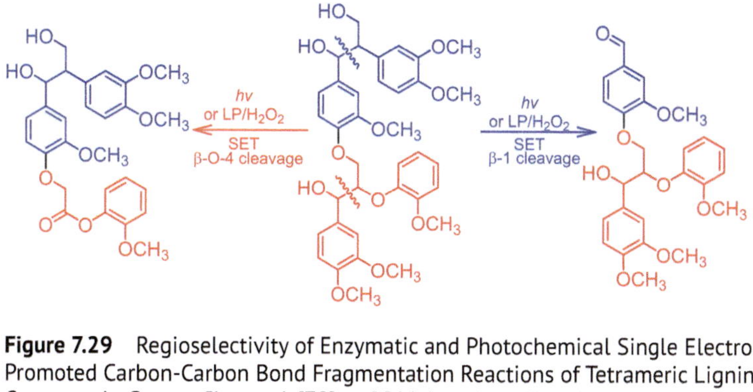

Figure 7.29 Regioselectivity of Enzymatic and Photochemical Single Electron Transfer Promoted Carbon-Carbon Bond Fragmentation Reactions of Tetrameric Lignin Model Compounds. *Source:* Cho et al. [70], © 2011 American Chemical Society.

In 2013, Mariano and co-workers [102] reported that the radical cations derived from less alkoxy-substituted dimeric β-O-4 models undergo more rapid C_α–C_β bond cleavage than those of more alkoxy-substituted analogs. These findings gained support from the results of DFT calculations, which demonstrate that C_α–C_β bond dissociation energies of β-O-4 radical cations decrease as the degree of alkoxy substitution decreases. In SET reactions of tetrameric compounds consisting of two β-O-4 units, containing different degrees of alkoxy substitution, regioselective radical cation C–C bond cleavage was observed to occur in one case at the C_α–C_β bond in the less alkoxy-substituted β-O-4 moiety. In addition, in 2015, Mariano and co-workers [105] presented a study about the effect of the number of methoxy substituents on arene rings on the C–C bond cleavage of β-1 linkage models with 9,10-dicyanoanthracene (DCA), which was used as an excited state electron acceptor to promote photochemical reactions design to explore SET promoted oxidation. Analysis of the relative quantum efficiencies of the DCA-promoted SET photochemical processes indicated that they do not depend greatly on the numbers and positions of arene ring methoxy substituents of the β-1 model compounds. On the basis of DFT calculation results, the energy changes involved in cation radical C_α–C_β bond cleavage are lower than those of the neutral analogs, reflected in BDEs, match the experimentally observed nature of SET promoted reactions of the β-1 model compounds. In addition, as shown in Figure 7.30, the cation radical intermediate generated from the SET oxidation of β-1 substrate has two different following routes for C_α–C_β bond cleavage, generating C_α cation/C_β radical or C_α radical/C_β cation, which further react with H_2O or O_2 and affect the distribution of the products. Following DFT calculations indicate that the choice between C_α cation/C_β radical or C_α radical/C_β cation production is influenced by the energies (stabilities) of the cations and radicals generated, which means the methoxy substituents on arene rings, connecting with C_α or C_β, can affect the distribution of the product.

Besides just using the organic photocatalyst 9,10-dicyanoanthracene (DCA), in 2018, Miura and co-workers [106] reported that the $CuSO_4$+NaOAc can enhance the photocatalytic oxidative cleavage of C_α–C_β bonds in the β-O-4 and β-1 dimeric

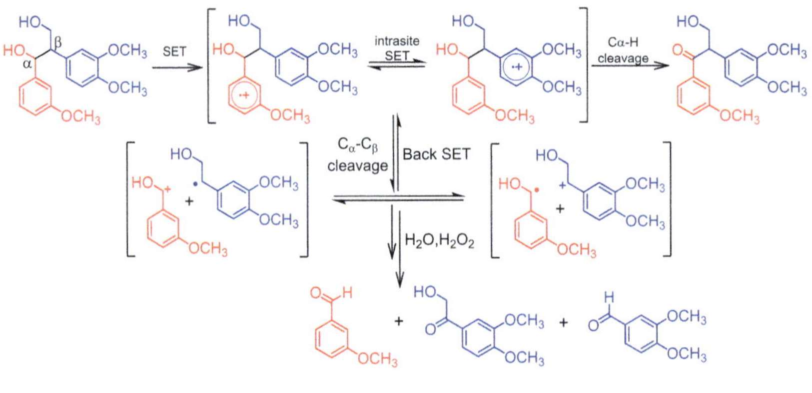

Figure 7.30 Depolymerization Route of β–1 Linkage via the First One-electron Oxidation of the Aromatic Rings. *Source:* Lim et al. [105], © 2015 Elsevier.

model compounds with photocatalyst Mes-$Acr^+BF_4^-$ or [Ir(III){dF(CF3) ppy}2(dtbpy)]PF_6. According to the proposed mechanism, the pathway consists of two steps, where the first one is catalyzed by the acridinium salt, and the second one is catalyzed by Cu(II) (Figure 7.31). The first step starts with the single electron transfer from the more electron-rich benzene ring (on the left) of **1** to the photo-excited acridinium cation (Mes-Acr^{+*}), which results in the formation of the cation radical **I** along with an acridinium radical (Mes-Acr$^\bullet$). The cation radical **I** then undergoes proton-coupled electron transfer (PCET) [107, 108] to furnish an alkoxy radical **II**. Then, the C_α–C_β bond of **II** is cleaved by β-scission to give rise to 3,5-dimethoxybenzaldehyde (2), along with aryloxy-stabilized carbon radical **III**. Molecular oxygen traps the radical **III** to form peroxide radical **IV**. The acridinium radical (Mes-Acr$^\bullet$) transfers one electron to **IV** to generate peroxide anion **V** with the regeneration of the acridinium cation (Mes-Acr^+), which initiates the next catalytic cycle of the first step. The second step is catalyzed by copper(**II**), which commences with ligand exchange of copper catalyst with the peroxide anion **V** to furnish copper peroxide **VI**. The highly energetic O–O bond undergoes homolytic cleavage, inducing the second C_β–C_γ bond cleavage to afford the formate **3**, formaldehyde, and generating the copper(II) species (**VII**).

In 2019, Zhang and co-workers [109] reported a redox-neutral photocatalytic strategy for selectively cleaving the C_α–C_β bonds in β-O-4 and β-1 linkages via a PCET process, which can provide aromatic aldehydes (up to 97%) and phenyl ethers (up to 96%) as the products at room temperature. By adjusting the alkalinity of the base to obtain a lignin models/base PCET pair with a bond dissociation free energy close to 102 kcal·mol^{-1}, this proton-coupled electron transfer (PCET) process does not need any pretreatments of the substrate. For the detail of the process, the Ir^{III} photocatalyst ([Ir(dF(CF3)ppy)2(5,5′-d(CF3)bpy)](PF_6)) was first excited to $^*Ir^{III}$ under the illumination of blue light. In the presence of a base, an intermolecular hydrogen bond was formed with lignin model, and then electron was transferred from the phenyl moiety (adjacent to the α-OH group) of the lignin model **1** to $^*Ir^{III}$ to produce an Ir^{II} containing compound along with the generation of a radical cationic intermediate **2**, which went through PCET process to furnish an alkoxy radical intermediate **3**. The generated alkoxy radical **3** without C–C bond cleavage will weaken the neighboring C_α–C_β bond, thus achieving the β-scission of C_α–C_β bond to produce benzaldehyde **4** and a carbon radical **5** (Figure 7.32). The carbon radical **5** could be trapped by hydrogen atom transfer (HAT) of thiophenol to form anisole **6**

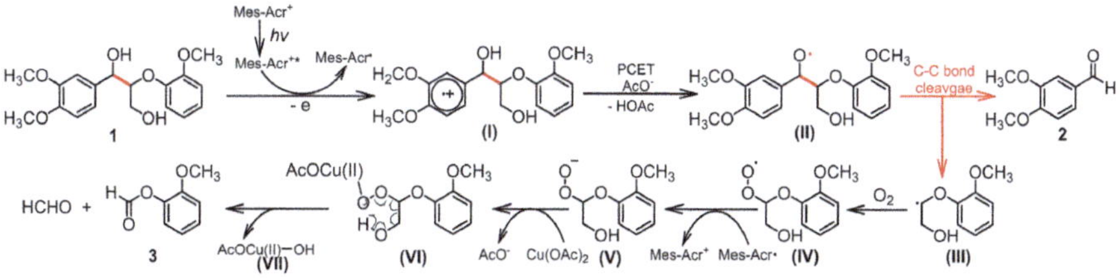

Figure 7.31 Light/Copper Relay for Aerobic Degradation of Lignin Model Compounds. *Source:* Zhou et al. [106] with Permission, © 2018 Wiley-VCH.

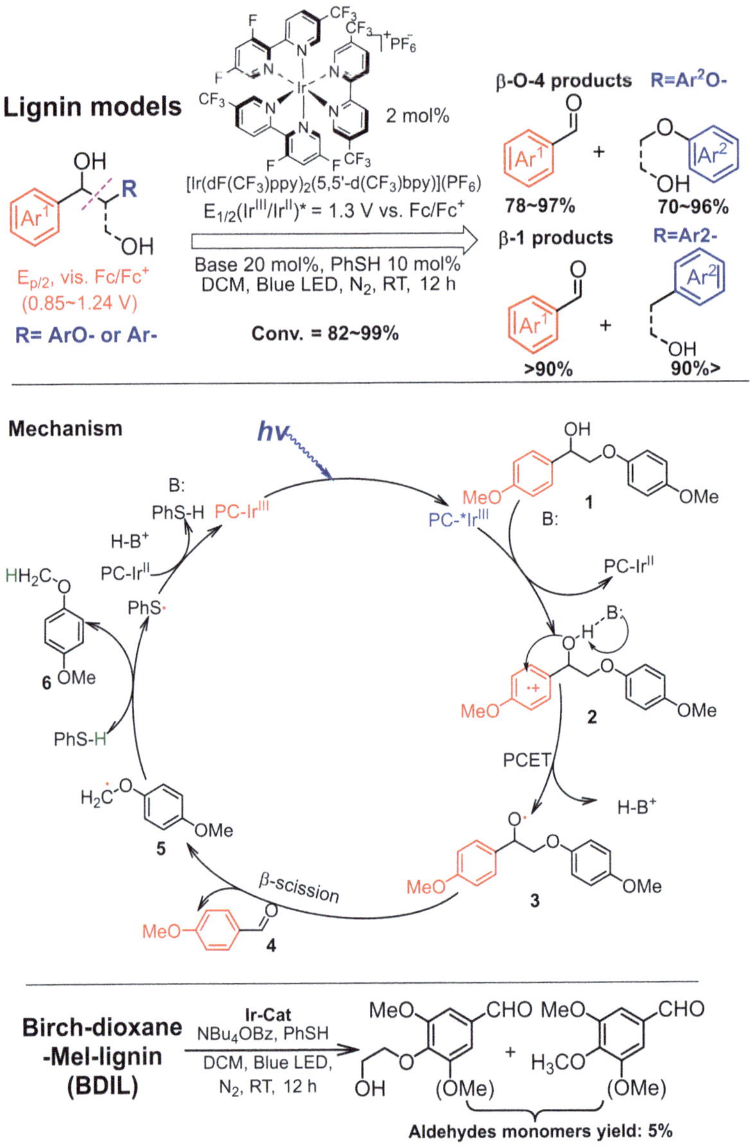

Figure 7.32 The mechanism for Selective Cα–Cβ Bond Cleavage by Ir Catalyst via a PCET Process. *Source:* Wang et al. [109] with Permission, © 2019 Science China Press.

along with the production of a PhS• radical. Then, the PhS• radical oxidized the Ir^{II} to Ir^{III} with base for the next catalytic cycle. In addition, their preliminary studies indicated that this PCET strategy is effective for the depolymerization of native lignin at room temperature, and 5% aromatic aldehyde monomers were obtained from birch-dioxane-Mel-lignin (BDIL).

In 2020, Knowles and co-workers [110] applied the Ir-catalyzed photocatalytic protocol in the catalytic cleavage of lignin C_α–C_β bonds. The system with [Ir(dF(CF3) ppy)$_2$(5,5-d(CF_3)-bpy)]PF_6, 2,4,6-triisopropylbenzene thiol (TRIP thiol), and tetrabutylphosphonium dimethylphosphate upon blue light irradiation exhibited good efficiency and excellent selectivity for the activation and fragmentation of the β-O-4 and β-1 linkages, even in the presence of numerous other PCET-active functional groups. Different from the above mechanism with the first SET oxidation of the aromatic ring to the cation radical, the reaction began with the first PCET oxidation of the C_α–OH to the $C_\alpha O^\bullet$ radical, and more discussion is provided in Chapter 9.7.

Besides the separate photocatalytic process, in 2020, Li and co-worker [111] studied the one-step photoelectrochemical decomposition of lignin model compounds at the surface of $BiVO_4$ (and $BiVO_4|V_2O_5$) photoanode with simulated sunlight and an applied bias of 2.0 V. The β-O-4 lignin model compounds were photoelectrochemically converted into the corresponding aryl aldehyde and phenol derivatives in a single step with the conversion up to 64% over 20 h.

Besides the β-O-4 and β-1 linkages with the first SET oxidation of the aromatic rings, in 2021, Jiang and co-workers [112] reported the uranyl-photocatalyzed hydrolysis of diaryl ethers to two phenols at room temperature under normal pressure (Figure 7.33). The single-electron transfer process was disclosed by a radical quenching experiment and Stern–Volmer analysis between diphenyl ether and uranyl cation catalyst, followed by oxygen atom transfer process between radical cation of diphenyl ether and uranyl peroxide species. The ^{18}O labeling experiment precisely demonstrates that the oxygen source is water. As shown in the proposed mechanism (Figure 7.33), the uranyl photoredox catalysis was stimulated by blue light through the LMCT (ligand-to-metal charge transfer) process, generating $^*UO_2^{2+}$. The SET process between $^*UO_2^{2+}$ and diaryl ether **1** produced UO_2^+ along with radical cation **A**, which was attacked by the uranyl peroxide dimer afforded via water-splitting [113, 114], yielding the radical cation of phenol **B** and phenyl oxygen anion **C**, accompanied by cleavage of the original C_{Ar}–O bond and formation of the

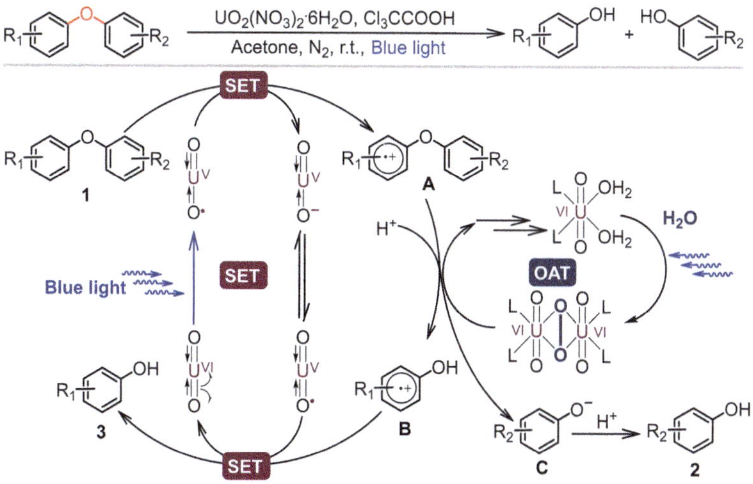

Figure 7.33 Uranyl-Photocatalyzed Hydrolysis of Diaryl Ethers. *Source:* Zhou et al. [112] with Permission, © 2021 American Chemical Society.

new one. The protonation of **C**, together with the SET process between intermediate **B** and UO_2^+ [115–117], afforded desired phenolic product **2** and regenerated the catalyst.

7.3 Lignin C_{Ar}–OC/C_{Ar}–C Bonds Cleavage via First Partly-Hydrogenation or Partly-Addition of the Neighbouring Aromatic Ring

As we have discussed in Chapter 4, direct breaking C_{Ar}–OR bonds is an effective but challenging means for the depolymerization of the lignin. In order to effectively break the C_{Ar}–OR bond, besides via the direct C_{Ar}–OR bond cleavage or via the first adjacent functional group modification of the R structure to reduce the BDE of ether bond or to introduce the substrate site to promote the breaking of the ether bond, the first transformation of the Ar structure group is also a potential approach to regulate the cleavage of lignin C_{Ar}–OR bonds. This subsection will mainly summarize the strategies for the lignin C_{Ar}–OC bonds cleavage via first partly-hydrogenation or partly-addition of the neighboring aromatic ring.

As a common organic chemistry knowledge point, the enol ether bond is unstable which is prone to occur hydration to form unstable hemiacetals (hemiketal), which further trend to transform to the corresponding carbonyl compounds with original ether bond cleavage [118–120].

In 2017, Han and co-workers [118] reported that bromide salt-modified Pd/C (m-Pd/C) in H_2O/CH_2Cl_2 could efficiently catalyze the transformation of aromatic ethers at 90~140 °C, 2 MPa H_2 to cyclohexanone and its derivatives via hydrogenation and hydrolysis processes. The yield of cyclohexanone from anisole can reach 96%, and the yields of cyclohexanone derivatives produced from the aromatic ethers, which can be extracted from plants or derived from lignin, are also satisfactory. Detailed study shows that the Pd, bromide salt, and H_2O/CH_2Cl_2 work cooperatively to promote the desired reaction and inhibit the side reaction. According to the proposed mechanism (Figure 7.34), the benzene ring of anisole is first partially hydrogenated to form 1-methoxycyclohexene on Pd catalyst in Step 1, which can be further hydrogenated to cyclohexyl methyl ether (Step 2) or hydrolyzed to

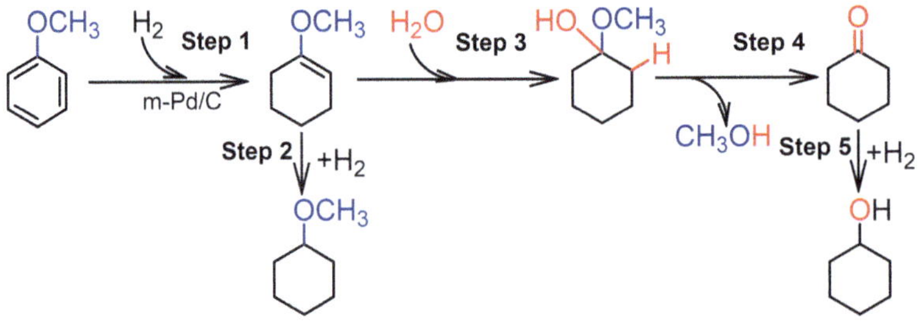

Figure 7.34 Synthesis of Ketones from Lignin-derived Methyl Aromatic Ether. *Source:* Meng et al. [118] with Permission, © 2017 Springer Nature.

cyclohexanone (Steps 3–4). Cyclohexanone can then be further hydrogenated to form cyclohexanol (Step 5). The m-Pd/C and H_2O/CH_2Cl_2 catalytic system promoted the production of cyclohexanone and inhibited the production of cyclohexyl methyl ether and cyclohexanol.

This hemiacetal (hemiketal) strategy is also effective in the conversion of 4-O-5 lignin linkage. In 2017, Lercher and co-workers [119] reported that metallic palladium surfaces are highly selective in promoting the reductive hydrolysis of aromatic ethers in an aqueous phase at relatively mild temperatures and pressures of H_2 (Figure 7.35). By analysis of the evolution of products with and without incorporation of $H_2{}^{18}O$, the pathway was concluded to be initiated by Pd-catalyzed partial hydrogenation of the phenyl group to an enol ether. Water then rapidly adds to the enol ether to form a hemiacetal, which then undergoes elimination to cyclohexanone and phenol/alkanol products.

Similar to the hemiacetal (hemiketal) mechanism, the enol ether bond (RCH = CR′–OR^1) can also react with the alcohol molecule (R^{2}OH) to provide a ketal (RCH$_2$CR′(OR^2)–OR^1), which can further transform to another enol ether (RCH = CR′–OR^2) after releasing the fragment of R^{1}OH. The process can be regarded as the exchange of the ether bond but it finally causes the cleavage of the original RCH = CR′–OR^1 ether bond. In terms of this mechanism, in 2018, Lercher and co-workers [120] reported that palladium on carbon can catalyze C_{Ar}–O bond cleavage of aryl ethers (diphenyl ether and cyclohexyl phenyl ether) by alcohols (R–OH) in H_2. The aromatic C_{Ar}–O bond is cleaved by reductive solvolysis (Figure 7.36), which is initiated by Pd-catalyzed partial hydrogenation of one phenyl ring of **1** to form an enol ether **2**. The enol ether **2** reacts rapidly with alcohols to form a ketal **4**, which generates 1-cyclohexenyl–O–R **5** by eliminating phenol or an alkanol. Subsequent hydrogenation leads to cyclohexyl–O–R **6**.

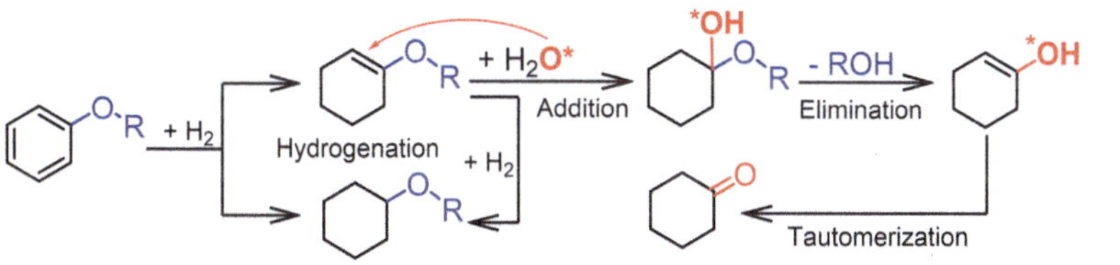

Figure 7.35 Palladium-Catalyzed Hydrolytic Cleavage of Aromatic C–O Bonds. *Source:* Wang et al. [119] with Permission, © 2017 Wiley-VCH.

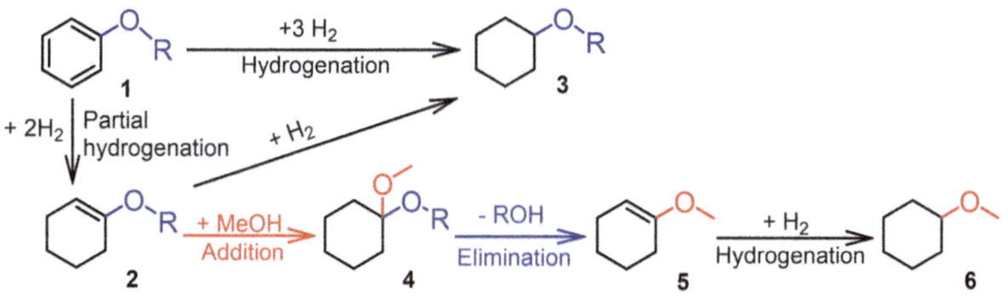

Figure 7.36 Palladium-Catalyzed Reductive Insertion of Alcohols into Aryl Ether Bonds. *Source:* Wang et al. [120] with Permission, © 2018 Wiley-VCH.

To cleave the aromatic ether bonds in 4-O-5 linkage, the above methods involve the first semi-hydrogenation of one aromatic ring to cyclohexene. As a result, the proportion of aromatic chemicals in the product is reduced to half or even lower. In order to maintain the two aromatic rings and use the strategy with the first activation of the double bond C = C bind in one aromatic ring, a feasible solution is to add a cleavage reagent (AB) to the C = C bond of the aromatic ring which connects with the Ar–OR^2 structure (Figure 7.37), and then the addition intermediate **2** occurs elimination reaction which can release the fragment B–OR^2 with Ar–OR^2 bond cleavage and rearomatizes the ring of **3** with the new C = C(–A) bond generation (Figure 7.37, Route I). Otherwise, the addition-intermediate **2** can even occur the elimination of A–OR^2 (Figure 7.37, Route II), after during which the left fragment intermediate **4** need occur the migration of group **B** to the carbon that originally connects with the –OR structure to achieve the rearomatization (**5→6**).

Compared with the above two routes, although Route II is feasible on chemistry principle, Route I is more easily to be accepted, while the process with the Route II mechanism really exists in the aromatic ether bond cleavage. In 2013, Martin and co-workers [121] reported a detailed mechanistic and computational study on the reductive cleavage of C_{Ar}–OMe bonds catalyzed by $Ni(COD)_2/PCy_3$ with silanes as reducing agents. Specifically, they demonstrate that the mechanism for this transformation did not proceed via oxidative addition of the Ni(0) precatalyst into the C–OMe bond. In the absence of an external reducing agent like R_3Si–H, the in-situ-generated oxidative addition complexes rapidly undergo β-hydride elimination at room temperature, ultimately leading to either Ni(0)– carbonyl- or Ni(0)–aldehyde-bound complexes, which have been summarized in Chapter 6.2. Characterization of Ni-centered complexes by X-ray crystallography, isotopic-labeling experiments, kinetic isotope effects, in-situ monitoring experiments by ^{1}H NMR, EPR, IR spectroscopy, and computational studies unambiguously suggested a different mechanism when silanes are present in the reaction media, during which the Ni(I) species work as the key catalytic species that are generated via comproportionation of initially formed Ni(II) species. Martin et. al. believe that the mechanism (Figure 7.38) is initiated by Ni(I)–$SiEt_3$ (**1**) that are generated from $Ni(COD)_2$, PCy_3, and Et_3SiH via a comproportionation event. Then the initially generated Ni(I) **1** coordinates to substrate **2** in a η^2-fashion and after a migratory insertion event, a benzyl nickel

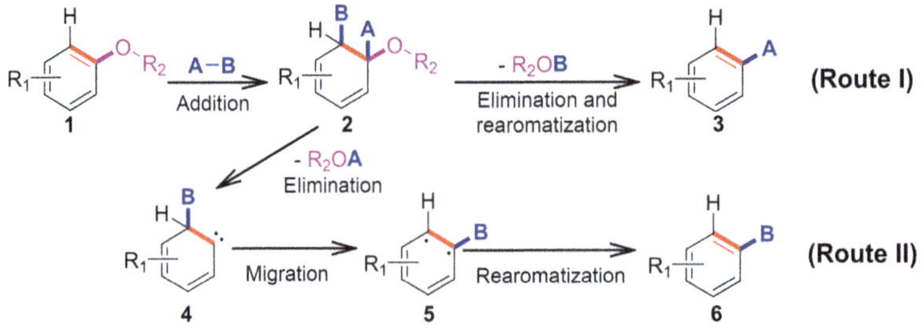

Figure 7.37 Cleaving Lignin Ether Bond via First Partly-addition of the Neighboring Aromatic Ring and Two Different Rearomatization Mechanisms.

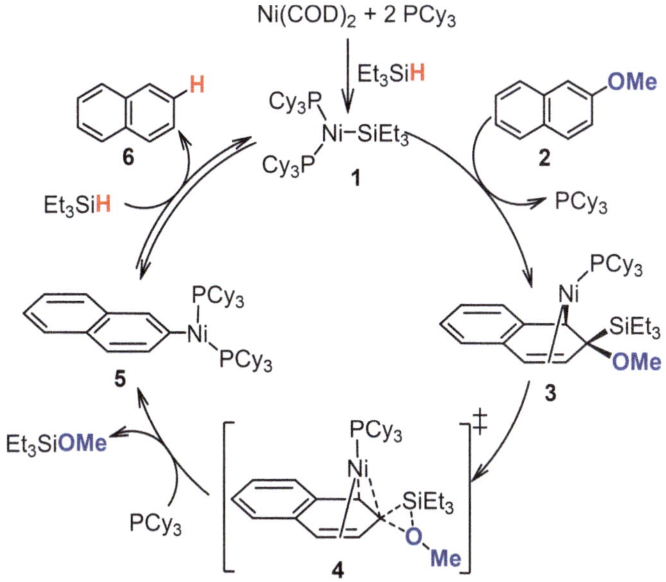

Figure 7.38 Ni-catalyzed Reductive Cleavage of Inert C_{Ar}–O Bonds with the Existence of Silanes. *Source:* Cornella et al. [121] with Permission, © 2013 American Chemical Society.

species **3** is generated. Elimination of $MeOSiR_3$ with the concomitant migration of the Ni atom results in **5** that undergoes a reversible σ-bond metathesis, delivering the final product **6** while generating back the Ni(I)–$SiEt_3$ species **1** to close the catalytic cycle.

Returning to the Route I mechanism shown in Figure 7.37, in 2020, John Ralph and co-workers [122] provided a detailed mechanism study of the palladium-catalyzed ether bond cleavage and rearrangement of 4-O-5 lignin models under a hydrogenolytic condition. Different from the previous study, the 4-O-5 lignin model contains a phenolic hydroxyl group, which makes the transformation of the 4-O-5 lignin model over the Pd surface present regioselectivity. As shown in Figure 7.39, hydrogenolysis reaction mechanisms begin with dissociative adsorption of H_2 onto the palladium catalyst surface, generating surface-bound hydrogen atoms (Step a). The reaction of a C5′–C6′-bond with the Pd–H species will partially saturate the B-ring (**Int-I**), with the regioselectivity for the addition of Pd–H to the aromatic ring rationalized similarly to Markovnikov's rule for charged species. The Pd–O bond could help stabilize the substrate on the catalyst surface via a Pd-coordinated intermediate (**Int-I**) (Step b). Different from the above hemiketal/ketal decomposition mechanism with the critical enol ether intermediate, migration of a second hydrogen atom to the palladium-B-ring complex will release the partially hydrogenated product (**Int-II**) and regenerate the Pd catalyst (Step c). Then, an elimination reaction from **Int-II** will cleave the ether bond to generate monomeric products while rearomatizing the B-ring (Step d). The products of this hydrogenolysis pathway are two identical 1,3,4-substituted monomers (**Cycle I**). According

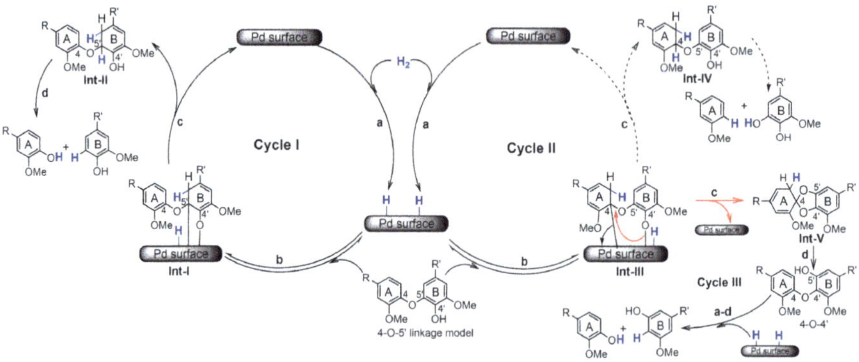

Figure 7.39 Pd-catalyzed Ether Bond Cleavage and Rearrangement of 4-O-5 Lignin Models under a Hydrogenolytic Condition. *Source:* Li et al. [122] with Permission, © 2020 Wiley-VCH.

to the products distribution analysis, another hydrogenolysis cycle (**Cycle II**) with the first reaction of the A-ring is not the main route, which involves the hydrogen addition to the A-ring to the Pd-coordinated **Int-III**, the addition of the second hydrogen atom to **Int-III** to release the partially saturated A-ring product (**Int-IV**), and subsequent elimination cleavage of the aryl ether bond to provide two different monomers with 1,3,4,5- and 1,3-substitution patterns. Nevertheless, a variant of the secondary pathway provides a rationale for the unique rearrangement products (**Cycle III**), during which **Int-III** could undergo rearrangement, rather than the addition of a second hydrogen atom. Intramolecular addition of the Pd-coordinated phenoxide onto the B-ring can give the spirocyclic cyclohexadiene (**Int-V**). Formation of the new C4–O bond could arise from the rotation of the B-ring to orient parallel to the Pd surface, leading to orbital overlap between the C4 and O atoms. An elimination reaction involving ring-opening of the 5-membered dioxolane ring will rearomatize the A-ring and either generate the 4-O-4′ rearrangement product or regenerate the 4-O-5′ starting material, depending on which C4–O bond is cleaved in Int-V. The 4-O-4′ products can then enter the ether cleavage cycle (Cyclic III, referring to step a–d) to afford the 1,3,4- and the novel 1,3,5-substituted (rearrangement) products.

Besides the above methods with the first semi-addition of one aromatic C = C bond with AB (two H* or Ni–Si species), the addition with two oxygen-centered groups for the oxidation transformation of the lignin unite is also feasible on principle, such as the methyltrioxorhenium (MTO)/H_2O_2-catalyzed aromatic ring oxidation with C_{Ar}–OC/C_{Ar}–C bonds cleavage (Figure 7.40) [123], during which MTO/H_2O_2 can first oxidize the aromatic ring and provide a 1,2-arene oxide intermediate. The 1,2-arene oxide intermediate can provide *p*-benzoquinones via a hydration-hydration-oxidation process (Route I). Otherwise, the 1,2-arene oxide can occur nucleophilic ring-opening with H_2O_2, following which the decomposition and further oxidation of the peroxide intermediate to the *o*-benzoquinones with cleavage of the C_{Ar}–C bond. As the side reactions, the overoxidation of the benzoquinones can cause the fragmentation of the original aromatic rings to various organic acids.

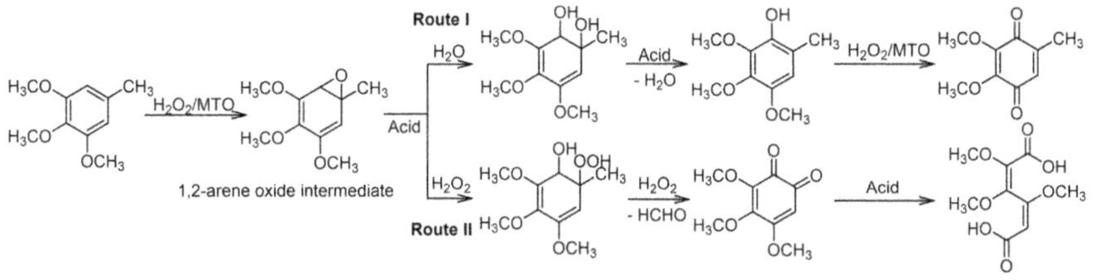

Figure 7.40 Methoxybenzenes Oxidation to Benzoquinones Catalyzed by MTO/H_2O_2 and HBF_4. *Source:* Adam et al. [123] with Permission, © 1995 Elsevier.

7.4 Lignin C(sp^2)–C(sp^2) σ Bond and C(sp^2)–OAr Bonds Cleavage via Adjacent Aromatic Groups Activation or Extra Radicals Attack

Due to the high bond dissociation energy, low proton acidity, and highly nonpolar character of the C–C bond, activation or cleavage of the bonds, especially the C(sp^2)–C(sp^2) σ bond, is one of the most challenging tasks in synthetic chemistry. In the lignin structure, the 5-5′ linkage contains the C(sp^2)–C(sp^2) σ bond. Compared with various C–O linkages, the 5-5′ linkage seems more difficult to be cleaved, which is regarded as the hard bone for the lignin depolymerization research.

To achieve the cleavage of the lignin 5-5′ C–C bond, breaking one aromatic ring and keeping another one could be a compromise solution, such as the Ru/$NbOPO_4$-catalyzed hydrogenation system reported in 2019 by the Wang Group (discussed in Chapter 6.1) [124], during which the biphenyl could be cleaved into one benzene and one cyclohexane at 280 °C under 0.5 MPa H_2. Before Wang's report [124], in 2016, Liu and co-workers [125] presented a potential strategy for the cleavage of C5–C5′ σ bond between two phenyl groups under mild reaction conditions during the construction of zinc(II) organic frameworks. According to the results and proposed mechanism (Figure 7.41), the coordination of Zn^{2+} with the biphenyl diacid ligand **A** and deprotonation of DMF in **B** (**step a** and **b**) generates a $[Zn–H]^+$ intermediate **C** and an iminium cation intermediate. Then, intramolecular nucleophilic attack of the hydride anion to the electron-deficient aromatic ring (step c) gives the anion intermediate **D**. Next, isomerization along with C–C bond cleavage affords the intermediate **E** (**step d**). Rapid protonation of the intermediate **E** (**step e**) generates the terephthalate **G** and benzoate **F**.

Inspired by the $[Zn–H]^+$ mechanism which involves the introduction of cleavage reagent [Zn–H] at the substrate site (adjacent aromatic carboxylic acid group) for the C5–C5′ σ bond cleavage, in 2019, Li and co-workers [126] developed a photoinduced dearomatization of nonphenolic biaryl compounds to generate spirolactones via a carboxyl radical. The reaction can be performed on substrates blocked in the *ortho*-position without *ipso* substitution using an acridinium catalyst under aerobic conditions. Importantly, due to the suppressed *ipso* substitution, spirolactones with labile groups (OMe) can be generated (Figure 7.42), which could be used in synthesizing several bioactive molecules. In addition,

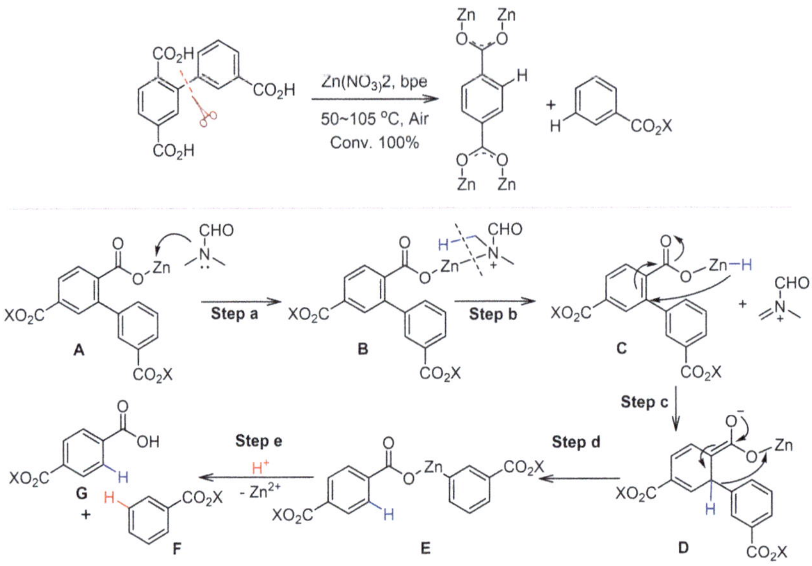

Figure 7.41 Cleavage of a C(sp^2)–C(sp^2) σ Bond between Two Phenyl Groups under Mild Conditions during the Construction of Zn(II) Organic Frameworks. *Source:* Liu et al. [125] with Permission, © 2016 The Royal Society of Chemistry.

a complementary methodology in which commercial 2,3-dichloro-5,6-dicyano-1, 4-benzoquinone (DDQ) is used as a photo oxidant to generate the spirolactones under aqueous conditions is disclosed to tolerate anaerobic synthetic conditions. Based on the experimental data and previous literature reports, reaction pathways under aerobic and anaerobic conditions were proposed (Figure 7.42). The formation of key carboxyl radical intermediate A occurs as the first step in both catalytic mechanisms (Method **A** and **B**) [127]. Carboxyl radical intermediate **A** induces dearomatization by an intramolecular cyclization to form intermediate **B** in both systems, where the twisted conformation promotes the cyclization [128]. From this point, the two pathways diverge. Under aerobic *Condition A*, nucleophilic attack of intermediate **B** by O_2 could deliver peroxy radical **C**, then reaction between **C** and **B** gives radical **E** [129, 130], which could further deliver product **2** via hydrogen atom transfer in the presence of TEMPO (2,2,6,6-tetramethylpiperidine-1-oxyl) [131]. The addition of TEMPO could quench the harmful superoxide radical anion and realize mild oxidative conditions [132]. Meanwhile, DABCO (1,4-diazabicyclo-[2.2.2]octane) additive has been proposed to facilitate carboxyl radical formation by deprotonation [133]. In the case of *Condition B*, intermediate **B** undergoes oxidation to give cation **D**, which then traps a molecule of water to deliver hydroxylated intermediate **F**. Intermediate **F** can then undergo further oxidation to give final product **2**.

Although the extra carboxyl radical introduction is not efficient for C5–C5′ σ bond cleavage, in 2020, Li and co-workers [134] from another group developed a visible-light photoredox-catalyzed strategy for the cleavage C_{Ar}–O bond in diaryl ethers (4-O-5 linkage model) by an acidolysis with an aryl carboxylic acid and a following

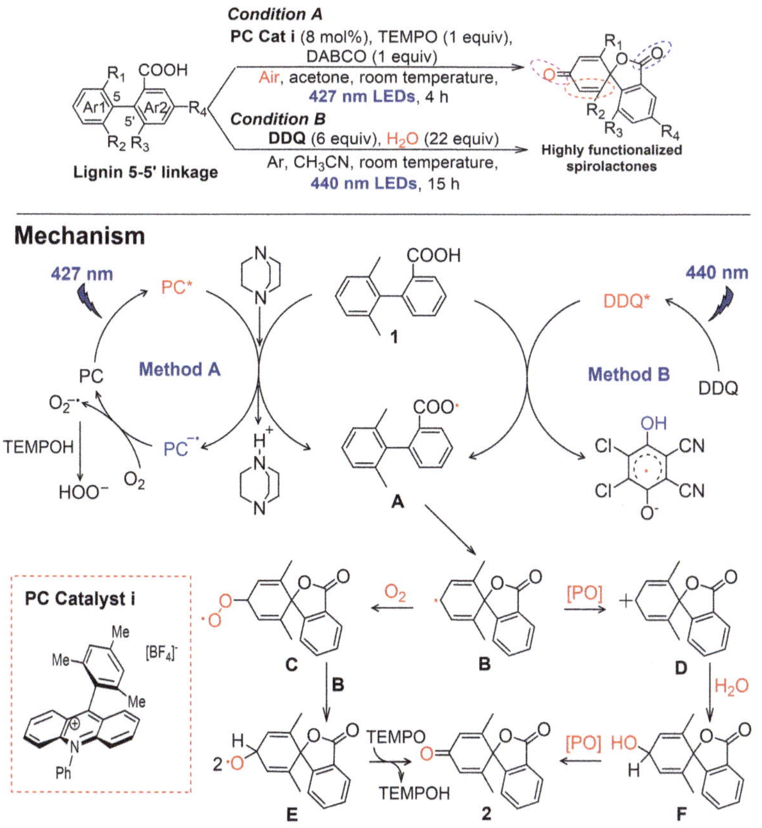

Figure 7.42 Functionalized Spirolactones by Photoinduced Dearomatization of Biaryl Compounds. *Source:* Li et al. [126] with Permission, © 2019 The Royal Society of Chemistry.

one-pot hydrolysis. Two phenols are obtained from a diaryl ether at room temperature (Figure 7.43). The key to the success of the acidolysis is merging visible-light photo redox catalysis with a new acridinium photocatalyst and Lewis acid catalysis with Cu(TMHD)$_2$. Preliminary mechanistic studies indicate that the catalytic cycle occurs via a rare selective electrophilic attack of the generated aryl carboxylic radical on the electron-rich aryl ring of diphenyl ether. In detail, Cu(TMHD)$_2$ firstly, promotes the ionization of aryl carboxylic acid **1** to afford the carboxylic anion **2** and a proton. Meanwhile, irradiation of PC with blue LEDs leads to the excited state PC*. A SET occurs between PC* and carboxylic anion **2** to generate the carboxylic acid radical **3** and $PC^{-\bullet}$. An electrophilic attack of **3** occurs on the electron-rich aryl ring of diphenyl ethers 4 to form intermediate **5**. A SET between **5** and $PC^{-\bullet}$ in the presence of a proton, and the promotion of Cu(TMHD)$_2$ as a Lewis acid afford aromatic ester **6**, phenol **7**, with the regeneration of PC. In the following hydrolysis step, the aromatic ester further transforms to another phenol, with the recovery of the aryl carboxylic acid **1**. In their further gram-scale reaction, after the optimization of the reaction condition, the aryl carboxylic acid (*p*-fluorobenzoic acid) after acidolysis can be recovered by 88%.

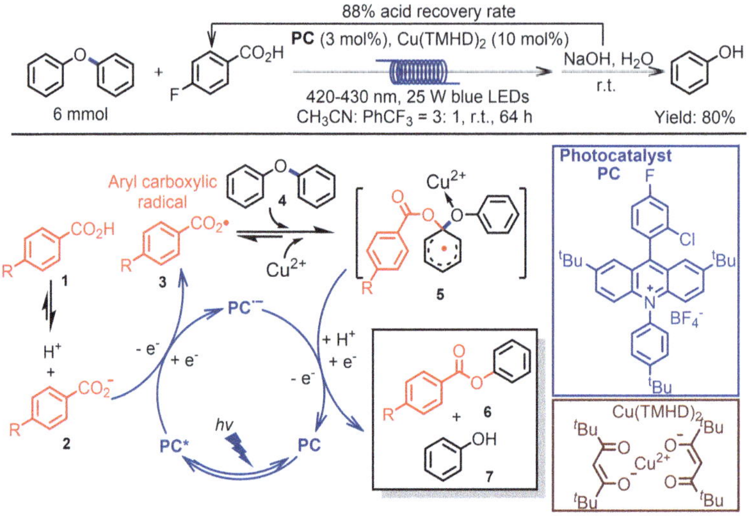

Figure 7.43 Visible-Light Photoredox-Catalyzed C_{Ar}–O Bond Cleavage of Diaryl Ethers by Acridinium Photocatalysts at Room Temperature. *Source:* Tan et al. [134] with Permission, © 2020 Springer Nature.

Let's return to cleavage of the stable C5–C5′ σ bond, in 2019, Dong and co-workers [135] reported a general approach for the catalytic activation of the unstrained C(aryl)–C(aryl) σ bonds in 2,2′-biphenols, which can be regarded as the fragment of the lignin 5-5′ linkage after the hydrogenolysis of dibenzodioxocin structure (Figure 7.44c). The success of this method is utilizing the phenol moiety as a handle to install phosphinites as a recyclable directing group (Figure 7.44a). Using H_2 as the reductant, monophenols are obtained with a low Rh catalyst loading and high functional group tolerance. A further mechanistic study suggests that the C5–C5′ σ bond activation step is mediated by a rhodium(I) monohydride species.

For the critical question as to how the C(aryl)–C(aryl) bond is cleaved during Rh-catalyzed reaction [135], based on previous organometallic studies on C–C activation [136–138], two pathways are possible. Path A involves a direct oxidative addition of the aryl–aryl bond to the Rh(I)–Cl species to generate a diaryl–Rh(III) intermediate; the further oxidative addition of H_2 and subsequent two C–H reductive elimination (a formal σ-bond metathesis) result in two monomer products (Figure 7.44b, Path A). Alternatively, the Rh(I)–Cl precatalyst can first react with H_2 to give one Rh(I) monohydride species, which is more electron-rich than the corresponding Rh(I)–Cl complex and can serve as the real catalyst in this reaction. In path B, the Rh(I)–H first undergoes oxidative addition with the aryl–aryl bond to give one Rh(III) intermediate, followed by C–H reductive elimination to give one monomer and one Rh(I)-aryl species. The following oxidative addition of H_2 converts the Rh(I)–aryl species back to the Rh–H catalyst and also produces another equivalent of the monomer (Figure 7.44b, Path B). Further results from the mechanistic study with control experiments, and in-situ characterizations, as well as the

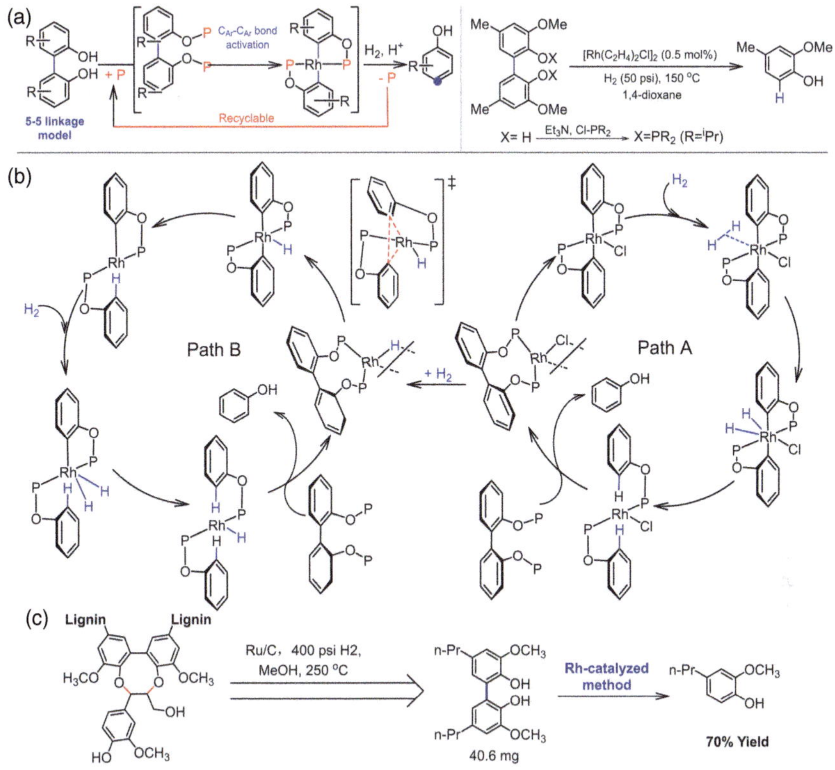

Figure 7.44 Strategy for the Catalytic Activation of C(aryl)–C(aryl) Bonds of 2,2′-Biphenols with Phenol Moieties as the Handle and P Atom in Phosphinite as the Directing Groups (a); Two Proposed Reaction Pathways for the Cleavage of C(aryl)–C(aryl) Bonds in 2,2′-Biphenols (b); The Application of Rh-catalyzed Method in the Transformation of Lignin Dibenzodioxocin linkage in Spruce (c). *Source:* Zhu et al. [135] with the Permission, © 2019 Springer Nature.

DFT calculation, support the C5–C5′ σ bond activation through Path B. Finally, Dong and co-workers [135] illustrated a preliminary study on breaking the inert biphenolic moieties in lignin models and spruces extract. Treatment of wet sawdust of spruce wood with Ru/C and H_2 at 250 °C in MeOH gave 82.9 mg of a crude mixture that contained ~40.6 mg of the lignin dimer 2,2′-biphenol. Further reaction of the crude mixture with Et_3N and $^iPr_2PCl$, followed by the Rh-catalysed C5–C5′ activation, afforded 28.6 mg of the pure phenol monomer in 70% yield based on extracted 2,2′-biphenol (Figure 7.44c).

In addition, besides the promotion effect of the first activation of adjacent aromatic groups, e.g. carboxylate, carboxylic acid radical, phenolic hydroxyl group, or the extra radical introduction for $C(sp^2)-C(sp^2)$ σ bond and $C(sp^2)$–OAr bonds cleavage, in 2021, Li and Vlachos [139] performed DFT calculations on 2-methoxy-1,1′-biphenyl, a 5-5′ lignin model compound, with various substitutions at all ring locations on Ru(0001). They analyzed the competition between the 5–5′ bond cleavage and the defunctionalization of the side functional groups at multiple degrees of depolymerization. They also discussed the role of ring functional groups in the adsorption of lignin oligomers and the 5–5 bond scission and, conversely, the effect of the aromatic group on the $-OCH_3$ decomposition. According to the most preferred steps in the reaction network, the $C(sp^2)$–$C(sp^2)$ σ bond cleavage begins with the first transformation of the $-OCH_3$ group to the active adsorbed species. Furthermore, they found that increasing the number of methoxy groups can decrease the energy barrier for C–C cleavage.

7.5 Chapter Summary

Based on the lignin linkage structure whether contains a phenolic hydroxyl group, Chapter 7 first summarises two kinds of methods for lignin C–C bonds cleavage: (1) Cleavage of lignin C_{Ar}–C bond after the first phenolic hydroxyl group dehydrogenation to the phenolic radical (Chapter 7.1); (2) Cleavage of lignin C–C bond via the first single-electron transfer from the aromatic ring (Chapter 7.2). Furthermore, as an extension of these methods, the first transformation of one aromatic ring to the active intermediate, such as semi-hydrogenation/addition to a cyclohexenyl ether (Chapter 7.3), or first inserting a cleavage reagent between the target bond with the assistance of the aromatic rings conjoint groups, metal catalyst or extra radical (Chapter 7.4), could also provide some useful methods for lignin linkage cleavage. The keywords of this chapter could be Lignin depolymerization, C–C Bond cleavage, Phenolic radical, Single-electron transfer, Ether bonds cleavage, Semi-hydrogenation/addition, Metal/radical inserting.

References

1 Zhang, C., Huang, Z., Lu, J. et al. (2018). Generation and confinement of long-lived N-oxyl radical and its photocatalysis. *J. Am. Chem. Soc.* 140: 2032–2035.

2 Deng, H.B., Lin, L., and Liu, S.J. (2010). Catalysis of Cu-doped Co-based Perovskite-type oxide in wet oxidation of lignin to produce aromatic aldehydes. *Energy Fuel* 24: 4797–4802.

3 Ponomarev, A.V. and Ershov, B.G. (2018). Radiation-thermal decomposition of lignin: Products and the mechanism of their formation (review). *High Energy Chem.* 52: 58–70.

4 Tarabanko, V.E., Hendogina, Y.V., Petuhov, D.V., and Pervishina, E.P. (2000). On the role of retroaldol reaction in the process of lignin oxidation into vanillin. Kinetics of the vanillideneacetone cleavage in alkaline media. *React. Kinet. Catal. Lett.* 69: 361–368.

5 Pardini, V.L., Smith, C.Z., Utley, J.H.P. et al. (1991). Electroorganic reactions. 38. Mechanism of electrooxidative cleavage of lignin model dimers. *J. Org. Chem.* 56: 7305–7313.

6 Reddy, G.V.B., Sridhar, M., and Gold, M.H. (2003). Cleavage of nonphenolic β-1 diarylpropane lignin model dimers by manganese peroxidase from Phanerochaete chrysosporium-Evidence for a hydrogen abstraction mechanism. *Eur. J. Biochem.* 270: 284–292.

7 Drago, R., Corden, B., and Barnes, C. (1986). Novel cobalt (II)-catalyzed oxidative cleavage of a carbon-carbon double bond. *J. Am. Chem. Soc.* 108: 2453–2454.

8 Bozell, J.J., Hames, B.R., and Dimmel, D.R. (1995). Cobalt-Schiff base complex catalyzed oxidation of para-substituted phenolics. Preparation of benzoquinones. *J. Org. Chem.* 60: 2398–2404.

9 Elder, T. and Bozell, J.J. (1996). Cobalt Schiff base complex catalyzed oxidation of para-substituted phenolics. Molecular orbital calculations on phenolic substrates. *Holzforschung* 50: 24–30.

10 Sippola, V., Krause, O., and Vuorinen, T. (2004). Oxidation of lignin model compounds with cobalt-sulphosalen catalyst in the presence and absence of carbohydrate model compound. *J. Wood. Chem. Technol.* 24: 323–340.

11 Badamali, S.K., Luque, R., Clark, J.H., and Breeden, S.W. (2009). Microwave assisted oxidation of a lignin model phenolic monomer using Co(salen)/SBA-15. *Catal. Commun.* 10: 1010–1013.

12 Zhou, X.-F., Quin, J., and Wang, S.-R. (2011). Oxidation of a lignin model compound of benzyl-ether type linkage in water with H_2O_2 under an oxygen atmosphere catalyzed by Co (salen). *Drewno. Pr. Nauk. Donies. Komunik.* 54: 15–25.

13 Cedeno, D. and Bozell, J.J. (2012). Catalytic oxidation of para-substituted phenols with cobalt-Schiff base complexes/O_2-selective conversion of syringyl and guaiacyl lignin models to benzoquinones. *Tetrahedron Lett.* 53: 2380–2383.

14 Lu, X.J. and Zhou, X.F. (2014). Co(Salen)-Catalysed oxidation of synthetic lignin-like polymer: H_2O_2 effects. *Oxid. Commun.* 37: 572–582.

15 Biannic, B., Bozell, J.J., and Elder, T. (2014). Steric effects in the design of Co-Schiff base complexes for the catalytic oxidation of lignin models to para-benzoquinones. *Green Chem.* 16: 3635–3642.

16 Riano, S., Fernandez, D., and Fadini, L. (2008). Oxidative cleavage of vic-diols catalyzed by manganese(III) complexes in ionic liquids. *Catal. Commun.* 9: 1282–1285.

17 Zhang, N. and Zhou, X.F. (2012). Salen copper (II) complex encapsulated in Y zeolite: An effective heterogeneous catalyst for TCF pulp bleaching using peracetic acid. *J. Mol. Catal. A: Chem.* 365: 66–72.

18 Zhang, G., Scott, B.L., Wu, R. et al. (2012). Aerobic oxidation reactions catalyzed by vanadium complexes of bis(phenolate) ligands. *Inorg. Chem.* 51: 7354–7361.

19 Basolo, F., Hoffman, B.M., and Ibers, J.A. (1975). Synthetic oxygen carriers of biological interest. *Acc. Chem. Res.* 8: 384–392.

20 Akira, N., Toshiyuki, K., and Teruo, M. (1985). Oxygenation of cobalt (II) Schiff base complexes in alcohols. *Chem. Lett.* 14: 905–908.

21 Bolzacchini, E., Chiavetto, L.B., Canevali, C. et al. (1996). Oxidation of propenoidic phenols catalysed by N,N′-ethylenebis(salicylideneiminato)cobalt(II) [Cosalen]: reactivity and spectroscopic studies. *J. Mol. Catal. A: Chem.* 112: 347–351.

22 Bolzacchini, E., Canevali, C., Morazzoni, F. et al. (1997). Spectromagnetic investigation of the active species in the oxidation of propenoidic phenols catalysed by [N,N′-bis(salicylidene)ethane-1,2-diaminato]cobalt(II). *J. Chem. Soc., Dalton Trans.* 4695–4700.

23 Canevali, C., Orlandi, M., Pardi, L. et al. (2002). Oxidative degradation of monomeric and dimeric phenylpropanoids: reactivity and mechanistic investigation. *J. Chem. Soc., Dalton Trans.* 3007–3014.

24 Zoia, L., Canevali, C., Orlandi, M. et al. (2008). Radical formation on TMP fibers and related lignin chemical changes. *BioResources* 3: 21–33.

25 Kervinen, K., Korpi, H., Mesu, J.G. et al. (2005). Mechanistic insights into the oxidation of veratryl alcohol with Co(salen) and oxygen in aqueous media: An in-situ spectroscopic study. *Eur. J. Inorg. Chem.* 2591–2599.

26 Zakzeski, J., Jongerius, A.L., and Weckhuysen, B.M. (2010). Transition metal catalyzed oxidation of Alcell lignin, soda lignin, and lignin model compounds in ionic liquids. *Green Chem.* 12: 1225–1236.

27 Zakzeski, J., Bruijnincx, P.C.A., and Weckhuysen, B.M. (2011). In situ spectroscopic investigation of the cobalt-catalyzed oxidation of lignin model compounds in ionic liquids. *Green Chem.* 13: 671–680.

28 Zhou, X.F. (2014). Selective oxidation of kraft lignin over zeolite-encapsulated Co(II) [H4] salen and [H2]salen complexes. *J. Appl. Polym. Sci.* 131: 40809.

29 Sippola, V.O. and Krause, A.O.I. (2003). Oxidation activity and stability of homogeneous cobalt-sulphosalen catalyst-Studies with a phenolic and a non-phenolic lignin model compound in aqueous alkaline medium. *J. Mol. Catal. A: Chem.* 194: 89–97.

30 Tarabanko, V.E., Petukhov, D.V., and Selyutin, G.E. (2004). New mechanism for the catalatic oxidation of lignin to vanillin. *Kinet. Catal.* 45: 569–577.

31 Wu, G.X., Heitz, M., and Chornet, E. (1994). Improved alkaline oxidation process for the production of aldehydes (vanillin and syringaldehyde) from steam-explosion hardwood lignin. *Ind. Eng. Chem. Res.* 33: 718–723.

32 Wu, G. and Heitz, M. (1995). Catalytic mechanism of Cu^{2+} and Fe^{3+} in alkaline O_2 oxidation of lignin. *J. Wood. Chem. Technol.* 15: 189–202.

33 Deng, H., Lin, L., Sun, Y. et al. (2008). Perovskite-type oxide $LaMnO_3$: An efficient and recyclable heterogeneous catalyst for the wet aerobic oxidation of lignin to aromatic aldehydes. *Catal. Lett.* 126: 106–111.

34 Deng, H., Lin, L., Sun, Y. et al. (2008). Activity and stability of $LaFeO_3$ catalyst in lignin catalytic wet oxidation to aromatic aldehydes. *Chin. J. Catal.* 29: 753–757.

35 Deng, H., Lin, L., Sun, Y. et al. (2009). Activity and stability of Perovskite-type oxide $LaCoO_3$ catalyst in lignin catalytic wet oxidation to aromatic aldehydes process. *Energy Fuel* 23: 19–24.

36 Deng, H., Lin, L., and Liu, S. (2010). Catalysis of Cu-doped Co-based Perovskite-type oxide in wet oxidation of lignin to produce aromatic aldehydes. *Energy Fuel* 24: 4797–4802.

37 Gao, P., Li, N., Wang, A. et al. (2013). Perovskite $LaMnO_3$ hollow nanospheres: The synthesis and the application in catalytic wet air oxidation of phenol. *Mater. Lett.* 92: 173–176.

38 Gao, P., Li, C., Wang, H. et al. (2013). Perovskite hollow nanospheres for the catalytic wet air oxidation of lignin. *Chin. J. Catal.* 34: 1811–1815.

39 Zhang, J., Deng, H., and Lin, L. (2009). Wet aerobic oxidation of lignin into aromatic aldehydes catalysed by a Perovskite-type oxide: $LaFe_{1-x}Cu_xO_3$ (x = 0, 0.1, 0.2). *Molecules* 14: 2747–2757.

40 Royer, S., Levasseur, B., Alamdari, H. et al. (2008). Mechanism of stearic acid oxidation over nanocrystalline La_1-xA′xBO_3 (A′ = Sr, Ce; B = Co, Mn): The role of oxygen mobility. *Appl. Catal., B* 80: 51–61.

41 Smith, C.Z., Utley, J.H.P., and Hammond, J.K. (2011). Electro-organic reactions. Part 60 [1]. The electro-oxidative conversion at laboratory scale of a lignosulfonate into vanillin in an FM01 filter press flow reactor: preparative and mechanistic aspects. *J. Appl. Electrochem.* 41: 363–375.

42 Zhu, H., Wang, L., Chen, Y. et al. (2014). Electrochemical depolymerization of lignin into renewable aromatic compounds in a non-diaphragm electrolytic cell. *RSC Adv.* 4: 29917–29924.

43 Wang, L., Chen, Y., Liu, S. et al. (2017). Study on the cleavage of alkyl-O-aryl bonds by in situ generated hydroxyl radicals on an ORR cathode. *RSC Adv.* 7: 51419–51425.

44 Zirbes, M., Quadri, L.L., Breiner, M. et al. (2020). High-temperature electrolysis of Kraft lignin for selective vanillin formation. *ACS Sustainable Chem. Eng.* 8: 7300–7307.

45 Zhang, Y.M., Peng, Y., Yin, X.L. et al. (2014). Degradation of lignin to BHT by electrochemical catalysis on Pb/PbO_2 anode in alkaline solution. *J. Chem. Technol. Biotechnol.* 89: 1954–1960.

46 Li, S., Kim, S., Davis, A.H. et al. (2021). Photocatalytic chemoselective C–C bond cleavage at room temperature in dye-sensitized photoelectrochemical cells. *ACS Catal.* 11: 3771–3781.

47 Subbotina, E., Rukkijakan, T., Marquez-Medina, M.D. et al. (2021). Oxidative cleavage of C—C bonds in lignin. *Nat. Chem.* 1118–1125.

48 Higuchi, T. (1990). Lignin Biochemistry - Biosynthesis and biodegradation. *Wood Sci. Technol.* 24: 23–63.

49 Higuchi, T. (1989). Mechanisms of lignin degradation by lignin peroxidase and laccase of white-rot fungi. *ACS Symp. Ser.* 399: 482–502.

50 Ander, P. and Eriksson, K.-E. (1977). Selective degradation of wood components by white-rot fungi. *Physiol. Plant.* 41: 239–248.

51 Schoemaker, H.E., Harvey, P.J., Bowen, R.M., and Palmer, J.M. (1985). On the mechanism of enzymatic lignin breakdown. *FEBS Lett.* 183: 7–12.

52 Harvey, P.J., Schoemaker, H.E., Bowen, R.M., and Palmer, J.M. (1985). Single-electron transfer processes and the reaction mechanism of enzymic degradation of lignin. *FEBS Lett.* 183: 13–16.

53 Paterson, A. and Lundquist, K. (1985). Radical breakdown of lignin. *Nature* 316: 575–576.

54 Kersten, P.J., Tien, M., Kalyanaraman, B., and Kirk, T.K. (1985). The ligninase of phanerochaete-chrysosporium generates cation radicals from methoxybenzenes. *J. Biol. Chem.* 260: 2609–2612.

55 Lundquist, K. and Kristersson, P. (1985). Exhaustive laccase-catalysed oxidation of a lignin model compound (vanillyl glycol) produces methanol and polymeric quinoid products. *Biochem. J.* 229: 277–279.

56 Dekker, R. and Higuchi, T. (1985). *Biosynthesis and biodegradation of wood components*. Academic Press Orlando.

57 Poulos, T.L., Edwards, S.L., Wariishi, H., and Gold, M.H. (1993). Crystallographic refinement of lignin peroxidase at 2 Å. *J. Biol. Chem.* 268: 4429–4440.

58 Higuchi, T. Biochemistry of wood components: Biosynthesis and microbial degradation of lignin. 2002.

59 Tien, M. and Kirk, T.K. (1984). Lignin-degrading enzyme from phanerochaete-chrysosporium-purification, characterization, and catalytic properties of a unique H_2O_2-requiring oxygenase. *Proc. Natl. Acad. Sci. USA* 81: 2280–2284.

60 Khindaria, A., Yamazaki, I., and Aust, S.D. (1995). Veratryl alcohol oxidation by lignin peroxidase. *Biochemistry* 34: 16860–16869.

61 Tien, M., Kirk, T.K., Bull, C., and Fee, J.A. (1986). Steady-state and transient-state kinetic studies on the oxidation of 3,4-dimethoxybenzyl alcohol catalyzed by the ligninase of Phanerocheate chrysosporium Burds. *J. Biol. Chem.* 261: 1687–1693.

62 Haemmerli, S.D., Schoemaker, H.E., Schmidt, H.W.H., and Leisola, M.S.A. (1987). Oxidation of veratryl alcohol by the lignin peroxidase of Phanerochaete chrysosporium Involvement of activated oxygen. *FEBS Lett.* 220: 149–154.

63 Glenn, J.K. and Gold, M.H. (1985). Purification and characterization of an extracellular Mn(II)-dependent peroxidase from the lignin-degrading basidiomycete, Phanerochaete chrysosporium. *Arch. Biochem. Biophys.* 242: 329–341.

64 Jensen, K.A., Bao, W.L., Kawai, S. et al. (1996). Manganese-dependent cleavage of nonphenolic lignin structures by Ceriporiopsis subvermispora in the absence of lignin peroxidase. *Appl. Environ. Microb.* 62: 3679–3686.

65 Hofrichter, M. (2002). Review: lignin conversion by manganese peroxidase (MnP). *Enzyme Microb. Technol.* 30: 454–466.

66 Marquez, L., Wariishi, H., Dunford, H.B., and Gold, M.H. (1988). Spectroscopic and kinetic properties of the oxidized intermediates of lignin peroxidase from Phanerochaete chrysosporium. *J. Biol. Chem.* 263: 10549–10552.

67 Baciocchi, E., Fabbri, C., and Lanzalunga, O. (2003). Lignin peroxidase-catalyzed oxidation of nonphenolic trimeric lignin model compounds: Fragmentation reactions in the intermediate radical cations. *J. Org. Chem.* 68: 9061–9069.

68 Cho, D.W., Parthasarathi, B.R., Pimentel, A.C. et al. (2010). Nature and kinetic analysis of carbon–carbon ond fragmentation reactions of cation radicals derived from SET-oxidation of lignin model Compounds. *J. Org. Chem.* 75: 6549–6562.

69 Bugg, T.D., Ahmad, M., Hardiman, E.M., and Rahmanpour, R. (2011). Pathways for degradation of lignin in bacteria and fungi. *Nat. Prod. Rep.* 28: 1883–1896.

70 Cho, D.W., Latham, J.A., Park, H.J. et al. (2011). Regioselectivity of enzymatic and photochemical single electron transfer promoted carbon–carbon bond fragmentation reactions of tetrameric lignin model compounds. *J. Org. Chem.* 76: 2840–2852.

71 Shimada, M., Habe, T., Higuchi, T. et al. (1987). Biomimetic approach to lignin degradation. 2. The mechanism of oxidative C–C bond cleavage reactions of lignin model compounds with natural iron(III) porphyrin chloride as a heme-enzyme model system. *Holzforschung* 41: 277–285.

72 Watson, P.A., Wright, L.J., and Fullerton, T.J. (1993). Reactions of metal ion complexes with lignin model compounds, Part II. Fe (TSPC) Catalyzed formation of oxidized products in the absence of oxygen. *J. Wood. Chem. Technol.* 13: 391–409.

73 Robinson, M.J., Wright, L.J., and Suckling, I.D. (2000). Fe(TSPc)-Catalysed benzylic oxidation and subsequent dealkylation of a non-phenolic lignin model. *J. Wood. Chem. Technol.* 20: 357–373.

74 Zakzeski, J., Bruijnincx, P.C.A., Jongerius, A.L., and Weckhuysen, B.M. (2010). The catalytic valorization of lignin for the production of renewable chemicals. *Chem. Rev.* 110: 3552–3599.

75 Lange, H., Decina, S., and Crestini, C. (2013). Oxidative upgrade of lignin-Recent routes reviewed. *Eur. Polym. J.* 49: 1151–1173.

76 Napoly, F., Jean-Gérard, L., Goux-Henry, C. et al. (2014). Fe(TAML)Li/(diacetoxyiodo) benzene-mediated oxidation of alcohol: evidence for mild and selective C–O and C–C oxidative cleavage in lignin model transformations. *Eur. J. Org. Chem.* 2014: 781–787.

77 Li, C., Zhao, X., Wang, A. et al. (2015). Catalytic transformation of lignin for the production of chemicals and fuels. *Chem. Rev.* 115: 11559–11624.

78 Zhu, W.M. and Ford, W.T. (1993). Oxidation of lignin model compounds in water with dioxygen and hydrogen-peroxide catalyzed by metallophthalocyanines. *J. Mol. Catal.* 78: 367–378.

79 Cui, F.T. and Dolphin, D. (1995). Metallophthalocyanines as possible lignin peroxidase models. *Bioorg. Med. Chem.* 3: 471–477.

80 Barbat, A., Gloaguen, V., Sol, V., and Krausz, P. (2010). Aqueous extraction of glucuronoxylans from chestnut wood: New strategy for lignin oxidation using phthalocyanine or porphyrin/H_2O_2 system. *Bioresour. Technol.* 101: 6538–6544.

81 Sheldon, R. (1981). *Metal-Catalyzed Oxidations of Organic Compounds*. New York: Academic Press.

82 Heiba, E.A.I., Dessau, R.M., and Koehl, W.J. Jr. (1969). Oxidation by metal salts. V. Cobaltic acetate oxidation of alkylbenzenes. *J. Am. Chem. Soc.* 91: 6830–6837.

83 Hirose, S., Kasuga, K., and Hatakeyama, H. (1985). Alkali-oxygen degradation of lignin catalyzed by cobalt(III). *Tappi J.* 39: 1151–1156.

84 Huynh, V.-B. (1986). Biomimetic oxidation of lignin model compounds by simple inorganic complexes. *Biochem. Biophys. Res. Commun.* 139: 1104–1110.

85 Sedai, B., Diaz-Urrutia, C., Baker, R.T. et al. (2011). Comparison of copper and vanadium homogeneous catalysts for aerobic oxidation of lignin models. *ACS Catal.* 1: 794–804.

86 Sedai, B. and Baker, R.T. (2014). Copper catalysts for selective C–C bond cleavage of β-O-4 lignin model compounds. *Adv. Synth. Catal.* 356: 3563–3574.

87 Mottweiler, J., Rauber, C., Bolm, C. et al. (2015). Copper- and vanadium-catalyzed oxidative cleavage of lignin using dioxygen. *ChemSusChem* 8: 2106–2113.

88 Dicosimo, R. and Szabo, H.C. (1988). Oxidation of lignin model compounds using single-electron-transfer catalysts. *J. Org. Chem.* 53: 1673–1679.

89 Fukuzumi, S. and Kochi, J. (1981). Electrophilic aromatic substitution. Charge-transfer excited states and the nature of the activated complex. *J. Am. Chem. Soc.* 103: 7240–7252.
90 Kurokawa, A., Osaki, N., and Shigeyasu, M. (1985). Studies on the oxidation of alkylbenzenes. 7. Synergistic effect between cobalt(II) and manganese(II) acetates catalysts in the liquid-phase autoxidation of para-xylene. *Nippon Kagaku Kaishi* 207–213.
91 Evtuguin, D.V., Daniel, A.I.D., Silvestre, A.J.D. et al. (2000). Lignin aerobic oxidation promoted by molybdovanadophosphate polyanion $[PMo7V5O_40]^{8-}$. Study on the oxidative cleavage of β-O-4 aryl ether structures using model compounds. *J. Mol. Catal. A: Chem.* 154: 217–224.
92 Walling, C., El-Taliawi, G.M., and Zhao, C. (1983). Oxidation of arylalkanols by peroxydisulfate-copper(II). *J. Org. Chem.* 48: 4914–4917.
93 Camaioni, D.M. and Franz, J.A. (1984). Carbon-hydrogen vs. carbon-carbon bond cleavage of 1, 2-diarylethane radical cations in acetonitrile-water. *J. Org. Chem.* 49: 1607–1613.
94 Maggioni, P. and Minisci, F. (1979). Electron-transfer processes-selective syntheses of aromatic-aldehydes of industrial interest by catalyzed peroxydisulfate oxidation. *Chim. Ind.* 61: 101–105.
95 Minisci, F., Citterio, A., and Giordano, C. (1983). Electron-transfer processes: Peroxydisulfate, a useful and versatile reagent in organic-chemistry. *Acc. Chem. Res.* 16: 27–32.
96 Bjørsvik, H.R. and Minisci, F. (1999). Fine chemicals from lignosulfonates. 1. Synthesis of vanillin by oxidation of lignosulfonates. *Org. Process. Res. Dev.* 3: 330–340.
97 Leopold, B. and Malmstrom, I.L. (1951). Constitution of resin phenols and their biogenetic relations. 15. Nitrobenzene oxidation of compounds of the lignan type. *Acta. Chem. Scand.* 5: 936–940.
98 Leopold, B. and Malmstrom, I.L. (1952). Studies on Lignin. 4. Investigation on the nitrobenzene oxidation products of lignin from different woods by paper partition chromatography. *Acta. Chem. Scand.* 6: 49–54.
99 Leopold, B. (1952). Studies on Lignin. 3. Oxidation of wood from picea-abies (L) karst (Norway spruce) with nitrobenzene and alkali. *Acta. Chem. Scand.* 6: 38–48.
100 Kavanagh, K.R. and Pepper, J.M. (1955). The Alkaline nitrobenzene oxidation of aspen wood and lignin model substances. *Can. J. Chem.* 33: 24–30.
101 Pardini, V.L., Vargas, R.R., Viertler, H., and Utley, J.H. (1992). Anodic cleavage of lignin model dimers in methanol. *Tetrahedron* 48: 7221–7228.
102 Lim, S.H., Nahm, K., Ra, C.S. et al. (2013). Effects of alkoxy groups on arene rings of lignin β-O-4 model compounds on the efficiencies of single electron transfer-promoted photochemical and enzymatic C–C bond cleavage reactions. *J. Org. Chem.* 78: 9431–9443.
103 Davis, H., Das, P., and Griffin, G. (1984). Electron-transfer sensitized carbon-carbon bond cleavage. Facile homolytic fission via geminate back electron transfer in photogenerated ion pairs. *J. Am. Chem. Soc.* 106: 6968–6973.

104 Okamoto, A. and Arnold, D.R. (1985). Radical ions in photochemistry. 16. The photosensitized (electron transfer) carbon–carbon bond-cleavage reaction of radical cations. *Can. J. Chem.* 63: 2341–2342.

105 Lim, S.H., Lee, W.S., Kim, Y.I. et al. (2015). Photochemical and enzymatic SET promoted C—C bond cleavage reactions of lignin β-1 model compounds containing varying number of methoxy substituents on their arene rings. *Tetrahedron* 71: 4236–4247.

106 Zhou, W., Nakahashi, J., Micra, T., and Murakami, M. (2018). Light/copper relay for aerobic fragmentation of lignin model compounds. *Asian. J. Org. Chem.* 7: 2431–2434.

107 Yayla, H.G., Wang, H., Tarantino, K.T. et al. (2016). Catalytic ring-opening of cyclic alcohols enabled by PCET activation of strong O–H bonds. *J. Am. Chem. Soc.* 138: 10794–10797.

108 Gentry, E.C. and Knowles, R.R. (2016). Synthetic applications of proton-coupled electron transfer. *Acc. Chem. Res.* 49: 1546–1556.

109 Wang, Y., Liu, Y., He, J., and Zhang, Y. (2019). Redox-neutral photocatalytic strategy for selective C–C bond cleavage of lignin and lignin models via PCET process. *Sci. Bull.* 64: 1658–1666.

110 Nguyen, S.T., Murray, P.R.D., and Knowles, R.R. (2020). Light-driven depolymerization of native ignin enabled by proton-coupled electron transfer. *ACS Catal.* 10: 800–805.

111 Li, T., Mo, J.Y., Weekes, D.M. et al. (2020). Photoelectrochemical decomposition of lignin model compound on a $BiVO_4$ photoanode. *ChemSusChem* 3622–3626.

112 Zhou, Y., Hu, D., Li, D., and Jiang, X. (2021). Uranyl-photocatalyzed hydrolysis of diaryl ethers at ambient environment for the directional degradation of 4-O-5 lignin. *JACS Au* 1: 1141–1146.

113 Petrus, E., Segado, M., Bandeira, N.A.G., and Bo, C. (2020). Unveiling a photoinduced hydrogen evolution reaction mechanism via the concerted formation of uranyl peroxide. *Inorg. Chem.* 59: 8353–8360.

114 McGrail, B.T., Pianowski, L.S., and Burns, P.C. (2014). Photochemical water oxidation and origin of nonaqueous uranyl peroxide complexes. *J. Am. Chem. Soc.* 136: 4797–4800.

115 Kimura, M. and Kaneko, Y. (1984). Kinetics of electron-transfer reactions of para-substituted phenols *p*-C_6H_4(X)OH with $[Fe(phen)3]^{3+}$(phen = 1,10-phenanthroline) and with $[IrCl6]^{2-}$ in aqueous acidic solutions: correlation between the Hammett constant of X and the one-electron redox potential of p-C_6H_4(X)OH. *J. Chem. Soc., Dalton Trans.* 341–343.

116 Li, C. and Hoffman, M.Z. (1999). One-electron redox potentials of phenols in aqueous solution. *J. Phys. Chem. B* 103: 6653–6656.

117 Capaldo, L., Merli, D., Fagnoni, M., and Ravelli, D. (2019). Visible light uranyl photocatalysis: Direct C–H to C–C bond conversion. *ACS Catal.* 9: 3054–3058.

118 Meng, Q.L., Hou, M.Q., Liu, H.Z. et al. (2017). Synthesis of ketones from biomass-derived feedstock. *Nat. Commun.* 8: 14190.

119 Wang, M., Shi, H., Camaioni, D.M., and Lercher, J.A. (2017). Palladium-catalyzed hydrolytic cleavage of aromatic C–O bonds. *Angew. Chem. Int. Ed.* 56: 2110–2114.
120 Wang, M., Gutierrez, O.Y., Camaioni, D.M., and Lercher, J.A. (2018). Palladium-catalyzed reductive insertion of alcohols into aryl ether bonds. *Angew. Chem. Int. Ed.* 57: 3747–3751.
121 Cornella, J., Gómez-Bengoa, E., and Martin, R. (2013). Combined experimental and theoretical study on the reductive cleavage of inert C–O bonds with silanes: ruling out a classical Ni(0)/Ni(II) catalytic couple and evidence for Ni(I) intermediates. *J. Am. Chem. Soc.* 135: 1997–2009.
122 Li, Y., Karlen, S.D., Demir, B. et al. (2020). Mechanistic study of diaryl ether bond cleavage during palladium-catalyzed lignin hydrogenolysis. *ChemSusChem* 13: 4487–4494.
123 Adam, W., Herrmann, W.A., Sahamoller, C.R., and Shimizu, M. (1995). Oxidation of methoxybenzenes to *p*-benzoquinones catalyzed by methyltrioxorhenium(VII). *J. Mol. Catal. A: Chem.* 97: 15–20.
124 Dong, L., Lin, L.F., Han, X. et al. (2019). Breaking the limit of lignin monomer production via cleavage of interunit carbon–carbon linkages. *Chem* 5: 1521–1536.
125 Liu, B., Zhou, H.F., Guan, Z.H. et al. (2016). Cleavage of a C-C σ bond between two phenyl groups under mild conditions during the construction of Zn(II) organic frameworks. *Green Chem.* 18: 5418–5422.
126 Li, H.J., Subbotina, E., Bunrit, A. et al. (2019). Functionalized spirolactones by photoinduced dearomatization of biaryl compounds. *Chem. Sci.* 10: 3681–3686.
127 Ramirez, N.P., Bosque, I., and Gonzalez-Gomez, J.C. (2015). Photocatalytic dehydrogenative lactonization of 2-arylbenzoic acids. *Org. Lett.* 17: 4550–4553.
128 Glover, S.A., Golding, S.L., Goosen, A., and McCleland, C.W. (1981). Intramolecular cyclisations of biphenyl-2-carboxyl radicals: evidence for a II-state aroyloxyl radical. *J. Chem. Soc., Perkin Trans.* 0: 842–848.
129 Keshari, T., Yadav, V.K., Srivastava, V.P., and Yadav, L.D.S. (2014). Visible light organophotoredox catalysis: a general approach to β-keto sulfoxidation of alkenes. *Green Chem.* 16: 3986–3992.
130 Yadav, A.K. and Yadav, L.D.S. (2015). Visible-light-mediated difunctionalization of styrenes: an unprecedented approach to 5-aryl-2-imino-1,3-oxathiolanes. *Green Chem.* 17: 3515–3520.
131 Romero, K.J., Galliher, M.S., Pratt, D.A., and Stephenson, C.R.J. (2018). Radicals in natural product synthesis. *Chem. Soc. Rev.* 47: 7851–7866.
132 Romero, N.A., Margrey, K.A., Tay, N.E., and Nicewicz, D.A. (2015). Site-selective arene C—H amination via photoredox catalysis. *Science* 349: 1326–1330.
133 Johnston, C.P., Smith, R.T., Allmendinger, S., and MacMillan, D.W.C. (2016). Metallaphotoredox-catalysed sp^3–sp^3 cross-coupling of carboxylic acids with alkyl halides. *Nature* 536: 322–325.
134 Tan, F.-F., He, X.-Y., Tian, W.-F., and Li, Y. (2020). Visible-light photoredox-catalyzed C–O bond cleavage of diaryl ethers by acridinium photocatalysts at room temperature. *Nat. Commun.* 11: 1–11.
135 Zhu, J., Wang, J.C., and Dong, G.B. (2019). Catalytic activation of unstrained C(aryl)—C(aryl) bonds in 2,2′-biphenols. *Nat. Chem.* 11: 45–51.

136 Gozin, M., Aizenberg, M., Liou, S.Y. et al. (1994). Transfer of methylene groups promoted by metal Complexation. *Nature* 370: 42–44.

137 Liou, S.Y., Gozin, M., and Milstein, D. (1995). Directly observed oxidative addition of a strong carbon-carbon bond to a soluble metal-complex. *J. Am. Chem. Soc.* 117: 9774–9775.

138 Liou, S.Y., van der Boom, M.E., and Milstein, D. (1998). Catalytic selective cleavage of a strong C–C single bond by rhodium in solution. *Chem. Commun.* 687–688.

139 Li, Q. and Vlachos, D.G. (2021). 5–5 Lignin linkage cleavage over Ru: A density functional theory study. *ACS Sustainable Chem. Eng.* 16143–16152.

8

Lignin Linkages Cleavage Beginning with $C_{\alpha}O$–H/ArO–H or C_{α}–OH Bond Heterolysis

8.1 Base-catalyzed C_{β}–OAr Bond Cleavage Beginning with $C_{\alpha}O$–H or ArO–H Heterolysis

8.1.1 The Mechanism of Base-mediated Lignin Depolymerization

In terms of industrial chemical utilization of lignocellulose, the most representative work is the classical kraft pulp process to depolymerize lignin. The target of this chemical pulping process is to separate cellulosic fibers from lignin, producing a pulp suitable for paper manufacture and other related products. In a conventional kraft cook, an aqueous solution of NaOH and Na_2S, also known as white liquor, is reacted with the wood chips in a large pressure vessel called a digester. The mixture of white liquor and the chips are heated to a cooking temperature of about 170 °C and are allowed to cook for about two hours [1]. During this treatment, the OH^- and HS^- anions react with the lignin [2, 3], causing the lignin polymer decomposition into smaller water/alkali-soluble fragments. Meanwhile, the bonds of cellulosic fiber are stable. For the lignin degradation, as typical lignin ether bonds such as the β-O-4 and α-O-4 bonds are the weakest bonds in the lignin structure, cleavage of these ether linkages is a dominant reaction in alkaline delignification processes [4, 5]. The fragmentation of the lignin macromolecule proceeds through the cleavage of the linkages holding the phenylpropane units together, with a concomitant generation of free phenolic hydroxyl groups [6]. The presence of these hydroxyl groups increases the hydrophilicity of the lignin and the lignin fragments. As a result, the solubility of the lignin in the cooking liquor can be increased. However, besides the β-O-4 and α-O-4 ether bonds cleavage, the carbon-carbon linkages, being more stable, tend to survive the pulping process. Like all lignin conversion processes, the types of lignin reactions that occur during kraft pulping can be classified into degradation and condensation reactions [7–14]. Degradation reactions are desirable since they lead to the depolymerization of lignin polymer to smaller fragments. Conversely, condensation reactions lead to the formation of alkali-stable C–C linkages, which inhibited the depolymerization of lignin and are therefore less desirable.

Lignin Conversion Catalysis: Transformation to Aromatic Chemicals, First Edition.
Chaofeng Zhang and Feng Wang.

The prevalent degradation reactions during kraft pulping are the cleavage of α-aryl ether and β-aryl ether bonds [10, 11, 14]. However, the reactivity of such linkages is sensitive to the type of moiety (i.e., free or etherified phenolic group) present at the para position relative to the propane side chain. If the substrate has a phenolic arylpropane structure, ether bonds of α-O-4 phenolic linkages are readily cleaved by the conversion of phenolate unit into the corresponding quinone methide intermediate (Figure 8.1a, **1**→**2**, R=Aryl). The phenolate unit of β-O-4 linkage can also be converted into the quinone methide intermediate (Figure 8.1a, **2**) if a suitable leaving group is present at the α-carbon. The hydrosulfide ion attacks the C_α of quinone methide intermediate and yields a benzyl mercaptide structure **4**. The mercaptide anion **4** then attacks the C_β to form a thiirane intermediate **6** and eliminates the β-aryloxy group **5** (Figure 8.1a, **4**→**6**). Alternatively, the quinone methide intermediate **2** can also be converted to an alkali-stable vinyl ether **3** by the elimination of the terminal hydroxymethyl group as formaldehyde (Figure 8.1a, **2**→**3**) [15]. The cleavage of β-aryl ether linkages in nonphenolic units **8** involves the attack of an ionized hydroxyl group present of C_α on C_β to form epoxy propane intermediate **9** and eliminate the β-aryloxy group **5** (Figure 8.1b, **8**→**9**) [16–18]. Then the epoxy propane intermediate **9** reacts with an OH^- and forms a trihydric alcohol **10** (Figure 8.1b, **9**→**10**). In 2016, Yu and co-workers [19] provided a detailed DFT calculations study about the base-catalyzed cleavage of the nonphenolic lignin β-O-4 linkage, with β-phenoxy-α-phenylethanol as a model. For the ether bond cleavage of the lignin β-O-4 linkage, in 2011, Lercher [20] proposed another different base-promoted mechanism (Figure 8.1c). It is suggested that cleavage of the β-O-4 ether bond takes place heterolytically via a six-membered transition state **11**, in which also the sodium cation and the hydroxide ion participate. The sodium cations catalyze the reaction by forming cation adducts with lignin and thus polarizing the ether bond. This leads to an increased negative partial charge on the oxygen atom and reduces the energy required for heterolytic bond cleavage. In addition, for the ether bond cleavage of the lignin β-O-4 linkage without a C_α–OH group, Lee [21] and Cronin [22] proposed another different base-catalyzed mechanism (Figure 8.1d). It begins with the attack of hydroxide ion at C_α–H in **15**, which further induces the heterolytic C_β–OAr bond cleavage and provides phenolate and styrene.

Besides the depolymerization reactions during base-mediated lignin depolymerization, the hydrogen sulfide and sodium hydroxide anions have to compete with the lignin nucleophiles (e.g., carbanions **3** from the resonance hybrid of phenolate **2**) for quinone methide intermediates **1** (Figure 8.2) [7, 12, 14, 23]. The addition of the nucleophile to the quinone methide is reversible, and the result of this competition will not only depend on the nucleophilicity of the species, but also on the ability of the addition product to undergo a fast, irreversible reaction. In the structures of **1** containing good leaving groups (e.g., R = OAr) at the β-carbon, the neighboring group participation reactions, resulting in the cleavage of β-aryl ether linkages, will predominate over condensation reactions (Figure 8.2a). Otherwise, for the lignin condensation reactions that occur during kraft pulping, they proceed via Michael addition, where the quinone methide acts as an acceptor and the phenolate ion or the resonant carbanions acts as the nucleophile (Figure 8.2b). The addition of the

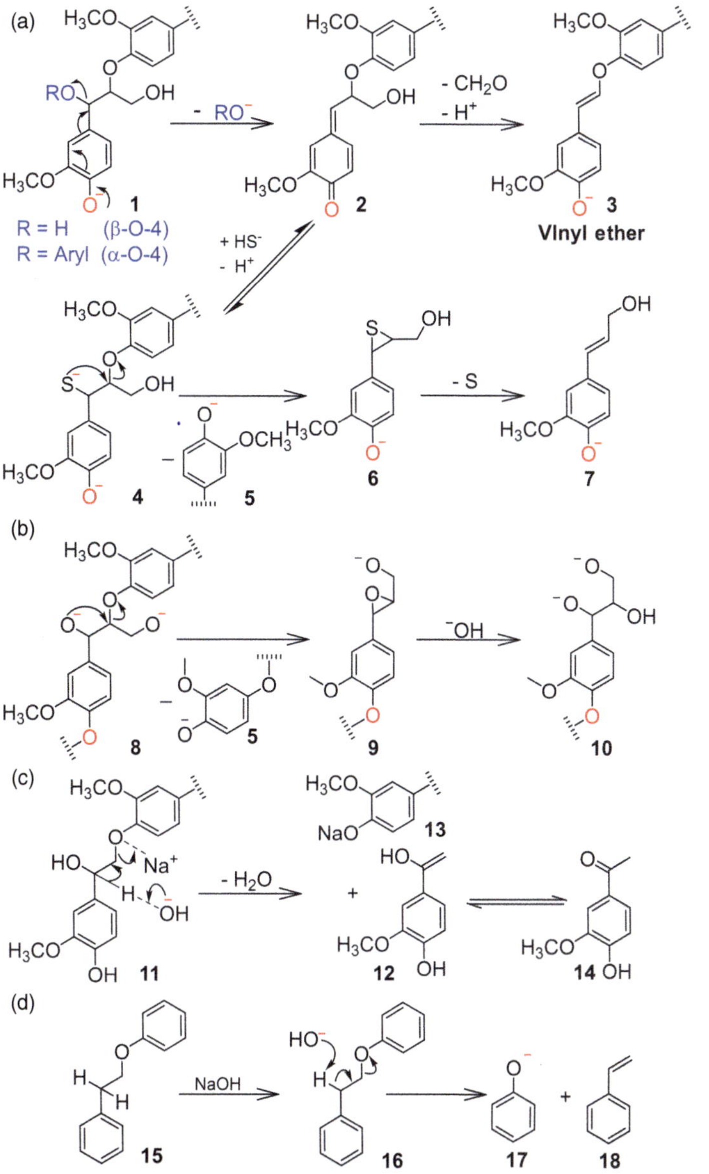

Figure 8.1 Alkaline Cleavage of α-aryl and β-aryl Ether Bonds in Phenolic Arylpropane (a) and Nonphenolic Arylpropane (b), and Two Base-catalysed Hydrolysis of Mechanisms with First Activation of the C_α–H (c and d).

nucleophile is followed by the abstraction of a proton and rearomatization, resulting in the new C–C bond generation.

Besides quinone methides, formaldehyde generated from C_γ releasing constitutes another type of acceptor [24]. Figure 8.3a illustrates the reaction between formaldehyde and two phenolate units (**1** or **2**). The primary addition product, a

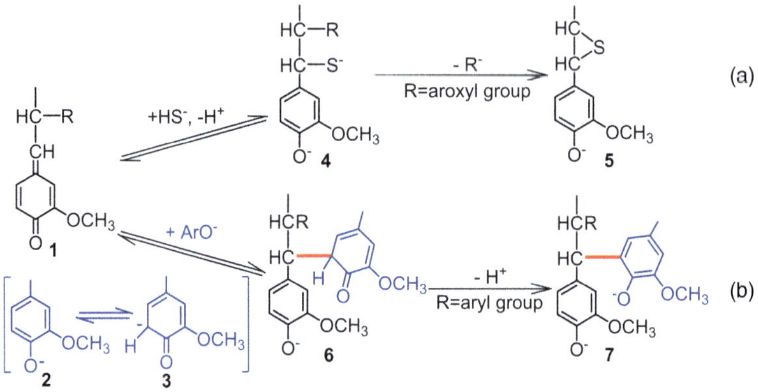

Figure 8.2 Competitive Addition of Nucleophiles Hydrosulfide Ion (a) or Phenolate to Quinone Methide Intermediates (b).

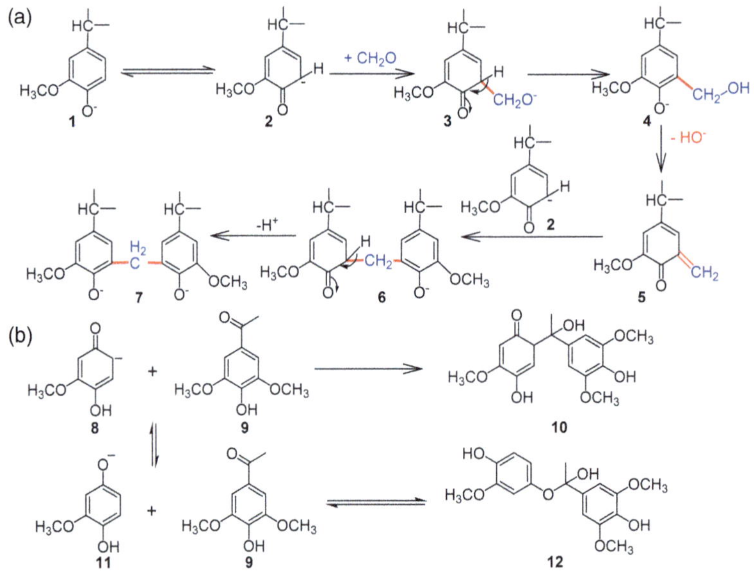

Figure 8.3 Alkali-promoted Condensation Reactions in Phenolic Units with Formaldehyde (a) and Ketone Fragments (b).

hydroxybenzyl alcohol **4**, generated from the addition of the phenolic ion **2** and formaldehyde, is converted to an o-quinone methide **5**. Another phenolic ion **2** reacts with **5** to generate a diarylmethane structure **7**. This process causes the generation of a new Ar-CH_2-Ar structure, which is alkali-stable and inhibits the further depolymerization of lignin fragments [17, 25]. In addition, during the base-catalyzed lignin depolymerization, the phenolate and its corresponding carbanions resonant hybrids can also work as the nucleophilic reagent to react with the ketones with new C–C or C–O bonds generations (Figure 8.3b) [20].

In the base-catalyzed lignin hydrolysis reaction, the selectivity and yield of the products are particularly dependent on pressure, temperature, time, the concentration of the base, and lignin/solvent ratio. Equally important, the oil yield and products composition vary strongly depending on the nature of the base. Besides Na_2S and $NaSO_3$, other cheap and commercially available bases such as LiOH, NaOH, KOH, Cs_2CO_3, and ^{t}BuOLi [20, 26–28] can also efficiently catalyze the lignin depolymerization to simple aromatic chemicals under mild conditions. Usually, a stronger base gives higher conversion since the polarization of the base governs the kinetics and the mechanism of the depolymerization reaction. The kinetics of reactions is faster in phenols or alcohols than in water due to the ether linkages solvolysis effect. A higher temperature and longer reaction time favor monomers generation. However, the formation of a solid residue increases due to the condensation/ repolymerization reactions of the degradation intermediates/products. Therefore, decreasing the rate of repolymerization and oligomerization reactions during base-catalyzed lignin depolymerization is the key issue to enhance the yields of the products. For the recondensation control, we provide the corresponding discussion in Chapter 11, this chapter will mainly focus on the issue of lignin depolymerization.

Treating lignin with aqueous NaOH solution at high temperatures leads to phenols and derivatives. The process is relatively simple, but its selectivity remains unsatisfactory. Lavoie and co-workers [5] have employed steam-exploited lignin from softwood and hemp with a 5 wt% solution of NaOH at a temperature between 300 and 330 °C under 9–13 MPa. Guaiacol, catechol, and vanillin, among a mixture of 26 products (identified by GC-MS), were the main low-M_w products under these conditions. When commercial organosolv lignin is treated under similar conditions, syringol, hydroxyacetophenone, and catechol are the main products. Contrarily, for the Kraft lignin, pyrocatechol was the main product. Toledano and co-workers [29] carried out the depolymerization of organosolv lignin with several bases such as KOH, NaOH, $Ca(OH)_2$, LiOH, and K_2CO_3 in water solutions above 270 °C, catechol being the main product with a yield ranging from 0.1 to 2.4 wt%. Besides the homogeneous base, in 2017, Dhepe and a co-worker [30] reported the depolymerization of high molecular weight lignin (60,000 Da) over various recyclable solid base catalysts at 250 °C over 1 h. Under these conditions, most of the zeolitic catalysts (NaX, NaY, NaP) showed very high yields of low molecular weight products compared to other catalysts (MgO, CaO, HT, HAP). In addition, among these zeolitic base catalysts, NaX zeolitic is capable of depolymerizing high molecular weight lignin into low molecular weight aromatic products (51% yield) in ethanol: water (1: 2 v/v) ratio solvent system.

In terms of making the lignin catalytic depolymerization conditions mild, the first approach is that the catalyst must have high activity; on the other hand, the lignin substrate or model molecule must have high solubility. In this regard, the solvent plays an important role. Recently, the ionic liquid (IL) has attracted wide attention in lignin conversion [31] because of its good solubility for lignin molecules, its strengthening effect on catalysts, especially the acid-base catalysis process, and the extraction ability [32]. In 2010, Ekerdt and co-workers [18] explored the combination of an Ionic liquid 1-butyl-2,3-dimethylimidazolium chloride ([BDMIm]Cl) and nonaqueous organic N-bases (Figure 8.4) in cleaving the β-O-4 bonds. 1,5,7-Triazabicyclo [4.4.0]dec-5-ene (TBD) showed the highest activity, leading to more than 40% β-O-4 ether bond cleavage. It was proposed that two exposed nitrogen atoms (N_1 and N_7) in TBD enable it to work as a dibasic nucleophile, both attacking the α- and β-carbon of quinone methide like hydrosulfide anion in the kraft pulping process and thereby assisting the cleavage of C_β–OAr bond.

Although the use of ionic liquids can make the transformation of lignin and its model compounds efficient, the excessively high cost often limits its wide range of applications. How to reduce the use of IL solvents, or develop a solvent-free method may be a very meaningful means for lignin depolymerization. In 2013, Bolm and co-workers [33] first applied solvent-free ball milling in the presence of a (solid) base for the degradation of lignin (Figure 8.5). It was found that sodium hydroxide or sodium tert-butoxide could effectively degrade various lignin model compounds via ether bond cleavage. The reaction mechanism was in agreement with the classical kraft pulp process [17]. The initial deprotonation at the primary hydroxyl group followed by intramolecular cleavage of the β-ether bond generates two products: an unstable epoxide and guaiacol. Although solvent-free mechanochemical transformation is proposed as a sustainable alternative solution of solvothermal chemical

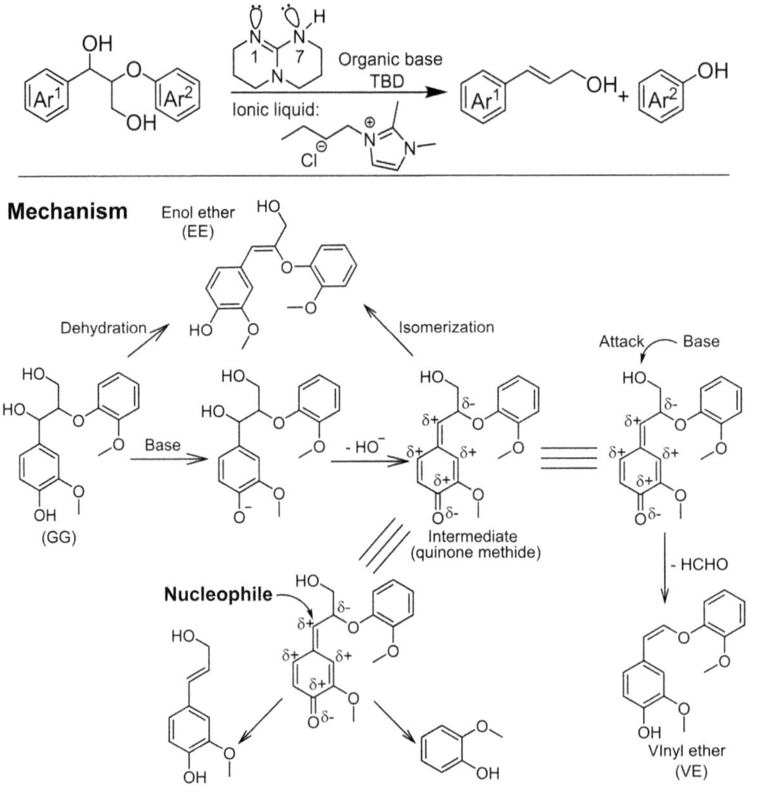

Figure 8.4 Decomposition of Phenolic Lignin Models over Organic N-bases in an Ionic Liquid. *Source:* Jia et al. [18] with Permission, © 2010 Walter de Gruyter.

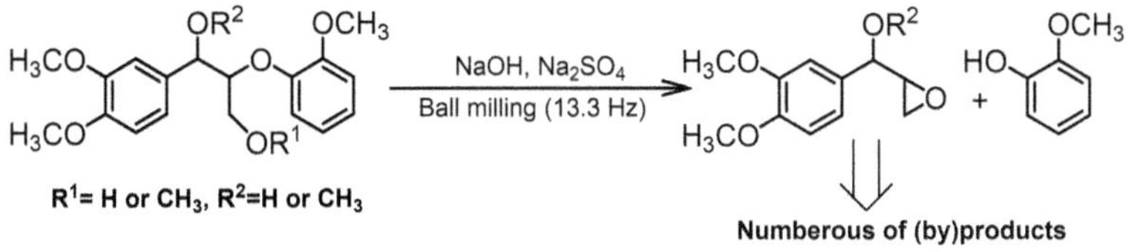

Figure 8.5 Degradation of β-O-4 Lignin Model Compounds by Solvent-free Grinding with NaOH.

processes, its energy efficiency should be improved; in addition, the base quantity is too large to be applied on a large scale.

8.1.2 Base-promoted Tandem Process for the Cleavage of C_β–OAr Bond

The above-reported strategies show that alkali-catalyzed means can effectively fragment the lignin. To effectively depolymerize and convert these fragments, according to the cascaded mode of IDA discussed in Chapter 5, tandem alkali hydrolysis with other methods, such as hydrogenation/reduction [34–38], may be a good choice to increase the depolymerization efficiency of real lignin.

In 2010, Barta and Ford [39] reported that a Cu-doped porous metal oxide catalyst can transfer hydrogen from methanol to an organosolv lignin at 300 °C, which leads to the complete hydrogenolysis of phenyl ether bonds, coupled with the hydrogenation of aromatic rings. The product is a complex mixture composed principally of monomeric substituted cyclohexyl derivatives with greatly reduced oxygen content and negligible aromatics. Next, in 2011, Ford and co-workers [34] reported quantitative catalytic conversions of wood and cellulosic solids to liquid and gaseous products in a single-stage reactor operating at 300–320 °C and 160–220 bar. The reaction medium is supercritical methanol (sc-MeOH) and the catalyst, a copper-doped porous metal oxide (Cu/Al_2O_3-MgO), is composed of earth-abundant materials. Little or no char is formed during this process, and the major liquid product is a mixture of C_2–C_6 aliphatic alcohols and methylated derivatives. In 2014, Barta and co-workers [40] then achieved the selective hydrodepolymerization of organosolv lignin to aromatic compounds over Cu-doped porous metal oxides (Cu-PMO) in methanol. The Cu-PMO catalyst was prepared by calcination of a hydrotalcite-like precursor resulting from modification of the hydrotalcite structure with Cu dopant keeping a 1:3 molar ratio of M^{3+}/M^{2+} by replacing 20 mol% of the Mg^{2+} ions with Cu^{2+}. In the temperature range 140–220 °C, C_9 catechols were obtained with high selectivity. These studies demonstrate that heterogeneous supported catalysts with basic sites and reduced metals under hydrogen pressure or in hydrogen-donor solvents show promise for lignin depolymerization.

In 2015, Ma and co-workers [35] reported an efficient base-catalyzed depolymerization and in situ hydrogenolysis process for lignin decomposition and char elimination was proposed using the synergic catalyst of NaOH coordinated with Ru/C, in which, lignin was first depolymerized to phenolic monomer and its oligomer with base, and then the oligomer was further hydrogenated to more stable aliphatic alcohols simultaneously. The results showed that more than 92.5% of lignin was converted at 240 °C for 4 h with 4 MPa H_2, giving 12.69% phenolic monomer, 6.12% aliphatic alcohol, and 14.03% residual solid. This residual solid selectivity was far lower than it from the single catalyst condition. In 2015, Yan and co-workers [36] presented a systematic investigation on the influence of pH in the hydrogenolysis of lignin over the bimetallic Ni_7Au_3 nanoparticles. Low pH is detrimental to both activity and aromatic product selectivity, whereas high pH, in particular in the range of pH 12-13, exhibits a significant promotional effect. The addition of a strong base such as NaOH significantly enhanced the activity and selectivity for C–O bond hydrogenolysis over undesired hydrogenation reactions, not only in lignin model compounds but also in real lignin conversion.

Different from the above systems and conclusion, in 2014, Beckham and co-workers reported [41] that a nickel supported hydrotalcite (Ni-HTC), a layered-double hydroxide catalyst (LDH), was active for β-O-4 bond cleavage in a lignin model dimer and that reduced the molecular weight of two insoluble lignin streams (Figure 8.6). The reaction carried out at 270 °C does not alter the overall catalyst microstructure, and nickel does not appreciably leach from the catalyst. The authors announce that nickel oxide on solid-basic support can function as an effective lignin depolymerization catalyst without the need for external hydrogen and reduced metal, and suggest that LDHs offer a novel, active support in multifunctional catalyst applications. However, the active site in the Ni-HTC catalysts was not determined in this work, and the authors speculated that nickel supported on HTC could depolymerize lignin via a base-catalyzed mechanism mediated by a strong interaction between the ether oxygen and the nickel metal.

In their following work, Beckham and co-workers [42] were motivated by these initial experiments to pursue a deeper understanding of catalyst activity and reaction parameters. To gain further understanding of the roles of the metals and anions, they explored a series of metal-free, ion-exchanged HTC as catalysts (Figure 8.7). HTC-based materials are effective catalysts for lignin depolymerization: namely, for cleavage ester bonds rather than the C–C bonds of β-O-4 linkages, but the catalyst

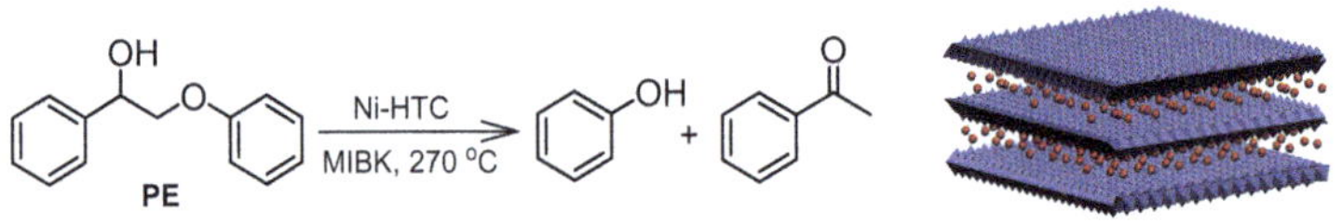

Figure 8.6 Lignin Depolymerization by Nickel-supported Layered Double Hydroxide Catalysts.

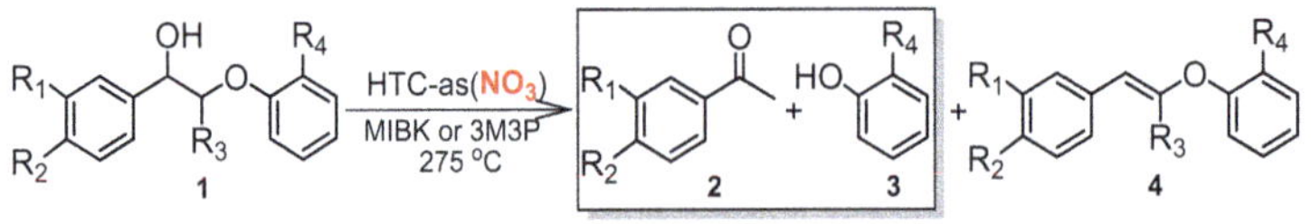

Figure 8.7 Lignin Depolymerization with Nitrate-Intercalated Hydrotalcite Catalysts.

is active only when a C_αH–OH group is present on the substrate. For this heterogeneous base catalysis process, controlled experiments show that the key component for catalytic activity is nitrate anions intercalated in the interlayer space. HTCs containing intercalated NO_3^- anions, even in the absence of transition metals, increase activity by 2–3-fold more than the HTC for some model compounds, but the high activity may benefit from neighboring basic anions, such as CO_3^{2-} and OH^-. For the role of NO_3^- anion on the catalytic mechanism, the authors proposed a nitration step of the benzylic hydroxyl group, but more detailed exploration was not carried out.

Besides above the normal base, basic salts, and metal oxide, in 2017, Zhang and co-workers [43] developed a mild and transition-metal-free protocol for the vinylation of epoxides and aryl ether linkage (β-O-4 lignin model compound) with "solid acetylene" calcium carbide (CaC_2) through the C_β–O bond cleavage, forming valuable vinyl ether products. Calcium carbide plays a vital role in the C_β–O bond activation and cleavage, and in providing an acetylide source for the formation of vinylated products. Based on the control experiments, a proposed reaction pathway is shown in Figure 8.8. It was postulated that isomerization of β-O-4 lignin model compound **1** could have taken place to form **10**, followed by an H_2O elimination, which gave product **11**. This side reaction [Route (I) in Figure 8.8] is slow as only a small amount of intermediate **10** and byproduct **11** (<10 %) were observed in the systems with or without calcium carbide. The initial detection of acetophenone and phenol could have occurred from direct β-elimination of **1**, followed by vinylation of the phenol to form **7** [Route (II)]. In addition, the O-vinylation of **1** took place as a fast and

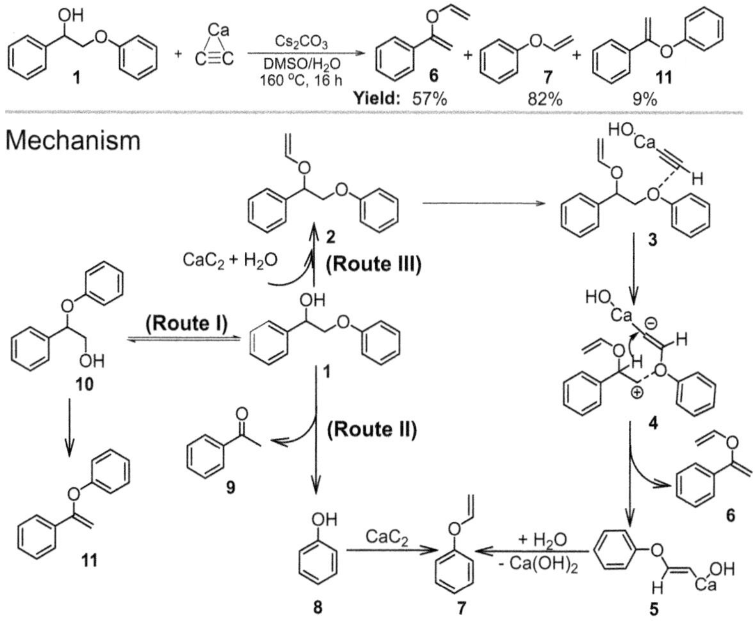

Figure 8.8 Vinylation of Lignin β-O-4 Linkage with Calcium Carbide through C–O Bond Cleavage. *Source:* Teong et al. [43] with Permission, © 2017 Wiley-VCH.

dominant reaction in this system to give intermediate **2**. Similar to the epoxide reactions, it was assumed that **2** was then activated by acetylide and went via intermediate **4** (C_β–O cleavage and hydrogen transfer) to give **5, 6,** and **7**. Route (III) is the fast and dominant reaction, whereas pathway (I) and (II) coexist in the system as complimentary reaction paths.

Besides the base-promoted methods targeting at the cleavage of lignin C_β–OAr bond, some base-promoted alkaline oxidation of lignin [44, 45] can achieve the lignin C–C bonds, especially for the C_α–C_β in β-O-4 linkage, which provide aromatic aldehydes/acids as the final product. Different mechanisms of alkaline catalyzed lignin oxidation to produce aldehyde have been proposed in the literature. The most acceptable one is the radical chain mechanism proposed by Tarabanko and co-workers [46, 47], which involves the generation of the phenoxyl anion with the assistance of one base and its further transformation to the phenoxyl radical. The corresponding discussion about the base-promoted lignin oxidation processes for the C_β–OAr/C–C bonds cleavage is summarized in Chapter 7.1.

8.2 Acid-catalyzed C_β–OAr Bonds Cleavage Beginning with C_α–OH Heterolysis

8.2.1 The Mechanism of Acid-mediated Lignin Depolymerization

For the lignin hydrolysis to cleave C_β–OAr bonds, besides the alkaline hydrolysis, acidolysis is another well-studied approach [48]. It is generally accepted that the weakest bonds in the lignin polymers are the α- and β-aryl-ether bonds followed by the aryl-aryl bond. Under acidic pretreatment conditions, the predominant reactions associated with lignin are fragmentation by acidolysis of α- and β-aryl ether linkages. This subsection will summarize the acid catalysis process, reaction mechanism, and the latest studies and advancements on the catalytic conversion of lignin via various types of acid catalysts, such as liquid acid, homogeneous metal salts, organic metal compounds, and solid acids metal oxides.

At the beginning of this section, the lignin acidolysis mechanism of various linkages is briefly summarized (Figure 8.9). In 2015, Barta and co-workers [49] identified two different reaction pathways for the cleavage of the β-O-4 linkage and modification of the β-5 linkage, the two most common intermonomer bonds in lignin, in which β-aryl ether bonds were readily cleaved via a benzylic cation reacting in one of two pathways, C2 cleavage or C3 cleavage (Figure 8.9a) [50, 51]. For the β-O-4 linkage, the C3 pathway provides the Hibbert ketones **10**, and the C2 pathway yields C_2-aldehydes **8** upon release of formaldehyde. During the lignin acidolysis, different from the traditional C2 cleavage pathway [52], the C3 cleavage pathway [53] is suggested to proceed via protonation of the hydroxyl group attached to the α-carbon of the β-ether subunit of **1** followed by dehydration to generate a stabilized benzylic carbocation **2**. Then the carbocation **2** transforms to the enol aryl ether (EE) intermediate **9**. This dehydration step is the rate-determining step for β-phenyl ether hydrolysis. EE intermediate **9** can be rapidly

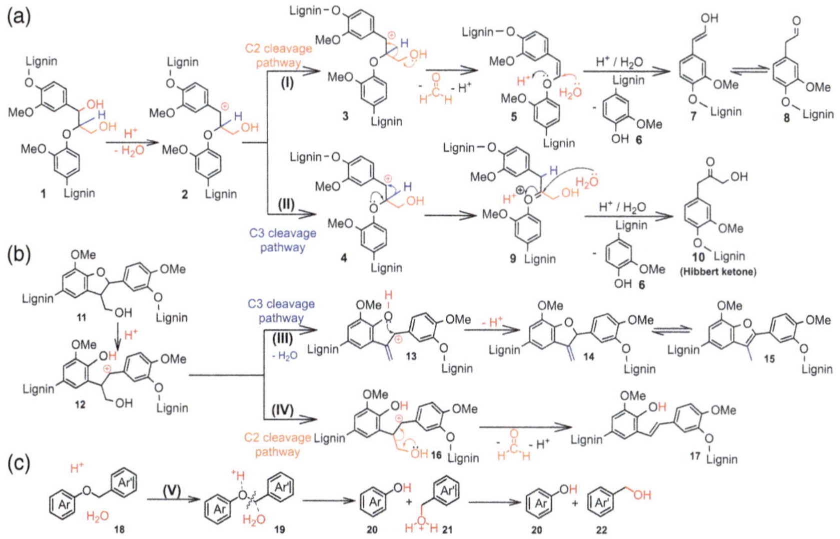

Figure 8.9 Pathways for the Acid-mediated Cleavage of the Lignin β-O-4 Linkage (a), β-5/cyclic α-O-4 (b), and α-O-4 (c)

hydrolyzed to guaiacol and α-ketocarbinol [54, 55]. The latter is gradually converted to a mixture called "Hibbert's ketones" via allylic rearrangement [56]. For the β-5/cyclic α-O-4 linkage acidolysis (Figure 8.9b), the C3 cleavage pathway just promotes the generation of benzofuran structure after the dehydration of **11** and following molecule transformation via the C_α cation (**11**→**15**), while the C2 cleavage pathway can lead to the cleavage of the C_5–OC_α ether bond after the elimination of formaldehyde and following molecule transformation via the C_α cation (**11**→**17**). Besides the α-aryl ether in β-5 linkage, in 2015, Pelzer and co-workers [57] provided a detailed study about the acidolysis of α-O-4 aryl-ether bonds in lignin model compounds with calculation results and kinetic study (Figure 8.9c). For α-aryl ether linkages, acid-catalyzed hydrolysis reactions follow predominantly a SN1-type mechanism [58] with first-order kinetics as outlined in the following two equations:

$$\text{Aryl}-\text{CH}_2\text{O}-\text{Aryl}'+\text{H}^+ \rightarrow \text{Aryl}'\text{OH}+\text{Aryl}-\text{CH}_2{}^+ \quad \text{(i)}$$

$$\text{Aryl}-\text{CH}_2{}^+ +\text{ROH} \rightarrow \text{Aryl}-\text{CH}_2\text{OR} \quad \text{(ii)}$$

(ROH is water or low molecular alcohol). The range of activation energies for the hydrolysis of the α-aryl ether bonds (80–118 $kJ{\cdot}mol^{-1}$) is much lower than for the β-aryl ether hydrolysis (148–151 $kJ{\cdot}mol^{-1}$) [59]. According to the report from Johansson and Miksche [60], the rate of hydrolysis of α-aryl ether was ~100 times faster than that of β-aryl ether. Thus, when lignin is subjected to acidic degradation, the α-ether linkages depolymerize before the β-aryl ether linkages. Furthermore, phenolic α-ethers hydrolyze faster than nonphenolic ones.

As can be seen from these phenomena, the formation of the benzylic carbocation is one way of acid-catalyzed lignin depolymerization, but it is also an important driver of the repolymerization reaction. As a result, the acid-catalyzed process needs to strike the right balance between the promotion of reaction activity of benzylic carbocation and the inhibition of repolymerization caused by benzylic carbocation [61]. This section is mainly focusing on the introduction of the depolymerization process, and the methods for restraining the condensation are provided in Chapter 11.

Different from other protonic acids, the lignin decomposition with the dry hydrogen iodide (HI) is a stoichiometric process [62–64]. In 2020, Shevchenko [65] reported that depolymerization of lignin in wood with hydrogen iodide in a nonpolar solvent is a selective, high-yield reaction, which can release a diiodide from the β-O-4 linkage (Figure 8.10). Dry HI works as the cleaving reagent to break the C–O bonds in ether and alcohol groups. Besides the HI, various iodide-based reagents such as trimethylsilyl iodide [66] and pivaloyl iodide [67] have been used for the cleavage of carbon-oxygen bonds under mild conditions.

Besides the first C_αH–OH transformation to C_α–I with HI and iodide-based reagents, some DFRC (derivatization followed by reductive cleavage) method [68–75] first use the reaction between C_αH–OR/C_γH–OR and acetyl bromide (AcBr) to achieve the C_αH–OH cleavage to C_αH–Br and the esterization protection of the C_γH–OH, the C_β–OAr bond in obtained bromide product can occur reductive cleavage with Zn powders in acetic acid (Figure 8.11).

In addition, the lignin thioacetolysis with the thioacetic acid and other sulfhydryl-containing compounds can be regarded as another type of acidolysis [76–80]. The principle of the three-step degradation method has been formulated by Nimz and Das [81] as shown in Figure 8.12. Treatment of wood with thioacetic acid and boron trifluoride converts the arylglycerol-β-aryl ether unit via the benzylium ion into the S-benzyl thioacetate (**1**→**3**). Subsequent saponification with 2N NaOH at 60 °C gives a benzyl thiolate ion (**3**→**4**), which loses the β-aryloxy group by nucleophilic

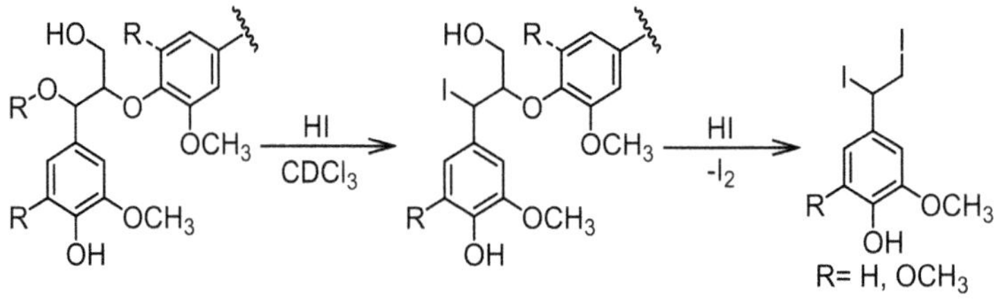

Figure 8.10 Depolymerization of Lignin Linkage with Hydrogen Iodide.

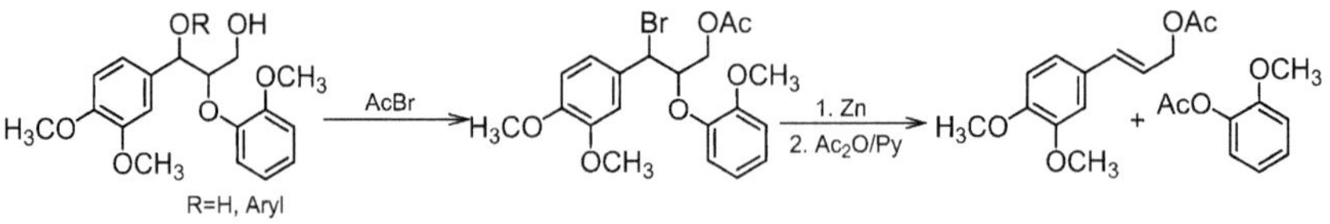

Figure 8.11 Selective Ether Bonds Cleavage in Lignins by the DFRC Method.

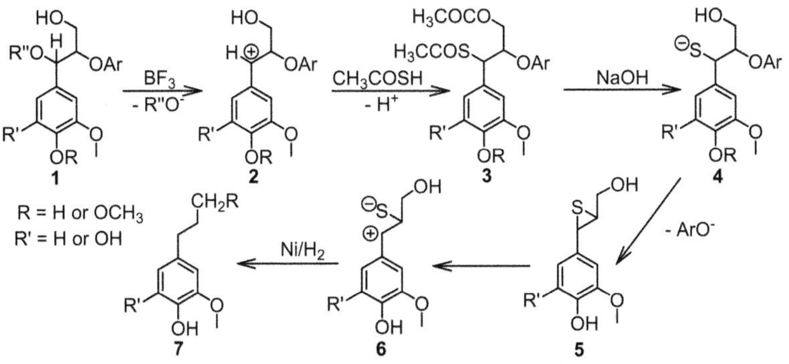

Figure 8.12 Degradation of Lignin by Thioacetolysis with Thioacetic Acid.

attack of the neighboring thiolate ion on the C_β atom to give an episulfide (**4→5**). In the final step, treatment with Raney nickel and alkali at 115 °C removes the sulfur and yields reduced phenolic reaction products (**7**).

8.2.2 Lignin Acidolysis with Liquid Acid Catalysts

As mentioned, discussed in Chapter 3, historically acid pulping was mainly used for the isolation of lignin fractions from the lignocellulose matrix, rather than in the depolymerization of lignin into valuable aromatic monomers. The first acid-catalyzed lignin hydrolysis reaction was reported in 1924 when Hägglund and Björkman [82] distilled lignin with 12% hydrochloric acid to obtain thiobarbituric acid, phloroglucinol, and barbituric acid. Besides the HCl [83, 84], the mineral acids H_2SO_4 [55, 85, 86], and H_3PO_4 [87] have also been used in lignin conversion studies. According to the corresponding research, the kinetics of reactions is usually faster in phenols [4] or alcohols [88] than in water due to solvent effects. However, phenol solvents should be carefully selected, as they are active species that can initiate nucleophilic substitution reactions with cations to form the corresponding phenolated byproducts [89–91]. In 2016, Yang and co-workers [92] proposed a novel method to depolymerize lignin under mild conditions which were conducted at 110 °C for 2 h in 60% LiBr solution with 0.3 M HCl as the catalyst, giving marked decreases in Mw and Mn of the lignins, indicating that the β-O-4 aryl ether bonds of the lignins were selectively cleaved, and both LiBr and HCl played crucial roles in catalyzing the cleavage of the ether bonds.

Besides the mineral acids, heteropoly acids (HPAs), as a class of Brønsted acids, have attracted much interest in biomass hydrolysis because of their potential to generate economic rewards and green benefits [93, 94]. HPAs are made up of a particular combination of hydrogen cations and polyoxometalate (POM) anions, among which the most stable and easy available Keggin HPAs comprise heteropoly anions of the formula $[XM12O40]^{n-}$, where X is the heteroatom (P^{5+}, Si^{4+}, etc.) and M the addendum atom (Mo^{6+}, W^{6+}, etc.). HAPs possess stronger acidity than conventional solid acid catalysts such as acidic oxides and zeolites [95, 96], while because of the redox properties of the HAPs, the acidolysis process usually contains the further oxidation of the lignin fragments [97–99]. Therefore, the corresponding discussion is put in Chapter 8.2.5 about the Acid-promoted tandem process for lignin depolymerization.

According to the recent research advances, the mineral acid-catalyzed conversion strategy has been adopted abstemiously, but the high corrosion and oxidation of mineral acids should not be ignored and avoided. In recent years, the substitution of inorganic catalysts by renewable organic catalysts is very interesting from an environmental and economic point of view. Different organic acid catalysts (formic acid, acetic acid, lactic acid, citric acid, and benzenesulfonic acid) were employed as an alternative to common catalysts used in lignin hydrolysis [100, 101]. Among these organic catalysts, the formic acid is special because of its insert properties as the acid catalyst and hydrogen donor [102]. In 2019, Wang and co-workers [103], found the recyclable mixture solvent of ethanol/1,4-dioxane/formic acid (FA) could efficaciously depolymerize industrial lignin to produce high-value-added phenolic monomers, in which 1,4-dioxane acts as a lignin solvent; ethanol acts as a solvent, reactant, and in situ hydrogen donor; and FA acts as an acid catalyst and in situ hydrogen donor. They proposed the preliminary reaction pathways: Initially, lignin was mostly dissolved in ethanol/1,4-dioxane to form a homogeneous reaction system. Moreover, numerous lignin fragments were generated due to the in situ hydrogen from ethanol; subsequently, FA brought about substantial acid-catalyzed cleavage of the linkages in lignin fragments. Meanwhile, atomic hydrogen produced in situ from FA decomposition led to a series of hydrogen-induced reactions of depolymerized intermediates, resulting in the increase of phenolic monomers. More discussion about the lignin acidolysis-hydrogenation systems with FA is put in Chapter 8.2.5. In addition, the formic acid can also promote the hydrolysis of pre-oxidized β-O-4 linkage [104], more discussion can be found in Chapter 9.5.

In addition, like formic acid with two roles in the lignin acidolysis, the conversion of lignin could be facilitated by using ionic liquids (ILs) as both solvent and catalyst, especially for acidic ILs [105–110]. Acidic ILs are not only able to dissolve lignin in a decent amount but also capable of activating the ether moieties in the lignin, resulting in a partial cleavage of C–O bonds. In 2010, Ekerdt and co-workers [111] employed 1-H-3-methylimidazolium chloride ([Hmim]Cl) for the treatment of β-O-4 lignin model GG and VG (Figure 8.13). These two model compounds underwent catalytic hydrolysis to produce guaiacol as the main product with yields higher than 70% at 150 °C. The reactivity and mechanism of the model compounds in the IL depend not only on the acidity, but also on the nature of IL cations and IL anions, and their interaction with the substrate [112]. For ILs with bromide, chloride, and hydrogen sulfate as the anions, these ILs could stabilize the hydroxyl group and prevent deprotonation through coordination. The hydrolysis follows a typical Brönsted acid-catalyzed reaction mechanism (Route A). Using IL as a solvent with a less coordinating anion such as BF_4^-, significant amounts of vinyl ether (VE) and formaldehyde were formed through deprotonation of the γ- hydroxyl group following break-off of the γ-carbon (Route B).

Besides the lignin models and just the lignin substrate, the ability of ILs to swell or dissolve lignocellulose makes it possible to obtain the direct conversion of raw

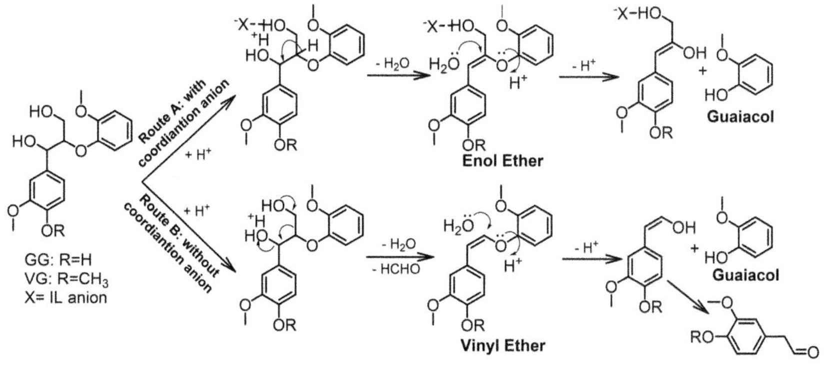

Figure 8.13 Acid-catalyzed Mechanism for Hydrolysis of the β-O-4 Bonds in Ionic Liquid. *Source:* Cox et al. [112] with Permission, © 2011 Elsevier.

biomass under mild conditions without pretreatment [113–115]. For example, in 2010, Argyropoulos and co-workers [114] showed that the acidic pretreatment of the raw wood species in IL [Amim]Cl resulted in not only the complete hydrolysis of cellulose and hemicellulose but also lignin depolymerization to obtain phenols. It should be noted that the lignin conversion in acidic ILs also suffers from the inevitable char formation and intermediates repolymerization via alky-aryl fragments coupling reaction [105]. Therefore, searching for appropriate acidic IL catalysts which afford excellent catalytic activities and char inhibition capabilities becomes a challenge. In 2017, Li and co-workers [116] reported a remarkable catalyst of cooperative IL pairs, consisting of [bSmim][HSO_4] (IL catalyst) and [bmim][CF_3SO_3] (IL additive, an excellent IL solvent for lignin dissolution), allowed the depolymerization of organosolv lignin to occur under mild conditions of 250 °C and 30 min, giving a 66.7% conversion of lignin, and a14.5% yield of phenolic monomer, and no obvious char formation. Moreover, it was noteworthy that the cooperative IL pairs both played important roles in the depolymerization of lignin in which the IL catalyst presented excellent ether linkage cleavage capability, whereas the IL additive exhibited efficient carbocation capping and scavenging.

8.2.3 Lignin Acidolysis with Homogeneous Metal Salts and Organic Metal Compounds

Besides the above protic acids, and acidic ionic liquids, some homogeneous Lewis metal salts [117, 118], and organic metal compounds can also catalyze the acidolysis of lignin depolymerization.

In 2015 and 2016, Barta and co-workers [49, 50] reported a series of triflic acid-catalyzed methods to achieve the depolymerization of lignin β-O-4 linkages with low recondensation, which were based on the rapid transformation of in-situ generated substituted phenylacetaldehyde. Capturing the unstable phenylacetaldehyde by reaction with diols leads to a high yield of acetal aromatics. Besides the triflic acid, during 2016-2017, Barta and co-workers further studied the promotion effect of metal triflates for the production of aromatics from lignin [119, 120]. Regarded as the typical method to restrain the lignin fragments' condensation, the corresponding discussion is provided in Chapter 11.2.3. For other metal salts for lignin acidolysis, in 2016, Güvenatam and co-workers [121] investigated the depolymerization of Protobind lignin with various metal salts including metal chlorides, acetates, and triflates, which was carried out in ethanol under supercritical conditions at 400 °C for 4 h. The results demonstrated that the metal salts led to high organic yields and the degree of depolymerization of lignin increased going from metal acetates to metal chlorides to metal triflates. Then they further investigated the depolymerization of soda lignin via alcoholysis using triflate salt $Al(OTf)_3$ as the catalyst, resulting in complete conversion of lignin without formation of char or insoluble residue. It is noteworthy that metal salts are observed to be soluble in strong polar solvents. For the depolymerization of lignin, this is possible using metal salts as catalysts due to the tendency to favor carbocation formation from the C_α–OH [122–124]. Moreover, ethanol was found to be involved in the alkylation of the lignin depolymerization products which was related to the strong catalytic activity of the triflate salts, increasing the product yield by inhibiting repolymerization of the products [125]. Besides the separate use of the Lewis metal salts in lignin depolymerization, the tandem process of metal salt+hydrogenation [126–129] has been discussed in Chapter 6.1 and Chapter 11.2.3 [49, 130], and one can check the discussion for more information.

For the organic metal compounds that can catalyze the lignin acidolysis, one representative homogeneous catalyst is methyltrioxorhenium (MTO) [132]. Recently, besides using it as the oxidation catalysts [133–136], MTO has also been used as a very efficient acid catalyst for deoxydehydration reactions of biomass-derived polyols [137–141]. MTO can also catalyze the dehydration of benzylic alcohol derivates [142], and the dehydration mechanism based on Re^{V} as the key species [143, 144]. In 2014, Kühn and co-workers [131] made an obvious progression for the application of MTO in lignin conversion. They reported that methyldioxorheniu (MDO), which was in situ generated by reduction of MTO, can effectively catalyze C–O bond cleavage of a variety of lignin β-O-4 model compounds in *p*-xylene without oxidant. Based on the observed and isolated intermediates, ^{17}O- and ^{2}H-isotope labeling experiments, DFT calculations, and several spectroscopic studies, a reaction mechanism is proposed (Figure 8.14). MTO is firstly reduced by substrate **1** to MDO (step I). MDO reacts subsequently with a second molecule of **1**, forming an alcoholate adduct **1'** (step II), from which enol ether **4** (step III) and a rhenium (V) bis(hydroxo) species **5'** eliminated. Then, **5'** adds an O and H atom to the double bond over **TS1** (step IV) to generate an acetal-like rhenium alcoholate **TS2**, thereby leading to C_β–OAr bond scission (step V) and liberation of MDO and cleavage products to close the catalytic cycle.

Whereas, in the above system lignin feedstock conversion still faces a challenge due to the poor dissolubility of lignin in *p*-xylene, and the relatively low reaction rate. In 2015, Zhang and co-workers [145] achieved the fast conversion of lignin β-O-4 model compounds and organosolv lignin to aromatics over methyltrioxorhenium (MTO) in ionic liquids under the assistance of microwave. This system did not

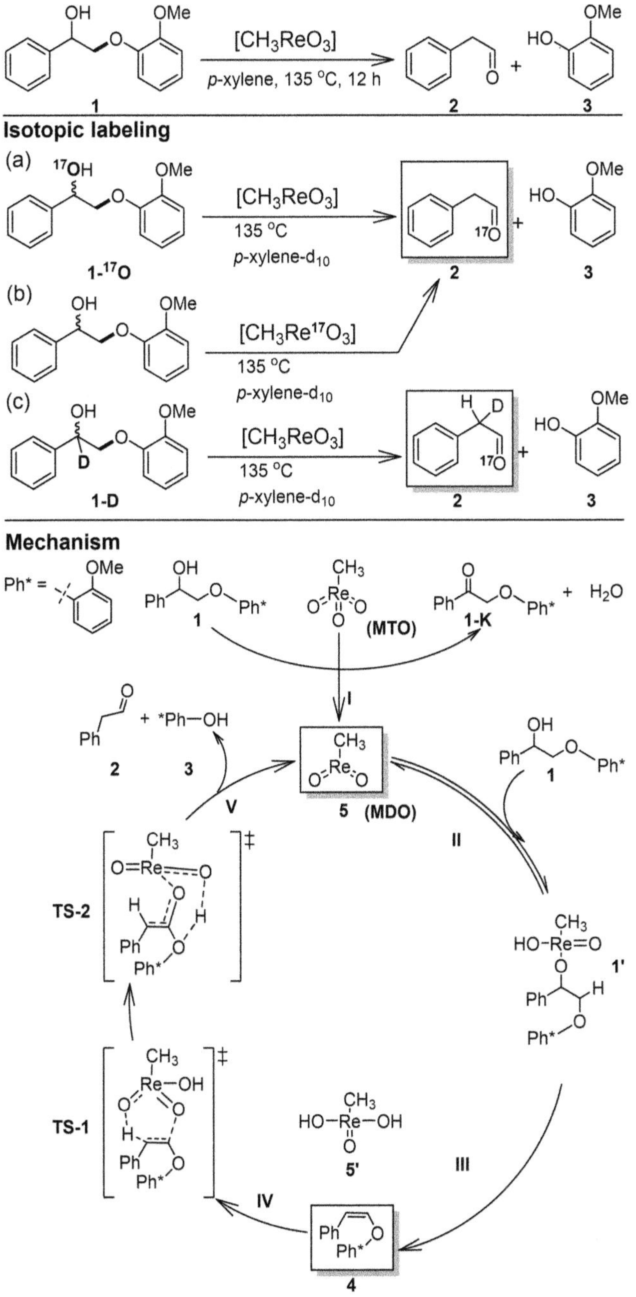

Figure 8.14 Cleavage of Lignin β-O-4 C–O Bonds Catalyzed by Methyldioxorhenium (MTO). *Source:* Harms et al. [131] with Permission, © 2014 Wiley-VCH.

need any oxidant and reducing agent. Mechanism study showed that the substrate was first converted into olefins-intermediates, which were then converted to the final products (Figure 8.15). This result is in agreement with the work of Kühn [131],

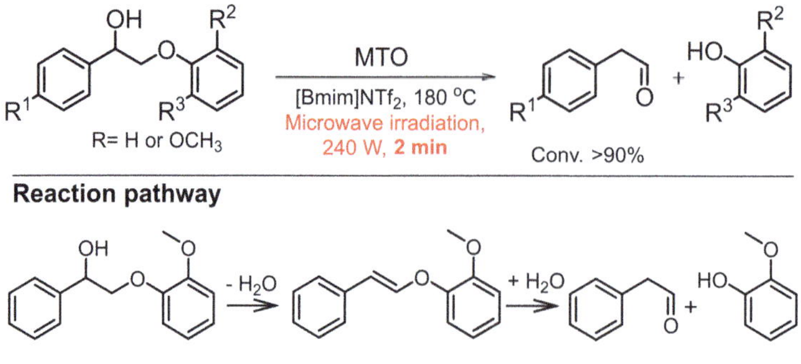

Figure 8.15 Microwave-assisted Fast Conversion of Lignin β-O-4 Linkage over Methyltrioxorhenium in Ionic Liquids.

suggesting that the substrate follows the same reaction pathway in ionic liquid compared to the reaction in organic solvent *p*-xylene.

8.2.4 Lignin Acidolysis with Solid Acid Catalysts

Traditionally, lignin can be readily depolymerized via liquid acids, which is mainly ascribed to the acidity, homogeneity, and permeability of the catalysts. However, corrosiveness and difficulty of catalyst recovery are unavoidable issues [48, 146]. Heterogeneous acid catalysis can provide numerous advantages, such as easier catalyst separation and reuse, easier separation of product, and less environmental damage. In this respect, different kinds of solid catalysts used in the catalytic conversion of lignin will be summarized in this subsection.

For the solid acid catalysts, liquid acids can be immobilized on various supports to form immobilized liquid acids. In 2017, Elfadly and co-workers [147] prepared various immobilized liquid acids with the combination of natural bentonite clay and several mineral acids (HCl, H_2SO_4, and H_3PO_4), which were further used in the production of aromatic hydrocarbons from catalytic pyrolysis of lignin. The results revealed that the immobilized liquid acids significantly improved the yields of BTX (benzene–toluene–xylenes) and naphthalene at 600 °C, explained by the activation of bentonite clay with mineral acids which resulted in the replacement of the exchangeable cations with H^+ ions of the acids. However, the stability of the immobilized liquid acids should be emphasized as the catalysts are particularly prone to losing their active groups under a high reaction temperature. In 2017, Li and co-workers [148] studied the catalytic performance of the ternary composite oxides $S_2O_8^{2-}/ZrO_2$-TiO_2-SiO_2 as an efficient solid super acid catalyst for depolymerization of lignin. In 2018, Ren and co-workers [149] reported that the tungstophosphoric acid (TPA), supported on zirconia materials with a high surface area, was shown to be an effective catalyst for the degradation of alkali lignin (AL). Catalyst loaded with 20% TPA and prepared at 200 °C demonstrated the highest catalytic efficiency when the catalyst-to-lignin ratio was 1:8. Under these conditions, total hydroxyl groups and phenolic hydroxyl groups increased by 203.9% and 148.4%, respectively. The catalyst could be reused at least three times, without significant loss of activity.

Besides these inorganic acid species, some solid acid catalysts with organic acid grafting or polymer structure sulfonation can be also used in lignin conversion.

In addition, metal oxides have been explored for the catalytic conversion of lignin [150]. In 2014 and 2015, Deepa and co-workers [151, 152] evaluated several types of solid acid catalysts ranging from crystalline to amorphous, such as zeolites, clay, sulfated zirconia, and metal oxides (Nb_2O_5, MoO_3–SiO_2, and SiO_2–Al_2O_3) for the depolymerization of lignin into aromatic monomers. Interestingly, compared to other acid catalysts, the SiO_2–Al_2O_3 catalyst gave exceptionally high yields of about 60% for organic solvent-soluble extracted products in the depolymerization of six different types of lignins at 250 °C within 30 min in the methanol solution. In 2015, Jin and co-workers [153] reported that well-dispersed niobic acid catalysts supported on activated carbon in which niobic acid was transformed from amorphous to crystalline Nb_2O_5 exhibited good C–O ether bond cleavage activity and acid-catalyzed dehydration activity when used to break the linkages of lignin-derived 4-O-5 aryl ether bonds under mild conditions due to the coexisting Brønsted and Lewis acid sites on the hydrated niobic oxide. Moreover, the activity toward the removal of oxygen gradually expanded with an increase in the loading of niobic acid, leading to a 70% selectivity to C–O bond cleavage and a 65% selectivity to O-free alkane at 200 °C under 3.0 MPa H_2. In 2018, Santana Junior and co-workers [154] achieved a higher selectivity of 92% for hydrocarbons, mainly alkanes, with niobic acid as a catalyst in the pyrolysis of industrial kraft lignin. Besides the single metal oxide, Huang and coworkers [155] prepared a complex metal oxide, Cu_xMgAl_y, for the depolymerization of soda lignin in supercritical ethanol, yielding 36% monomers without formation of char after reaction at 340 °C for 4 h.

Additionally, aluminosilicate zeolites, with abundant Brønsted acid sites (Si–OH–Al) in their framework, are promising catalysts in lignin pyrolysis, because of the large surface area and well-ordered pore structures [156]. To date, zeolite and modified zeolite catalysts have received much attention for the catalytic pyrolysis of lignin or upgrading of pyrolysis vapors. ZSM-5/HZSM-5 has acid sites and appropriate pore diameter (~ 0.5 nm) and shape selectivity and presents good deoxygenating properties to produce aromatic hydrocarbons from the catalytic pyrolysis of lignin [157–161]. In 2014, Adhikari and co-workers [162] investigated the effect of zeolite acidity on catalytic pyrolysis of lignin, demonstrating that ZSM-5 with a relatively higher acidity was more efficient in eliminating oxygenates and promoting aromatic hydrocarbon production due to the catalytic cleavages of the methoxy groups and aromatic C–O bonds in the lignin polymer. Kim and co-workers [163] reported that the yield of aromatic hydrocarbons increased from 0.76 (Si/Al = 280) to 2.62% (Si/Al = 30) in proportion with increasing catalyst acidity during milled wood lignin pyrolysis. In addition to acidity, pore size, internal pore volume, and the crystal structures of the zeolite catalyst play an important role in the migration of pyrolysis vapors into the catalyst and the sequent reactions during the catalytic process [164]. In 2012, Ma and co-workers [165] tested a series of acidic zeolites such as Na-ZSM-5, ASA, silicalite, HZSM-5, H-beta, and H-USY for the catalytic fast pyrolysis of alkaline lignin. The results revealed that H-USY with a larger pore size (0.74 nm) and lower Si/Al ratio (7) presented the highest catalytic activity; furthermore, the yield of liquid and the selectivity to desired products could be controlled by tuning the acidity and pore size of the catalyst. In 2017, Nandiwale and co-workers [166] reported a breakthrough for the depolymerization of dealkaline lignin, in which predominantly Lewis acidic mesoporous silicate Zr-KIT-5 showed a marvelous catalytic activity, providing a 7.5% yield of aromatic monomers and a 77% yield of tetrahydrofuran-soluble aromatic species, which were much higher than those observed with H-ZSM-5. More excitingly, the yield of catalyst deposits was lower in Zr-KIT-5 (14%), compared to HZSM-5 (27%). The predominant Lewis acidity of Zr-KIT-5 appeared to reduce aldehyde condensation reactions promoted by the Brønsted acid sites present in zeolites. Unfortunately, a lack of hydrothermal stability of the mesoporous silicates was a drawback that must be overcome for practical viability. In 2019, Jiang and co-workers [167] found that when the acid-treated mesoporous MCM-41 molecular sieves were adopted as catalysts for degradation of dealkalized lignin dissolved in acidic ionic liquid 1-methyl-3-benzylimidazolium chloride (BnMIMCl), the efficiency of lignin degradation was as high as 74.02% at 200 °C for 4 h; meanwhile, over 80% of the liquid products were tested as phenols.

8.2.5 Acid-promoted Tandem Process for the Cleavage of Lignin $C_{\beta}-OAr$ Bonds

The above strategies show that acidolysis and acid-catalyzed pyrolysis can effectively fragment the lignin. To more effectively depolymerize and convert these fragments, according to the cascaded mode of IDA, combining acidolysis with hydrogenation may be a good choice to increase the depolymerization efficiency of real lignin.

In 2009, Tang and co-workers [168] synthesized an immobilized liquid acid Pd/SO_4^{2-}/ZrO_2/SBA-15 to upgrade bio-oil with the combination of hydrotreatment, esterification, and cracking, resulting in a trace amount of tar or coke produced and a large number of organic components kept in the upgraded bio-oil in which aldehydes and ketones produced from lignin pyrolysis were almost completely removed. In 2011, Zhao and co-workers [169] reported a dual-functional catalyst system, phosphoric acid, and Pd/C, for the aqueous-phase hydrodeoxygenation of bio-derived phenols to cycloalkanes. In 2012, Leahy and co-worker [170] studied the hydropyrolysis of biomass with Ni-ZSM-5 and Ni-MCM-41.

In 2015, Ma and co-workers [171] reported the simultaneous conversion of cellulose, hemicellulose, and lignin from raw biomass into gasoline alkanes (hexanes and pentanes) and monophenols and related hydrocarbons over layered $LiTaMoO_6$ and Ru/C in the aqueous phosphoric acid medium at 330 °C and 6 MPa H_2 for 24 h. Specifically, gasoline alkanes were directly yielded from the carbohydrate components, based on hemicellulose and cellulose, and the total yield could be up to 82.4%. Notably, the lignin fraction could also be transformed into monophenols, related alcohols and hydrocarbons by the one-pot reaction. The total yield of volatile products was 53% based on the lignin fraction. In 2016, Sels and co-workers [37] studied the influence of acidic (H_3PO_4) additives on the Pd/C-catalyzed reductive processing of poplar wood in methanol (MeOH) at 150-250 °C. It was found that the addition of small quantities of H_3PO_4 results in three rather than two product streams, since under acidic conditions both delignification and alcoholysis of hemicellulose and lignin are promoted, leaving behind a cellulose-rich pulp. In 2016, Anderson and co-workers [87] present a more inexpensive system consisting of phosphoric acid and Ni/C was for the reductive catalytic fractionation (RCF) of corn stoves. After 6 h at 200 °C in methanol under 30 bar of H_2, the yields of monomer with Ni/C alone and with the addition of acid co-catalyst to the Ni/C system were 28.3% and 38%, respectively, indicating that the acid cocatalyst could improve lignin depolymerization. In 2017, Wang and co-workers [172] developed an efficient catalytic system for the liquefaction of kraft lignin by hydrocracking, in which the immobilized catalyst $S_2O_8^{2-}$-KNO_3/TiO_2 with active acidic and basic sites acted as the catalyst in lignin hydrocracking with Ru/C, leading to a high liquid product yield of 93% and an extremely low solid product yield of less than 1% at 320 °C for 6 h. In 2018, Jin and co-workers [173] employed the catalyst $HTaMoO_6$ with strong Brønsted acidity and the hydrogenation catalyst Rh/C for the depolymerization of kraft lignin in dioxane-water systems, giving a liquid product yield of 95.6% at 290 °C and 2 MPa H_2 for 2 h which was higher than those catalyzed by an independent catalyst.

Besides the normal solvent, ionic liquids are attractive reaction media for the solubilisation and acidolysis of lignin into value-added products. In 2010, Yan and co-workers [174] explored a Ru nanoparticle (Ru NP)–Brønsted acidic IL system to convert lignin-derived phenol into cyclohexane with excellent selectivity at 130 °C for 4 h under 4 MPa H_2. In 2016, Barta and co-workers [175] presented a detailed study of the scission of the most abundant lignin β-O-4 motif in Brønsted acidic ionic liquids followed by conversion of reactive aldehyde intermediates using Ru-NPs and H_2. In 2016, by using a choline-derived acidic IL ([Ch][MeSO3]) as both a solvent and an acid catalyst and Pd/C as a metal catalyst, Liu and co-workers [176] selectively convert lignin to phenol or catechol. The conversion of kraft lignin and the selectivity to phenol and catechol reached 20.3%, 18.4%, and 18.1%, respectively, at 200 °C for 5 h under 2.0 MPa of H_2. In 2016, Zhang and co-workers [177] developed a pseudo-homogeneous catalyst system, consisting of uniformly stabilized noble metal nanoparticles (NPs) in ionic liquids with H_3PO_4 for the selective reductive cleavage of lignin C–O bond. Compared with Ru, Rh, and Pd NPs, Pt NPs in $[Bmim]PF_6$ were identified as the best catalytic system for the transformation of lignin monomeric and dimeric model compounds without alcohol hydroxyl group to substituted cyclohexanes.

Besides using (noble) metal NPs as the hydrogenation catalysts, in 2016, Li and co-workers [178, 179] reported that W_2C/AC exhibited promising potential for the substitution of noble metals in the cleavage of the major aryl ether bonds of lignins without destroying the aromatic rings. High yields of C–O cleavage products (up to 96.8%) from model compounds and liquid oils (up to 70.7%) from lignin feedstock were obtained under low hydrogen pressure (0.69 MPa) in methanol. Taking into account all intermediates and products formed in the above reactions, the overall reaction pathway for the W_2C catalyzed conversion of the β-O-4 model compound is proposed in Figure 8.16. It proceeds by initial dehydration of β-O-4 compound **1**

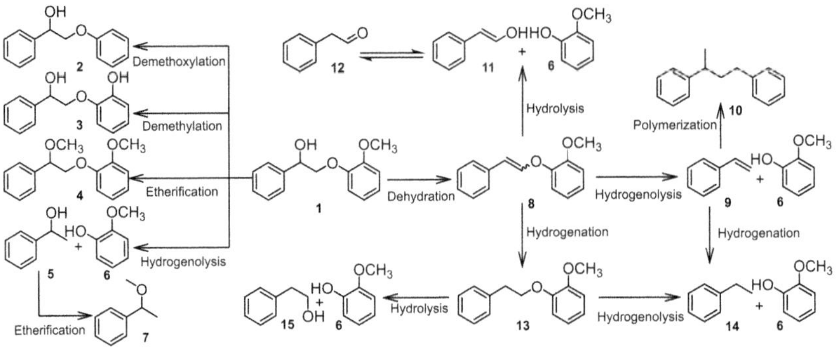

Figure 8.16 Proposed Reaction Pathway for the W_2C/AC-catalyzed Deconstruction of 2-(2-methoxyphenoxy)-1-phenylethanol (β-O-4 compound 1) in Alcohol Solvent. *Source:* Guo et al. [178] with Permission, © 2014 Wiley-VCH.

to form the phenyl vinyl ether intermediate **8** (including the E and Z isomers), which is followed by hydrogenolysis of **8** to produce styrene and guaiacol and the hydrogenation of styrene to obtain ethylbenzene. According to the product distribution, the hydrogenolysis of **8** is proposed as the major route to the target products, while the hydrogenation of **8** to **13** followed by hydrogenolysis of **13** to ethylbenzene and guaiacol is a possible minor route. Moreover, compound **13** can produce 2-phenylethanol by hydrolysis over the WC_x catalyst. In addition, the hydrolysis of **8** can generate guaiacol and **11**, which tends to isomerize into aldehyde **12**. Further, direct hydrogenolysis of compound **1** occurs to produce guaiacol and 1-phenylethanol, and the latter is then etherified to **7**. Other side reactions for the mechanism include polymerization of styrene to **10** and the demethoxylation, hydrogenolysis, and etherification of compound **1** to give **2**, **3**, and **4**, respectively.

For the cleavage of the C_β–OAr bond in β-O-4 compound **8** (Figure 8.16), although Li and co-workers declare that the hydrogenolysis is the main process to provide the styrene and the phenol [178], the more detailed mechanism is not provided. Focusing on this reaction mechanism, the direct cleavage of the target ether in phenyl vinyl ether may be difficult, because its BDE increase by about 62 $kJ\cdot mol^{-1}$ from the value of compound **1**. Besides the hydrolysis route (**8→11+6**), as another potential choice, the semi-hydrogenation of the C=C bond with the generation of the $C_\alpha^\bullet$ radical can obviously decrease the BDE of the C_β–OAr bond in **8**. A more detailed discussion can be found from the dehydroxylation-(reduction)-hydrogenation process reported by Zhang [180] which will be discussed in Chapter 9.6.2.

Currently, a promising and relatively new lignin conversion methodology involves the use of a hydrogen donor solvent instead of molecular hydrogen. Besides the commonly used alcohols, a well-known hydrogen donor is a formic acid, which is converted in situ, either thermally or catalytically, to molecular hydrogen and CO/CO_2 [181]. When employed in lignin depolymerization, formic acid has a triple effect of promoting lignin decomposition, improving the chemical properties of the depolymerized products as well as serving to inhibit the repolymerization reaction due to its excellent hydrogen-donating ability [100, 102], and acid-catalytic activity [182].

For instance, formic acid was added to improve the yield of lignin oil and inhibit repolymerization during supercritical treatment. In 2016, Park [183] reported that the lignin oils produced in ethanol medium with 7% formic acid showed the best properties, yielding the highest total monomeric content and the lowest O/C value. It is worth noting that the accession of catalysts provides an increase in the rate of hydrogen production from formic acid and promotes the lignin depolymerization to monomers, which was confirmed by Kristianto [184] in 2017. Besides the use of hydrogen donors, formic acid can also be involved in the depolymerization reaction by providing methyl to stabilize lignin depolymerization intermediates.

As a class of Brønsted acids, heteropoly acids (HPAs) are made up of a particular combination of hydrogen cations and polyoxometalate (POM) anions, among which the most stable and easy available Keggin HPAs comprise heteropoly anions of the formula $[XM12O40]^{n-}$, where X is the heteroatom (P^{5+}, Si^{4+}, etc.) and M the addendum atom (Mo^{6+}, W^{6+}, etc.). HPAs possess stronger acidity than conventional solid acid catalysts such as acidic oxides and zeolites. When the HPAs are used in the lignin depolymerization, besides the acidolysis mechanism, the redox property of the POM can cause the further oxidation of the lignin fragments. In 2010, Marques et al. [97] reported a two-stage POM delignification method, in which reduced Mn-substituted POM, $SiW_{11}Mn^{II}$, was oxidized by versatile peroxidase, resulting in a high selectivity for delignification. To improve the selectivities of degradation products from the delignification process, Voitl and co-workers [98] proposed a process to produce vanillin and methyl vanillate from oxidation depolymerization of industrial kraft lignin in aqueous methanol with $H_3PMo_{12}O_{40}$ as a homogeneous catalyst, resulting in a stable yield of 7% vanillin plus methyl vanillate, which was much higher than that by alkaline oxidation (< 4%). Soon afterward, Zhao and co-workers [99] further optimized the oxidation depolymerization reaction conditions with the catalyst of $H_5PMo_{10}V_2O_{40}$. A high yield of 65.2% of the oxidative degradation products that mainly consisted of aromatic compounds and dimethyl succinate was achieved from pyrolytic lignin in a 90% methanol solution at 190 °C for 10 min. Although the heteropoly acids and their salts display good catalytic activities for the depolymerization of lignin, however, they are hard to be recovered because of their high solubility in water. Satisfactorily, various methods have been proposed by researchers to overcome this drawback [93, 94], including the selection of suitable solvents and the introduction of larger monovalent cations to make the modified POM catalyst a heterogeneous one.

8.3 Chapter Summary

In this chapter, strategies for lignin linkages cleavage beginning with $C_\alpha O$–H/ArO–H or C_α–OH bonds heterolysis are summarized, primarily focusing on lignin alkaline hydrolysis, acidolysis, and their corresponding tandem processes. Chapter 8.1 focuses on the base-catalyzed cleavage of the C_β–OAr bond beginning with $C_\alpha O$–H or ArO–H heterolysis to oxygen anion, which further induces the cleavage of the lignin C_β–OAr bond. Chapter 8.2 summarizes the various acid-catalyzed methods for the C_β–OAr bond cleavage beginning with C_α–OH heterolysis, which further induce the C_β–OAr bond cleavage via a C_α=C_β–OAr intermediate. The keywords of this chapter could be Lignin depolymerization, Lignin alkaline hydrolysis, Lignin acidolysis, Ether bond cleavage.

References

1 Smook, G.A. (1992). Handbook for pulp and paper technologists vol. 11, seconde. Angus Wilde Publications.

2 Brink, D.L., Hossfeld, R.L., and Sandstrom, W.M. (1949). Products from the degradation of lignin by sodium hydrosulfide. *J. Am. Chem. Soc.* 71: 2275–2275.

3 Löfstedt, J., Dahlstrand, C., Orebom, A. et al. (2016). Green diesel from kraft lignin in three steps. *ChemSusChem* 9: 1392–1396.

4 Yuan, Z.S., Cheng, S.N., Leitch, M., and Xu, C.B. (2010). Hydrolytic degradation of alkaline lignin in hot-compressed water and ethanol. *Bioresour. Technol.* 101: 9308–9313.

5 Lavoie, J.M., Bare, W., and Bilodeau, M. (2011). Depolymerization of steam-treated lignin for the production of green chemicals. *Bioresour. Technol.* 102: 4917–4920.

6 Gellerstedt, G. and Lindfors, E.L. (1984). Structural-changes in lignin during kraft cooking. 4. Phenolic hydroxyl-groups in wood and kraft pulps. *Sven. Papperstidn.* 87: R115–R118.

7 Gierer, J. (1970). The reactions of lignin during pulping. *Svensk Papperstidning* 18: 571–596.

8 Gierer, J. and Ljunggren, S. (1979). Reactions of lignin during sulfate pulping. 17. Kinetic Treatment of the formation and competing reactions of quinone methide intermediates. *Sven. Papperstidn.* 82: 503–512.

9 Gierer, J. and Ljunggren, S. (1979). Reactions of lignins during sulfate pulping. 16. Kinetics of the cleavage of β-Aryl ether linkages in structures containing carbonyl groups. *Sven. Papperstidn.* 82: 71–81.

10 Gierer, J. (1980). Chemical aspects of kraft pulping. *Wood Sci. Technol.* 14: 241–266.

11 Gierer, J., Ljunggren, S., Ljungquist, P., and Noren, I. (1980). Reactions of lignin during sulfate pulping. 18. Significance of alpha-carbonyl groups for the cleavage of β-Aryl ether structures. *Sven. Papperstidn.* 83: 75–82.

12 Gierer, J. (1982). The chemistry of delignification - a general concept. *Holzforschung* 36: 43–51.

13 Gierer, J. (1982). The chemistry of delignification - a general concept. 2. *Holzforschung* 36: 55–64.

14 Gierer, J. (1985). Chemistry of delignification. 1. General concept and reactions during pulping. *Wood Sci. Technol.* 19: 289–312.

15 Pasco, M.F. and Suckling, I.D. (1994). Lignin removal during kraft pulping-an investigation by thioacidolysis. *Holzforschung* 48: 504–508.

16 Hubbard, T.F., Schultz, T.P., and Fisher, T.H. (1992). Alkaline-hydrolysis of nonphenolic-β-O-4 lignin model dimers - Substituent effects on the leaving

phenoxide in neighboring group vs direct nucleophilic-attack. *Holzforschung* 46: 315–320.

17 Chakar, F.S. and Ragauskas, A.J. (2004). Review of current and future softwood kraft lignin process chemistry. *Ind. Crops Prod.* 20: 131–141.

18 Jia, S., Cox, B.J., Guo, X. et al. (2010). Decomposition of a phenolic lignin model compound over organic N-bases in an ionic liquid. *Holzforschung* 64: 577–580.

19 Lankau, T. and Yu, C.H. (2016). Intermediate oxiranes in the base-catalyzed depolymerisation of lignin. *Green Chem.* 18: 1590–1596.

20 Roberts, V.M., Stein, V., Reiner, T. et al. (2011). Towards quantitative catalytic lignin depolymerization. *Chem. Eur. J.* 17: 5939–5948.

21 Eom, H.J., Lee, D.W., Hong, Y.K. et al. (2014). Effects of a sodium carbonate (Na_2CO_3) additive on the conversion of phenethyl phenyl ether (PPE) in high-temperature water. *Appl. Catal., A* 472: 152–159.

22 Cronin, D., Moghaddam, L., Rackemann, D. et al. (2016). Degradation of phenethoxybenzene in sodium hydroxide. *RSC Adv.* 6: 57889–57901.

23 Liu, Y., Carriero, S., Pye, K., and Argyropoulos, D.S. (2000). A comparison of the structural changes occurring in lignin during Alcell and kraft pulping of hardwoods and softwoods. *Lignin: Historical, Biological, and Materials. Perspectives* 742: 447–464.

24 Gierer, J. and Lindeberg, O. (1979). Studies on the condensation of lignins in alkaline media. 3. Formation of stilbenes, arylcoumarans and diarylmethanes on treatment of spruce wood meal with alkali and white liquor in the presence of xylenols. *Acta Chem. Scand. B* 33: 580–582.

25 Okuda, K., Umetsu, M., Takami, S., and Adschiri, T. (2004). Disassembly of lignin and chemical recovery—rapid depolymerization of lignin without char formation in water-phenol mixtures. *Fuel Process. Technol.* 85: 803–813.

26 Dabral, S., Mottweiler, J., Rinesch, T., and Bolm, C. (2015). Base-catalysed cleavage of lignin β-O-4 model compounds in dimethyl carbonate. *Green Chem.* 17: 4908–4912.

27 Katahira, R., Mittal, A., McKinney, K. et al. (2016). Base-catalyzed depolymerization of biorefinery lignins. *ACS Sustainable Chem. Eng.* 4: 1474–1486.

28 Dabral, S., Engel, J., Mottweiler, J. et al. (2018). Mechanistic studies of base-catalysed lignin depolymerisation in dimethyl carbonate. *Green Chem.* 20: 170–182.

29 Toledano, A., Serrano, L., and Labidi, J. (2012). Organosolv lignin depolymerization with different base catalysts. *J. Chem. Technol. Biotechnol.* 87: 1593–1599.

30 Chaudhary, R. and Dhepe, P.L. (2017). Solid base catalyzed depolymerization of lignin into low molecular weight products. *Green Chem.* 19: 778–788.

31 Zhang, Z.R., Song, J.L., and Han, B.X. (2017). Catalytic transformation of lignocellulose into chemicals and fuel products in ionic liquids. *Chem. Rev.* 117: 6834–6880.

32 Garron, A., Arquilliere, P.P., Al Maksoud, W. et al. (2015). From industrial black liquor to pure phenolic compounds: A combination of catalytic conversion with ionic liquids extraction. *Appl. Catal., A* 502: 230–238.

33 Kleine, T., Buendia, J., and Bolm, C. (2013). Mechanochemical degradation of lignin and wood by solvent-free grinding in a reactive medium. *Green Chem.* 15: 160–166.

34 Matson, T.D., Barta, K., Iretskii, A.V., and Ford, P.C. (2011). One-pot catalytic conversion of cellulose and of woody biomass solids to liquid fuels. *J. Am. Chem. Soc.* 133: 14090–14097.

35 Long, J.X., Xu, Y., Wang, T.J. et al. (2015). Efficient base-catalyzed decomposition and in situ hydrogenolysis process for lignin depolymerization and char elimination. *Applied Energy* 141: 70–79.

36 Konnerth, H., Zhang, J., Ma, D. et al. (2015). Base promoted hydrogenolysis of lignin model compounds and organosolv lignin over metal catalysts in water. *Chem. Eng. Sci.* 123: 155–163.

37 Renders, T., Schutyser, W., Van den Bosch, S. et al. (2016). Influence of acidic (H_3PO4) and alkaline (NaOH) additives on the catalytic reductive fractionation of lignocellulose. *ACS Catal.* 6: 2055–2066.

38 Qi, S.C., Hayashi, J., Kudo, S., and Zhang, L. (2017). Catalytic hydrogenolysis of kraft lignin to monomers at high yield in alkaline water. *Green Chem.* 19: 2636–2645.

39 Barta, K., Matson, T.D., Fettig, M.L. et al. (2010). Catalytic disassembly of an organosolv lignin via hydrogen transfer from supercritical methanol. *Green Chem.* 12: 1640–1647.

40 Barta, K., Warner, G.R., Beach, E.S., and Anastas, P.T. (2014). Depolymerization of organosolv lignin to aromatic compounds over Cu-doped porous metal oxides. *Green Chem.* 16: 191–196.

41 Sturgeon, M.R., O'Brien, M.H., Ciesielski, P.N. et al. (2014). Lignin depolymerisation by nickel supported layered-double hydroxide catalysts. *Green Chem.* 16: 824–835.

42 Kruger, J.S., Cleveland, N.S., Zhang, S.T. et al. (2016). Lignin depolymerization with nitrate-intercalated hydrotalcite catalysts. *ACS Catal.* 6: 1316–1328.

43 Teong, S.P., Lim, J., and Zhang, Y. (2017). Vinylation of aryl ether (lignin β-O-4 linkage) and epoxides with calcium carbide through C-O bond cleavage. *ChemSusChem* 10: 3198–3201.

44 Walde, A.W. and Hixon, R.M. (1934). Alkaline oxidation of lignin. *J. Am. Chem. Soc.* 56: 2656–2659.

45 Lee, T.W. and Yang, J.W. (2018). Transition-metal-free conversion of lignin model compounds to high-value aromatics: scope and chemoselectivity. *Green Chem.* 20: 3761–3771.

46 Tarabanko, V.E., Hendogina, Y.V., Petuhov, D.V., and Pervishina, E.P. (2000). On the role of retroaldol reaction in the process of lignin oxidation into vanillin. Kinetics of the vanillideneacetone cleavage in alkaline media. *React. Kinet. Catal. Lett.* 69: 361–368.

47 Tarabanko, V.E., Petukhov, D.V., and Selyutin, G.E. (2004). New mechanism for the catalytic oxidation of lignin to vanillin. *Kinet. Catal.* 45: 569–577.

48 Wu, Z., Hu, L., Jiang, Y. et al. (2020). Recent advances in the acid-catalyzed conversion of lignin. *Biomass Convers. Biorefin.* 1–21.

49 Deuss, P.J., Scott, M., Tran, F. et al. (2015). Aromatic monomers by in situ conversion of reactive intermediates in the acid-catalyzed depolymerization of lignin. *J. Am. Chem. Soc.* 137: 7456–7467.

50 Lahive, C.W., Deuss, P.J., Lancefield, C.S. et al. (2016). Advanced model compounds for understanding acid-catalyzed lignin depolymerization: Identification of renewable aromatics and a lignin-derived solvent. *J. Am. Chem. Soc.* 138: 8900–8911.

51 Wan, G.G. and Frazier, C.E. (2017). Lignin acidolysis predicts formaldehyde generation in pine wood. *ACS Sustainable Chem. Eng.* 5: 4830–4836.

52 Agarwal, A., Rana, M., and Park, J.-H. (2018). Advancement in technologies for the depolymerization of lignin. *Fuel Process. Technol.* 181: 115–132.

53 Dutta, T., Isern, N.G., Sun, J. et al. (2017). Survey of lignin-structure changes and depolymerization during ionic liquid pretreatment. *ACS Sustainable Chem. Eng.* 5: 10116–10127.

54 Jia, S.Y., Cox, B.J., Guo, X.W. et al. (2011). Hydrolytic cleavage of β-O-4 ether bonds of lignin model compounds in an ionic liquid with metal chlorides. *Ind. Eng. Chem. Res.* 50: 849–855.

55 Sturgeon, M.R., Kim, S., Lawrence, K. et al. (2014). A mechanistic investigation of acid-catalyzed cleavage of aryl-ether linkages: Implications for lignin depolymerization in acidic environments. *ACS Sustainable Chem. Eng.* 2: 472–485.

56 Sarkanen, K.V. and Hoo, L.H. (1981). Kinetics of hydrolysis of erythro-guaiacylglycerol β-(2-methoxyphenyl) ether and its veratryl analogue using hcl and aluminum chloride as catalysts. *J. Wood. Chem. Technol.* 1: 11–27.

57 Pelzer, A.W., Sturgeon, M.R., Yanez, A.J. et al. (2015). Acidolysis of α-O-4 aryl-ether bonds in lignin model compounds: A modeling and experimental study. *ACS Sustainable Chem. Eng.* 3: 1339–1347.

58 Jasiukaitytė-Grojzdek, E., Huš, M., Grilc, M., and Likozar, B. (2020). Acid-catalysed α-O-4 aryl-ether bond cleavage in methanol/(aqueous) ethanol: understanding depolymerisation of a lignin model compound during organosolv pretreatment. *Sci. Rep.* 10: 1–12.

59 Wongsiriwan, U., Noda, Y., Song, C.S. et al. (2010). Lignocellulosic biomass conversion by sequential combination of organic acid and base treatments. *Energy Fuel* 24: 3232–3238.

60 Johansson, B. and Miksche, G. (1972). Über die Benzyl-arylätherbindung im Lignin. II. Versuche an Modellen. *Acta Chem. Scand.* 26: 289–301.

61 Fraile, J.M., Garcia, J.I., Hormigon, Z. et al. (2018). Role of substituents in the solid acid-catalyzed cleavage of the β-O-4 linkage in lignin models. *ACS Sustainable Chem. Eng.* 6: 1837–1847.

62 Shevchenko, S.M., Akim, L.G., Pranovich, A.V., and Zarubin, M.Y. (1991). Dry hydrogen iodide as a mild cleaving reagent for lignins. *Tappi J.* 74: 257–262.

63 Akim, L.G., Shevchenko, S.M., and Zarubin, M.Y. (1993). ^{13}C NMR-Studies on lignins depolymerized with dry hydrogen iodide. *Wood Sci. Technol.* 27: 241–248.

64 Shevchenko, S.M., Paszner, L., and Saddler, J.N. (2001). Treatment of organosolv and steam explosion lignins with molecular hydrogen iodide: A structural probe. *Cellul. Chem. Technol.* 35: 295–318.

65 Shevchenko, S.M. (2000). Depolymerization of lignin in wood with molecular hydrogen iodide. *Croat. Chem. Acta.* 73: 831–841.

66 Meshitsuka, G., Kondo, T., and Nakano, J. (1987). Cleavage of β-Aryl ether bonds in lignin by trimethylsilyl iodide. *J. Wood. Chem. Technol.* 7: 161–178.

67 Fukagawa, N., Meshitsuka, G., and Ishizu, A. (1992). 2D NMR study of residual lignin in beech kraft pulp combined with selective cleavage with pivaloyl iodide. *J. Wood Chem. Technol.* 12: 425–445.

68 Lu, F. and Ralph, J. (1997). Derivatization followed by reductive cleavage (DFRC method), a new method for lignin analysis: Protocol for analysis of DFRC monomers. *J. Agr. Food. Chem.* 45: 2590–2592.

69 Lu, F.C. and Ralph, J. (1997). DFRC method for lignin analysys. 1. New method for β-aryl ether cleavage: Lignin model studies. *J. Agr. Food. Chem.* 45: 4655–4660.

70 Lu, F.C. and Ralph, J. (1998). The DFRC method for lignin analysis. 2. Monomers from isolated lignins. *J. Agr. Food. Chem.* 46: 547–552.

71 Lu, F.C. and Ralph, J. (1998). The DFRC method for lignin analysis. 3. NMR studies. *J. Wood. Chem. Technol.* 18: 219–233.

72 Peng, J.P., Lu, F.C., and Ralph, J. (1998). The DFRC method for lignin analysis. 4. Lignin dimers isolated from DFRC-degraded loblolly pine wood. *J. Agr. Food. Chem.* 46: 553–560.

73 Peng, J.P., Lu, F.C., and Ralph, J. (1999). The DFRC method for lignin analysis. 5. Isochroman lignin trimers from DFRC-degraded pinus taeda. *Phytochemistry* 50: 659–666.

74 Ralph, J. and Lu, F.C. (1998). The DFRC method for lignin analysis. 6. A simple modification for identifying natural acetates on lignins. *J. Agr. Food. Chem.* 46: 4616–4619.

75 Lu, F.C. and Ralph, J. (1999). The DFRC method for lignin analysis. 7. Behavior of cinnamyl end groups. *J. Agr. Food. Chem.* 47: 1981–1987.

76 Rolando, C., Monties, B., and Lapierre, C. (1992). Thioacidolysis. In: *Methods in Lignin Chemistry* (ed. S.Y. Lin and C.W. Dence), 334–349. Berlin, Heidelberg: Springer Berlin Heidelberg.

77 Yamamura, M., Hattori, T., Suzuki, S. et al. (2012). Microscale thioacidolysis method for the rapid analysis of β-O-4 substructures in lignin. *Plant Biotechnol. (Tokyo)* 29: 419–423.

78 Yue, F., Lu, F., Sun, R.-C., and Ralph, J. (2012). Syntheses of lignin-derived thioacidolysis monomers and their uses as quantitation standards. *J. Agr. Food. Chem.* 60: 922–928.

79 Yue, F., Lu, F., Regner, M. et al. (2017). Lignin-derived thioacidolysis dimers: reevaluation, new products, authentication, and quantification. *ChemSusChem* 10: 830–835.

80 Matsushita, Y., Yagami, S., Kato, A. et al. (2020). Combinations of the aromatic rings in β-1 structure formation of lignin based on quantitative analysis by thioacidolysis. *J. Agr. Food. Chem.* 68: 9245–9251.

81 Nimz, H. and Das, K. (1971). Low molecular weight degradation products of lignin. II. Dimeric phenolic degradation products obtained by degradation of beech lignin with thioacetic acid. *Chem. Ber.* 104: 2359–2380.

82 Hägglund, E. and Björkman, C.B. (1924). Lignin Hydrochloride. *Biochem. Z.* 147: 74–89.

83 Phillips, M. and Goss, M.J. (1932). The hydrolysis of lignin with 12% hydrochloric acid. *J. Am. Chem. Soc.* 54: 3374–3377.

84 Zhang, G., Wen, Y., Liu, Z. et al. (2017). Acid-catalyzed hydrolysis of conifer lignosulfonate in black liquor for the production of value-added chemicals. *Appl. Catal., A* 542: 1–9.

85 Zhang, L., Yan, L., Wang, Z. et al. (2015). Characterization of lignin derived from water-only and dilute acid flowthrough pretreatment of poplar wood at elevated temperatures. *Biotechnol. Biofuels* 8: 203.

86 Mahmood, N., Yuan, Z.S., Schmidt, J., and Xu, C.B. (2015). Hydrolytic depolymerization of hydrolysis lignin: Effects of catalysts and solvents. *Bioresour. Technol.* 190: 416–419.

87 Anderson, E.M., Katahira, R., Reed, M. et al. (2016). Reductive catalytic fractionation of corn stover lignin. *ACS Sustainable Chem. Eng.* 4: 6940–6950.

88 Miller, J.E., Evans, L., Littlewolf, A., and Trudell, D.E. (1999). Batch microreactor studies of lignin and lignin model compound depolymerization by bases in alcohol solvents. *Fuel* 78: 1363–1366.

89 Lin, L.Z., Yao, Y.G., Yoshioka, M., and Shiraishi, N. (1997). Liquefaction mechanism of lignin in the presence of phenol at elevated temperature without catalysts - Studies on β-O-4 model compound. 1. Structural characterization of the reaction products. *Holzforschung* 51: 316–324.

90 Lin, L.Z., Yoshioka, M., Yao, Y.G., and Shiraishi, N. (1997). Liquefaction mechanism of lignin in the presence of phenol at elevated temperature without catalysts - Studies on β-O-4 lignin model compound. 2. Reaction pathway. *Holzforschung* 51: 325–332.

91 Lin, L.Z., Yoshioka, M., Yao, Y.G., and Shiraishi, N. (1997). Liquefaction mechanism of lignin in the presence of phenol at elevated temperature without catalysts - Studies on β-O-4 lignin model compound. 3. Multi-condensation. *Holzforschung* 51: 333–337.

92 Yang, X.H., Li, N., Lin, X.L. et al. (2016). Selective cleavage of the aryl ether bonds in lignin for depolymerization by acidic lithium bromide molten salt hydrate under mild conditions. *J. Agr. Food. Chem.* 64: 8379–8387.

93 Park, H.W., Kim, J.K., Hong, U.G. et al. (2013). Catalytic decomposition of lignin model compounds to aromatics over acidic catalysts. *Catal. Surv. Asia* 17: 119–131.

94 Wu, X., Jiao, W., Li, B.-Z. et al. (2017). Decomposition of a β-O-4 lignin model compound over solid Cs-substituted polyoxometalates in anhydrous ethanol: acidity or redox property dependence? *Chin. J. Catal.* 38: 1216–1228.

95 Alsalme, A.M., Wiper, P.V., Khimyak, Y.Z. et al. (2010). Solid acid catalysts based on $H_3PW_{12}O_{40}$ heteropoly acid-Acid and catalytic properties at a gas-solid interface. *J. Catal.* 276: 181–189.

96 Omwoma, S., Gore, C.T., Ji, Y.C. et al. (2015). Environmentally benign polyoxometalate materials. *Coordin. Chem. Rev.* 286: 17–29.

97 Marques, G., Gamelas, J.A.F., Ruiz-Duenas, F.J. et al. (2010). Delignification of eucalypt kraft pulp with manganese-substituted polyoxometalate assisted by fungal versatile peroxidase. *Bioresour. Technol.* 101: 5935–5940.

98 Voitl, T. and Rohr, P. R. v. (2010). Demonstration of a process for the conversion of kraft lignin into vanillin and methyl vanillate by acidic oxidation in aqueous methanol. *Ind. Eng. Chem.* 49: 520–525.

99 Zhao, Y., Xu, Q., Pan, T. et al. (2013). Depolymerization of lignin by catalytic oxidation with aqueous polyoxometalates. *Appl. Catal. A: Gen* 467: 504–508.
100 Holmelid, B., Barth, T., Brusletto, R., and Kleinert, M. (2017). Production of monomeric phenols by formic acid assisted hydrous liquefaction of lignin. *Biomass Bioenergy* 105: 298–309.
101 da Silva, S.H.F., Egues, I., and Labidi, J. (2019). Liquefaction of kraft lignin using polyhydric alcohols and organic acids as catalysts for sustainable polyols production. *Ind. Crops Prod.* 137: 687–693.
102 Huang, S.H., Mahmood, N., Zhang, Y.S. et al. (2017). Reductive de-polymerization of kraft lignin with formic acid at low temperatures using inexpensive supported Ni-based catalysts. *Fuel* 209: 579–586.
103 Wu, Z., Zhao, X.X., Zhang, J. et al. (2019). Ethanol/1,4-dioxane/formic acid as synergistic solvents for the conversion of lignin into high-value added phenolic monomers. *Bioresour. Technol.* 278: 187–194.
104 Rahimi, A., Ulbrich, A., Coon, J.J., and Stahl, S.S. (2014). Formic-acid-induced depolymerization of oxidized lignin to aromatics. *Nature* 515: 249–252.
105 Cox, B.J. and Ekerdt, J.G. (2012). Depolymerization of oak wood lignin under mild conditions using the acidic ionic liquid 1-H-3-methylimidazolium chloride as both solvent and catalyst. *Bioresour. Technol.* 118: 584–588.
106 Liu, S.W., Shi, Z.L., Li, L. et al. (2013). Process of lignin oxidation in an ionic liquid coupled with separation. *RSC Adv.* 3: 5789–5793.
107 da Costa Lopes, A.M. and Bogel-Lukasik, R. (2015). Acidic ionic liquids as sustainable approach of cellulose and lignocellulosic biomass conversion without additional catalysts. *ChemSusChem* 8: 947–965.
108 De Gregorio, G.F., Weber, C.C., Grasvik, J. et al. (2016). Mechanistic insights into lignin depolymerisation in acidic ionic liquids. *Green Chem.* 18: 5456–5465.
109 Thierry, M., Majira, A., Pegot, B. et al. (2018). Imidazolium-based ionic liquids as efficient reagents for the C–O bond cleavage of lignin. *ChemSusChem* 11: 439–448.
110 Bodachivskyi, I., Kuzhiumparambil, U., and Williams, D.B.G. (2018). Acid-catalyzed conversion of carbohydrates into value-added small molecules in aqueous media and ionic liquids. *ChemSusChem* 11: 642–660.
111 Jia, S.Y., Cox, B.J., Guo, X.W. et al. (2010). Cleaving the β-O-4 bonds of lignin model compounds in an acidic ionic liquid, 1-H-3-methylimidazolium chloride: An optional strategy for the degradation of lignin. *ChemSusChem* 3: 1078–1084.
112 Cox, B.J., Jia, S.Y., Zhang, Z.C., and Ekerdt, J.G. (2011). Catalytic degradation of lignin model compounds in acidic imidazolium based ionic liquids: Hammett acidity and anion effects. *Polym. Degrad. Stab.* 96: 426–431.
113 Li, C.Z., Wang, Q., and Zhao, Z.K. (2008). Acid in ionic liquid: An efficient system for hydrolysis of lignocellulose. *Green Chem.* 10: 177–182.
114 Li, B., Filpponen, I., and Argyropoulos, D.S. (2010). Acidolysis of wood in ionic liquids. *Ind. Eng. Chem. Res.* 49: 3126–3136.
115 Yu, H.M., Hu, J., Fan, J., and Chang, J. (2012). One-pot conversion of sugars and lignin in ionic liquid and recycling of ionic liquid. *Ind. Eng. Chem. Res.* 51: 3452–3457.

116 Li, Y.W., Cai, Z.P., Liao, M.C. et al. (2017). Catalytic depolymerization of organosolv sugarcane bagasse lignin in cooperative ionic liquid pairs. *Catal. Today* 298: 168–174.

117 Schwiderski, M., Kruse, A., Grandl, R., and Dockendorf, D. (2014). Comparison of the influence of a Lewis acid $AlCl_3$ and a Bronsted acid HCl on the organosolv pulping of beech wood. *Green Chem.* 16: 1569–1578.

118 Wang, H.L., Zhang, L.B., Deng, T.S. et al. (2016). $ZnCl_2$ induced catalytic conversion of softwood lignin to aromatics and hydrocarbons. *Green Chem.* 18: 2802–2810.

119 Deuss, P.J., Lahive, C.W., Lancefield, C.S. et al. (2016). Metal triflates for the production of aromatics from lignin. *ChemSusChem* 9: 2974–2981.

120 Deuss, P.J., Lancefield, C.S., Narani, A. et al. (2017). Phenolic acetals from lignins of varying compositions via iron(III) triflate catalysed depolymerisation. *Green Chem.* 19: 2774–2782.

121 Güvenatam, B., Heeres, E.H.J., Pidko, E.A., and Hensen, E.J.M. (2016). Lewis-acid catalyzed depolymerization of Protobind lignin in supercritical water and ethanol. *Catal. Today* 259: 460–466.

122 Hepditch, M.M. and Thring, R.W. (2000). Degradation of solvolysis lignin using Lewis acid catalysts. *Can. J. Chem. Eng.* 78: 226–231.

123 Beauchet, R., Monteil-Rivera, F., and Lavoie, J.M. (2012). Conversion of lignin to aromatic-based chemicals (L-chems) and biofuels (L-fuels). *Bioresour. Technol.* 121: 328–334.

124 Zhu, G., Jin, D., Zhao, L. et al. (2017). Microwave-assisted selective cleavage of C_α–C_β bond for lignin depolymerization. *Fuel Process. Technol.* 161: 155–161.

125 Guvenatam, B., Heeres, E.H.J., Pidko, E.A., and Hensen, E.J.M. (2016). Lewis acid-catalyzed depolymerization of soda lignin in supercritical ethanol/water mixtures. *Catal. Today* 269: 9–20.

126 Assary, R.S., Atesin, A.C., Li, Z. et al. (2013). Reaction pathways and energetics of etheric C–O bond cleavage catalyzed by lanthanide triflates. *ACS Catal.* 3: 1908–1914.

127 Lohr, T.L., Li, Z., and Marks, T.J. (2015). Selective ether/ester C–O cleavage of an acetylated lignin model via tandem catalysis. *ACS Catal.* 5: 7004–7007.

128 Huang, X., Zhu, J., Koranyi, T.I. et al. (2016). Effective release of lignin fragments from lignocellulose by Lewis acid metal triflates in the lignin-first approach. *ChemSusChem* 9: 3262–3267.

129 Huang, X.M., Gonzalez, O.M.M., Zhu, J.D. et al. (2017). Reductive fractionation of woody biomass into lignin monomers and cellulose by tandem metal triflate and Pd/C catalysis. *Green Chem.* 19: 175–187.

130 Jastrzebski, R., Constant, S., Lancefield, C.S. et al. (2016). Tandem catalytic depolymerization of lignin by water-tolerant Lewis acids and rhodium complexes. *ChemSusChem* 9: 2074–2079.

131 Harms, R.G., Markovits, I.I., Drees, M. et al. (2014). Cleavage of C–O bonds in lignin model compounds catalyzed by methyldioxorhenium in homogeneous phase. *ChemSusChem* 7: 429–434.

132 Felixberger, J.K., Kuchler, J.G., Herdtweck, E. et al. (1988). Alkyne coordination to organorhenium oxides with rhenium in high oxidation states. *Angewandte Chemie International Edition in English* 27: 946–948.

133 Crestini, C., Pro, P., Neri, V., and Saladino, R. (2005). Methyltrioxorhenium: a new catalyst for the activation of hydrogen peroxide to the oxidation of lignin and lignin model compounds. *Bioorg. Med. Chem.* 13: 2569–2578.

134 Crestini, C., Caponi, M.C., Argyropoulos, D.S., and Saladino, R. (2006). Immobilized methyltrioxo rhenium (MTO)/H_2O_2 systems for the oxidation of lignin and lignin model compounds. *Bioorg. Med. Chem.* 14: 5292–5302.

135 Korstanje, T.J. and Gebbink, R.J.M.K. (2012). Catalytic oxidation and deoxygenation of renewables with rhenium complexes. In: *Top Organometal Chem* (ed. M.A.R. Meier, B.M. Weckhuysen and P.C.A. Bruijnincx), 129–174. Berlin, Heidelberg: Springer Berlin Heidelberg.

136 Lange, H., Decina, S., and Crestini, C. (2013). Oxidative upgrade of lignin-Recent routes reviewed. *Eur. Polym. J.* 49: 1151–1173.

137 Ziegler, J.E., Zdilla, M.J., Evans, A.J., and Abu-Omar, M.M. (2009). H_2-driven deoxygenation of epoxides and diols to alkenes catalyzed by methyltrioxorhenium. *Inorg. Chem.* 48: 9998–10000.

138 Shiramizu, M. and Toste, F.D. (2012). Deoxygenation of biomass-derived feedstocks: Oxorhenium-catalyzed deoxydehydration of sugars and sugar alcohols. *Angew. Chem. Int. Ed.* 51: 8082–8086.

139 Korstanje, T.J., de Waard, E.F., Jastrzebski, J.T.B.H., and Klein Gebbink, R.J.M. (2012). Rhenium-catalyzed dehydration of nonbenzylic and terpene alcohols to olefins. *ACS Catal.* 2: 2173–2181.

140 Denning, A.L., Dang, H., Liu, Z. et al. (2013). Deoxydehydration of glycols catalyzed by carbon-supported perrhenate. *ChemCatChem* 5: 3567–3570.

141 Li, X., Wu, D., Lu, T. et al. (2014). Highly efficient chemical process to convert mucic acid into adipic acid and DFT studies of the mechanism of the rhenium-catalyzed deoxydehydration. *Angew. Chem. Int. Ed.* 53: 4200–4204.

142 Korstanje, T.J., Jastrzebski, J.T.B.H., and Klein Gebbink, R.J.M. (2010). Catalytic dehydration of benzylic alcohols to styrenes by rhenium complexes. *ChemSusChem* 3: 695–697.

143 Zhu, Z.L. and Espenson, J.H. (1996). Organic reactions catalyzed by methylrhenium trioxide: Dehydration, amination, and disproportionation of alcohols. *J. Org. Chem.* 61: 324–328.

144 Korstanje, T.J., Jastrzebski, J.T.B.H., and Gebbink, R.J.M.K. (2013). Mechanistic insights into the rhenium-catalyzed alcohol-to-olefin dehydration reaction. *Chem. Eur. J.* 19: 13224–13234.

145 Zhang, B., Li, C., Dai, T. et al. (2015). Microwave-assisted fast conversion of lignin model compounds and organosolv lignin over methyltrioxorhenium in ionic liquids. *RSC Adv.* 5: 84967–84973.

146 Bouxin, F., Baumberger, S., Pollet, B. et al. (2010). Acidolysis of a lignin model: Investigation of heterogeneous catalysis using montmorillonite clay. *Bioresour. Technol.* 101: 736–744.

147 Elfadly, A.M., Zeid, I.F., Yehia, F.Z. et al. (2017). Production of aromatic hydrocarbons from catalytic pyrolysis of lignin over acid-activated bentonite clay. *Fuel Process. Technol.* 163: 1–7.

148 Wang, H., Li, W., Wang, J. et al. (2017). A ternary composite oxides $S_2O_8^{2-}$/ ZrO_2-TiO_2-SiO_2 as an efficient solid super acid catalyst for depolymerization of lignin. *RSC Adv.* 7: 50027–50034.

149 Liu, X., Zhao, G., Ren, S. et al. (2018). Zirconia-supported 12-tungstophosphoric acid: Characterization and catalytic behavior in degradation of alkali lignin. *Ind. Crops Prod.* 125: 496–504.

150 Nakasaka, Y., Yoshikawa, T., Kawamata, Y. et al. (2017). Fractionation of degraded lignin by using a water/1-butanol mixture with a solid-acid catalyst: A potential source of phenolic compounds. *ChemCatChem* 9: 2875–2880.

151 Deepa, A.K. and Dhepe, P.L. (2014). Solid acid catalyzed depolymerization of lignin into value added aromatic monomers. *RSC Adv.* 4: 12625–12629.

152 Deepa, A.K. and Dhepe, P.L. (2015). Lignin depolymerization into aromatic monomers over solid acid catalysts. *ACS Catal.* 5: 365–379.

153 Jin, S.H., Xiao, Z.H., Chen, X. et al. (2015). Cleavage of lignin-derived 4-O-5 aryl ethers over nickel nanoparticles supported on niobic acid-activated carbon composites. *Ind. Eng. Chem. Res.* 54: 2302–2310.

154 Santana, J.A., Carvalho, W.S., and Ataide, C.H. (2018). Catalytic effect of ZSM-5 zeolite and HY-340 niobic acid on the pyrolysis of industrial kraft lignins. *Ind. Crops Prod.* 111: 126–132.

155 Huang, X., Atay, C., Korányi, T.I. et al. (2015). Role of Cu–Mg–Al mixed oxide catalysts in lignin depolymerization in supercritical ethanol. *ACS Catal.* 5: 7359–7370.

156 Chen, P., Zhang, Q., Shu, R. et al. (2017). Catalytic depolymerization of the hydrolyzed lignin over mesoporous catalysts. *Bioresour. Technol.* 226: 125–131.

157 Foster, A.J., Jae, J., Cheng, Y.-T. et al. (2012). Optimizing the aromatic yield and distribution from catalytic fast pyrolysis of biomass over ZSM-5. *Appl. Catal. A: Gen* 423-424: 154–161.

158 Yu, Y., Li, X., Su, L. et al. (2012). The role of shape selectivity in catalytic fast pyrolysis of lignin with zeolite catalysts. *Appl. Catal. A: Gen* 447-448: 115–123.

159 Dorado, C., Mullen, C.A., and Boateng, A.A. (2014). H-ZSM5 Catalyzed co-pyrolysis of biomass and plastics. *ACS Sustainable Chem. Eng.* 2: 301–311.

160 Li, X., Li, J., Zhou, G. et al. (2014). Enhancing the production of renewable petrochemicals by co-feeding of biomass with plastics in catalytic fast pyrolysis with ZSM-5 zeolites. *Appl. Catal. A: Gen* 481: 173–182.

161 Xue, Y., Zhou, S., and Bai, X. (2016). Role of hydrogen transfer during catalytic copyrolysis of lignin and tetralin over HZSM-5 and HY zeolite catalysts. *ACS Sustainable Chem. Eng.* 4: 4237–4250.

162 Adhikari, S., Srinivasan, V., and Fasina, O. (2014). Catalytic pyrolysis of raw and thermally treated lignin using different acidic zeolites. *Energy Fuel* 28: 4532–4538.

163 Kim, J.-Y., Lee, J.H., Park, J. et al. (2015). Catalytic pyrolysis of lignin over HZSM-5 catalysts: Effect of various parameters on the production of aromatic hydrocarbon. *J. Anal. Appl. Pyrolysis* 114: 273–280.

164 Shen, D.K., Liu, N.N., Dong, C.J. et al. (2015). Catalytic solvolysis of lignin with the modified HUSYs in formic acid assisted by microwave heating. *Chem. Eng. J.* 270: 641–647.

165 Ma, Z., Troussard, E., and van Bokhoven, J.A. (2012). Controlling the selectivity to chemicals from lignin via catalytic fast pyrolysis. *Appl. Catal. A: Gen* 423-424: 130–136.

166 Nandiwale, K.Y., Danby, A.M., Ramanathan, A. et al. (2017). Zirconium-incorporated mesoporous silicates show remarkable lignin depolymerization activity. *ACS Sustainable Chem. Eng.* 5: 7155–7164.

167 Kang, S., Bu, D., Xue, L. et al. (2019). Degradation of lignin catalyzed by nanochannels solid acids in ionic liquid. *Polym. Degrad. Stab.* 166: 1–7.

168 Tang, Z., Lu, Q., Zhang, Y. et al. (2009). One step bio-oil upgrading through hydrotreatment, esterification, and cracking. *Ind. Eng. Chem. Res.* 48: 6923–6929.

169 Zhao, C., He, J., Lemonidou, A.A. et al. (2011). Aqueous-phase hydrodeoxygenation of bio-derived phenols to cycloalkanes. *J. Catal.* 280: 8–16.

170 Melligan, F., Hayes, M.H.B., Kwapinski, W., and Leahy, J.J. (2012). Hydro-pyrolysis of biomass and online catalytic vapor upgrading with Ni-ZSM-5 and Ni-MCM-41. *Energy Fuel* 26: 6080–6090.

171 Liu, Y., Chen, L.G., Wang, T.J. et al. (2015). One-pot catalytic conversion of raw lignocellulosic biomass into gasoline alkanes and chemicals over $LiTaMoO_6$ and Ru/C in aqueous phosphoric acid. *ACS Sustainable Chem. Eng.* 3: 1745–1755.

172 Wang, J., Li, W., Wang, H. et al. (2017). Liquefaction of kraft lignin by hydrocracking with simultaneous use of a novel dual acid-base catalyst and a hydrogenation catalyst. *Bioresour. Technol.* 243: 100–106.

173 Jin, L., Li, W., Liu, Q. et al. (2018). Liquefaction of kraft lignin over the composite catalyst $HTaMoO_6$ and Rh/C in dioxane-water system. *Fuel Process. Technol.* 178: 62–70.

174 Yan, N., Yuan, Y.A., Dykeman, R. et al. (2010). Hydrodeoxygenation of lignin-derived phenols into alkanes by using nanoparticle catalysts combined with bronsted acidic ionic liquids. *Angew. Chem. Int. Ed.* 49: 5549–5553.

175 Scott, M., Deuss, P.J., de Vries, J.G. et al. (2016). New insights into the catalytic cleavage of the lignin β-O-4 linkage in multifunctional ionic liquid media. *Catal. Sci. Technol.* 6: 1882–1891.

176 Liu, F., Liu, Q.Y., Wang, A.Q., and Zhang, T. (2016). Direct catalytic hydrogenolysis of kraft lignin to phenols in choline-derived ionic liquids. *ACS Sustainable Chem. Eng.* 4: 3850–3856.

177 Chen, L., Xin, J.Y., Ni, L.L. et al. (2016). Conversion of lignin model compounds under mild conditions in pseudo-homogeneous systems. *Green Chem.* 18: 2341–2352.

178 Guo, H., Zhang, B., Li, C. et al. (2016). Tungsten carbide: A remarkably efficient catalyst for the selective cleavage of lignin C–O bonds. *ChemSusChem* 9: 3220–3229.

179 Guo, H.W., Zhang, B., Qi, Z.J. et al. (2017). Valorization of lignin to simple phenolic compounds over tungsten carbide: Impact of lignin structure. *ChemSusChem* 10: 523–532.

180 Zhang, C., Lu, J., Zhang, X. et al. (2016). Cleavage of the lignin β-O-4 ether bond via a dehydroxylation-hydrogenation strategy over a NiMo sulfide catalyst. *Green Chem.* 18: 6545–6555.

181 Wang, Q., Guan, S., and Shen, D. (2017). Experimental and kinetic study on lignin depolymerization in water/formic acid system. *Int. J. Mol. Sci.* 18: 2082.

182 Shao, L.P., Zhang, Q.L., You, T.T. et al. (2018). Microwave-assisted efficient depolymerization of alkaline lignin in methanol/formic acid media. *Bioresour. Technol.* 264: 238–243.

183 Park, S.Y., Hong, C.Y., Jeong, H.S. et al. (2016). Improvement of lignin oil properties by combination of organic solvents and formic acid during supercritical depolymerization. *J. Anal. Appl. Pyrolysis* 121: 113–120.

184 Kristianto, I., Limarta, S.O., Lee, H. et al. (2017). Effective depolymerization of concentrated acid hydrolysis lignin using a carbon-supported ruthenium catalyst in ethanol/formic acid media. *Bioresour. Technol.* 234: 424–431.

9

Lignin Linkages Cleavage Beginning with C_α–H, C_α–OH, or C_αO–H Bond Non-ionized Activation

9.1 Lignin C_β–OAr Bond Cleavage via a Transfer Hydrogenation or Dehydrogenation-hydrogenation Process Beginning with the First Activation of C_α–H(O–H) to C_α=O

According to the BDE change of C_β–OAr bonds (Chapter 5), the first dehydrogenation of a β-O-4 alcohol can not only provide two equivalents of H atoms but can also generate a β-O-4 ketone intermediate with an active C_β–OAr bond. It is an atom-economical process to use these H atoms in the cleavage of C_β–OAr to prepare phenylacetone derivatives and phenols. Given the fact that the properly increasing H species can benefit the reaction kinetics and the potential hydrogenation ketone products could consume H extra species, appropriate addition of H_2 or other hydrogen donor reagents could promote the subsequent hydrogenation. However, the extra hydrogen donor can restrain the dehydrogenation process and cause the hydrogenolysis of C_α–OH to $C_\alpha H_2$, restraining the C_β–OAr bond hydrogenation. Therefore, the appropriate catalytic systems should have a balance of selectivity between the dehydrogenation of the C_αH–OH and the hydrogenation of the C_β–OAr bond.

9.1.1 Homogeneous Catalytic Systems

In 2010, Bergman and co-workers [1] developed a ruthenium-catalyzed redox neutral method to selectively cleave the C_β–OAr bond in 2-aryloxy-1-arylethanols (β-O-4 model compound). $PhCOCH_3$ and PhOH were the main products. As shown in Figure 9.1, they propose a mechanism: It begins with a well-known Ru-catalyzed dehydrogenative equilibrium between a benzylic alcohol (**1**) and the corresponding aryl ketone (**2**). This is followed by the loss of an HX from the catalyst precursor and the formation of the Ru(0) complex. Then, the C–O activation of **2** with Ru catalyst in **3** leads to Ru-enolate (**4**). Hydrogenation of **4** yields an aromatic ketone (**5**) and a Ru-alkoxide (**6**) followed by reductive elimination of phenol (**7**) and association with **2** to close the cycle.

Lignin Conversion Catalysis: Transformation to Aromatic Chemicals, First Edition.
Chaofeng Zhang and Feng Wang.

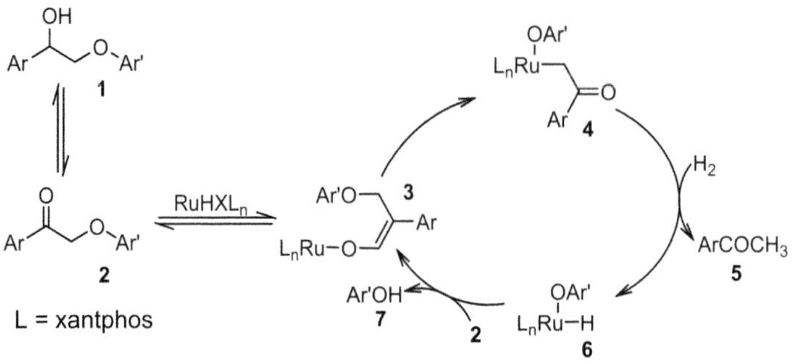

Figure 9.1 The Mechanism for Ru Catalyzed C–O Bond Cleavage in 2-Aryloxy-1-arylethanol. *Source:* Nichols et al. [1] with Permission, © 2010 American Chemical Society.

Then, in 2012, James and co-workers [2] reported more additional details and results on a wider substrate range for the same Ru-xantphos catalyst and confirmed that the catalytic cycle likely proceeds via a ketone intermediate. In 2013, Paton and co-workers [3] employed density functional theory (DFT) calculations and kinetics measurements to understand the mechanism of a xantphos-containing molecular ruthenium catalyst acting on an alkyl aryl ether linkage similar to that found in lignin to produce acetophenone and phenol (Figure 9.2). The DFT calculations reveal several interesting features, including an unusual 5-membered transition state structure for oxidative insertion in contrast to the typically proposed 3-membered transition state (**6** → **7**), a preference for an O-bound over a C-bound Ru–enolate, and a significant kinetic preference for the order of product release from the catalyst. The experimental measurements confirm that the reaction proceeds via a free ketone intermediate, but also suggest that the conversion of the intermediate ketone to acetophenone and phenol does not necessarily require ketone dissociation from the catalyst. Overall, this work elucidates the kinetically and thermodynamically preferred reaction pathways for tandem alcohol dehydrogenation and reductive ether bond cleavage by the ruthenium-xantphos catalyst.

In addition, although a base, especially KOH, can be used in place of xantphos for the Ru-complex-catalyzed C–O bond cleavage of β-O-4 lignin model compounds as reported by Huo in 2014 [4], the ligand participation can promote and regulate the reaction. In 2017, Kamer and co-workers [5] studied the bite angle and electronic effects on the ruthenium–diphosphine catalyzed ether bond cleavage of the lignin β-O-4 model compound 2-phenoxy-1-phenylethanol. Enhanced conversion of the substrate was observed with the increasing σ-donor capacity of the ligands. Further kinetic and thermodynamic data suggest that the oxidative addition of the dehydrogenated model compound to the diphosphine Ru(0) complex is to be rate-limiting. Besides the xantphos ligand, in 2013, Jürgen Klankermayer and co-workers [6] demonstrated that a Ru(II)-complex with a trimethylenemethane ligand ([Ru(P3)(TMM)]) was a promising catalytic system for hydrogen transfer reactions with unprecedented high activity in the C_β–OAr cleavage of β-O-4-glycerolarylether linkages of lignin by means of internal hydrogen transfer (Figure 9.3).

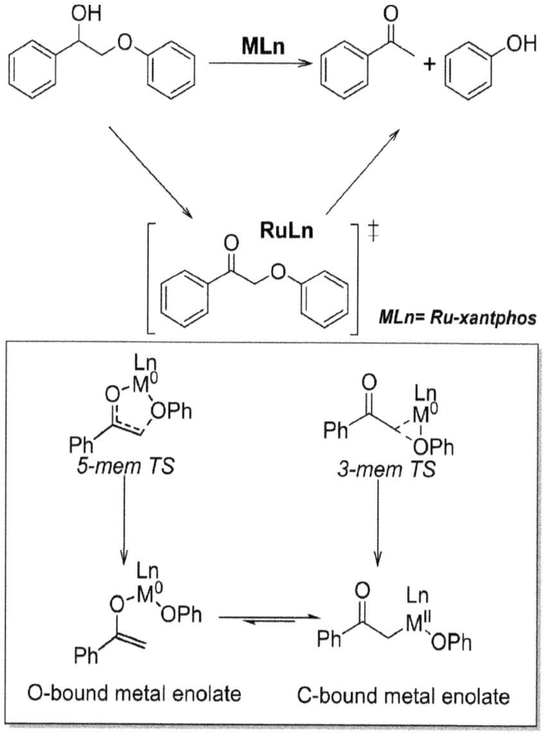

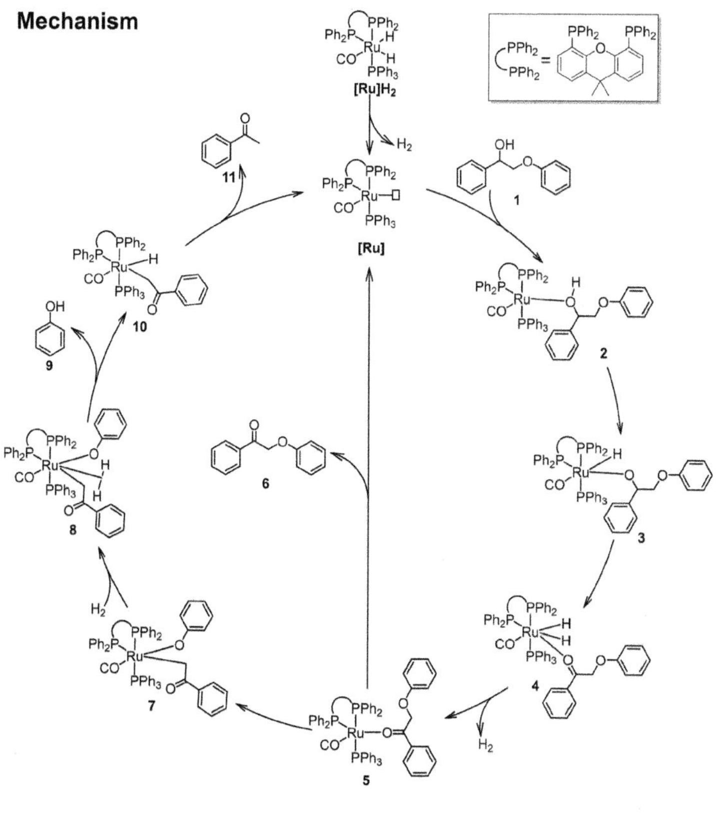

Figure 9.2 Mechanism of 2-Aryloxy-1-arylethanol Catalytic Deconstruction by the $[Ru]H_2$. *Source:* Chmely et al. [3] with Permission, © 2013 American Chemical Society.

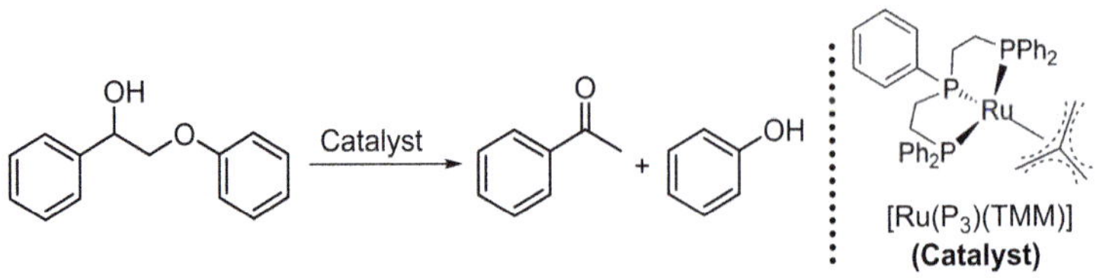

Figure 9.3 Dehydrogenation and C–O Cleavage by Intermolecular Hydrogen Transfer Catalyzed by the [Ru(P_3)(TMM)].

Besides the Ru-based homogeneous catalysts, Wilson and co-workers [7] provided a detailed study in 2015 of the cleavage process of the β-O-4 linkage model (2-phenoxy-1-phenylethan-1-ol, with a C_α–OH) over various group 8 pincer complexes (M= Fe, Ru, and Os; ligand=PNP, NNN, PCP, PCN, or NCN). Their study suggests that C_β–OAr bond activation by transition metal catalysts is essentially controlled by the ability of the metal center to donate electrons to the C_β–OAr and that ligands that donate electrons to the metal center tend to facilitate the cleavage of the C_β–OAr bond. Between the two proposed reaction pathways, routes A and B (Figure 9.4), route A showed lower effective barriers than route B for Fe and Ru systems, but no significant differences in the thermodynamics between the two routes were obtained for Os systems.

Besides the single metal homogeneous catalyst approach, Li and co-workers [8] developed a mild redox-neutral lignin depolymerization system for lignin depolymerization in 2019. This approach features a water-soluble binuclear Rh complex (Figure 9.5). The catalytic system could be successfully applied to the depolymerization of a lignin-like polymer, alkaline lignin, as well as to raw lignocellulose samples. One typical alkaline lignin extracted from basswood was converted and provided 88 wt % oil products, during which the quantified monomers were about

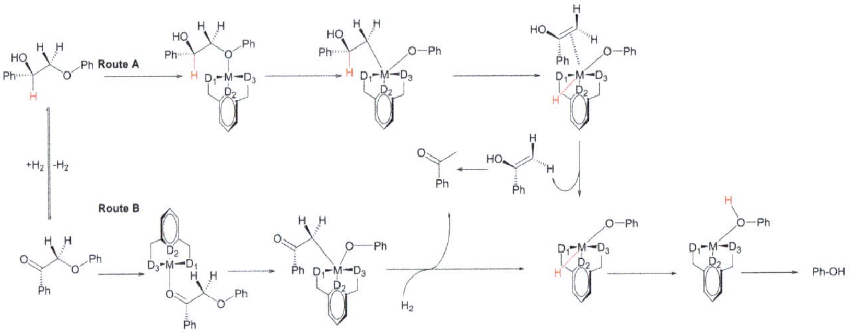

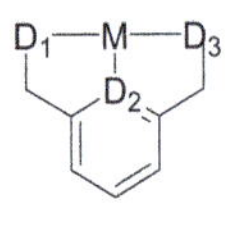

D_1= PR_2,NR_2, SR,OR, etc.
D_2= N, O, C, etc.
D_3= PR_2,NR_2, SR,OR, etc.
M= Metal center
R= Me, Ph, ᵗBu, etc

Ligand	D_1	D_2	D_3	Metal valence
PNP	PH_2	N	PH_2	0
NNN	NH_2	N	NH_2	0
PCP	PH_2	C	PH_2	+2
PCN	PH_2	C	NH_2	+2
NCN	NH_2	C	NH_2	+2

Figure 9.4 Cleavage of the β-O-4 Linkage with Pincer Complexes. *Source:* Liu and Wilson [7] with Permission, © 2015 Elsevier.

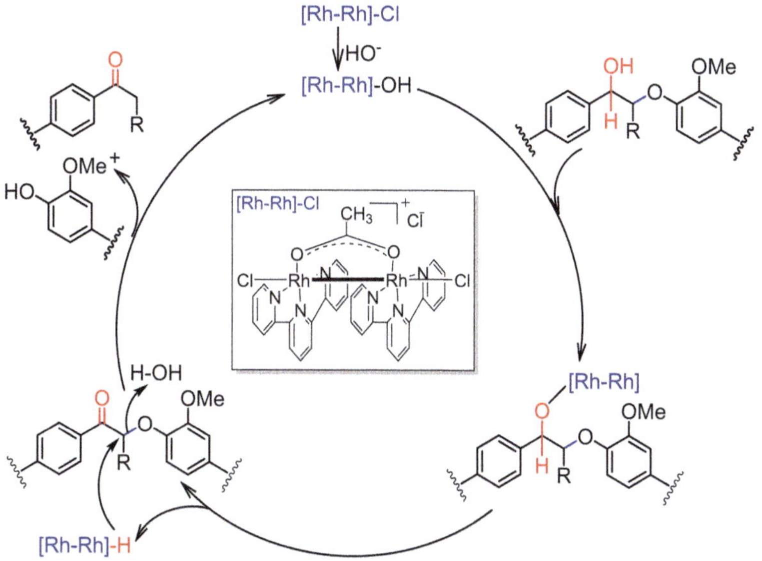

Figure 9.5 Binuclear Rh-catalyzed Redox-neutral System for Lignin Depolymerization in Water. *Source:* Liu et al. [8] with Permission, © 2019 American Chemical Society.

11 wt % as measured by GC-FID and aromatic ketones were the major products. Mechanistic studies on the model substrate suggest that the reaction proceeds via a metal-catalyzed dehydrogenation step to afford a carbonyl intermediate (C_α=O), followed by C—O bond cleavage to afford ketone and phenol products and this was confirmed by the deuterium labeling study.

Li and co-workers [9] then studied the redox-neutral lignin depolymerization catalyzed by an Rh-terpyridine catalyst in water (Figure 9.6), during which the first dehydrogenation of the C_α–OH to C_α=O could induce the release of $-C_\gamma$HOH as HCHO via a base-promoted Retro-Aldol process. The C_β–OAr bond in the generated ketone intermediate can further occur reductive cleavage with the [Rh–H] species. As another route, the $-C_\gamma$H–OH of the ketone intermediate can occur dehydration

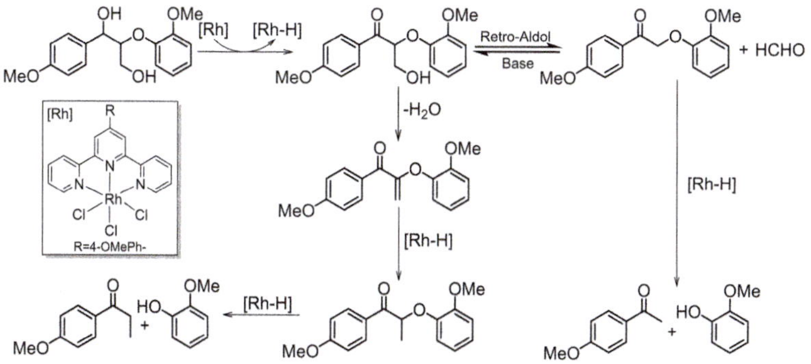

Figure 9.6 Ru-terpyridine Catalyzed Redox-neutral Depolymerization of Lignin. *Source:* Liu et al. [9] with Permission, © 2020 The Royal Society of Chemistry.

and provide a C_β=C_γ. The following C=C bond hydrogenation can consume the limited [Rh–H] species and compete with the following reductive cleavage of the C_β–OAr bond in the intermediate with $-C_\beta H-C_\gamma H_3$.

9.1.2 Heterogeneous Catalytic Systems

The dehydrogenation-hydrogenation or internal hydrogen transfer strategies for cleaving the β-O-4 C_β–OAr ether bond are also used over heterogeneous catalysts. However, the complex active center of heterogeneous catalysts usually consists of multiple metal sites and the reaction condition of the heterogeneous process has a more obvious influence on the reaction route than homogeneous ones. Furthermore, the heterogeneous catalytic processes usually contain the activation of the C_β–H bond or C_α=O bond before cleaving the ether bonds in the second hydrogenation process, which makes the detailed mechanism quite different from the homogeneous processes via an oxidative addition to generate an R-M-OAr intermediate.

As a widely used and commercial catalyst, Pd/C has been widely used in research of the lignin model and of lignin conversion. Pd/C was firstly used in direct hydrogenation of lignin models and lignin with extra H_2, but the high activity of the palladium center and the harsh reaction conditions can make the catalytic process uncontrollable, which leads to the complex products distribution and hydrogenation of aromatic rings [10]. In 2014, Rauchfuss and co-workers [11] reported that the commercially available Pd/C could catalyze the cleavage of β-O-4 ether bonds in lignin model compounds to acetophenone and phenol without hydrogen. A redox neutral process that utilizes the internal hydrogen source for the cleavage is proposed (Figure 9.7). The mechanistic study indicates that the activation of C_α–H in the substrate plays a critical role. In this catalytic system, the solvent dioxane can

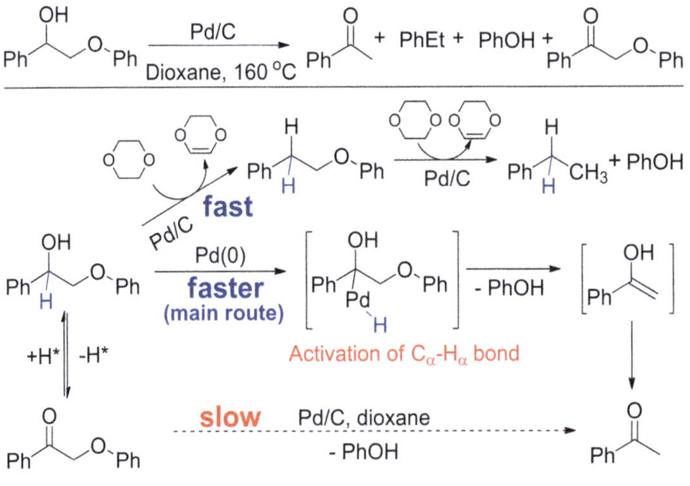

Figure 9.7 Proposed Mechanism of β-O-4 Ether Bond Cleavage via the C_α–M–H Intermediate over the Pd/C in 1,4-Dioxane. *Source:* Zhou et al. [11] with Permission, © 2014 Wiley-VCH.

also work as the hydrogen donor and promote the transfer hydrogenation of the β-O-4 model compound to ethylbenzene and phenol.

Before Rauchfuss's work, in 2014, Samec and co-workers [12] reported that Pd/C can catalyze C–O bond cleavage of 2-aryloxy-1-arylethanols with formic acid as the reductant in air. Mechanistic studies under kinetic control indicate that an initial palladium-catalyzed dehydrogenation of the β-O-4 C_α alcohol is followed by insertion of palladium to an enol equivalent. A palladium-formate complex reductively cleaves the palladium–enolate complex to generate the ketone (Figure 9.8). Samec and co-workers [13] then reported that Pd/C can catalyze the transfer hydrogenolysis of in-situ generated organosolv wood to 2-propenyl aryls (Figure 9.9). The addition of HCOOH can increase the yields of products with C_β–OAr ether bond cleavage. When the reaction was performed in ethanol without extra HCOOH as the active hydrogen donor, only endogenous hydrogen from wood was consumed (mechanism in Figure 9.9).

Then, in 2015, Samec and co-workers [14] found that Pd/C could also catalyze the redox neutral C_β–OAr bond cleavage of 2-aryloxy-1-arylethanols to aryl ketones in EtOAc at a lower temperature (80 °C). Their mechanism studies support a transfer hydrogenolysis reaction mechanism that proceeds through an initial dehydrogenation followed by an enol adsorption to Pd/C and a reductive C_β–OAr bond cleavage,

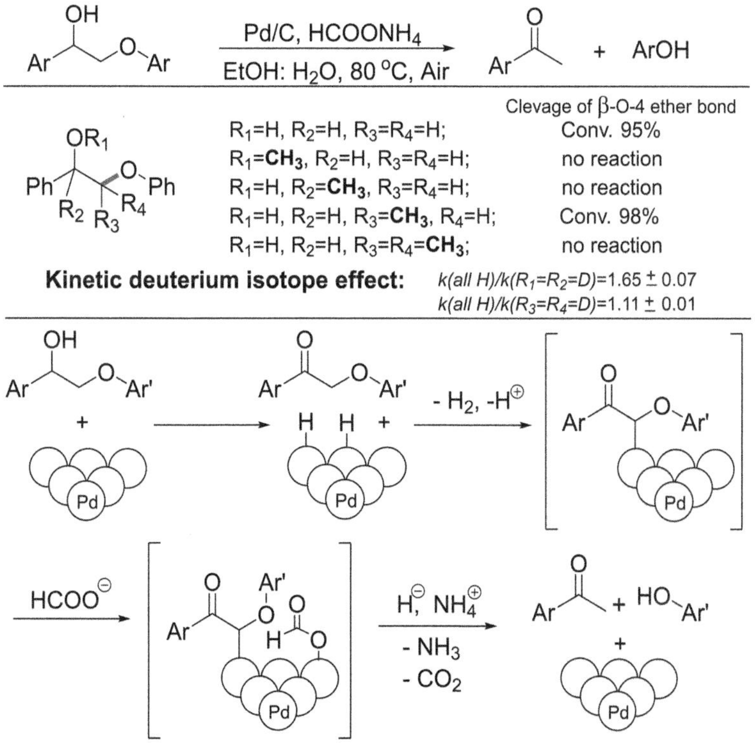

Figure 9.8 Proposed Mechanism for the Transformation of β-O-4 Ether Bond Cleavage over the Pd/C with $HCOONH_4$. *Source:* Galkin et al. [12] with Permission, © 2014 Wiley-VCH.

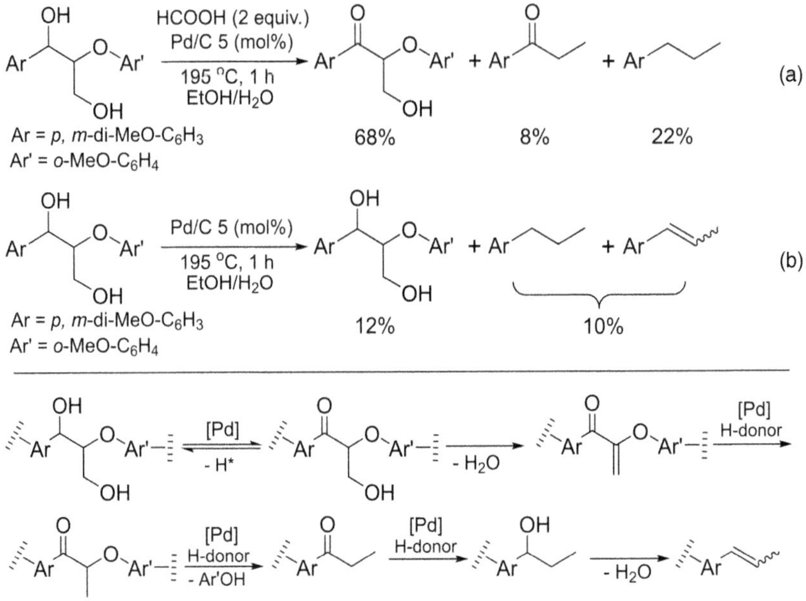

Figure 9.9 Proposed Mechanism of Aryl Propene Formation from Wood over the Pd/C. *Source:* Galkin and Samec [13] with Permission, © 2014 Wiley-VCH.

which is distinctly different from the mechanism [11] proposed by Rauchfuss (Figure 9.7). The addition of catalytic amounts of a hydrogen source ($NaBH_4$ or H_2) to the reaction mixture activates the catalyst surface to proceed through a low energy barrier pathway (12 kcal · mol^{-1}). Hydrogens at both the C_α and C_β positions of β-O-4 ethanolaryl ethers participate in the reaction when performed at 80 °C. The proposed reaction mechanisms are as follows (Figure 9.10): The oxide on the palladium surface is removed by hydrogen or $NaBH_4$. The reaction starts with a reversible dehydrogenation of **1** to generate **4** and chemically adsorbed hydrogen on Pd/C. The enolate form of **4** adsorbs to palladium to generate **C**. Following a Horiuti–Polanyi-type mechanism [15, 16], the atomic hydrogen may insert into either the α or β position of intermediate **C**. In Route **I**, atomic hydrogen inserts into the β position to generate intermediate **D**. C–O bond cleavage generates intermediate **E** where the enol equivalent is chemically adsorbed together with the aryloxide. Intermediate **E** decomposes to products **2** and **3**. In Route II, atomic hydrogen inserts into the position to give intermediate **F**. Palladium inserts into the C–O bond to generate **G**, in which a carbene and aryloxide are chemically adsorbed on palladium. Reductive elimination of aryloxide and the carbine tautomerization lead to products **2** and **3**.

Combined with density functional theory (DFT) calculations and constrained minima hopping, in 2016, Samec and co-workers [17] found that the keto tautomer was the most stable for the gas phase molecules. Interestingly, the enol tautomer has a higher affinity to the Pd catalyst than the keto form and becomes the most stable molecular form when adsorbed on the Pd(111) surface. The global minimum complex found on the metal surface corresponds to an enolate structure formed when

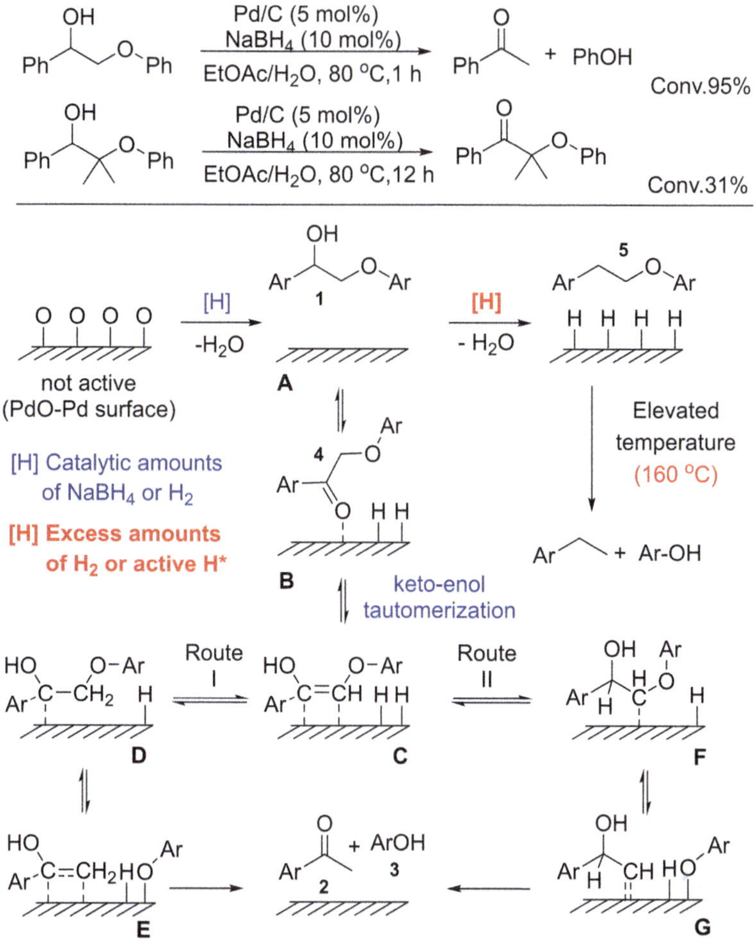

Figure 9.10 Proposed Reaction Mechanisms for the Redox Neutral Transfer Hydrogenolysis of $ArCH(OH)CH_2OAr$ over the Pd/C Surface. *Source:* Galkin et al. [14] with Permission, © 2015 Wiley-VCH.

the enol tautomer chemisorbs onto the surface and donates its π-electrons from the C–C region to two adjacent palladium atoms. The adsorbed enolate is the key intermediate for the C_β–OAr ether bond cleavage. While the β-O-4-ketone model compound is simplified to 1-methoxypropan-2-one, and the strong affinity effect between aryl groups of β-O-4-ketone and Pd(111) surface is ignored (Figure 9.11).

Focusing on the critical intermediate for the self-decomposition of β-O-4-alcohol with a C_α–OH over the Pd surface, in 2016, Lu and co-workers [18] performed detailed density functional theory calculations to investigate the cleavage mechanism of the C_β–OAr ether bond in the lignin model compound 2-phenoxy-1-phenylethanol with a β-O-4 linkage over a Pd(111) catalyst surface model. They propose that the favorable reaction pathway (Figure 9.12) proceeds as follows: the β-O-4 lignol reactant **2a** is first dehydrogenated on the C_α and then on the –OH group to

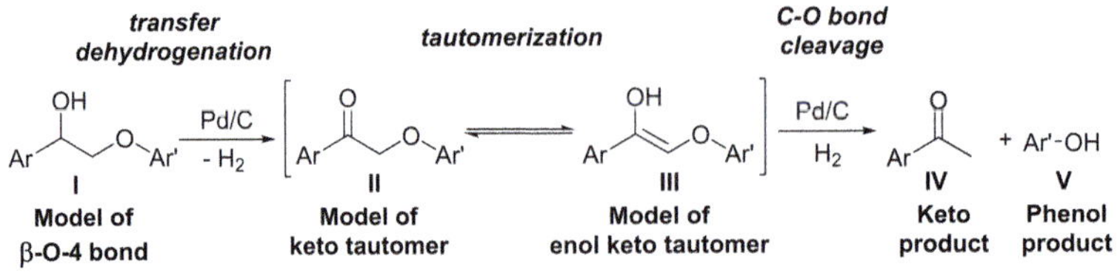

Figure 9.11 Pd/C-Catalyzed Redox Neutral Cleavage of the Model I to Products IV and V, via an Initial Transfer Dehydrogenation to Generate Intermediates Tautomers II and III.

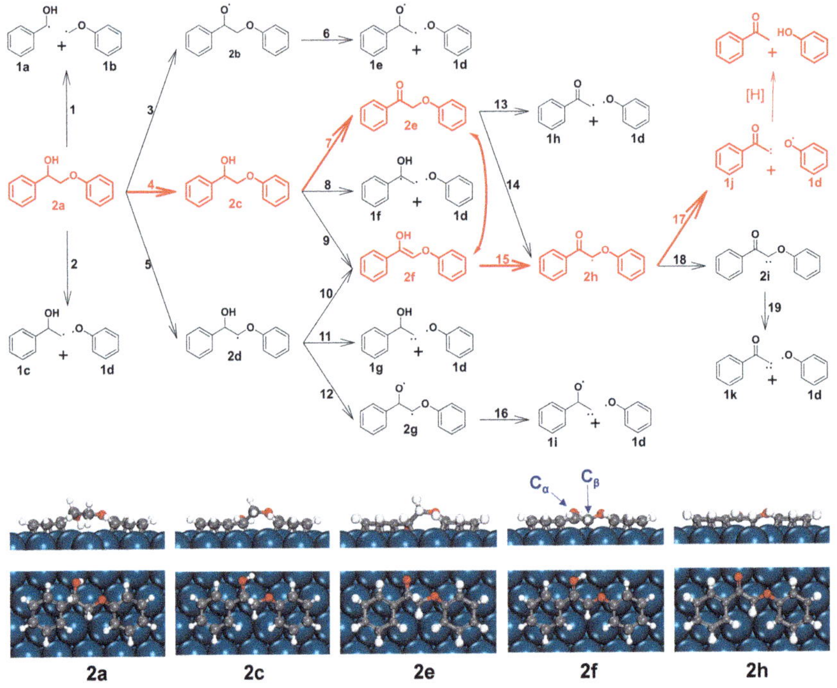

Figure 9.12 Reaction Network for the β-O-4 Bond Cleavage of 2-Phenoxy-1-phenylethanol over Pd(111). Note: The Main Route is Marked with Bold Red Arrows. *Source:* Permission from Lu et al. [18], © 2016 American Chemical Society.

generate its corresponding ketone **2e**; the ketone continues to get dehydrogenated on the C_β by first an equilibrated keto–enol tautomerization to its enol form **2f** and then –OH dehydrogenation; the C_β–OAr ether bond cleavage happens afterward, leading to one-aromatic-ring surface intermediates (**1j** and **1d**) followed by hydrogenation to yield acetophenone and phenol.

As shown above, experimental results in the literature have suggested that the keto-enol tautomerization is an important reaction step in the depolymerization of lignin β-O-4 linkage. Earlier computational results in the literature showed that the keto form is more stable on the Pd surface than enol by 0.16 eV, which thermodynamically makes the reaction unlikely to occur. For more detailed information about the keto-enol tautomerization during the β-O-4 linkage cleavage, in 2020, de Andrade and co-workers [19] explored the bonding and interactions of keto and enol conformers of a lignin model molecule on Pd(111) using DFT. Considerable care was taken to achieve tight optimizations, including dispersion correction, exploring the convergence of the *k*-point grid, and applying zeropoint energy corrections. Their main results are as follows:

(1) New low-energy enol conformers were found. The most stable enol conformer binds more strongly on the Pd(111) surface than the keto conformer: Eads(enol)= 4.23 eV and Eads(keto)= 4.01 eV (Figure 9.13). The result is that the enol form is at least as stable as the keto form when adsorbed on the Pd (111) surface, which strengthens and highlights the likelihood of keto-enol tautomerization in the lignin fragmentation;

(2) An energy decomposition analysis gives insights concerning the origin of the fine energy balance between the keto and enol forms, where the molecule–surface interaction (–7 eV) and the molecular strain energy (+3 eV) are the main contributors to the adsorption energy;

(3) Keto conformers bind to the Pd surface through the C_α and O1 atom (O1 connecting with the C_α), forming two new chemical bonds, namely C_α–σ and O1–σ bonds. The enol–surface interaction results in a weakening of the C_α–C_β bond as a result of the newly formed di–σ bond.

In addition to Pd/C, other noble catalysts are also used in the self-hydrogen transfer hydrogenolysis of β-O-4 linkages and the transfer hydrogenation with extra hydrogen donors. In 2012, the Jones group [20] reported that the organosolv switchgrass lignin in ethanol solvent could be depolymerized and hydrodeoxygenated with a formic acid hydrogen source over the 20 wt% Pt/C, and the lignin was significantly depolymerized to form liquid products with a 76% reduction in the weighted

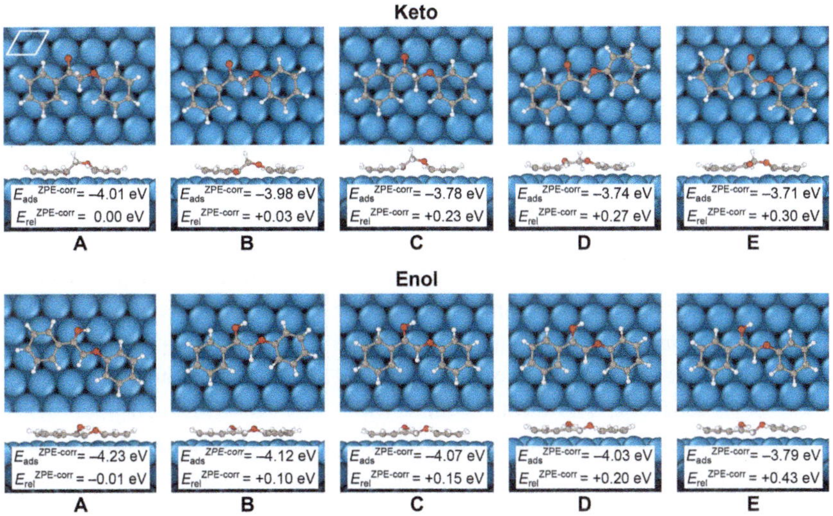

Figure 9.13 The Optimized $_\beta$-O-4 Molecules/Pd(111) Structures. In Each Row, the Structures Are Ordered from the Most Stable to the Least Stable. *Source:* de Andrade et al. [19], © 2020 Wiley-VCH.

average molecular weight after 20 h reaction. Then, in 2017, Hamou and co-workers [21] studied the adsorption and decomposition mechanism of lignin on Pt(111) by using 2-phenoxyethanol as the β-O-4 model.

In addition, the alloy catalysts with noble metal are also used in the hydrogen transfer hydrogenolysis of β-O-4 linkages. In 2017, Zhang and co-workers [22] reported that bimetallic nanoparticles with a Pd_1Ni_4 composing proportion supported by porous metalorganic frameworks MIL-100(Fe) can achieve the catalytic self-hydrogen transfer hydrogenolysis of β-O-4 linkage models to phenols and functionalized acetophenones. For the more complex mechanism study over the alloy catalysts, in 2018, Li and López workers [23] employed DFT calculations to investigate lignin decomposition on pure and Ru-doped Ni(111) surface to unravel the key issues that limit performance. The reaction network was screened by using complex coniferyl dimer models containing different stereocenters to simulate the most abundant β-O-4 linkages in lignin, thus presenting a better representation of the complexity of the parent compound. Their results show that, on nickel, both adsorption geometries and the preferred reaction paths are different depending on the chirality of the dimer reactants. In addition, the rigidity of the β-O-4 link is also responsible for the reactivity found. Finally, the presence of small amounts of Ru on the surface simplifies the reaction as they act as preferential points for β-O-4 bond cleavage via the weakened C—O bond that is induced by increasing metal—O bond strength. For a more detailed mechanism discussion, one who is interested can check their research article [23].

Although the self-hydrogen-transfer process is an atom-economical method for β-O-4 lignin ether bond cleavage, the reaction temperature of the reported thermal catalytic process is always above 170 °C, which can cause lignin fragment recondensation and an over-hydrogenation process with the active noble metal catalyst. To avoid high reaction temperatures, photocatalytic strategies at mild conditions should be a good choice to explore. In addition to the thermal Pd-catalyzed self-hydrogen transfer hydrogenolysis process [18]. in 2017, Wang and co-workers [24] reported a photocatalytic protocol with a heterogeneous $ZnIn_2S_4$ catalyst (Figure 9.14). In this process, alcoholic groups (C_αH—OH) of lignin act as hydrogen donors. The proliferation of phenolic products to dark substances is suppressed under visible light illumination (450 nm LED) at low temperature (below 50 °C). The formation of a light and transparent reaction solution allows visible light to be absorbed by the catalyst $ZnIn_2S_4$. With this strategy, 71–91% yields of phenols in the conversion of lignin β-O-4 models are achieved. Mechanistic studies revealed that the C_αH—OH group of β-O-4 linkage was initially dehydrogenated on $ZnIn_2S_4$ holes (h^+) to a ketone intermediate without C_β—OAr bond cleavage. The electron and proton form a "hydrogen pool" with active H* rather than H_2 at the CB (conduction band) structure, then the C_β—OAr bond of ketone intermediate was subsequently cleaved to two monomers by active H*. Thus, the dehydrogenation and hydrogenolysis reaction are integrated in one-pot with lignin itself as a hydrogen donor. Where after, $ZnIn_2S_4$ was used to depolymerize dioxanesolv poplar lignin under the photocatalytic transfer hydrogenolysis strategy, providing a ~10% monomers yield.

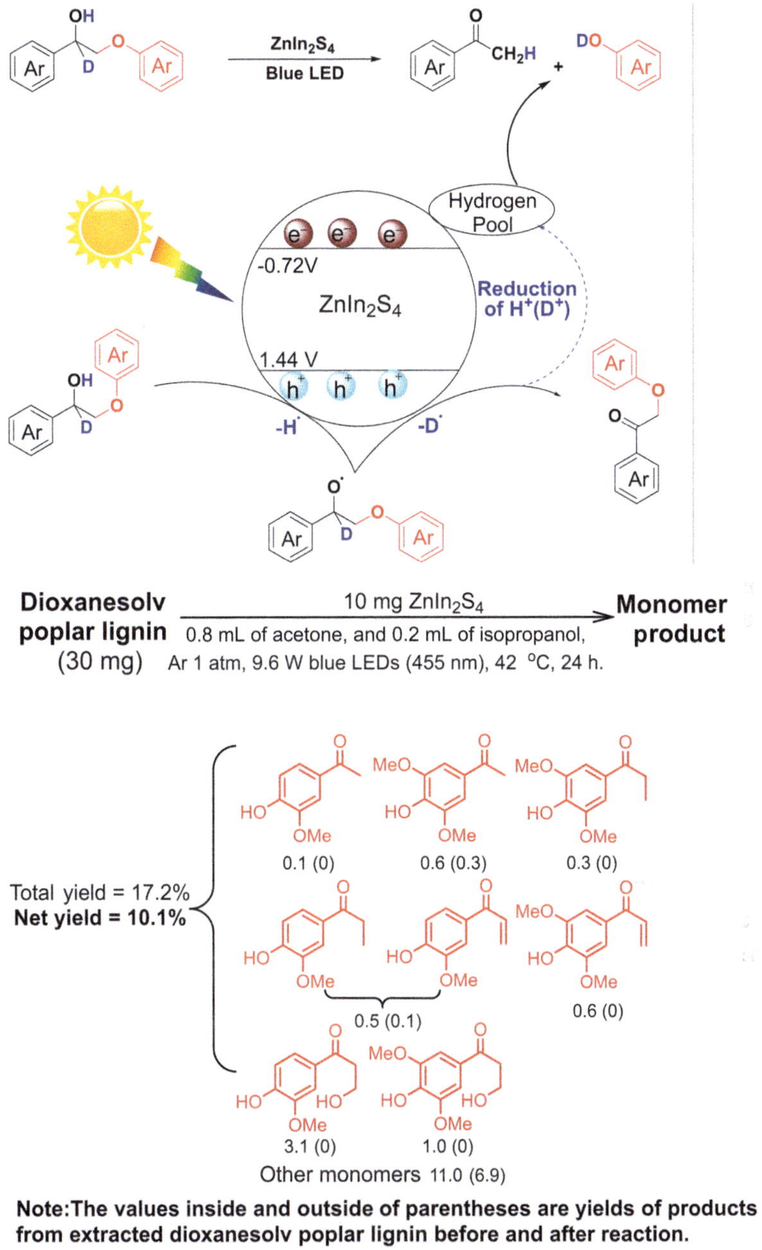

Figure 9.14 $ZnIn_2S_4$-catalyzed Photocatalytic Self-transfer Hydrogenolysis of Lignin. *Source:* Luo et al. [24] with Permission, © 2017 American Chemical Society.

Besides the oxygenated chemicals prepared from lignin via the self-hydrogen transfer process shown above, in 2021, Li and co-workers [25] designed a one-pot two-step strategy for the direct transformation of lignin β-O-4 model compounds into benzylamines and phenols in the presence of organic amines and

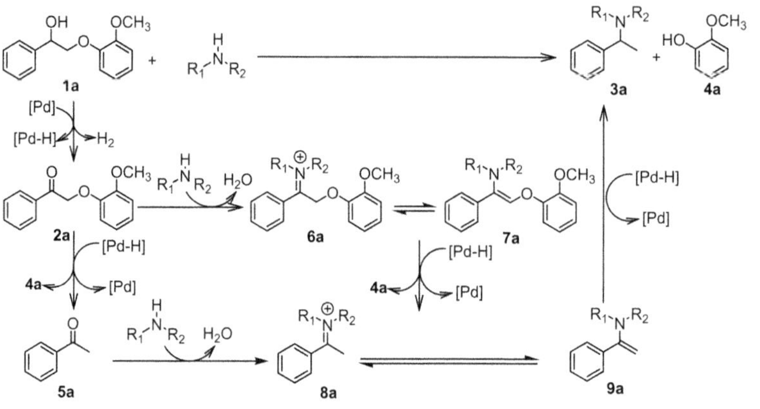

Figure 9.15 The Transformation of Lignin β-O-4 Models to Benzylamines. *Source:* Zhang et al. [25] with Permission, © 2021 Wiley-VCH.

Pd/C. The transformation involves the dehydrogenation of C_α–OH, hydrogenolysis of the C_β–OAr bond, and reductive amination in the presence of the Pd/C catalyst (Figure 9.15). Experimental data suggest that the dehydrogenation reaction proceeds over the other two reactions and secondary amines serve as both reducing agents and amine sources in the transformation.

9.2 Lignin C_β–OAr Bond Cleavage in the Dehydrogenation/Oxidation-Hydrogenation (Reduction) Process Beginning with the First Activation of C_α–H(OH) to C_α=O

While, given the existence of some ether linkages (α-O-4, 4-O-5, etc.) without adjacent hydroxyl group that cannot provide the hydrogen species for linkage ether bonds cleavage and the extra consumption of H* species caused by some unsaturated bonds during the self-hydrogen transfer hydrogenolysis, converting protolignin via the two-step oxidation-hydrogenation (reduction) strategy with an extra reductant is still a feasible method. As mentioned above, in these two-step methods, the C_αH–OH of β-O-4-alcohol is first selectively oxidized to C_α=O, then the chemical reduction methods with a stoichiometric reducant, thermal catalytic systems with H_2 or photocatalytic/electrocatalytic system with different sacrificial agents are employed to achieve the further C_β–OAr bond cleavage (Figure 9.16).

9.2.1 Chemical Reduction with Stoichiometric Reductant for the C_β–OAr Bond Cleavage in the Pre-oxidized Lignin

In 2015, Westwood and co-workers [26] reported a two-step depolymerization process to isolate functionalized phenolic monomers from an organosolv lignin (Figure 9.17). They first achieve the chemoselective catalytic oxidation of

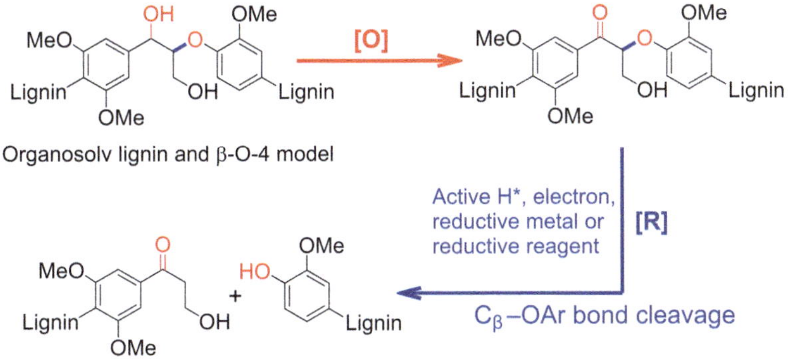

Figure 9.16 Dehydrogenation/Oxidation-hydrogenation (Reduction) Strategies for Lignin C_β–OAr Bond Cleavage.

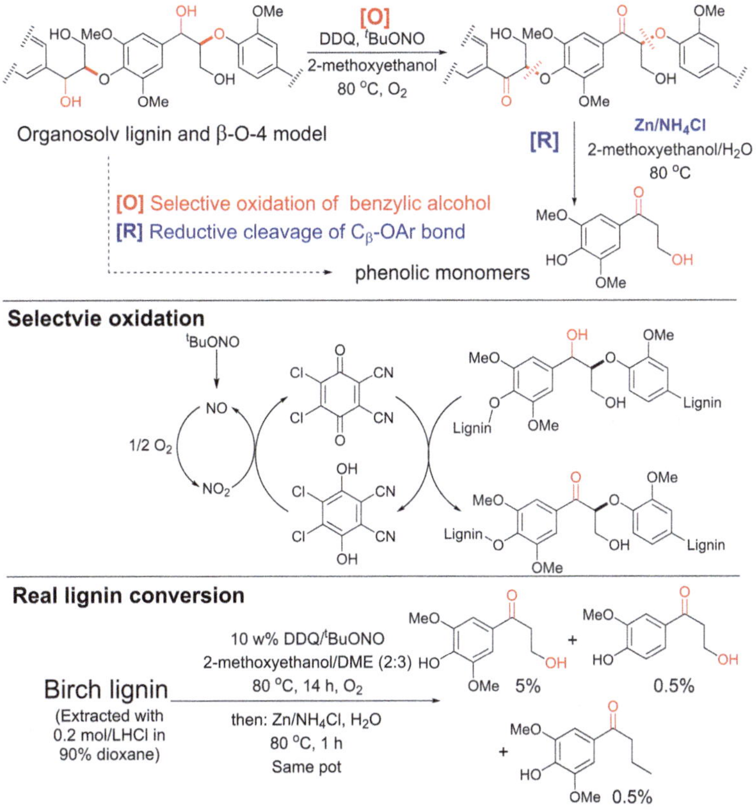

Figure 9.17 Isolation of Functionalized Phenolic Monomers through Selective Oxidation and C–O Bond Cleavage of the β-O-4 Linkages in Lignin. *Source:* Lancefield et al. [26] with Permission, © 2015 Wiley-VCH.

C_αH–OH to C_α=O in β-O-4 linkages in model compounds, polymeric models, and in real lignin with the DDQ/^{t}BuONO/O_2 system. Then, the C_β–OAr bond of oxidized β-O-4 linkages was further cleaved with Zn/NH_4Cl [27] in

2-methoxyethanol. When the dioxane-extracted birch lignin was used as subtract, the above two-step method can provide an ~6 w% functionalized phenolic monomers (Figure 9.17).

To achieve the first C_αH–OH conversion to C_α=O in β-O-4 linkage, other than using oxidation methods, in 2016, Fu and co-worker [28] reported an Ir-based catalytic system that can achieve the first selective dehydrogenation of C_αH–OH to C_α=O and the byproduct hydrogen has been trapped by 1-decene successfully, which can keep the C_β–OAr unconverted. Then the C_β–OAr bonds in the dehydrogenation substrate can be reductively cleaved with the combination of Zn/NH_4Cl (Figure 9.18), which was supported by the GPC of the samples that showed an obvious molecule weight decrease after the dehydrogenation-reduction treatment.

To achieve the reductive cleavage of the ether bond in the pre-oxidized lignin with C_α=O, thiols can also play key roles in biochemical pathways in the natural process, rather than just the active metal. An example for ether cleavage is a biomimetic approach, mimicking etherases such as the glutathione-dependent enzymes from wood-digesting bacteria like *Sphingobium* sp strain SYK-6 [29–31]. Glutathione (GS) is a tripeptide used by the Lig enzymes E, P, F, and BaeAB (etherases), and Lig enzyme G and GSTNu (lyase) [29–33]. This cofactor contains a cysteine (thiol) residue that performs a protein-assisted SN_2 reaction on an α-keto lignin dimer, forming a thioether. The thioether is then reduced by another glutathione resulting in the formation of a disulfide bond and liberation of an enolate that is protonated to form the ketone. The net result is the addition of two hydrogen atoms across the β-O-4-aryl ether C_β–O bond, reductively fragmenting the alpha-keto β-O-4 linkage, as summarized in Figure 9.19a.

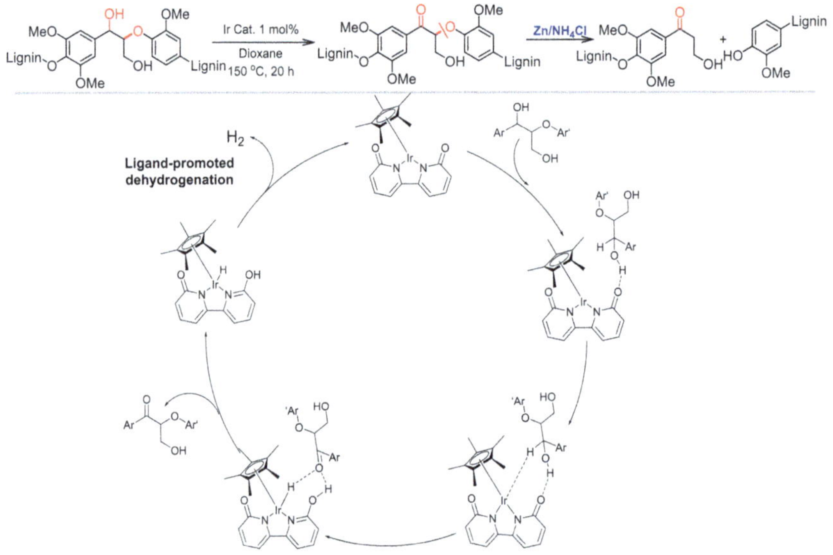

Figure 9.18 Chemoselective Oxidant-free Dehydrogenation of C_αH–OH in Lignin β-O-4 Linkage with a C_p*Ir Catalyst and Further Reductive Conversion of the Pre-oxidized Lignin. *Source:* Zhu et al. [28] with Permission, © 2016 The Royal Society of Chemistry.

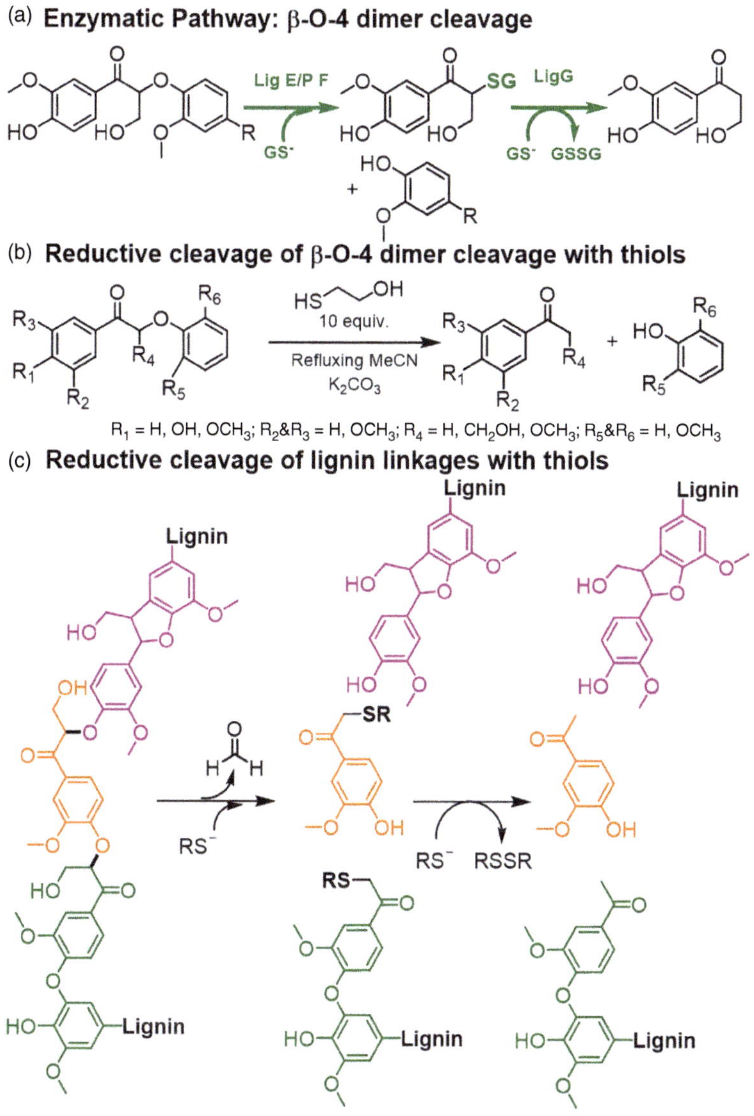

Figure 9.19 Nucleophilic ThiolsReductively Cleave Ether Linkages in Lignin Models and Lignin. *Source:* Klinger et al. [34], © 2020 Wiley-VCH.

Inspired by the chemistry harnessed by the Lig enzymes of *Sphingobium* to achieve reductive depolymerization of lignin dimers, in 2019, Hegg and co-workers [35] reported that organic thiols ($HOCH_2CH_2SH$), under alkaline polar aprotic solvent conditions, can cleave keto aryl ether bonds in model lignin systems with nearly 100% yields (Figure 9.19b). Then, in 2020, Hegg and co-workers [34] further examined the nucleophilic and reductive capabilities of free organic thiols to cleave ether bonds in polymeric lignin models, lignin extracted from poplar, and to other lignin-type linkages (Figure 9.19c). Synthetic β-O-4 linked polymers with C_α=O were treated with organic thiols, yielding up to 90% cleaved monomer products. Lignin extracted from poplar was also treated with organic thiols resulting in molecular weight reductions as high as 65% (Mn) in oxidized lignin. The activity difference between the lignin model and real lignin could be attributed to the low activity of the system in the cleavage of ether bonds of 4-O-5 than β-O-4. Furthermore, they also suggest that these findings open the door for future studies in which electrochemical 2-electron/2-proton reductions of the disulfide byproducts would regenerate the thiols, enabling net electrocatalytic lignin depolymerization.

For the recycling of the thiol from the disulfide via a biochemical pathway, in 2013, Sieber and co-workers [36] have supposed that three enzymes (LigD, a C_α-dehydrogenase; LigF, a β-etherase; and LigG, a glutathione lyase) were involved in lignin degradation. They cloned and recombinantly expressed these genes in *E. coli* and determined their pH and temperature optima on the lignin model substrate 1-(4-hydroxy-3-methoxyphenyl)-2-(2-methoxyphenoxy)-1,3-propanediol. Using an NAD^+ dependent glutathione reductase from *Allochromatium vinosum* (AVR), they established an efficient way to regenerate the co-substrates NAD^+ and glutathione allowing for a self-sufficient balanced enzymatic cascade with net internal hydrogen transfer (Figure 9.20). They also showed the capability of this enzyme system to release lignin monomers from complex lignin structures.

Besides the first transformation of C_αH–OH to C_α=O for activating the C_β–OAr bond, in 1999, Ralph and co-worker [37] reported a DFRC (derivatization followed by reductive cleavage) method for lignin analysis, which involved the cleavage of the C_β–OAr bond (Figure 9.21). Derivatization, accompanied by cell wall solubilization, is accomplished with acetyl bromide (AcBr); reductive cleavage of resulting β-bromo ethers utilizes zinc in acetic acid. Under the optimized conditions, the β-ether linkage (β-O-4) of all models was cleaved in a high yield (>92%). The DFRC

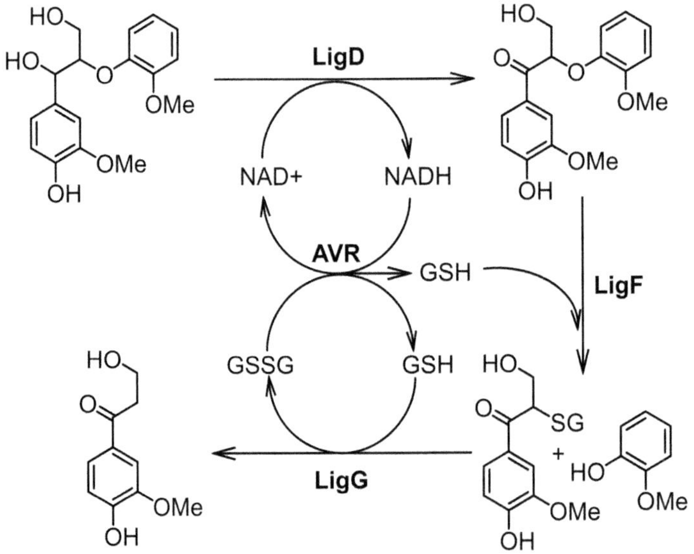

Figure 9.20 The reaction of the LigDFG Enzyme System with AVR. *Source:* Reiter et al. [36] with Permission, © 2013 The Royal Society of Chemistry.

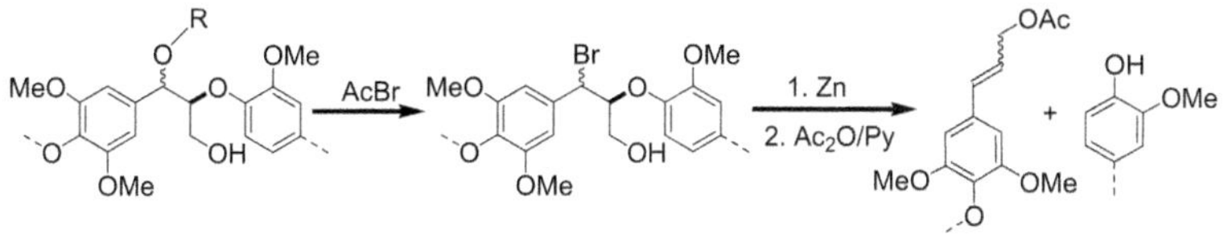

Figure 9.21 DFRC Method for Lignin Analysis with C_β–OAr Cleavage.

method produces simpler mixtures of monomers with higher yields than alternative hydrolytic methods. In addition, because of its relative simplicity, mild conditions, and exceptional selectivity, the DFRC method become a powerful analytical method for lignin depolymerization.

9.2.2 Thermal Catalytic Hydrogenation Methods for the C_β–OAr Bond Cleavage in the Pre-oxidized Lignin

Given the existence of other ether linkages (α-O-4, 4-O-5, etc.) without adjacent hydroxyl groups to provide active hydrogens, converting lignin via oxidation-hydrogenation strategy is still a feasible method. However, difficulties exist. The less controlled hydrogenation system can catalyze the hydrogenation of aromatic rings and the ketone could be hydrogenated back to alcohol before the C_β–OAr cleavage. Therefore, the hydrogenation catalyst should hydrogenate C_β–OAr rather than the C_α=O bond and aromatic rings.

In 2017, after the catalyst screening, Zhang and co-workers [38] found $NiMoS_x$, prepared by calcinating the precipitate of reaction between $(NH_4)_2MoS_4$ and $NiCl_2$, was an appropriate hydrogenation catalyst for the selective hydrogenation of pre-oxidized β-O-4 linkage with a C_α=O. On the basis of the XRD characterization, MoS_2, NiS, and $Ni_{0.7}Mo_3S_4$ phases exist in the $NiMoS_x$ catalyst. They further tested the catalytic performances of MoS_2 and NiS. When MoS_2 was used as the catalyst for the hydrogenation of β-O-4-Ketone, although the products with hydrogenation of aromatic could be ignored and the main products were target aromatic products with C_β–OAr bond cleavage (71%), a part of β-O-4-ketone was transformed to β-O-4-alcohol (7%) with a C_αH–OH but without C_β-OAr bond cleavage that could further undergo hydrolysis to β-O-4-H (18%) with a $C_\alpha H_2$, and etherification to β-O-4-E with a C_αH–OR (3%). When NiS is used as the catalyst, the β-O-4-Ketone conversion is <2%. The low activity indicates that the NiS_x species might just be the catalytic promoter to the $NiMoS_x$ catalyst. Then DFT calculations and further characterizations were carried out to check the reason. DFT calculations of β-O-4 compound adsorption models (Figure 9.22d–g) indicated the hydrogenative Mo sites on MoS_2 were in a linear arrangement at the edge of this 2D-layered material. Besides the potential Mo sites to cleave C_β–OAr bond, few Mo sites were working as the hydrogenation centers around the aromatic ring, which was different from the face-face adsorption models over bulk Pd(111) and Ni(111). As shown in Figure 9.22d, the EDX images show that Ni and Mo atoms are equally distributed at some zones of $NiMoS_x$, which might indicate the hydrogenative Mo sites in NiMo sulfide are not in a good linear arrangement like MoS_2 but interrupted by the Ni^{2+}. Therefore, by

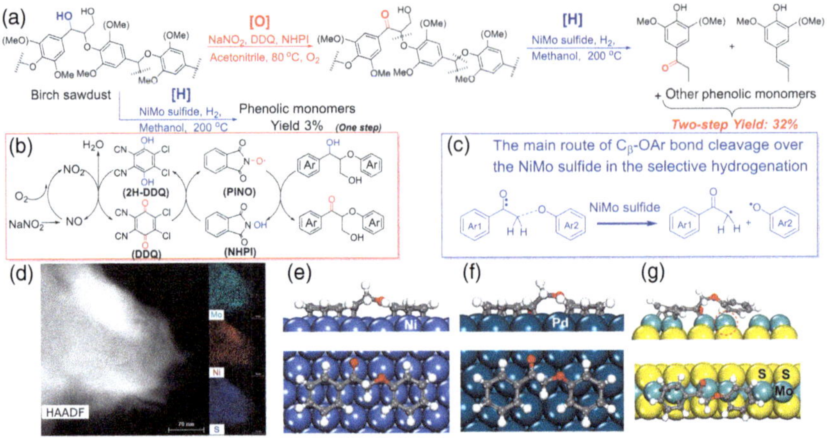

Figure 9.22 The Conversion of Birch Sawdust to Aromatic monomers via an Oxidation-hydrogenation Strategy (a-c); DFT Adsorption Models of 2-Phenoxy-1-phenylethan-1-one on Pd(111), Ni(111), and MoS_2 (d-g). *Source:* Zhang and Wang [40] with Permission, © 2017 American Chemical Society.

partly replacing edge Mo atoms with NiS structure with low hydrogenation activity, the $NiMoS_x$ could further restrain the hydrogenation of C_α=O and aromatic rings but focus on the C_β–OAr bond cleavage (Figure 9.22c), leading to an increase of monomers yield. Coordinating this NiMo hydrogenation system with an O_2/$NaNO_2$/DDQ/NHPI system [39] to selectively oxidize β-O-4 C_αH–OH to C_α=O (Figure 9.22b), they achieved the birch sawdust depolymerization with a 32 w% phenolic monomers yield, which was 10 times higher than the direct hydrogenation (Figure 9.22a).

Given the effect of hydrogenation catalyst modification by the heteroatom, in 2018, Wang and co-workers [41] then prepared a carbon modified Ni catalyst from the carbothermal reduction of Ni-doped layered double hydroxides (NiMgAl-LDH) with inserted lignosulfonate (LS), during which an LDH-LS composite was transformed into a MgAl-oxide sheet, and Ni cations were reduced by carbon to well-dispersed nanoparticles (Figure 9.23a). Further XPS, EXAFS, and DFT characterization showed that the high-temperature carbothermal reduction could facilitate a small number of carbon atoms residing in the Ni surface and greatly modified the nickel 3d electronic structure, which could make the obtained Ni catalyst (Ni/MgAlO-C) prefer to catalyze the selective hydrogenolysis of C_β–OAr bonds in pre-oxidized lignin models into aromatics rather than catalyze the hydrogenation of the C_α=O or aromatic rings. Moreover, they then focused on the real lignin as a substrate (Figure 9.23b). Birch lignin with 43% content of β-O-4 linkages was first oxidized by 2,3-dichloro-5,6-dicyano-1,4-benzoquinone (DDQ)/^{t}BuONO. Then, the oxidized birch lignin was further subjected to hydrogenolysis over the carbothermal reduced Ni/MgAlO-C and a control catalyst Ni/MgAlO which was obtained by further treating Ni/MgAlO-C in the H_2 flow to remove the resided carbon in the Ni nanoparticles. With Ni/MgAlO as a catalyst, only 5 wt % yields of aromatics were obtained. Comparatively, enhanced performance was observed in the presence of a Ni/MgAlO-C catalyst, and a mixture of aromatics was obtained in a 22 wt % monomers yield.

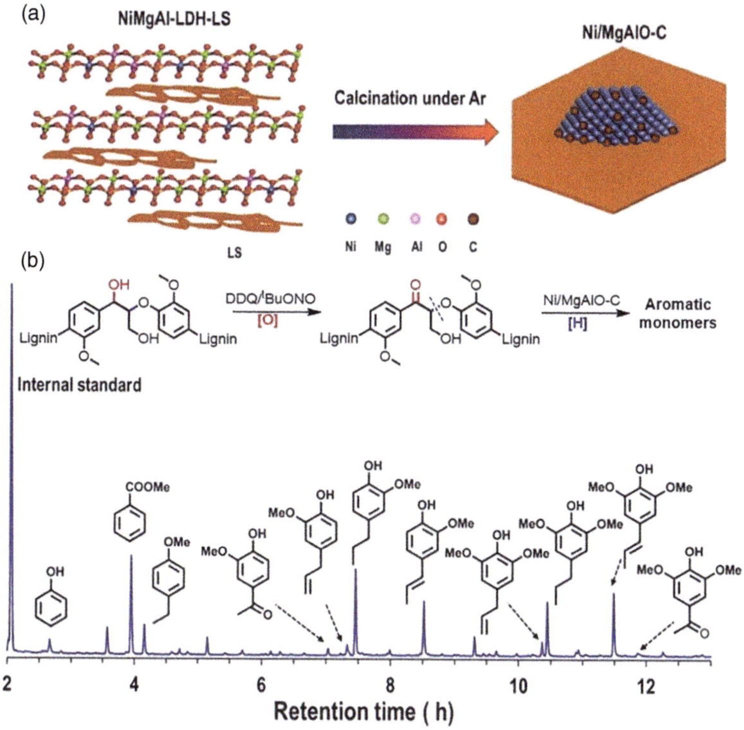

Figure 9.23 Carbon Modification of Nickel Catalyst Ni/MgAlO-C for Depolymerization of Oxidized Lignin to Aromatics. *Source:* Wang et al. [41] with Permission, © 2018 American Chemical Society.

In 2018, Beltramini and co-workers [42] further studied the two-step selective oxidation and hydrogenolysis sequence (Figure 9.24). During their catalytic process, the first oxidation of lignin was conducted using an O_2/HCl-HNO_3/TEMPO system, followed by the catalytic hydrogenolysis over the active and inexpensive TiN-Cu nanocatalyst. Compared with the direct hydrogenolysis of the Soda lignin with a 12 w% yield of aromatic monomers, the oxidation-hydrogenolysis provided a 56.7 w% yield of aromatic monomers over the TiN-Cu.

9.2.3 Photocatalytic and Electrocatalytic Reduction Methods for the C_β–OAr Bond Cleavage in the Pre-oxidized Lignin

Besides the above thermal catalytic processes, in 2014, Stephenson and co-workers developed a two-step photocatalytic lignin degradation strategy [43]. The strategy involved chemoselective benzylic oxidation to generate a C_α=O with a selective oxidant ([4-AcNHTEMPO]BF_4), and a catalytic reductive C–O bond cleavage system which utilized the photocatalyst [Ir(ppy)2(dtbbpy)]PF_6 and the sacrificial agent combination DIPEA/HCOOH (Figure 9.25, DIPEA means N,N-Diisopropylethylamine). This system was tested on relevant lignin model substrates

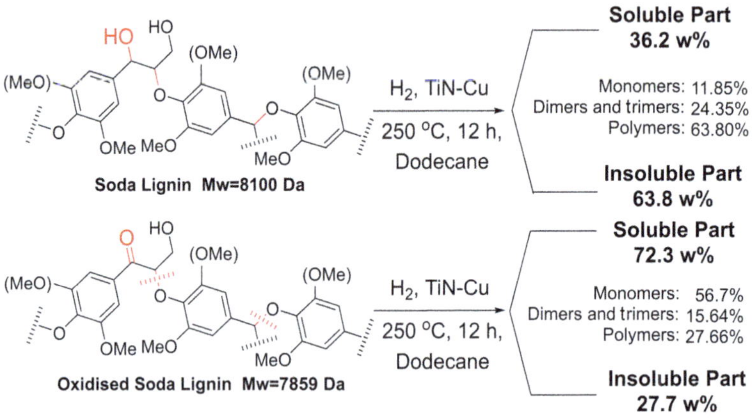

Figure 9.24 Catalytic Depolymerisation of Oxidized Lignin over TiN-Cu Nanocatalysts. *Source:* Reaction Data Abstracted from Rashidi et al. [42].

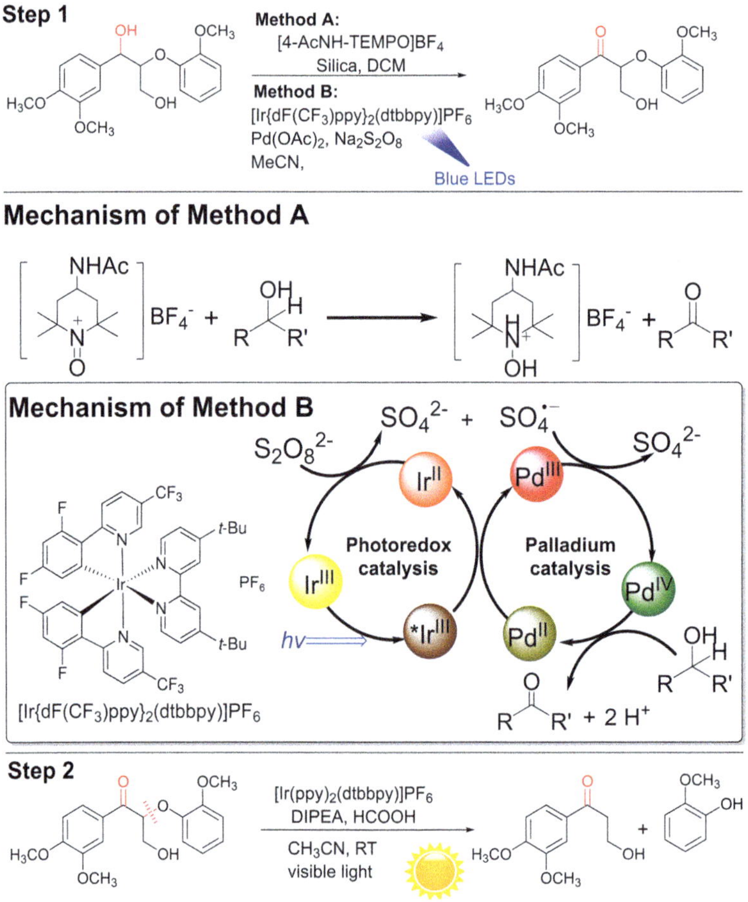

Figure 9.25 Two-step Visible-light Mediated Depolymerization Strategies for Lignin and Lignin Model Systems. *Source:* Karkas et al. [44] with Permission, © 2016 American Chemical Society.

containing β-O-4 linkages to generate fragmentation products in good to excellent yields. For the first selective oxidation, besides the oxidation system with stoichiometric [4-AcNHTEMPO]BF_4 (Figure 9.25, Step 1, Method A), in 2016, they further achieved the selective benzylic oxidation by merging Ir-photoredox and Pd catalysis (Figure 9.25, Step 1, Method B), [44] which make the lignin conversion with C_β–OAr bond cleavage a two-step photocatalytic process. The mechanism of the oxidation step is initiated with irradiation of the Ir^{III} photocatalyst and results in metal-to-ligand charge transfer to produce a long-lived excited state $*Ir^{III}$. Then, $*Ir^{III}$ can undergo reductive quenching with $Pd(OAc)_2$ to generate a Pd^{III} species. The reduced Ir^{II} photocatalyst is a strong reductant and subsequently reacts with $S_2O_8^{2-}$ to regenerate the ground-state Ir^{III} photocatalyst and produce SO_4^{2-} and $SO_4^{\bullet-}$. The sulfate radical, $SO_4^{\bullet-}$, is a potent oxidant and has sufficient potential to facilitate the oxidation of Pd^{III} to Pd^{IV}. The high-valent Pd^{IV} species is expected to oxidize the substrate to afford the desired oxidized product, regenerating Pd^{II} to complete the proposed dual catalytic cycle.

In 2019, to remove the peroxide used in the above two-step degradation approach, Stephenson and co-workers [45] further modified the method involving Pd-catalyzed aerobic oxidation and visible-light photoredox Ir-catalyzed reductive fragmentation for the chemoselective cleavage of the β-O-4 linkage-the predominant linkage in lignin-for the generation of lower-molecular-weight aromatic building blocks (Figure 9.26). The developed strategy affords the β-O-4 bond

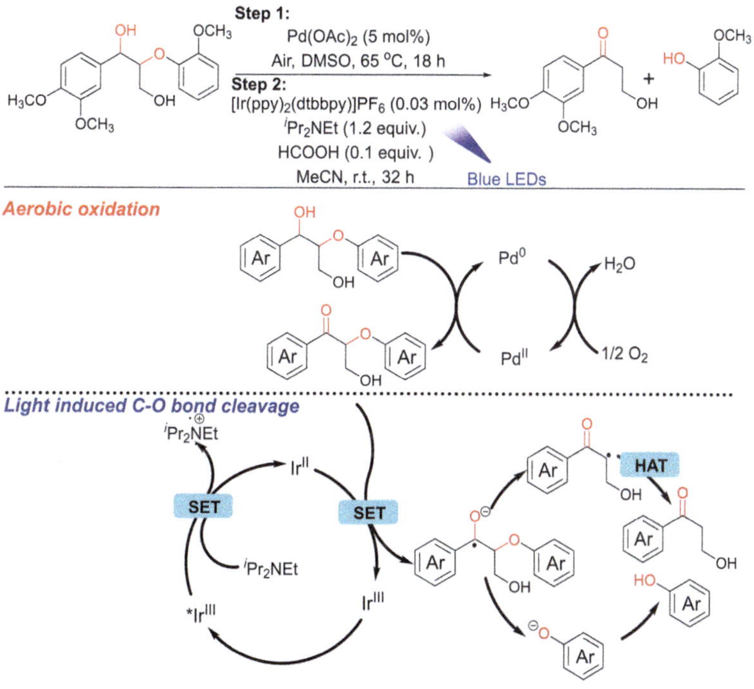

Figure 9.26 Selective C–O Bond Cleavage of Lignin Systems and Polymers Enabled by Sequential Pd-catalyzed Aerobic Oxidation and Visible-light Photoredox Catalysis. *Source:* Magallanes et al. [45] with Permission, © 2016 American Chemical Society.

cleaved products with high chemoselectivity and in high yields, is amenable to continuous flow processing, operates at ambient temperature and pressure, and is moisture- and oxygen-tolerant.

Besides the thermal process, photocatalytic system and their combinations, electrochemical oxidations and reductions have long been recognized as a more environmentally benign alternative tool for chemical reactions, allowing for specific reactivity thanks in part to the ability to control the desired reaction potential. However, as reported by previous research, the direct electroreduction of lignin backbone has been reported to be poorly efficient and to give complicated mixtures [46] and attempts of carrying out electrooxidation of lignin structure have proven to require high oxidation potentials, necessitating the use of expensive and toxic electrodes (IrO_2, Pt, Ru/V oxides, Au, or Pb). In 2017, on the basis of their previous two-step strategies used in lignin conversion, Stephenson and co-workers [47] described an efficient and selective one-pot two-step method for fragmentation of the β-O-4 linkages, which combined electrocatalytic oxidation and photocatalytic fragmentation in a seamless batch to flow processing at ambient temperature (Figure 9.27). They use NHPI/2,6-lutidine as a hydrogen atom transfer (HAT) mediator, which significantly reduces the necessary applied potential (from 0.85 to 0.35 V) and thus offers milder oxidation conditions and enhanced efficiency and selectivity. The secondary alcohol (C_αHOH) is selectively oxidized to ketone (C_α=O)

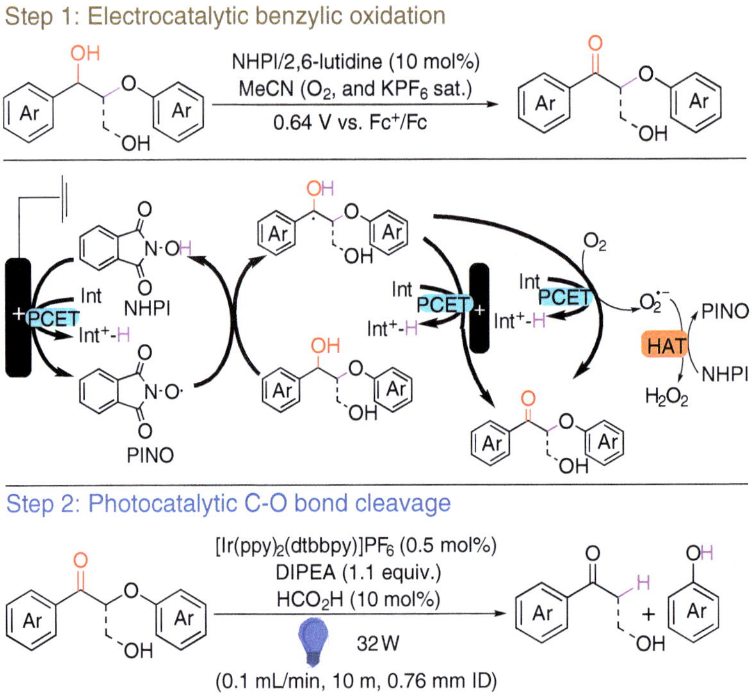

Figure 9.27 The One-pot Electrocatalytic Oxidation–reductive Cleavage of β-O-4 Linkage. *Source:* Process Summarized According to Bosque et al. [47].

rather than the oxidation of $C_\gamma H_2OH$ (Figure 9.27, Step 1). Then, kinetic isotope effect (KIE) experiments suggest that the abstraction of the benzylic hydrogen might be the rate- or turnover-determining step. The mechanism of the electrocatalytic oxidation is proposed to be initiated by the anodic oxidation of NHPI to PINO, a process facilitated by 2,6-lutidine (Int) via PCET. HAT from the β-O-4 substrate C_αHOH would regenerate the NHPI (rate-determining step) and provide the α-hydroxy radical C_α(•)OH [48]. Then, PCET-based oxidation of C_α(•)OH (2,6-lutidine as the base) with O_2 as oxidant (or the anode) would directly provide β-O-4 ketone (C_α=O) and superoxide. Then the Ir-based catalyst can achieve the photocatalytic fragmentation of the oxidized β-O-4 linkages to ketones and phenols (Figure 9.27, Step 2) [43].

To promote the activity in ether bonds cleavage of pre-oxidized lignin, the quantum dots materials were then used. In 2019, Cossairt and co-workers [49] demonstrated the versatility of CdSe quantum dots in cleaving the C_β–OAr bond of the β-O-4-Ketone model with DIPEA/HCOOH. Quantum dots outperform state-of-the-art molecular iridium catalysts with up to a factor of 15 times greater turnover frequency at equivalent catalyst loading. Additionally, the CdSe quantum dot system required 333 times less catalyst to achieve the same extent of product conversion previously reported for the optimized iridium-catalyzed systems over the same time period.

Besides the homogeneous photocatalysts and quantum dot systems, the heterogeneous photocatalyst, represented by the typical TiO_2, has been used in lignin conversion from the 1990s under UV light (>310 nm), which mainly focused on the direct degradation of lignin as the waste of the pulp industry [50]. However, the reaction selectivity and catalytic performance are not good and the aromatic rings are usually destroyed. Given the first oxidation of C_α–OH to C_α=O can reduce the BDE of C_β–OAr bond, the semiconductor, with appropriate conduction band (CB) potential, could achieve the reductive cleavage of the pre-oxidized β-O-4 linkage more easily with some sacrificial agent at valence band (VB) under a mild condition. In 2016, Luo and co-workers [51] reported a photocatalytic oxidation-reduction tandem method for cleaving C–O bonds of β-O-4 alcohols. After catalyst screening, they found TiO_2 with abundant surface Ti^{3+} could efficiently catalyze the photocleavage of C_β–OAr bond in β-O-4 lignin ketones with the assistance of EtOH/NaOAc system as the sacrificial agent. The oxidation–hydrogenolysis tandem reaction could be accomplished in one pot to offer ketones and phenols via a dual light wavelength switching (DLWS) strategy from lignin models. With this strategy, 93~99% yields of phenols in the conversion of lignin β-O-4-ketone models are achieved. In detail about the mechanism (Figure 9.28), the overall conversion of the β-O-4 model to monomer products begins with photooxidation of C_α–OH within β-O-4 linkage to C_α=O with 455 nm light over a Pd/$ZnIn_2S_4$ catalyst with O_2. Then, the 365 nm illumination induced the generation of holes (h^+) and electrons (e^-) on TiO_2. Holes of valence band oxidized sacrificial reagent ethanol while electrons reduced surface Ti^{4+} to Ti^{3+}. EPR characterizations indicated the adsorption of β-O-4 ketone to the in situ formed Ti^{3+} weakened the C_β–OAr bond and finally resulted in the ether bond cleavage to phenol and a ketone.

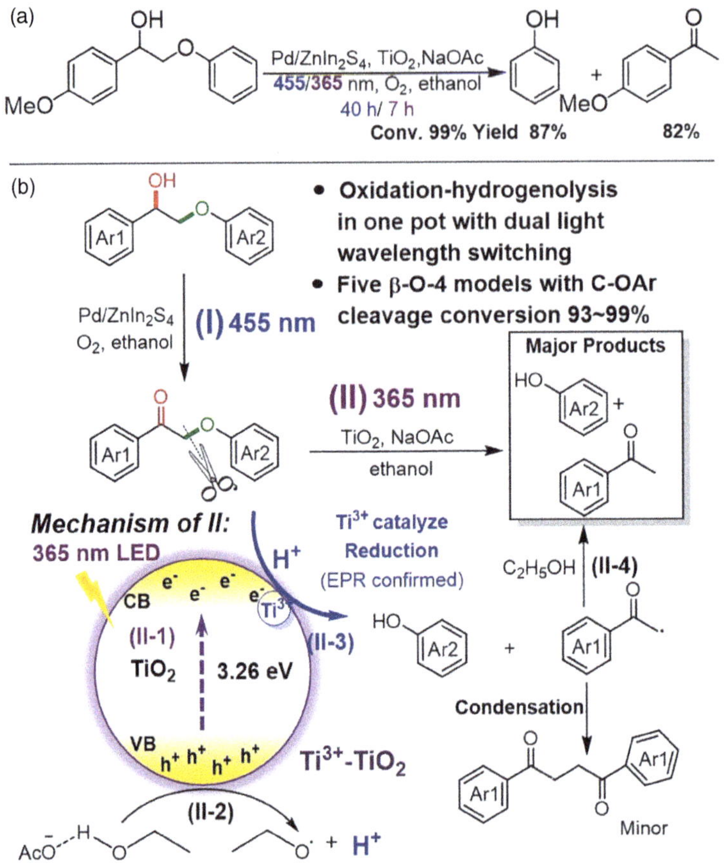

Figure 9.28 Proposed Reaction Mechanism for Photocatalytic C–O Bond Cleavage of β-O-4 Ketones via Hydrogenolysis over the TiO_2. *Source:* Luo et al. [51] with Permission, © 2016 American Chemical Society.

To accelerate the photocatalytic reductive cleavage of the C_β–OAr bond within pre-oxidized β-O-4 linkage, the Wang group [52] and other groups [53–55] then loaded some noble metal on the surface TiO_2, targeting at cleaving the C_β–OAr bond with active H* which generates from the combination of H^+ and photo-induced electron at CB of semiconductor. Besides the catalytic cleavage of the C_β–OAr bond to ketones and phenols, their further tests with optimized Pd/TiO_2 system could efficiently catalyze the photocatalytic deoxygenation of aromatic ketones to alkyl arenes at room temperature [52], which utilized alcohols as green hydrogen donors and no other additive was needed. No hydrogen evolution occurred during this transformation, suggesting a mechanism of direct hydrogen transfer from alcohols. Control experiments with additives indicated the role of acid sites in this transfer hydrogenolysis and further characterization indicated that a combination of hydrogen transfer sites and acidic sites were responsible for the efficient deoxygenation. More specifically (Figure 9.29), under irradiation of 365 nm light,

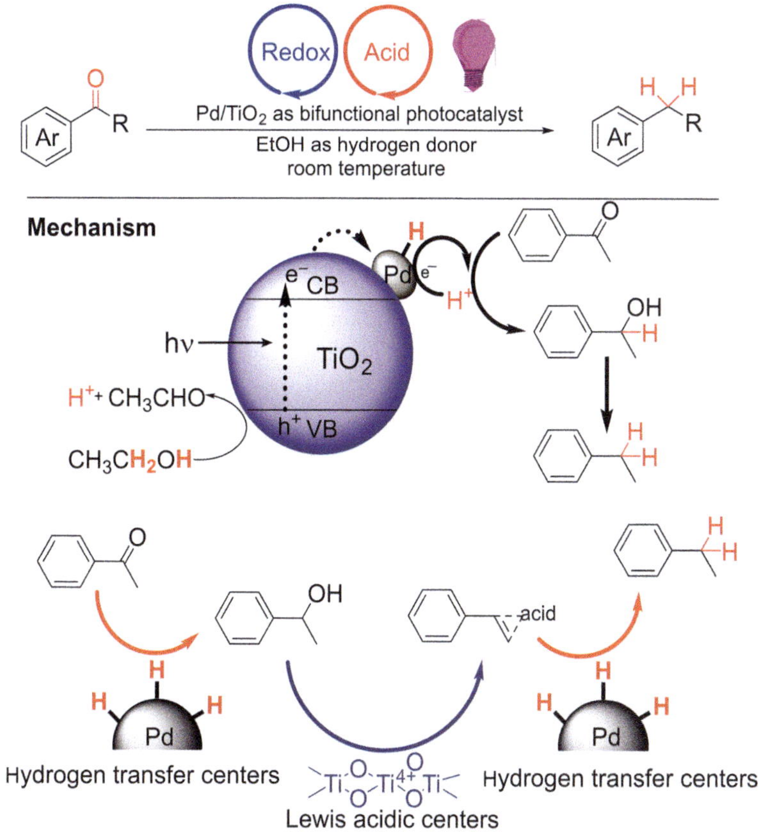

Figure 9.29 Further Deoxygenation of Aromatic Ketones to Alkyl Arenes over the Pd/TiO_2. *Source:* Li et al. [52] with Permission, © 2020 The Royal Society of Chemistry.

the photogenerated holes on TiO_2 oxidized the ethanol to acetaldehyde and proton. The photogenerated electrons on TiO_2 migrated to Pd^0 particles and reduced the proton to activated hydrogen species (Pd–H). Then, the ketone substrate was reduced to an alcohol intermediate by Pd–H. In the presence of abundant Lewis acidic sites on TiO_2, the alcohol intermediate was dehydrated and further hydrogenated to ethylbenzene by Pd–H over the Pd/TiO_2.

As shown above, the TiO_2-based photocatalytic systems can work well in lignin models conversion, but the conversion and selectivity in native lignin conversion are not good. For the reasons, besides the complex lignin fragments condensation, the competitive light absorption between photocatalysts and lignin with dark color and the hard contract between solid TiO_2 and lignin polymer fragment are two non-negligible reasons. To overcome these points, decreasing the concentration of the lignin extractive can work a little but will make the photocatalytic process impractical. Besides the method by changing the concentration, the Vinu group [56] reports that wet milling of lignin and TiO_2 using polar solvents improves the contact between lignin and TiO_2, thus facilitating lignin depolymerization to produce valuable phenolics and aromatic hydrocarbons. Furthermore, in 2020, Sherman and co-workers [57] reported the use of a DSPEC-type photoanode specifically in chemoselective oxidation of benzylic alcohol moieties in lignin model compounds and real lignin in organic media (Note: DSPECs, molecular-based dye-sensitized photoelectrochemical cells). The TiO_2-based photoanode incorporates a surface-bound Ru(II)-based photocatalyst and solution dissolved hydrogen atom transfer (HAT) co-catalyst to perform solar-driven photocatalytic oxidation of the lignin substrates (Figure 9.30). During the actual operation, the photoanode can be put on the shell of the pipe not in the middle of the solution, which can overcome the question of light competitive absorption. In addition, the photoanode can quickly consume the photo-induced electron from the activation photosensitizer Ru^{II} to Ru^{III}, which can further promote the following HAT process with NHPI/PINO. Although this photoanode system is not for ether bond reductive cleavage, the method used can provide some guidance for the modification of heterogeneous photocatalytic systems. Of course, some photocatalysts with stronger light absorption ability can be

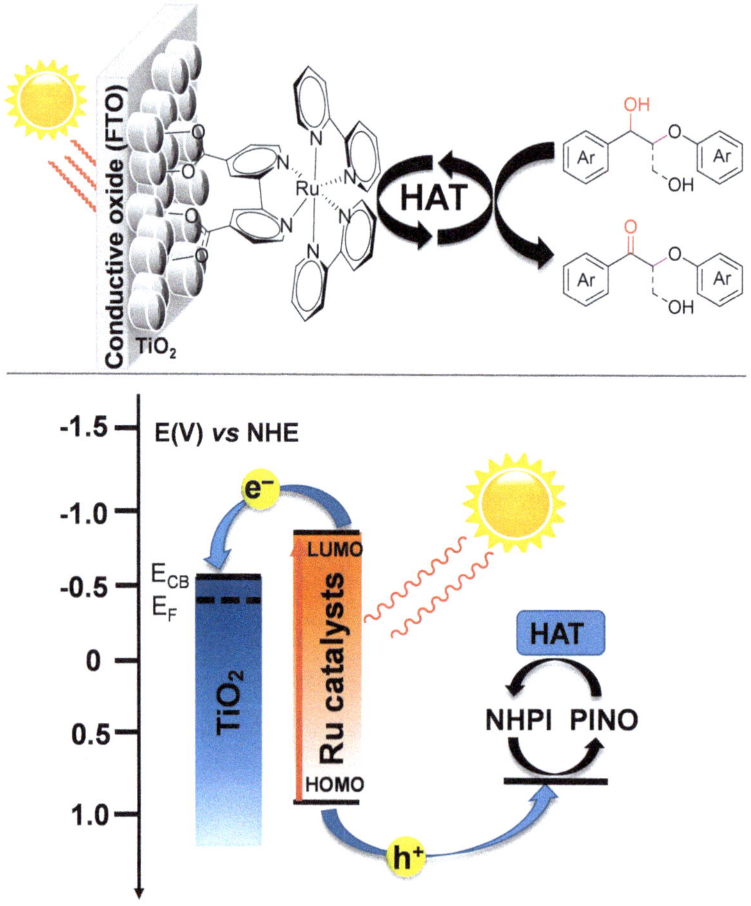

Figure 9.30 Solar-driven Lignin Oxidation via Hydrogen Atom Transfer with a Dye-sensitized TiO_2 Photoanode.

employed in pre-oxidized depolymerization, such as the yellow solid $ZnIn_2S_4$ [24] shown above (Figure 9.14), or preparing the color-less lignin substrate.

Besides the homogeneous photocatalysts with a metal center like Ru, Ir, etc., and solid metal oxide semiconductors used in oxidized lignin conversion, organic (visible light) photocatalysts (OPCs) have received much attention owing to their distinct properties such as commercial availability, low toxicity, and structural tunability. In 2017, Zhang and co-workers [58] reported a facile approach to fine tune the redox potentials of π-conjugated porous organic frameworks (POFs) by copolymerizing carbazoles electron donor (D) and electron acceptor (A) based comonomers at different ratios. The resulting carbazolic copolymers (CzCPs) exhibit a wide range of redox potentials and are used in the stepwise photocatalytic degradation of lignin

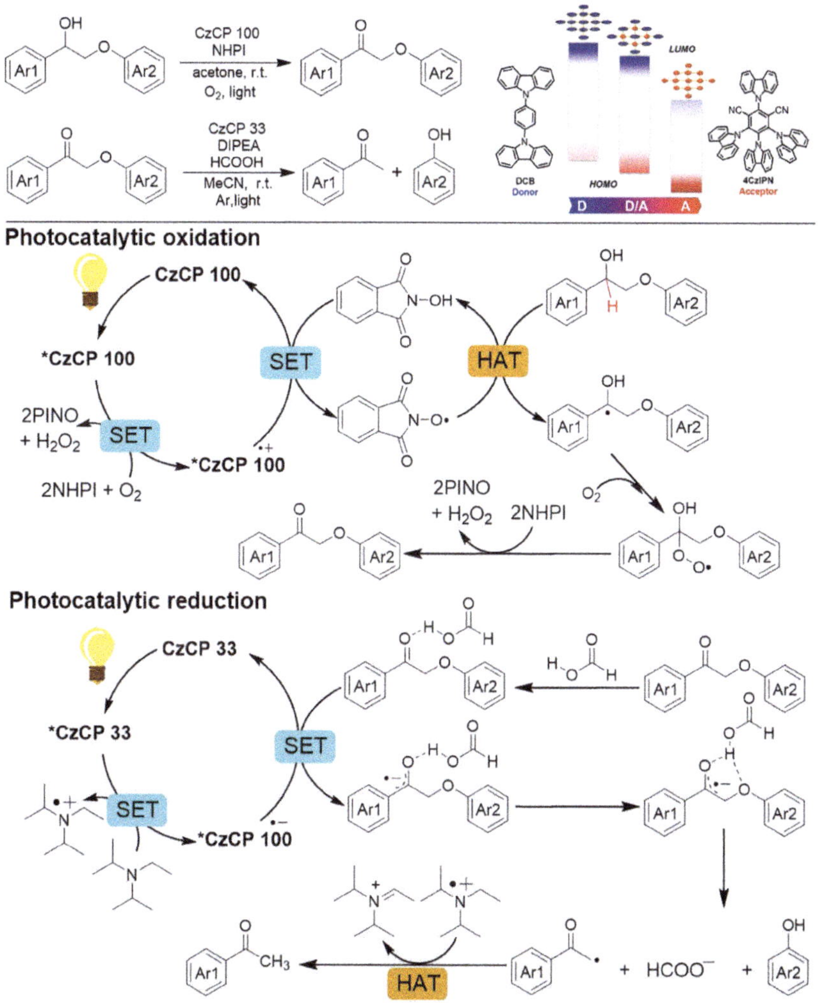

Figure 9.31 Fine Tuning the Redox Potentials of Carbazolic Porous Organic Frameworks for Photocatalytic Degradation of Lignin β-O-4 Models. *Source:* Luo et al. [58] with Permission, © 2017 American Chemical Society.

β-O-4 models (Figure 9.31). With the strongest oxidative capability, CzCP100 (D: A =0: 100) exhibits the highest efficiency for the oxidation of benzylic β-O-4 alcohols to a ketone with NHPI as a HAT co-catalyst (Figure 9.31, Photocatalytic oxidation), while the highly reductive CzCP33 (D: A=66: 33) gives the highest yield for the reductive cleavage of β-O-4 ketones with formic acid (Figure 9.31, Photocatalytic reduction).

In 2020, Zhang and co-workers [59] provided direct evidence that $^2(\mathrm{PDI}^{-\bullet})^*$ (Perylene diimide, PDI, structure in Figure 9.32) undergoes efficient PET (photoinduced electron transfer) with electron acceptors with reduction potential > −1.7 V. Extrapolation of this result to anion radicals that are more strongly reducing suggests an avenue to develop more reactive photosensitizers with the ability to catalyze reactions that require very negative potentials. Then, they demonstrated that perylene diimide (PDI) can act as an efficient metal-free photocatalyst for the depolymerization of pre-oxidized β-O-4 lignin models with a C_α=O and real organosolv lignin at room temperature under visible light illumination [60] in which the pre-oxidized lignin linkage was transformed to phenolic and ketone fragmentation molecules with very high selectivity (Figure 9.32). A mechanism study indicated that superior

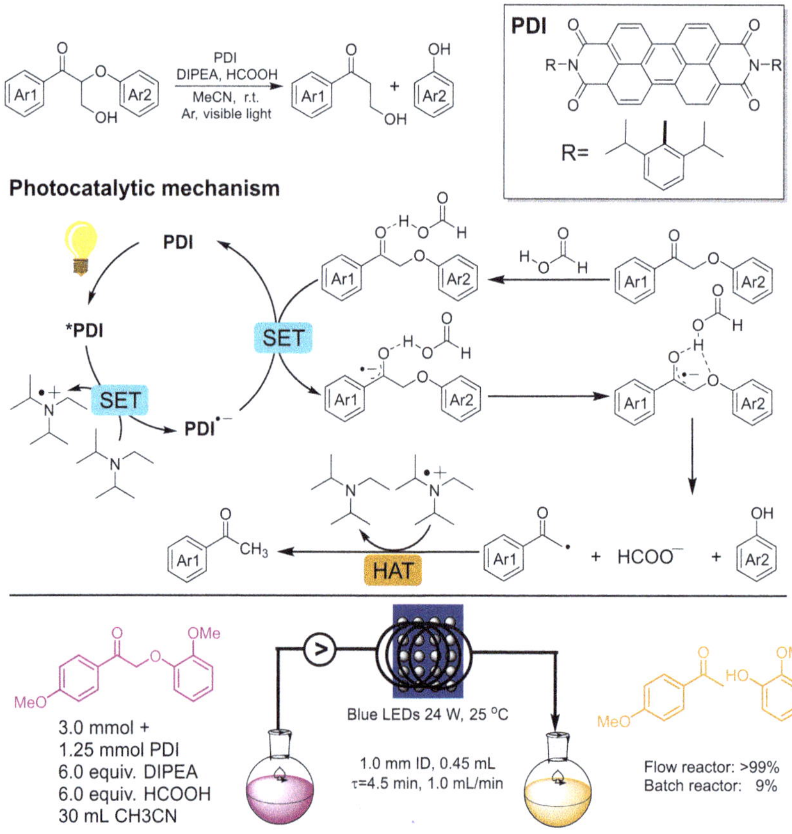

Figure 9.32 Visible Light-enabled Selective Depolymerization of Oxidized Lignin by An Organic Photocatalyst PDI. *Source:* Li et al. [60] with Permission, © 2020 The Royal Society of Chemistry.

performances were realized by single-electron transfer (SET) from the photoexcited strongly reducing $PDI^{\bullet-}$ anion to the ketone groups of the β-O-4 linkage in the lignin. Compared with the PDI catalyst with a 99% C_β–OAr bond conversion, the organic photocatalysts of 4CzIPN and acridinium salts provided 84% and 20% conversion, respectively. The author speculated that the inferior catalytic performances of 4CzIPN and acridinium salts were probably attributed to the low reduction potential (−1.21 eV and −0.57 eV), respectively. For the mechanism of this process, in the PDI/DIPEA/HCOOH catalytic system, PDI was a photocatalyst, DIPEA provided the electron and HCOOH was a proton source. After visible light illumination, the generated photoexcited *PDI oxidized DIPEA to create the strongly reducing $PDI^{-\bullet}$ anion. Single-electron transfer (SET) from the $PDI^{-\bullet}$ anion to the lignin/formic acid complex produced the intermediate product and meanwhile, this process regenerated the ground state PDI. The intermediate product underwent selective C–O bond cleavage to give the desired phenol fragmentation product through protonation reaction by HCOOH and the other ketone radical product. Meanwhile, the radical product went through a hydrogen atom transfer process with the oxidized DIPEA to afford the ketone product. Furthermore, they further efficiently shortened the reaction time within an hour by using a homemade photocatalytic continuous-flow reactor. Even for organosolv lignin, nearly 86% mass ratio of lignin was degraded to low-molecular-mass monoaromatic or diaromatic products.

9.3 Lignin C_α–C_β/C_β–OAr Bonds Cleavage via Multiple Oxidation Process Beginning with the First Activation of C_α–OH to C_α=O

To cleave the complex and stable C–C bonds in the lignin structure, oxidation methods are the potential choice. For the direct oxidation of lignin linkage, Hanson and co-workers did a series of studies about vanadium-catalyzed aerobic oxidation of alcohols [61–70]. They found that vanadium catalysts containing ligands with pyridine structure can effectively catalyze the oxidation of β-O-4 and β-1 linkages [63, 65, 66, 69]. The remarkably different selectivities observed for different vanadium catalysts show the complexity of vanadium-mediated oxidations and support the viability of homogeneous catalysts for controlling selectivity in the aerobic oxidation of lignin. Besides the V-based catalysts, the Cu-based [71] catalytic system is also used in lignols oxidation. Although the V and Cu catalytic can promote the transformation of lignin, the uncontrolled radical process makes products are complex. Besides the above one-step methods, recently, more efficient two-step strategies have been explored in the lignin and model compounds transformation. In some certain cases, the pre-modification can increase targeted bonds BDE, but these methods may generate a new substrate site on the substrate to introduce cleavage reagent for further efficient conversion. As discussed in Chapter 5, the first C_α–OH oxidation to C_α=O can cause the increase of C_α–C_β BED by 30 $kJ \cdot mol^{-1}$, but the new C_β–H can be easily abstracted as $H^\bullet$ or H^+ and the rest C_β can work as the "substrate site" to react with active O_2* or other O* species, causing the cleavage of C_α–C_β

bond. In this section, the strategies for the lignin C_α–C_β bond cleavage in the multiple oxidation process which begin with the first activation of C_α–OH to C_α=O will be summarized.

9.3.1 Thermal Catalytic Oxidation Cleavage of C_α–C_β Bonds in the Pre-oxidized Lignin

In 2016, Wang and co-workers [72] reported a two-step strategy for oxidative cleavage of lignin C–C bond to aromatic acids and phenols with molecular oxygen as oxidant. In the first step, lignin β-O-4 alcohol was oxidized to β-O-4 ketone over a $VOSO_4$/TEMPO [(2,2,6,6-tetramethylpiperidin-1-yl)oxyl)] catalyst. In the second step, the C_α–C_β bond of β-O-4 linkages was selectively cleaved to acids and phenols by oxidation over a Cu/1,10-phenanthroline catalyst (Figure 9.33). Phenol formate was detected in the initial stage but disappeared after the reaction, suggesting that the C_α–C_β in **2** was first cleaved to phenol formate, followed by C_β–O cleavage to phenol. Benzoylformic acid gave no benzoic acid (Eq 2), indicating that it was not the reaction intermediate. A kinetic isotopic experiment (Eq 1) showed a large value of k_H/k_D (3.1), indicating the C_β–H oxidation could be the rate-determining step for the C_α–C_β cleavage. Single crystals of the $Cu(OAc)_2$/1,10-phenanthroline complex, ESR experiments, and computational investigations suggested a copper-oxo-bridged dimer was the catalytically active site for hydrogen-abstraction from C_β–H bond. DFT calculation shows the most favorable way for C_β–H bond oxidation is that the H is abstracted by one oxygen in the copper–oxygen center, and the remaining part is bound to the other oxygen via C–O bond formation (Figure 9.33, C_β–H cleavage). The H-abstraction step is exothermic with −2.18 eV. The activation of the C_α–C_β bond in the form of hydroxyl ketone structure-like intermediate significantly decreases its bond energy from 307.7 kJ mol^{-1} to 205.5 kJ mol^{-1}, which enables the C_α–C_β bond to be easily broken, resulting in the formation of benzoic acid and phenol formate. Further transfer of a hydroxyl group generates benzoic acid, restoring the initial copper complex. Finally, oxidative decarboxylation of phenol formate generates phenol and CO_2.

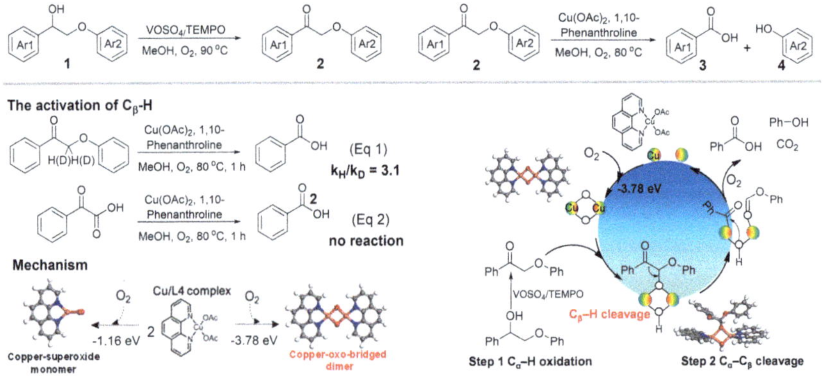

Figure 9.33 Two-step Oxidation Strategy for β-O-4 Cleavage. *Source:* Wang et al. [72] with Permission, © 2016 American Chemical Society.

Then, in 2017, Wang and co-workers [73] tried to introduce Lewis acid to assist the activation of the C_{β}–H for lignin C–C bonds oxidative cleavage. They further found that $Cu(OAc)_2/BF_3 \cdot OEt_2$ can efficiently catalyze the cleavage of C_{α}–C_{β} bond in β-O-4 ketone to esters and phenols, and cleave C–C bonds in pre-oxidized β-1 ketone to produce esters and aromatic aldehydes (Figure 9.34). In-depth studies showed that C–H bond activation was the rate-determining step for C_{α}–C_{β} bond cleavage, and $BF_3 \cdot OEt_2$ could promote the reaction via activating the C_{β}–H bond (BDE change of C_{β}–H). Mechanism studies suggest the C_{β} radical after C_{β}–H abstraction reacted with active O_2 to an oxygen-centered radical, which led to the cleavage of C_{α}–C_{β} bond. More detailedly, in the initial step, $BF_3 \cdot OEt_2$ activates the C_{β}–H bond via coordination with the oxygen atom in a carbonyl group. The C_{β}–H bond is oxidized by the activated oxygen to give a hydroperoxide intermediate (**A**). Then, O–O bond cleavage in **A** generates oxygen-centered radicals (**B**) which facilitates C_{α}–C_{β} bond cleavage via single electron transfer, generating benzoic acid. $BF_3 \cdot OEt_2$ catalyzes the esterification of acid in situ with alcohol solvent to an ester and the acetalization of aldehyde, which can restrain the overoxidation of the products.

Besides Lewis acid assistance, Wang and co-workers [74, 75] further found the KOH assistant $O_2/Cu(OAc)_2$/1,10-phenanthroline system could achieve oxidative cleavage of lignin model C–C bond to acids in one-step via the C_{α} ketone intermediate (Figure 9.35). DFT calculations show that the *Ea* value of the H_{β} abstraction from the ketone intermediate by the dimer (0.21 eV) is much lower than that of the monomer (1.37 eV), which indicate a copper-oxo-bridged oxygen dimer was the active species for the C–H bond cleavage which is the rate-determining step for C–C bond cleavage. For the mechanism, in the initial step, 1-phenylethanol is fast oxidized to acetophenone. The base favors the oxidation of alcohol to carbonyl compounds via deprotonation of the O–H bond. The addition of KOH facilitates the fast oxidation of 1-phenylethanol to acetophenone. C_{α}–H bond of acetophenone is oxidized by the copper-oxo-bridged oxygen dimer to give 2-hydroxy-1-phenylethanone which is further oxidized to phenylglyoxal. Phenylglyoxal can be oxidized to benzoylformic acid by-product, which is hardly converted to benzoic acid under the reaction conditions. Acetalation of phenylglyoxal intermediate with water forms α, α-bis(hydroxy)acetophenone. After a 1,2-hydride shift, C–C bond cleavage takes

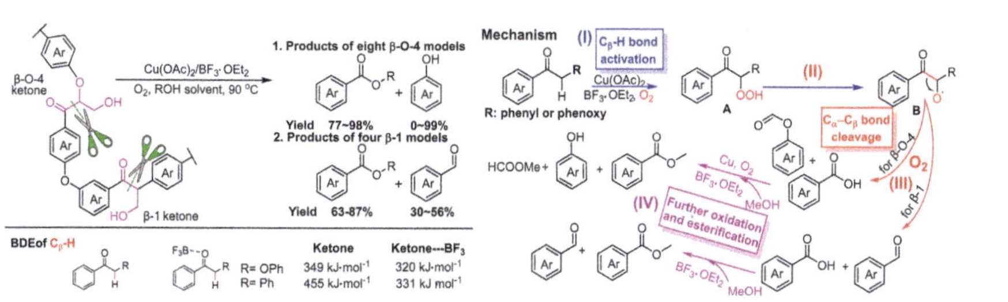

Figure 9.34 Possible Reaction Mechanism of the Copper ComplexCatalyzed Oxidative Cleavage of Oxidized Lignin Linkages. *Source:* Wang et al. [73] with Permission, © 2017 The Royal Society of Chemistry.

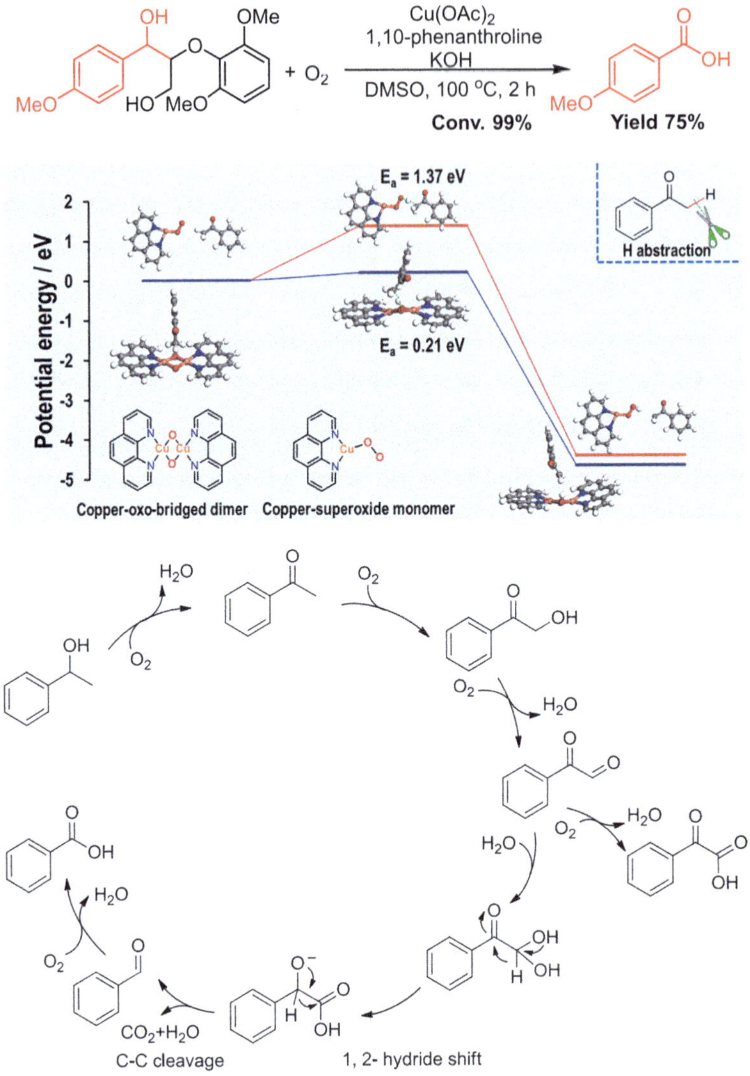

Figure 9.35 Oxidative C(OH)–C Bond Cleavage of Secondary Alcohols to Acids over a Copper Catalyst with the KOH. *Source:* Wang et al. [74] with Permission, © 2017 Elsevier.

place, leading to the formation of benzaldehyde and CO_2. Benzaldehyde is further oxidized by O_2 to benzoic acid. The oxygen of the acid product is derived from O_2.

Besides the homogeneous Cu systems for the C–C Bond cleavage via the multiple oxidations to oxygenated chemicals, the M-N-C catalyst (M=Fe, Co, Ni, Cu, etc.) prepared by the calcination of the metal salt with N-containing ligand are widely used in the selective oxidation [76–79]. In 2019, Liu and co-workers [80] reported that the single-atom Co catalyst (Co–N–C) prepared by the pyrolysis of $Mg(OH)_2$ and the $Co(phen)_2(OAc)_2$ (phen = 1,10-phenanthroline) complex could be used for the oxidative cleavage of the β-O-4 bonds of lignin model compounds at a low oxygen pressure. Under the optimized conditions (150 °C, 4 h, 1 MPa Air, methanol solvent), 95% conversion of 2-(2-methoxyphenoxy)-1-phenylethanol can be attained and the yield of guaiacol and methyl benzoate is 59 and 83%, respectively.

In 2020, Luo and co-workers [81] reported that the heterogeneous Co-NC catalysts, obtained by pyrolyzing the cobalt acetate and 1,10-phenanthroline at 700–900 °C, can efficiently catalyze the breaking of successive C–C bonds in alcohols to form esters with one or multiple carbon atoms less with dioxygen as the oxidant. According to the HAADF-STEM characterization, although the high-temperature pyrolysis caused most of the cobalt to aggregate into cobalt nanoparticles, there were still lots of atomically dispersed cobalt that could coordinate with nitrogen atoms to form Co-N_X. Further characterization, control experiment, and KSCN poisoning experiment disclose that graphene encapsulated Co nanoparticles indeed boost the reaction and are responsible for the successive cleavage of C–C bonds to achieve the excellent catalytic activity, while the presence of Co-N_X has just the opposite effect. Preliminary mechanistic studies reveal that a tandem sequence reaction (Figure 9.36). Initially, substrate **1a** is oxidized to **1b** through the reactive oxygen species produced by activation of oxygen by cobalt nanoparticles. Subsequent aerobic oxidation of **1b** generates **1c** which is further oxidized to **1d**. Then, there are two possible reaction pathways. On one hand, the generated **1d** is oxidized to **1e**, and then the cleavage of the C–C bond affords the ester **1p** with the release of formic acid followed by the decomposing to CO_2 and H_2. On the other hand, **1d** reacts with water to form the hydrated α, α-bis(hydroxy)acetophenone (**1f**). C–C bond cleavage occurs via 1,2-hydride shift, generating **1h** and CO_2. Then, nucleophilic addition

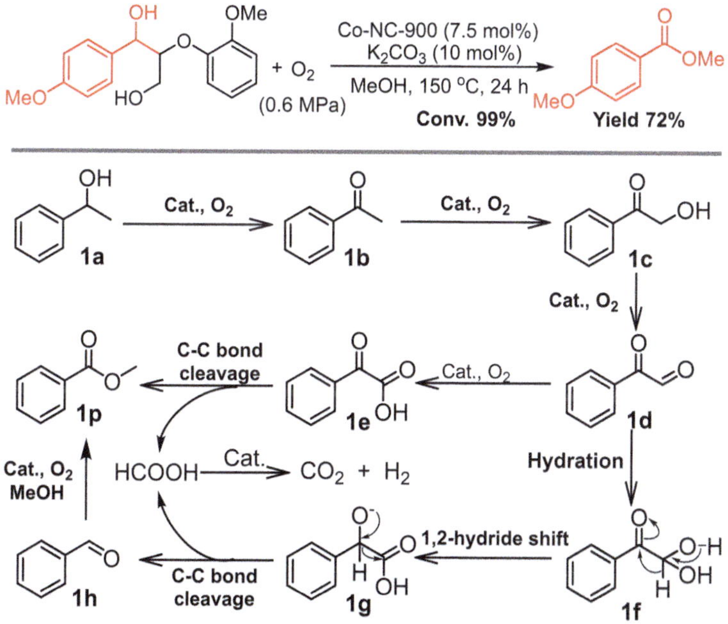

Figure 9.36 Cobalt Nanoparticles-Catalyzed Successive C–C Bond Cleavage in Alcohols to Access Esters. *Source:* Luo et al. [81] with Permission, © 2020 Wiley-VCH.

takes place between **1h** and MeOH to form the hemiacetal intermediate. Finally, the hemiacetal intermediate is transformed to **1p** catalyzed by the Co/O_2 system.

In addition, besides the model compounds conversion research, in 2021, Luo and co-workers from Stahl Group [82] demonstrates the oxidative fractionation method of lignocellulosic biomass which was conducted in the presence of a heterogeneous nonprecious-metal Co-N-C catalyst and O_2 in acetone as the solvent. The process affords a 15 wt% yield of phenolic products bearing aldehydes (vanillin, syringaldehyde) and carboxylic acids (*p*-hydroxybenzoic acid, vanillic acid, syringic acid).

Besides the metal-catalyzed lignin oxidative depolymerization to oxygenated chemicals, in 2015, Han and co-workers [83] reported that the ionic liquid (IL) 1-benzyl-3-methylimidazolium bis(trifluoromethylsulfonyl)-imide ([BnMIm] [NTf2]) with a catalytic amount of H_3PO_4 could promote the generation of the HOO• free radical and thereby efficiently transformed the β-O-4 lignin model compound 2-phenoxyacetophenone into benzoic acid and phenol using O_2 as the oxidant (Figure 9.37). According to the UV-vis spectra characterization of 2-phenoxyacetophenone in [BnMIm] [NTf2] at 353K for 12 h, the red shift of the maximum absorption from 243 nm under N_2 to 282 nm under O_2 atmosphere suggests the formation of CCTC (Contact charge transfer complex) between 2-phenoxyacetophenone and oxygen via van der Waals force [84]. Further EPR characterization with DMPO as the radical trapping agent proved the generation of HOO• free radical, which provided a typical EPR signal ($g = 2.01 \pm 0.002$). During the proposed mechanism, first, 2-phenoxyacetophenone forms CCTC with oxygen by the interaction between oxygen and the benzene ring conjugated with the carbonyl group. Then, the CCTC can be oxidized into the ROOH transition state, which is the key step in the oxidation process. After the dissociation of CCTC into an oxygen-free radical with a partial negative charge and 2-phenoxyacetophenone

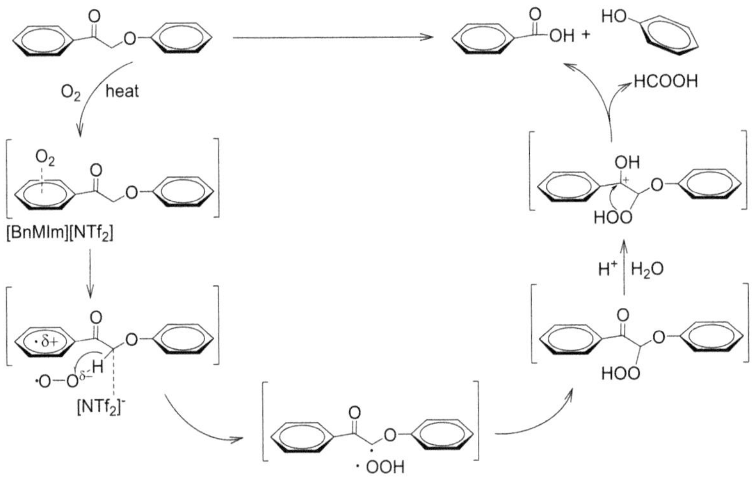

Figure 9.37 Possible Mechanism of Oxidation of Lignin Model Compound in Ionic Liquid. *Source:* Yang et al. [83] with Permission, © 2015 The Royal Society of Chemistry.

with a partial positive charge, the C–H bond adjacent to the ether-oxygen atom is activated. On one hand, the lone pair electrons of oxygen atom in oxygen-free radical can activate the hydrogen atom by forming a hydrogen bond. On the other hand, the strong electronegativity of $[NTf2]^-$ is also in favor of the delocalization of electrons between the C–H bond and thus promotes the cleavage of the C–H bond. As a result, the dissociated hydrogen atom combines with O_2^- to form $HOO^\bullet$ free radical, and the ROOH transition state is obtained. Finally, under the joint attack of H^+ and water, the intermediate was transformed, releasing formic acid and producing benzoic acid and phenol. They also studied the oxidation of organosolv lignin employing the IL-based system. After 20 hours of reaction, an orange oil fraction (75 wt%) was obtained. The products contained benzoic acid (43 wt%), benzaldehyde, and some other compounds.

Besides the oxygenated chemicals, introducing amines/ammonia into lignin catalytic oxidation can promote the C–C/C–O bond cleavage and also provide some N-containing products [85]. In 2020, Li and co-workers [86] systematically studied the catalytic oxidation methods of amines/ammonia-mediated bond cleavage in oxidized lignin β-O-4 models using a copper catalyst ($Cu(OAc)_2$) at room temperature in DMSO, which demonstrated the effect of the amine source on the selectivity of products (Figure 9.38). For primary and secondary aliphatic amines, lignin ketone models underwent oxidative C_α–C_β bond cleavage and C_α–N bond formation to generate aromatic amides. For ammonia, the competition between oxygen and ammonia determined the selectivity between C_α–N and C_β–N bond formation, generating amides and a-keto amides, respectively. For tertiary amines, the lignin models underwent oxidative C_α–C_β bond cleavage to benzoic acids. In addition, besides the aliphatic amine, in 2018, Liu and co-workers [87] reported an approach for the synthesis of benzanilides via the reaction of 2-phenoxyacetophenones with anilines catalyzed by $CuCl_2$ in DMSO at 120 °C under an air atmosphere.

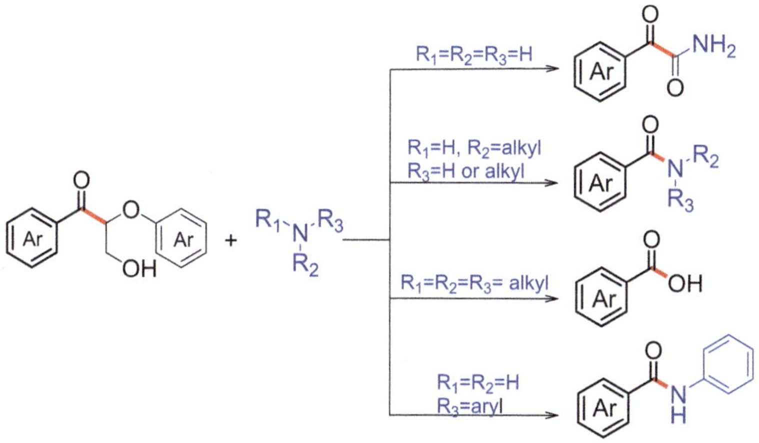

Figure 9.38 Amine-Mediated Bond Cleavage in Oxidized Lignin Models.

9.3.2 Photocatalytic and Electrocatalytic Oxidation Cleavage of C_α–C_β/C_β–OAr Bonds in the Pre-oxidized Lignin

For the further oxidation of the pre-oxidized lignin, targeting at the cleavage of the C_α–C_β or C_β–OAr bonds, besides the thermal methods shown above, the photocatalytic and electrocatalytic oxidation methods for C_β–OAr bond cleavage have also drawn much attention in lignin utilization.

In 2019, Kang and co-workers [88] reported a photochemical method for the oxidative cleavage of C_α–C_β bond in pre-oxidized lignin β-O-4 model by the sequent participation of two functional ionic liquids (ILs) (Figure 9.39). The bond rupture initiates with C_β–H bond activation of pre-oxidized β-O-4 ketone (**1**) to deliver C_β radical (**3**) with the aid of [Pmin][NTf2] IL (1-propenyl-3-methylimidazolium bis [(trifluoromethyl)sulfonyl] imide) under UV light irradiation. The strong electronegativity of this IL is vital to this initiation step. Subsequently, the C_β radical (**3**) reacts with atmospheric oxygen to form the peroxy radicals (**4**), which further abstracts protons from water to form hydroperoxide intermediate (**5**) and hydroxyl radical. The hydroperoxide (**5**) is eventually transformed into benzoic acid and phenol via C–C bond scission (**6**) catalyzed by an acidic IL, [PrSO$_3$HMim][OTf] (1-propylsulfonic-3-methylimidazolium trifluoromethanesulfonate).

During the multiple oxidation, besides the cleavage of lignin C_α–C_β bond, the C_β–OAr can occur cleavage. In 2017, Gao and co-workers [89] studied the selective electrochemical C_β–OAr bond oxidative cleavage of β-O-4 lignin models mediated by iodide ion. As illustrated in Figure 9.40, the chemistry starts with the anodic oxidation of iodide ion to form its oxidation state, I^+; meanwhile, the solvent

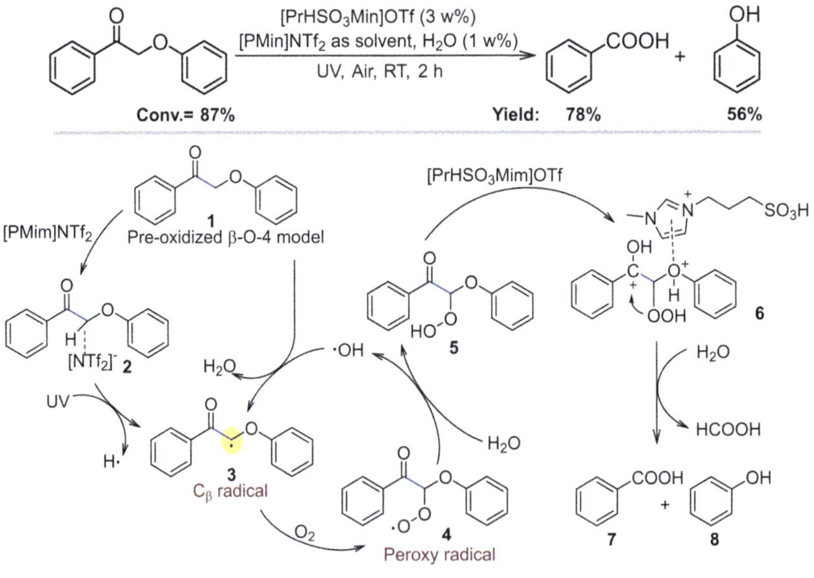

Figure 9.39 Dual ILs-promoted Photochemical Degradation of Pre-oxidized Lignin β-O-4 Model. *Source:* Kang et al. [88] with Permission, © 2019 Wiley-VCH.

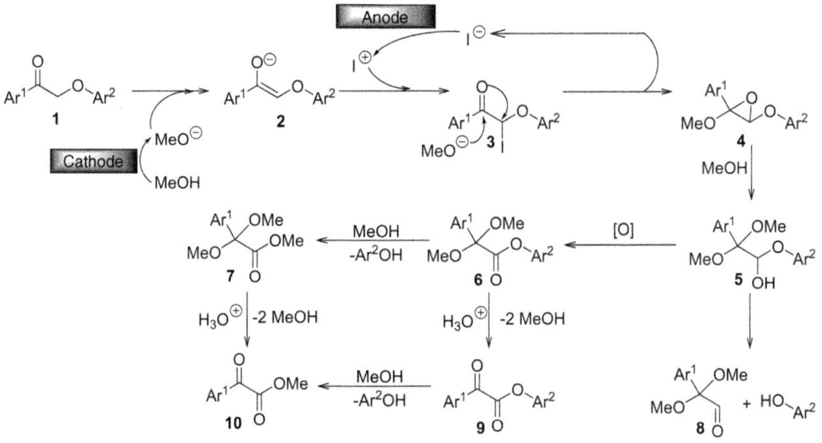

Figure 9.40 Selective Electrochemical C–O Bond Oxidative Cleavage of β-O-4 Lignin Models Mediated by Iodide Ion. *Source:* Gao et al. [89] with Permission, © 2017 Elsevier.

methanol is reduced at the cathode to generate methoxide anion, MeO^-. The methoxide serves as a base to deprotonate the starting 2-aryloxyarylethanone **1** followed by the subsequent reaction with electrochemically in situ generated I^+ to generate β-iodoketone **3**. The addition of the methoxide anion to the carbonyl group of α-iodoketone (C_α=O), followed by an intramolecular cyclization leads to an epoxide intermediate **4**. Subsequent ring-opening of the epoxide intermediate induced by the second equivalent of methoxide anion results in the formation of α-hydroxy ketal (**5** hemiacetal at C_β). Once the key hemiacetal is generated, it undergoes hydrolysis with C_β–OAr bond cleavage, removing the Ar_2OH unit to form α, α-dimethoxy aryl acetoaldehyde **8**. On the other hand, oxidation of hemiacetal to corresponding ester intermediate **6**, followed by trans-esterification and hydrolysis affords product **10**; alternatively, ester **6** first undergoes hydrolysis to give intermediate **9**, followed by a trans-esterification to generate methyl α-oxo arylacetate **10**.

9.4 Lignin C_α–C_β, C_{Ar}–C_α, or C_βO–C_{Ar} Bonds Cleavage by Inserting an O- or N-containing Fragment after the First Oxidation of C_α–OH to C_α=O

As shown in Figure 9.41, the lignin C_α–C_β or C_{Ar}–C_α bond can be cleaved by wedging an O or N atom. In this 'Wedging' mode, the C_α–OH was first oxidized to C=O bond, then the generated ketone intermediate reacted with peroxide to ester via the Baeyer-Villiger oxidation [90–94]. or reacted with hydroxylamine to amide via the Backman rearrangement [95, 96]. The C–O and C–N bonds in the generated ester and amide could be further cleaved by the subsequent hydrolysis. The corresponding strategies and their derivative methods for lignin depolymerization are summarized in this section.

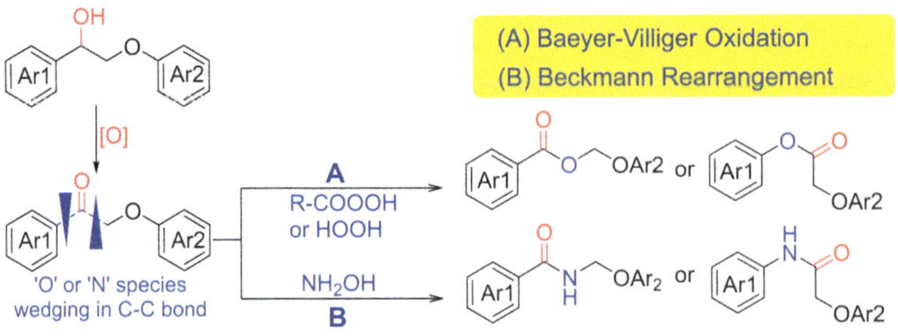

Figure 9.41 The β-O-4 Models Transformation via a 'Wedging' Mode.

9.4.1 Baeyer-Villiger Oxidation Methods

Besides the degradation of lignin model compounds with peroxymonophosphoric acid [91] and other organic peroxides, in 2005 and 2006, Crestini and co-workers [97–99] found that methyltrioxorhenium ($MeReO_3$, MTO) was an efficient catalyst for the oxidation of lignin and lignin model compounds with H_2O_2, and some monomer products are suggested coming from the Baeyer-Villiger Oxidation. Then, in 2013, Stahl and co-workers [100] identified an efficient organocatalytic method for chemoselective aerobic oxidation of secondary benzylic alcohol (C_α–OH) within lignin β-O-4 model compounds and natural lignins. The optimal catalyst system consisted of 4-acetamido-TEMPO in combination with HNO_3 and HCl. Their studies also highlighted the prospect of combining this method with subsequent Baeyer-Villiger oxidation and alkaline hydrolysis or Darkin-like reaction to achieve C_α–C_β bond cleavage (Figure 9.42). Then various two-step BV oxidation strategies were founded for the cleavage of lignin C–C bonds.

In 2015, Meier and Crocker [101] observed the activity of several catalytic systems on the selective oxidation of benzylic hydroxyl groups, followed by Baeyer–Villiger

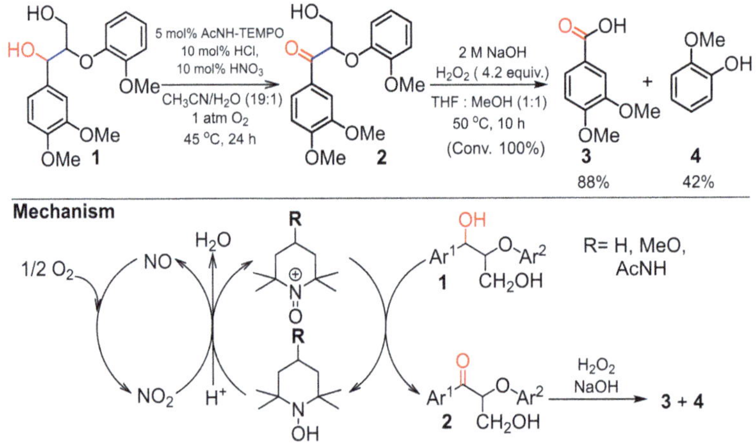

Figure 9.42 Chemoselective Aerobic Alcohol Oxidation in Lignin and C_α–C_β Bond Cleavage.

oxidation of the resulting ketones with H_2O_2 and other peroxy acids. In 2017, Crocker [102] achieved the stepwise depolymerization of lignin involves oxidation of the benzylic alcohol group in β-O-4 and β-1 linkages with tin beta zeolite catalyst and hydrogen peroxide, followed by Baeyer–Villiger oxidation of the resulting ketones and subsequent ester hydrolysis. In the same year, Wang and co-workers [103] modified the Baeyer–Villiger oxidation for the C–C bond cleavage of lignin model compounds. The inert C_α–C_β or C_α–Cphenyl bond of oxidized lignin model compounds was first converted to an active ester bond, and thus acetal esters and aryl esters were produced in high yields (up to 99%) at room temperature. Next, K_2CO_3 catalyzed the alcoholysis of the resulting ester products at 45 °C, affording various useful chemical platforms in excellent yields (up to 99%), such as phenols and multifunctional esters. In 2018, Andrade and co-workers [104] achieved the Baeyer-Villiger oxidation of β-O-4 lignin models with a continuous flow reactor. The supported Au/SiO_2 was first used as the catalyst to achieve the oxidation of C_α–OH to C=O with H_2O_2, the generated ketones were further transformed to esters over a phenylseleninic acid resin (PAR) catalyst. In 2020, Hao and co-workers [105] reported that the Mo-V-based bifunctional catalysts can achieve the catalytic conversion of lignin dimer model compounds at room temperature with H_2O_2, which contains the first C_α–OH to ketone and the following Baeyer–Villiger oxidation for the C–C bond cleavage.

9.4.2 Beckmann Rearrangement Methods and Its Derivative Methods

Turning to the N atom inserting, in 2018, Wang and co-workers [95] achieved the highly effective cleavage of C–C bonds in pre-oxidized β-O-4 lignin model compounds with C_α=O for the production of N-substituted aromatics, including benzonitriles and amides, via oxime formation followed by Beckmann Rearrangement (BR) (Figure 9.43). In this strategy, $NH_2OH \cdot HCl$ reacts with the carbonyl group of the oxidized lignin model compound to form a mixture of Z/E ligninoxime. In the

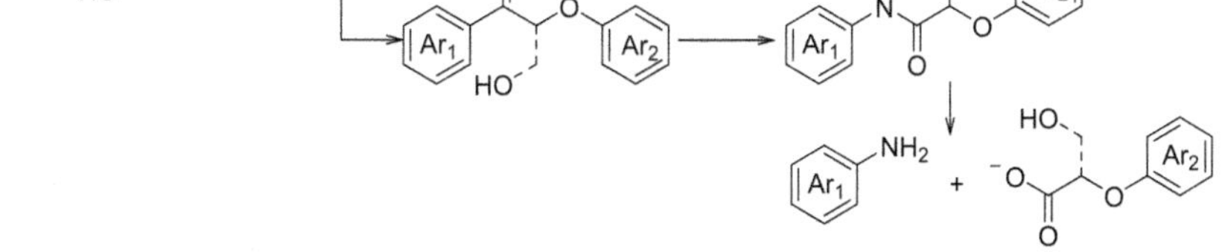

Figure 9.43 Transformation of Lignin Model Compounds to N-substituted Aromatics via Beckmann Rearrangement. *Source:* Wang et al. [95] with Permission, © 2018 The Royal Society of Chemistry.

presence of an acid catalyst, the Z-ligninoxime with or without γ-OH rearranged to generate an amide product via the BR reaction, which could be hydrolyzed to anilines and carboxylic acids. For E-ligninoxime without γ-OH, nitriles and phenols were obtained; while for E-ligninoxime with γ-OH, C-2 substituted oxazole was also available, except for nitriles and phenols.

Little earlier than Wang's report [95], in 2018, Li and co-workers [96] tried to introduce NH_2OH to achieve the depolymerization of oxidized lignin to N-containing monomers (Figure 9.44). While, different from the target amides as the products [95], they reported the first protocol that converts lignin models and pre-oxidized lignin to isoxazole and aromatic nitrile [96]. Catalyst screening indicated magnesium oxide could promote an oximation reaction and an intramolecular condensation. DFT calculations showed that the formation of oxime reduced the C_β–O bond dissociation energy of the β-O-4 model by 14.2 $kJ \cdot mol^{-1}$. Therefore, the pathway via oximation facilitates the breakage of the β-O-4 linkage to N-containing monomers. The isoxazole motif was constructed by condensation of β-hydroxyl ketone, containing a $C_\gamma H_2OH$, with hydroxylamine. Aromatic nitriles and esters are obtained via Beckmann rearrangement or an acidolysis reaction depending on the selected additive. The product difference was caused by the cis-trans isomerism of oximation intermediates. Furthermore, the hydroxylamine-mediated strategy can also work well for the pre-oxidized lignin conversion to aromatic isoxazole, nitrile, and ester monomers with up to a 7.6% yield.

Besides cleaving aliphatic lignin C_β–OAr bond by introducing NH_2OH at C_α=O to react with a base-catalyzed epoxide fragment and form isoxazole ring, in 2019, Li

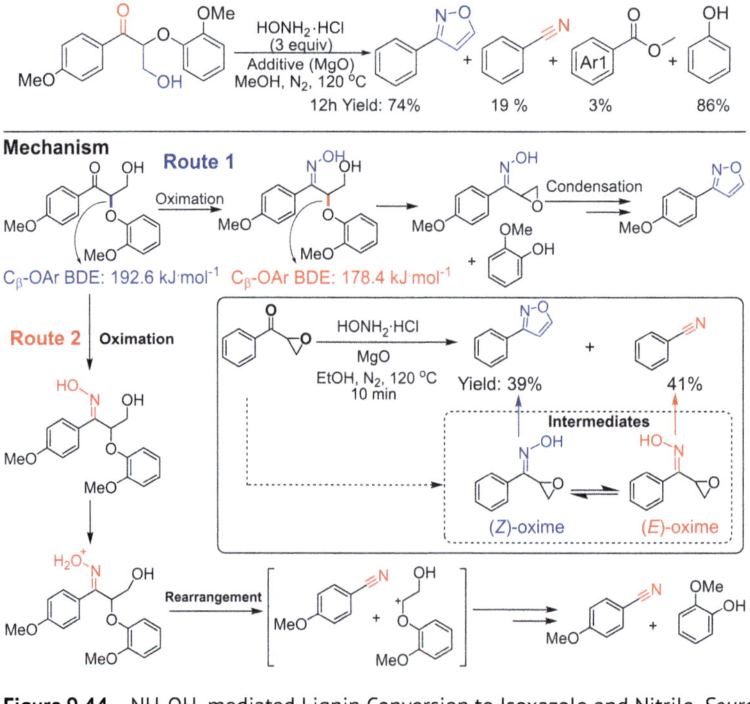

Figure 9.44 NH_2OH-mediated Lignin Conversion to Isoxazole and Nitrile. *Source:* Li et al. [96] with Permission, © 2018 American Chemical Society.

and co-workers [106] further found introducing an iminyl radical at pre-oxidized lignin C_α could directly cleave the aromatic lignin C_βO–Ar bond (Figure 9.45). In detail, they firstly modified the α-alcohol β-aryl ether structure to the α-oximated β-aryl ether structure in lignin. Then, the phenothiazine (PTH), working as the photocatalyst, could catalyze the decomposition of the α-oximated β-aryl ethers into iminyl radicals. The iminyl radical then induced the aryl migration from O atom to N atom via spiro cyclization. DFT calculations suggested that *ortho*-methoxy could facilitate the spiro cyclization due to its steric effect. After aryl migration, further imine intermediate hydrolysis and O-radical reduction released the primary arylamine and α-hydroxy ketone as the non-phenolic aromatics. The application of this method on N-modified lignin provides a 1.9% yield of primary arylamine monomers and a 2.7% yield of phenol monomers. The moderate yields of desired arylamine monomers could be ascribed to the low content of α-oximated β-aryl ethers after N-modification (21 per 100 aromatic units) and more restricted polymerized structures in real lignin than free dimer models.

Similar to the cleavage mechanism of C_βO–Ar bond that provides N-Ar products from β-O-4 linkages, before Li's report [106], In 2017, Nicewicz and co-workers [107] described a cation radical accelerated nucleophilic aromatic substitution using methoxy- and benzyloxy-groups as nucleofuges, which can be also used to cleave the aryl C–O bond in pre-oxidized lignin β-O-4 model to deliver N-contained

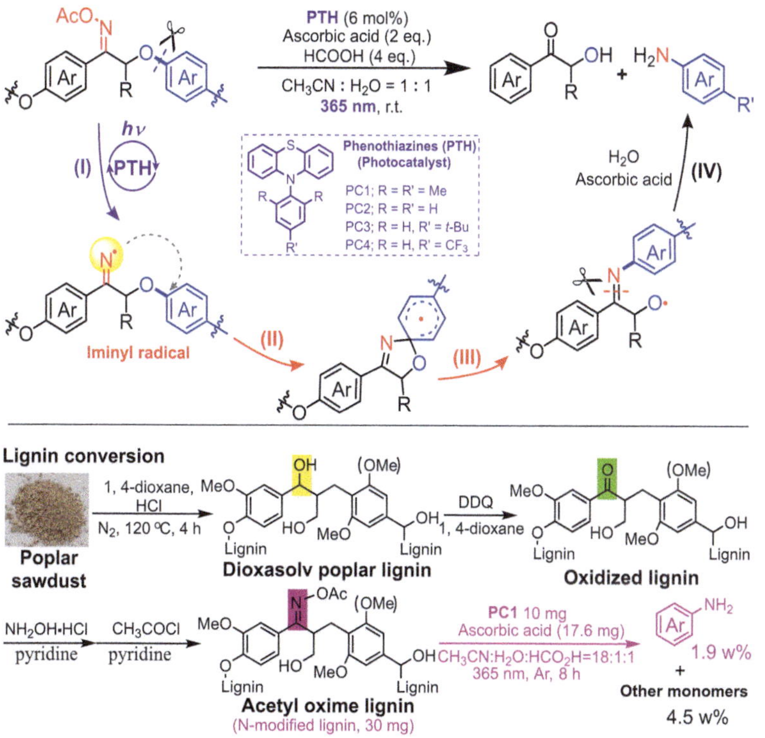

Figure 9.45 Introducing N-radical at C_α Mediated Lignin Aryl Ether Cleavage to Non-phenolic Aromatics. *Source:* Li et al. [106] with Permission, © 2019 American Chemical Society.

aromatics (Figure 9.46). Upon irradiation, the electron-rich arene moiety is first oxidized into an open-shell radical cation by the excited acridinium photocatalyst (Mes-Acr^{+*}, E^*_{red} = +2.15 V vs SCE), which is pivotal to this reaction. Subsequently, the nucleophilic attack on the carbon connecting with the OR structure by imidazole species leads to the ether bond cleavage under redox-neutral conditions, during which the radical intermediate can also oxidize the Mes-Acr$^{\bullet}$ to the ground Mes-Acr$^+$ for the next catalytic cycle.

Like Beckmann Rearrangement methods that can cleave the lignin C–C/C–O bonds by introducing the cleavage reagent at the C_α, in 2019, the Jiao group [108] reported a site-directed C–C bond primary amination of alkylarenes or benzyl alcohols for the direct preparation of anilines (Figure 9.47). The chemistry involves a novel C–C bond transformation and offers a versatile protocol for the synthesis of substituted anilines. As shown in Figure 9.47, the mechanism involves: (1) the first oxidation at benzylic CH_2 with DDQ or O_2 or the dehydroxylation of benzylic alcohol without any oxidant offer a carbocation intermediate; (2) the carbocation intermediate then further reacts with the nucleophile NaN_3 to the -N_3 substituted benzylic intermediate; (3) with the assistance of the H^+, the active N_3-containing intermediate occurs molecule arrangement which releases an N_2 and inserts an imide N atom into the Ar-alkyl bond; and (4) the hydrolysis of imide cation offers the aniline and the corresponding carbonyl compound. Then, they reported the present chemistry could be applied in the cleavage of the β-1 and β-O-4 lignin model compounds. As described in Figure 9.47, β-1 lignin model compounds could be

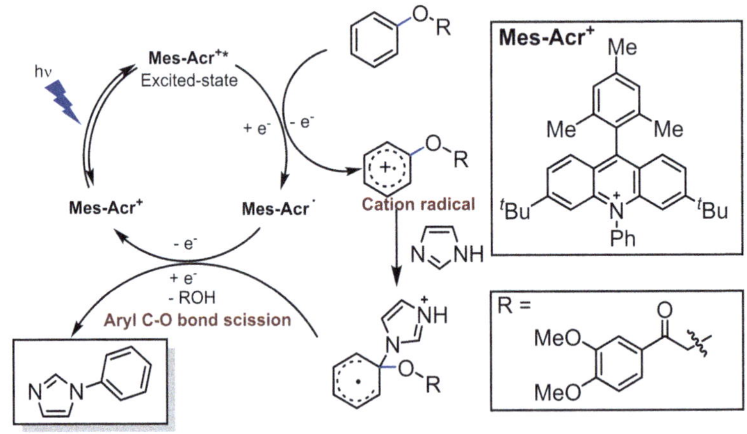

Figure 9.46 Nucleophilic Aromatic Substitution for the Cleavage of Aryl C–O Bond in Pre-oxidized Lignin Model. *Source:* Tay and Nicewicz [107] with Permission, © 2017 American Chemical Society.

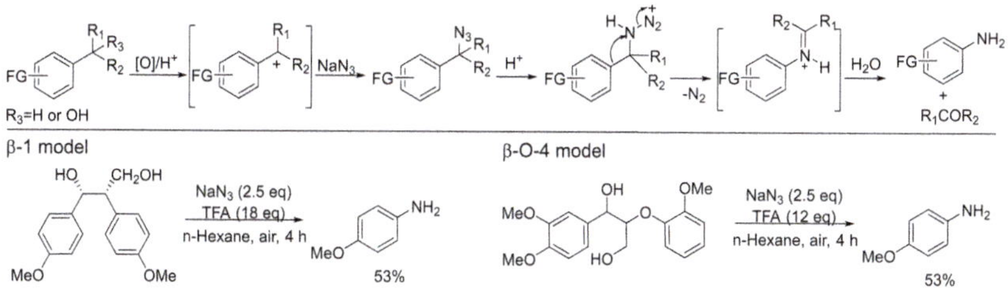

Figure 9.47 From Alkylarenes to Anilines via Site-directed Carbon-carbon Amination. *Source:* Liu et al. [108] with Permission, © 2019 Springer Nature.

converted into the corresponding 4-methoxyaniline in a 53% yield, and the β-O-4 lignin model could be converted into 3,4-dimethoxyaniline in a 40% yield.

Besides inserting an N atom between the two carbon atoms of C–C bond to cleave the lignin structure, given the success of the Baeyer-Villiger oxidation, inserting a CO in the ether structure to provide an ester structure that can occur hydrolysis cleavage could be another amazing process. In 2017, the Han group [109] proposed a strategy for the transformation of the methoxy group (-OCH_3) connecting with the aromatic rings to acetic acid. In their catalytic system, the methoxy group in lignin could react with CO and water to generate acetic acid over $RhCl_3$ in the presence of a promoter LiI. As shown in Figure 9.48, $LiBF_4$ catalyzed the reaction of methoxy

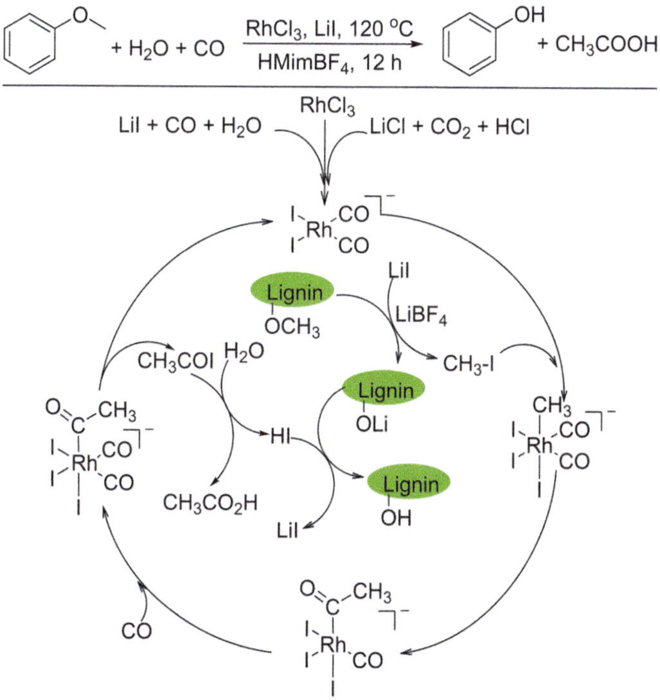

Figure 9.48 Selective Utilization of the Methoxy Group in Lignin to Produce Acetic Acid. *Source:* Mei et al. [109] with Permission, © 2017 Wiley-VCH.

group and LiI to form CH_3I and lithium phenolate. The sequence involves the oxidative addition of CH_3I to Rh species, CO insertion, then reductive elimination of acetyl iodide, and its subsequent hydrolysis to acetic acid and HI. Finally, HI reacts with lithium phenolate and regenerates LiI. In this system, iodide is a necessary component for the carbonylation of methanol to acetic acid, and iodide serves as a promoter. LiI has two roles in this process. One role is breaking the ether bond and the other is promoting the carbonylation of methyl iodide to acetic acid. Then they used this method in the transformation of lignin. Further experiments showed that the conversions of the methoxy group in the kraft lignin and organosolv lignin reached 87.5% and 80.4%, respectively, and no by-product was generated. In addition, in 2019, the Han Group [110] further reported that LiI in the ionic liquid 1-hexyl-3-methylimidazolium tetrafluoroborate could efficiently catalyze the reaction between lignin and aniline to provide the N-methylation products. Although the lignin was not efficiently depolymerized in the reaction but rich in hydroxyl groups, which was suitable for the preparation of functional materials. In addition, although the Ru-catalyzed real reaction mechanism does not involve the direct CO inserting into the ether bond, we think the Wedging Mode (discussed in Chapter 5) with a molecule or fragment inserting into the lignin linkage bonds is still efficient for the design of lignin depolymerization strategy.

9.5 Embellishing Lignin β-O-4 Linkages Hydrolysis Involving the C_α–OH First Oxidation and C_γ–OH Transformation

As discussed above, the first activation of -C_αHOH to ketones can make a weaker C_β–OArr bond, or generate the "substrate sites" for introducing cleavage reagents to cleave lignin C_α–C_β/C_β–OAr bonds. After the two-step oxidation-BV oxidation method reported by the Stahl group [100], the instability of the guaiacol under the reaction conditions prompted them to consider reductive cleavage methods (Figure 9.49) [111]. In 2014, the reactivity of oxidized β-O-4 model compound **1** was tested in the presence of different reducing metals, including zinc, aluminum, magnesium, iron, and manganese, in aqueous formic acid at 110 °C (Figure 9.49a, entries 1-5). Four different major products were identified from these reaction conditions. In the presence of zinc, the depolymerization afforded small amounts of the O-formylated product 3 (6%), together with good yields of the aryl ethyl ketone **3** (76%) and guaiacol (69%) (entry 1). When the reaction was performed with metal sources other than Zn (entries 2–5), the diketone product **4** was observed instead of **3**. Good yields of **4** (74%) and guaiacol (63%) were obtained from the reaction with manganese. Given the fact that the conversion of **1** into the diketone product **4** and guaiacol is a redox-neutral process. They therefore tested the reaction in the absence of a reducing metal, and products **4** and **5** were obtained in yields comparable to the best results obtained with a stoichiometric metal source (Figure 9.49a, entry 6). Even better yields of **4** and **5** resulted (96% and 87%, respectively) when 3 equivalents of sodium

(a)

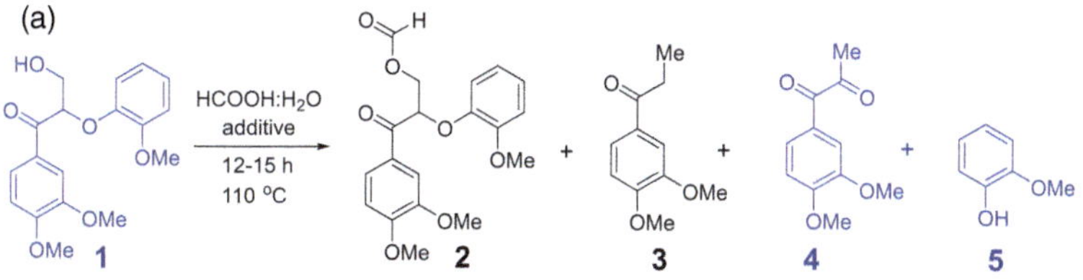

C_β-OAr cleavage of lignin model compounds with formic acid

Entry	M (2 equiv.)	Additive	2 (%)	3 : 4 (%)	5 (%)
1	Zn	---	6	76:0	69
2	Al	---	49	0:39	36
3	Mg	---	51	0:45	24
4	Fe	---	49	0:31	24
5	Mn	---	19	0:74	63
6	---	---	18	0:77	64
7	---	HCOONa (3 equiv)	0	0:96	87

(b)

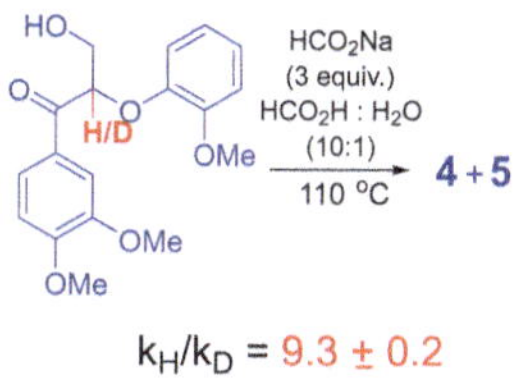

(c)

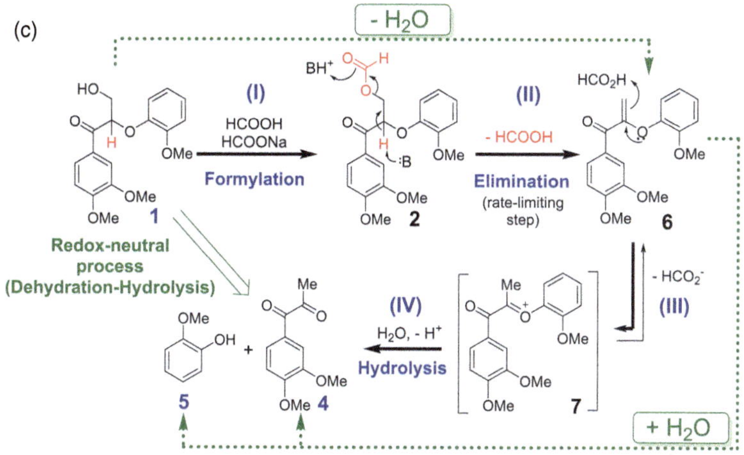

Figure 9.49 C_β–OAr Cleavage of Oxidized Lignin Model Compounds with Formic Acid/Formate (a); Kinetic Isotope Affect for Cleavage of the Oxidized $_\beta$-O-4 Model Compound 1 with the Reaction Condition of Entry 7 (b); Proposed Redox-neutral Mechanism for C_β–OAr Cleavage (c). *Source:* Rahimi et al. [111] with Permission, © 2014 Springer Nature.

formate were included in the reaction mixture (Figure 9.49a, entry 7). These conditions proved to be effective with several other oxidized lignin model compounds including those derived from *p*-hydroxyphenyl (H)-, guaiacyl (G)- and syringyl (S)-type lignin units. For the redox-neutral conversion mechanism, the C_α ketone was crucial to the success of the reaction, because no cleavage products were observed when the non-oxidized model compound with a C_αHOH was subjected to the formic acid/formate reaction conditions.

Mechanistic insights into the depolymerization process were obtained from additional studies of model compound **1**. Reaction time-course data, obtained by ^{1}HNMR spectroscopy, revealed that oxidized lignin model **1** converts quickly into the alcohol formylation product **2**, followed by a slower conversion of **2** into products **4** and **5** (Figure 9.49c). Formylation of **1** into **2** is rapid, even at room temperature. Direct investigation of the reaction of **2** revealed a small steady-state concentration of the aryl vinyl ether intermediate **6** during the formation of diketone product **4**. Independent testing of **6** confirmed that this intermediate is consumed more rapidly than the overall conversion of **2** into **4**. A large deuterium kinetic isotope effect, observed when the independent reaction rates of **1** and **1-d1** with a C_βD were compared (k_H/k_D=9.3±0.2; (Figure 9.49b), indicates that the elimination of formic acid from **2** is the rate-limiting step in the transformation. This kinetic isotope effect is larger than the semiclassical limit associated with the cleavage of a C–H bond, but it aligns with a previous study providing evidence for proton tunneling in concerted E2 elimination reactions [112].

According to the mechanism in Figure 9.49c, the beneficial effect of lignin oxidation may be attributed to the ability of the benzylic carbonyl group to polarize the C–H bond and lower the barrier for the rate-limiting E2 elimination reaction. Both the increased C–H acidity and orbital overlap between the carbonyl C=O π system and the developing π bond of the alkene are expected to contribute to a lower barrier for this step. The mechanism also accounts for the beneficial effect of using a buffered reaction medium, containing both formate and formic acid (Figure 9.49a, entries 6 and 7). The rate-limiting elimination step III is proposed to involve both a base (formate) to remove the proton and an acid (formic acid) to assist in the loss of the formate as a leaving group (Figure 9.49c). Overall, the sequence is a redox-neutral process that results in no net consumption of formic acid. This feature distinguishes the present approach from other lignin conversion methods that employ formic acid as a source of H_2 in transfer hydrogenation/hydrogenolysis reactions with heterogeneous catalysts.

Although the author did not mention the reaction type, from the analysis of the overall reaction process, this process could be regarded as a new type of dehydration-hydrolysis process or esterification-elimination –hydrolysis process. Different from the traditional acidolysis beginning with the dehydration at C_α–OH and following hydrolysis with a C_α=C_β–OAr center, the dehydration position in preoxidized β-O-4 is at the C_γ–OH, and the hydrolysis center becomes the vinyl ether of C_γ=C_β–OAr (Figure 9.49c, **6→7→4+5**). The C_α oxidation here should have two effects in promoting lignin depolymerization: (1) the C_α oxidation can reduce the BDE of the C_β–OAr bond; promoting the lignin linkage cleavage; (2) The vinyl ether center change can prevent the generation of C_α cation, restraining the potential lignin fragments condensation (more discussion is provided in Chapter 11.1). When the preoxidation-(formic acid-induced)-hydrolysis is used in the lignin transformation, the yield of monomeric aromatic compounds varies depending on the lignin plant source and pretreatment method [111, 113]. The best results are obtained from poplar lignin isolated via an acidolysis pretreatment method, which gives a 42 wt% yield of low-molecular-weight aromatics. Use of other pretreatment methods and/or use of maple and maize lignins afford yields of aromatics ranging from 3 to 31 wt%.

For the energetic and geometric details of the proposed unusual elimination step, in 2015, Wang and co-workers [114] presented a further DFT mechanistic study. They observed that the remarkable transformation adopted an unconventional elimination mechanism (termed as E1H-3c4e), which was different from the known ones such as E2 and E1cb. The E1H-3c4e mechanism passes through an intermediate stabilized by proton-shared 3c–4e bond ($HCOO^{\ominus}\cdots H^{\oplus}\cdots {}^{\ominus}O{=}C_\alpha$) and by shifting the 3c–4e $HCOO^{\ominus}\cdots H\oplus\cdots {}^{\ominus}O{=}C_\alpha$ bond to the 3c–4e ($HCOO^{\ominus}\cdots H\oplus\cdots {}^{\ominus}OOCH$) bond in the joint leaving group that is originally a regular H-bond (HCOO–H···OOC(H)–). The root cause of the E1H-3c4e elimination is that the poor leaving group (formate) is less competitive in stabilizing the negative charge resulting from H_β abstraction by $HCOO^-$ than the nearby carbonyl group (C_α=O) that can utilize the negative charge to form a stabilizing 3c–4e bond with a formic acid molecule. These results indicate the importance of preoxidation and support the E1H-3c4e mechanism for the elimination in the β-O-4 ketone case. In addition, they further studied the effect of HCO_2H/HCO_2Na medium in the whole transformation of the oxidized β-O-4 linkage model, one can check the article [114] for more detailed information.

After the Stahl group's work, in 2016, Chan and co-workers [115] successfully carried out depolymerization of commercial kraft lignin with the aid of formic acid and sodium formate to generate a mixture of monomeric and dimeric lignin, which was further chemically modified by alkylation and subsequent esterification to afford high-quality lignin oil. Similar strategies were also adopted by Gasser [116] in 2017, who reported a sequential lignin depolymerization process by the combination of biocatalytic and formic acid treatment steps, yielding 45% of soluble products which were mainly low molecular weight aromatic monomers and oligomers. In addition, besides using formic acid as the catalyst, in 2017, Patti and co-workers [117] reported a novel degradation method for preoxidized lignin using a formate ionic liquid. The β-O-4 bonds in lignin were oxidized by a TEMPO/$Cu(OAc)_2$ system and then depolymerized by a recyclable ionic liquid, 2-hydroxy ethylammonium formate prepared from formic acid and 2-aminoethanol. 2-Hydroxyethylammonium formate, which could dissolve lignin acting as a solvent, was found to also catalyze the depolymerization of the oxidized lignin. Besides the promotion effect of formic acid in the pre-oxidized lignin depolymerization, in 2017, Oregui-Bengoechea and co-workers [118] studied the role of formic acid together with the effect of the solvent type and their synergic interactions with a NiMo catalyst for the conversion of lignin into bio-oil in an alcohol/formic acid media. The replacement of formic acid with H_2 or isopropanol decreased the oil yield to a considerable degree, increased the solid yield. Control experiments confirmed that formic acid not only acts as an in situ hydrogen source but also reacts with lignin C_γ–OH through the formylation–elimination–hydrogenolysis mechanism reported by Stahl [111] that leads to the depolymerization of the lignn polymer with C_β–OAr bond cleavage.

9.6 Lignin C_β–OAr Bond Cleavage after the First Activation of C_α–H, C_α–OH, or C_αO–H to $C_\alpha^\bullet$ Radical

As shown in Figure 9.50, to cleave the β-O-4 C_β–OAr bond, besides the first transformation of the C_α–OH to C_α=O, the methods with the first generation of C_α radical

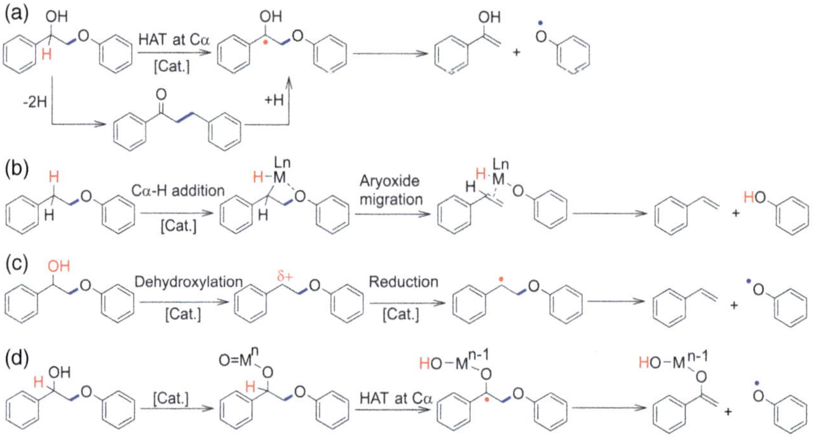

Figure 9.50 C_β–OAr Bond Cleavage after the First Activation of C_α–OH(–H) to C_α Radical.

can also achieve this goal by obviously reducing the BDE of the C_β–OAr bond [119–122]. For the generation C_α radical from lignin β-O-4 linkage, the methods could be divided into the following kinds: (1) The selective HAT oxidation process, consisting the direct hydrogen atom abstraction from C_α–H and following oxidation of the cleavage fragments (Figure 9.50a); (2) The (dehydroxylation)-hydrogenation process, containing C_α–OH dehydroxylation and following C_α cation reduction to the C_α radical and following cleavage fragments hydrogenation (Figure 9.50c); (3) Transfer hydrogenation process with the C_α radical intermediate, beginning with first hydrogen atom abstraction from C_α–H or an active metal insertion to provide the C_α–M–H intermediate (Figure 9.50a and b); (4) The redox-neutral process with first C_αO–H activation by M=O species, inducing the further HAT at C_αH–OH to the C_α radical and following C_β–OAr bond cleavage (Figure 9.50d). In addition, as another choice, the ketone after the first dehydrogenation or oxidation of the C_α–OH bond may occur semi-hydrogenation to provide a C_α radical (Figure 9.50a, bottom). While, by now, this mechanism has not been reported. This subsection will summarize the lignin C_β–OAr bond cleavage strategies which contain the first activation of C_α–H, C_α–OH, or C_αO–H to the C_α• radical.

9.6.1 The Selective HAT Oxidation Process

For the C_α–H HAT oxidation, the laccase/mediator systems (LMS) are widely studied [123, 124]. Past studies have shown that laccase can catalyze the C_α–C_β cleavage, C_α-oxidation, β-ether cleavage, and aromatic ring cleavage of a nonphenolic β-O-4 lignin model dimer, in the presence of 1-hydroxybenzotriazole (1-HBT) [125, 126]. In 2002, Kawai and co-workers [127] investigated the detailed degradation mechanism for these pathways of substrate 1,3-dihydroxy-2-(2,6-dimethoxyphenoxy)-1-(4-ethoxy-3-methoxyphenyl)propan (β-O-4-I) by the laccase/1-HBT couple with H_2O_2 and O_2, based on incorporation experiments from $H_2{}^{18}O$ and ${}^{18}O_2$. As shown in Figure 9.51, the laccase/1-HBT can cause the first SET oxidation of the aromatic ring to an aryl cation radical, which could be further attacked by the

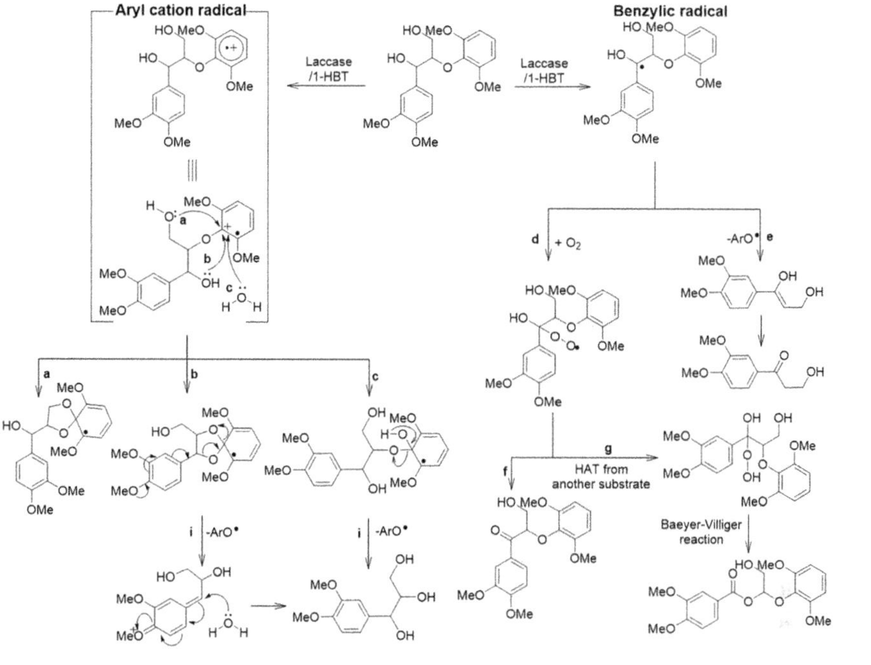

Figure 9.51 Possible Radical Intermediates for the Degradation Reactions of β-O-4 linkage by Laccase/1-Hydroxybenzotriazole (1-HBT) Couple. *Source:* Kawai et al. [127] with Permission, © 2002 Elsevier.

hydroxyl groups on its adjacent propyl side chain (pathways a, b) or with water (pathway c), and the resulting intermediates degrade via β-ether and aromatic ring cleavages. Otherwise, the laccase/1-HBT system can cause the first HAT of the C_α–H to the active C_α radical. The C_α can work as the substrate site to combine with the O_2 to the peroxide intermediate which can further release a HOO• species and provide the C_α=O ketone (pathway f) without the cleavage of the C–C/C–O bonds. Or the benzylic peroxide radical can cause the cleavage of C–C via a kind of Baeyer-Villiger reaction to give the benzoic acid (pathway g). Meanwhile, the C_α radical intermediate can release a phenol radical and induce the cleavage of the C_β–OAr bond (pathway e), followed by the further oxidation of the fragments.

As shown above, the enzymatic process can provide various radicals, and these radicals react further (nonenzymatically) via several routes, depending on the structure of the substrate. Studies on LMS treatments of nonphenolic β-O-4' linked lignin model compounds mainly report C_α-oxidation and bond cleavage as outcomes [126, 128–131] and it has been suggested that their formation occurs via competing routes [124]. Typically, C_α-oxidation is the major reaction outcome, whereas cleavage products are only formed in minor amounts [129–131]. Therefore, to increase the efficiency of LMS-catalyzed lignin degradation, two strategies could be followed: (1) maximizing the total extent of oxidation by increasing the catalytic performance of the LMS; or (2) steering the nonenzymatic follow-up reactions toward bond cleavage, such as the first C_α-oxidation to the active C_α radical [132–136].

In 2020, Kabel and co-workers [137] reported a modified laccase/mediator system (LMS) for oxidation cleavage of the lignin β-O-4 structure. They studied the influence of buffer properties on the product profile of a β-O-4 linked lignin model dimer upon conversion by a laccase/hydroxybenzotriazole (HBT) system, which targeted the cleavage of the C_β–OAr bond not the traditional C_α oxidation. They showed that the ratio between β-O-4 ether cleavage and C_α-oxidation can be substantially increased by using unconventional buffer properties (i.e., highly concentrated buffers at near-neutral pH). Whereas <10% ether cleavage was obtained in conventional buffer (i.e., weak buffer at pH 4), as much as 80% ether cleavage was obtained in highly concentrated buffers at pH 6. Combining with the DFT calculation results, they proposed that increasing buffer pH or increasing buffer strength enhances H-bonding between the lignin model and buffer anions, which drives the overall reaction outcome toward ether cleavage (Figure 9.52).

9.6.2 The Dehydroxylation-hydrogenation Process

In 2013, Abu-Omar and co-workers [138] described a bimetallic catalyst system, based on metallic Pd nanoparticles (NP) on carbon with Zn(II) Lewis acid ($ZnCl_2$), can selectively cleave C–O bonds in a variety of monomeric, dimeric, and polymeric lignin molecules under relatively mild conditions, 150 °C and 2 MPa of H_2 (Figure 9.53). Control experiments with different combinations of Lewis acid Sn(II), Al(III), or Zn(II) with Pd/C showed that the catalytic enhancement of Zn(II) for HDO selectivity over this Zn(II)/Pd/C catalyst was specific, which can restrain the hydrogenation of aromatic rings and the etherification at C_α–OH with solvent methanol with Pd/C or Zn(II) as the catalyst. Elemental analysis of the prepared Zn/Pd/C catalyst confirmed that the Pd/C was indeed picking up and retaining the Zn^{2+} added to the reaction mixture. Spectroscopic characterization of the Zn/Pd/C catalyst (EXAFS and XANES) under relevant reaction conditions points to a synergistic effect between Zn(II) and Pd without the formation of a Zn–Pd alloy. In addition, they observed no evidence in Zn/Pd/C catalyst of close Pd–Zn neighbors, and authentic Pd–Zn alloy catalyst on carbon support gave low conversion of the model dimer, which negated the hypothesis that a bimetallic Pd–Zn alloy was the catalyst. The chemical observations in conjunction with XAS are consistent with Zn

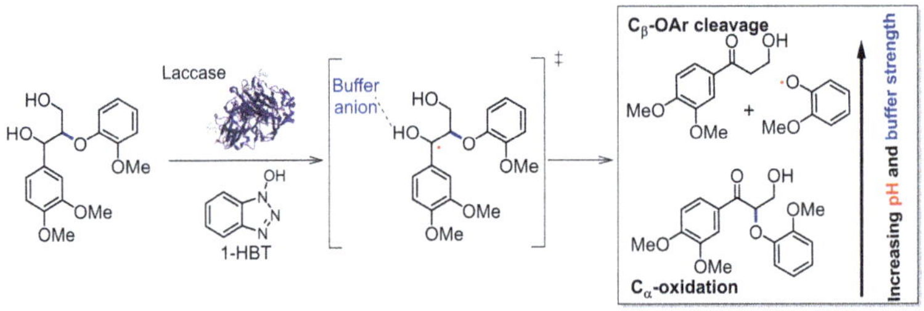

Figure 9.52 Boosting Laccase/HBT-Catalyzed the Cleavage of β-O-4′ Linkage in the Highly Concentrated Buffers at Near-neutral pH. *Source:* Hilgers et al. [137] with Permission, © 2020 American Chemical Society.

adsorbing on carbon, most likely onto –OH sites. A mechanistic hypothesis on the synergy between Pd^0 and Zn(II) was presented (Figure 9.53). The Zn^{2+} ions could participate in one of two ways. Activate the substrate via binding to OH groups inducing reactivity with Pd–H on the surface via hydrogen spillover. An alternative mechanistic hypothesis is that the Zn^{2+} ions are desorbed into the solution at the reaction temperature of 150 °C. Zinc binds the substrate and activates its cleavage upon encountering Pd–H sites on the catalyst surface. Once the reaction mixture is cooled, the Zn^{2+} ions are reabsorbed onto the C surface.

As discussed above, Abu-Omar and co-workers [138] give strict research about the active phase of the Zn(II)/Pd/C catalyst, while the detailed process about the HDO reaction containing the cleave order and type of the β-O-4 was not involved. In 2016, Abu-Omar and co-workers [139] present a detailed mechanistic study of the hydrodeoxygenation (HDO) of lignin by using a synergistic Pd/C and Zn(II) catalyst through the use of both lignin model compounds and lignocellulose. Their spectroscopic data coupled with the study of lignin model compounds suggest that Zn^{II} can activate and facilitate the removal of the primary OH at C_γ before the benzylic OH at C_α in β-O-4 ether linkage. Activation is proposed to occur through the formation of a six-membered ring complex of Zn^{II} coordinated to the oxygen atoms at C_α and C_γ of the lignin model compound guaiacylglycerol-β-guaiacy (Figure 9.54, step A). The aromatic substrates adsorb at the Pd metal surface forming a π-complex with the metal d-orbitals. This π-interaction (Figure 9.54, step B) brings Pd–H into close proximity to the β-O-4 ether linkage. Hydride transfer from Pd–H to the β-O-4 oxygen initiates C–O cleavage concurrently with a concerted removal of the OH

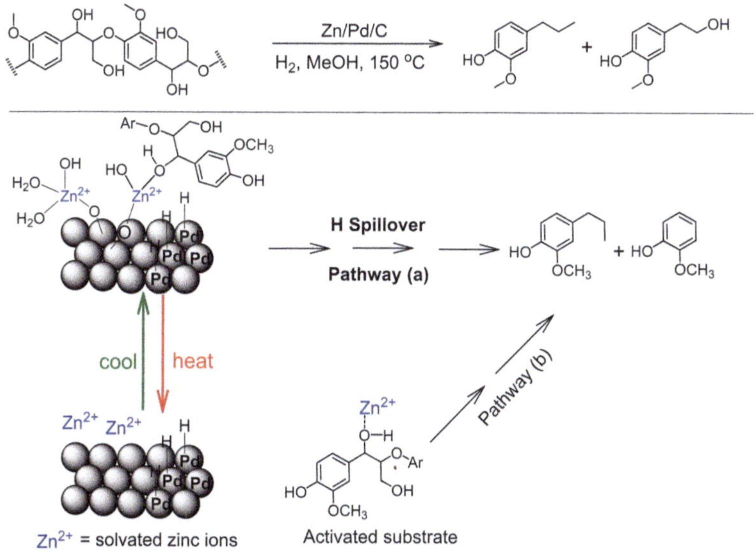

Figure 9.53 Proposed Pathways for HDO of β-O-4 Lignin Substrate over Zn(II)-Pd/C Catalyst. Pathway (a) Involves Substrate Binding to Zn^{2+} Sites and Hydrogen Spillover from Pd Sites. Pathway (b) Represents Desorption of Zn Ions into Solution. *Source:* Parsell et al. [138] with Permission, © 2013 The Royal Society of Chemistry.

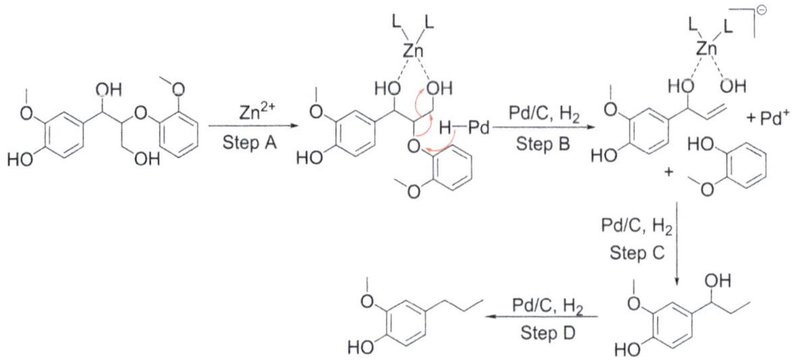

Figure 9.54 Zn(II)/Pd/C Catalyzed Cleavage and Hydrodeoxygenation of Lignin Model. *Source:* Klein et al. [139] with Permission, © 2016 The Royal Society of Chemistry.

group at the C_γ position (Figure 9.54, Step B), generating a Pd^+ on the surface and a $[Zn^{I}I–OH]^-$ anion. In this step, the coordination of Zn^{II} makes the OH a better leaving group. Catalyst regeneration from this point is envisaged to take place by reaction of Pd^+ with H_2 to form Pd–H and solvated [H+]. The later protonates $[Zn^{I}I–OH]^-$ anion to give H_2O and a Zn^{II} complex that can bind another lignin molecule. Then, the generated fragment from step B which contains a terminal $C_\beta{=}C_\gamma$ bond and a benzylic OH is readily hydrogenated by Pd/ C (Figure 9.54, Step C and D).

Besides the homogeneous Zn(II) promoting the hydrogenolysis of the lignin ether bonds over solid metal catalyst Pd/C [138, 139], in 2020, Li and co-workers [140] reported that the Co-Zn/Off-Al H-beta catalyst could expedite the reaction rate of C–O bonds cleavage in the β-O-4 ether linkages (Figure 9.55). For the catalyst preparation, briefly, H-beta was first treated with 13 M HNO_3 at 100 °C for 20 h to obtain dealuminized beta material (Off-Al H-beta). Then the Off-Al H-beta powder was finely ground with cobalt acetate and zinc acetate to get an intimate mixture. Finally, it was calcined at 550 °C for 5 h in the air to get the Co-Zn/Off-Al H-beta catalyst. Based on their results, the reaction of the β-O-4 model (**1**) most likely involves two distinct pathways, the reductive cleavage (Route I and II) and the acid catalyzed cleavage (Route III) as shown in Figure 9.55. Both pathways give rise to the degradation of the β-ether linkage and the formation of guaiacol. In the reductive cleavage, the formation of a six-membered ring complex structure between Zn and the two hydroxyl groups at the C_α and C_γ of **1** facilitates the direct attack of a hydride anion on the phenolic oxygen with a concomitant elimination of the α-hydroxyl group (Route I) or the γ-hydroxyl group (Route II). Different from the reaction mechanism proposed by Abu-Omar and co-workers (Route II) [138], they suggest that Route I is the dominant one as **5** derived from **4** is the dominant product, and **4** is a more stable product than **6** as the result of conjugation. Furthermore, the acid-catalyzed cleavage of **1** (Route III), which competes with the aforementioned reductive cleavage (Route I and II), produces only a single compound **11**, likely via **10** through a reversed aldol reaction (Route IIIA). Alternatively, a vinyl ether intermediate **12** may be formed via the well-known elimination of formaldehyde (Route IIIB). The vinyl ether underwent the addition of water giving intermediated compound **13**.

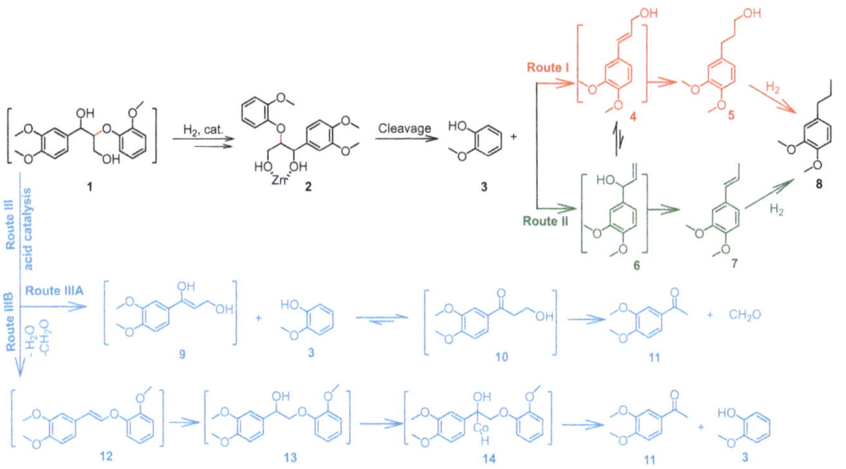

Figure 9.55 Catalytic Cleavage of Aryl–ether Bonds in Lignin Model Compounds with a Co–Zn-beta Catalyst. Zn Works as the Lewis Acid and the Co Works as the Hydrogenation Center. *Source:* Dou et al. [140] with Permission, © 2020 The Royal Society of Chemistry.

The compound further underwent reductive cleavage, forming the product **11**. Oxidative addition across the benzylic C–H bond of **13** by a Co metal center would yield a β-phenoxyalkyl metal hydride **14**, which underwent transfer hydrogenation to give a ketone and phenol as reported by Zhou with Pd/C [11] or by Goldman with a homogeneous Ir catalyst [141]. According to the experiments carried out at different temperatures, they reported that the acid-catalyzed cleavage appears to compete well with the reductive cleavage at temperatures below 200 °C, above which the reductive cleavage becomes the dominant reaction. Before this work, in 2020, Li and co-workers [142] discovered that the Co-Zn/Off-Al H-beta catalyst displayed superior activity in kraft lignin depolymerization. The corresponding process gave an 81 wt% yield of petroleum ether soluble product (mainly monomers and dimers) with a low molecular weight of ca. 330 $g{\cdot}mol^{-1}$ at 320 °C for 24 h.

For the transformation of β-O-4 lignin models, different catalytic routes contain different intermediaries, which can cause variations in the C_β–OAr ether bond cleavage. While the dominating factor for the C_β–OAr bond cleavage should not only contain the target bond BDE of the original substrate but also the one of the activated intermediates' phase [119]. In an ideal route of cleavage of the β-O-4 bond, an activated intermediate is generated after a series of activation steps with low activation energy, whose target bond has lower bond energy and easily breaks in the subsequent transformations. In 2016, Zhang and co-workers [119] carried out the hydrogenation study of lignin β-O-4 alcohols models over a NiMo sulfide catalyst, which provided unique styrene and ether as the products from cleavage of the C_β–OAr. Besides, the results of the competitive conversion of β-O-4 model compounds with C_αH–OH and C_α=O group in one reactor showed that NiMo sulfide preferred to catalyze the hydrogenative cleavage of C_β–OAr bond connecting with a C_αH–OH structure rather than catalyze the cleavage of the weaker C_β–OAr bond connecting with a C_α=O structure, which was on the contrary of the BDE of target C_β–OAr bond and indicated a different reaction mechanism. Based on the mechanism study and lignin models conversion, they proposed an efficient dehydroxylation-hydrogenation strategy for the C_β–OAr bond cleavage (Figure 9.56), which was not via C_β–OAr bond direct cleavage or C_α–OH dehydrogenation to C_α=O before the subsequent transformation. The acid site on the NiMo catalyst played a critical role in the lignin model hydrogenolysis cleavage but showed low efficiency in the lignin model hydrolysis cleavage without H_2. Also, further control experiments with radical trap with TEMPO (2,2,6,6-tetramethyl-1-piperidinyloxy) suggested a critical radical intermediate involved in the process. Further mechanism study, including the generation analysis of styrene and ether products, and DFT calculation of the ether products generation, suggested that the β-O-4-A model $PhCH(OH)CH_2OPh$ first lost a hydroxyl group (C_α–OH) at medium strong acid sites of NiMo catalyst. Then the generated $PhCH\delta^+CH_2OPh$ transformed to $PhCH{\bullet}CH_2OPh$ (β-O-4-R) by obtaining an electron from the NiMo catalyst redox cycle. DFT calculation suggested the dehydroxylation of C_α–OH to a β-O-4-R radical could greatly decrease the C_β–OAr BDE from 274.0 $kJ \cdot mol^{-1}$ to 66.9 $kJ \cdot mol^{-1}$ and thereby facilitate C_β–OAr cleavage, which further provided styrene, ethylbenzene, phenol, and various ethers under a hydrogenation condition in methanol. The cooperation between the acid-redox sites and hydrogenation sites played a crucial role in this route.

9.6.3 The Transfer Hydrogenation (Reduction) Process

For the self-decomposition or β-scission of the 1-phenyl-2-phenoxypropyl radical ($PhCH{\bullet}CH_2OPh$) shown above, Franz et. al. [121] had provided detailed kinetic research in 1991. Then, in 2002, Franz and co-workers [122] studied further the phenoxyl radical ($PhO{\bullet}$) expression rate from the β-scission of 1-phenyl-2-phenoxyethanol-1-yl radical ($PhC{\bullet}(OH)CH_2OPh$), a β-O-4 radical fragment which lost an H at C_α but containing a hydroxy group. The remaining part $PhC{\bullet}(OH)CH_2{\bullet}$ or $PhC(OH){=}CH_2$ further transformed to the aromatic ketone $PhCOCH_3$. The observed scission rates were consistent with a conventional scission pathway. In addition, the electronic structure calculations showed that intramolecular hydrogen-bond formation, between C_α–OH and –OPh, stabilized both transition and ground states. The combined experimental and computational results suggested that intramolecular hydrogen bonding resulted in slightly greater stabilization of the TS for scission, resulting in a slight net decrease in the β-scission barrier for the lignin model radical. In 2003, Müller and co-worker [143] provided a study of yellowing and IR-changes of spruce wood as a result of UV-irradiation, during which the UV light could induce the generation of the lignin fragment $ArC{\bullet}(OH)CHROAr$ and the further transformation of the phenolic fragment to quinones. In 2005, Lanzalunga et. al. [144] reported a detailed study of the decomposition of $ArC{\bullet}(OH)CHROAr$ and the possible recondensation of the fragments from the β-scission of $ArC{\bullet}(OH)CHROAr$, which could explain the reason for the low monomer yields in these UV-induced processes.

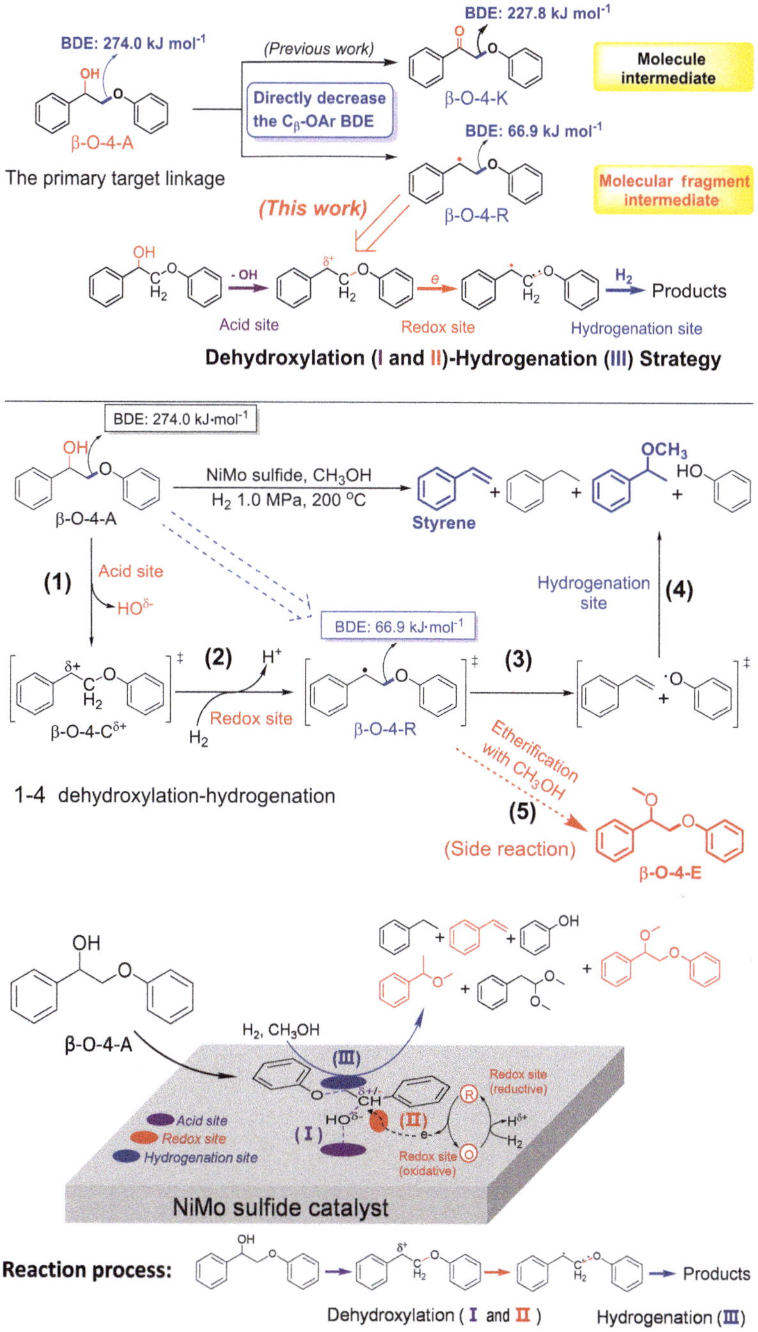

Figure 9.56 The Potential Dehydroxylation-hydrogenation Strategy for β-O-4-A Conversion over the NiMo Sulfide Catalyst. *Source:* Zhang et al. [119] with Permission, © 2016 The Royal Society of Chemistry.

To make the lignin conversion a catalytic process containing PhC•(OH)CH_2OPh as the critical radical, in 2018, Wang and co-workers [120] reported a photocatalytic lignin transformation methods (Figure 9.57), during which the photo-induced hole (h^+) over the photocatalyst CdS quantum dots (QDs) could selectively abstract the H from the carbon of C_αH–OH and generate a C_α carbon radical (PhC•(OH)CH_2OPh), which could decrease the BDE of C_β–OAr bond from 55 kcal · mol^{-1} to 7.8 kcal · mol^{-1} (1 kcal · mol^{-1} = 4.18 kJ · mol^{-1}) and induce the reductive cleavage of C_β–OAr bond in the intermediate PhCH•CH_2OPh with the assistance of electron (e^-) and proton. They further found that the CdS quantum dots could catalyze the cleavage of β-O-4 bonds in lignin models quantitatively but also were efficient for the conversion of native lignin within biomass into functionalized aromatics under visible light, while cellulose/hemicellulose remain almost intact. Under the optimized condition, functionalized aromatic monomers, mainly ketones, were obtained with a yield of about 27 wt% from birch woodmeal, corresponding to 84% of the theoretical maximum by assuming the selective cleavage of β-O-4 linkages.

Based on the proposed CdS QDs mechanism (Figure 9.57), the C_α radical intermediate is first formed via the oxidative dehydrogenation (ODH) of the C_α–H bond driven by the hole, and then the intermediate undertakes a reductive cleavage of the β-O-4 bond triggered by the electron. DFT calculations suggested that the valence band maximum (VBM) of CdS (1.7 V versus SHE) was more positive than the C_α–H bond ODH potential (1.1 V versus SHE), which was even higher than the ODH potential of C_αO–H (1.5 V versus SHE). To further modify the reaction selectivity and enhance the intimate contact between CdS QDs and lignin, leading to high catalytic activity, Wang and co-workers [145] reported a strong dependence of photocatalytic conversion of native lignin into high-value functionalized aromatics on the ligand ($HS(CH_2)_nCOOH$) attached on CdS quantum dots (QDs) (Figure 9.58). The surface engineering by manipulating the anchor group and the length of ligands demonstrated that the ligand participates in the electron-transfer (ET) process, a crucial step of photocatalysis. The anchor group and the alkyl length of the ligand determine the charge transfer from the core of QDs to lignin, which is mediated by the organic ligand. The kinetic study of the electron decay process reveals that the

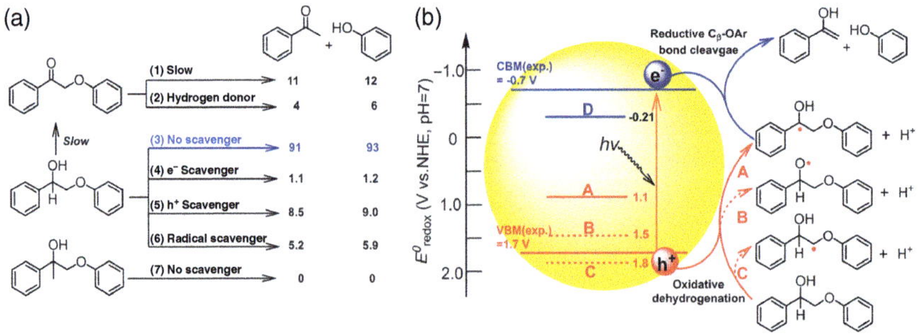

Figure 9.57 Solar Energy-driven Lignin-first Approach to Full Utilization of Lignocellulosic Biomass under Mild Conditions. *Source:* Wu et al. [120] with Permission, © 2018 Springer Nature.

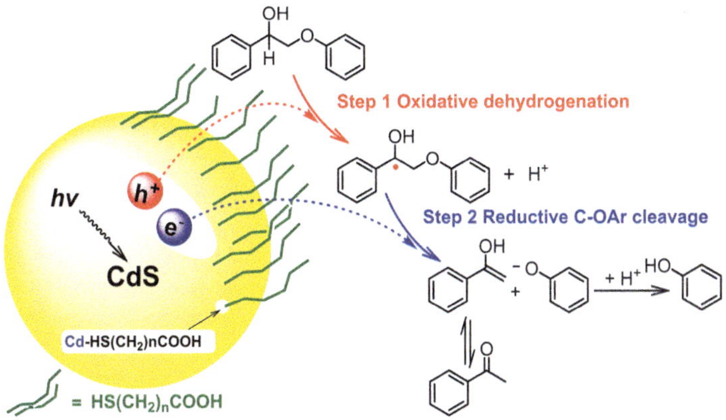

Figure 9.58 Ligand-controlled Photocatalysis of CdS Quantum Dots for Lignin Valorization under Visible Light. *Source:* Wu et al. [145] with Permission, © 2019 American Chemical Society.

ET through ligands proceeds via an electron-tunneling pathway and that the rate decreases exponentially with the length of ligands. They found that CdS QDs with 3-mercaptopropionic acid ligand could catalyze the conversion of various technical lignins into functionalized aromatics with a higher yield, and the yield of aromatics increases with the content of β-O-4 linkages. Dioxasolv lignin provided a 15.4 w% yield of aromatic monomers.

Besides the VB structure change of CdS via surface engineering by manipulating the anchor group ligand, in 2020, Lee and co-workers [146] reported that simple Ag^+ exchange of the CdS photocatalyst could significantly enhance the efficiency of lignin β-O-4 conversion, which results in an almost 100% conversion yield and the formation of only cleaved aromatic compounds. They proposed that suitable Ag^+-exchanged domains would dramatically shift the Fermi level of CdS and that the resulting Ag_2S@CdS would render photogenerated electrons and holes to effectively participate in the cleavage of β-O-4 bonds via the mechanism shown in Figure 9.59.

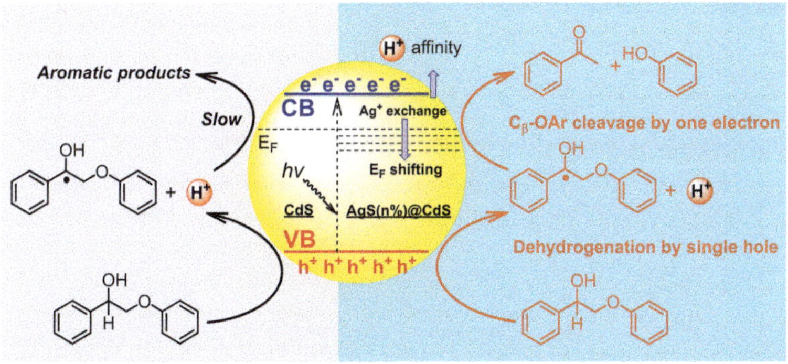

Figure 9.59 Enhancing Photocatalytic β-O-4 Bond Cleavage in Lignin Model Compounds by Ag-exchanged CdS. *Source:* Yoo et al. [146] with Permission, © 2020 American Chemical Society.

In detail, the Ag^+-exchanged CdS could achieve an easy photoexcited electron transfer to the oxidized intermediate (C_α• radical) that resulted in a marked increase in the conversion yield with high product selectivity under mild reaction conditions without any additives.

Besides the CdS-based photocatalyst, in 2019, Wang and co-workers [147] studied the photocatalytic activity of zinc–indium–sulfide ($Zn_mIn_2S_{m+3}$) for the cleavage of the lignol β-O-4 bond under mild conditions. After tuning the energy band structure of $Zn_mIn_2S_{m+3}$ by controlling the atomic ratio of Zn/In (Figure 9.60a), it was found that $Zn_4In_2S_7$ performed best for cleavage of the β-O-4 bond under visible-light irradiation, owing to its appropriate energy band structure for offering adequate visible-light absorption and suitable redox capability. In addition, functionalized aromatic monomers with near 18.4 wt% yields could be obtained from organosolv birch lignin with the catalyst $Zn_4In_2S_7$. Mechanistic studies revealed that the β-O-4 bond was efficiently cleaved mainly through a one-step redox-neutral pathway via a C_α radical intermediate (Figure 9.60b), which was different from the self-hydrogen transfer process with the C_α ketone intermediate over the $ZnIn_2S_4$ reported by Luo [24]. The thiol groups on the surface of $Zn_4In_2S_7$ played a key role in the process of β-O-4 bond cleavage.

For the cleavage of the C_β–OAr, besides the above photocatalytic transfer hydrogenation(reduction) system via the first dehydrogenation of C_α–H to a C_α• carbon radical, a similar reaction mechanism, beginning with first an active metal insertion into the C_α–H bond to provide the C_α–M–H intermediate, can also promote the cleavage of the target ether bond. In 2014, Haibach et. al. [141] reported a unique process whereby the C–O bonds of aromatic ethers are catalytically cleaved by pincer iridium catalysts with complete atom-economy (Figure 9.61). This dehydroaryloxylation reaction proceeds in moderate to excellent conversion for a variety of substituted alkyl aryl ethers. The initial step is the addition of a C–H bond with Ir to yield **1**, followed by aryloxide migration to form the six-coordinate intermediate **2**. Loss of the olefin and phenol from **2** regenerates the active (pincer)Ir species and completes the catalytic cycle.

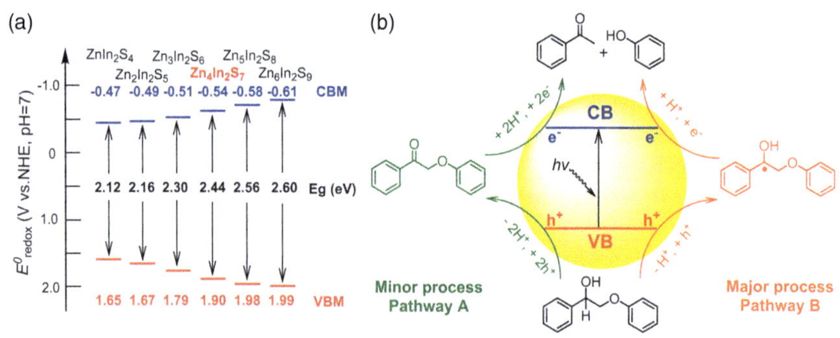

Figure 9.60 Visible-Light-Driven Cleavage of C–O Linkage for Lignin Valorization to Functionalized Aromatics. *Source:* Lin et al. [147] with Permission, © 2019 Wiley-VCH.

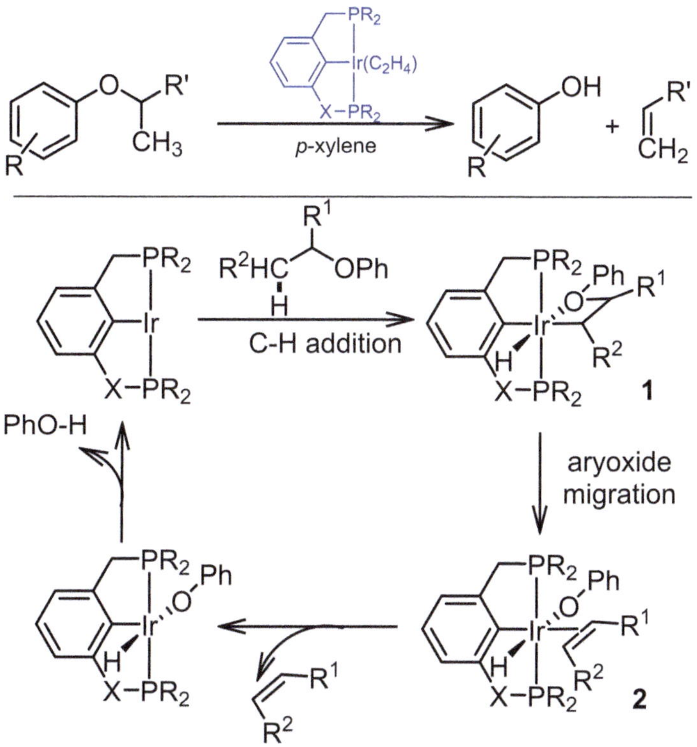

Figure 9.61 Catalytic Cleavage of Ether C–O Bonds by Pincer Iridium Complexes via a Dehydroaryloxylation Mechanism. *Source:* Haibach et al. [141] with Permission, © 2014 Wiley-VCH.

In 2020, the Jackson group [148] presented detailed mechanistic studies of electrocatalytic hydrogenation (ECH) in aqueous solution over skeletal nickel cathodes to probe the various paths of reductive catalytic C–O bond cleavage among functionalized aryl ethers. For the simple 2-phenoxy-1-phenylethane or its alcohol congener, 2-phenoxy-1-phenylethanol, they adsorb over the Ni surface via phenyl ring coordination followed by benzylic C–H activation, then the benzylic site is activated via Ni C–H insertion, followed by β-elimination of the phenoxide leaving group (Figure 9.62a). But in the case of the ketone, 2-phenoxyacetophenone, the polarized carbonyl π system apparently binds directly with the electron-rich Ni cathode surface without breaking the aromaticity of the neighboring phenyl ring, leading to rapid cleavage. In the mechanism discussion, they also discussed the potential semi-hydrogenation of C_α=O to provide a $C_\alpha^\bullet$ radical (Figure 9.50a, bottom). While, on the basis of the direct involvement of catalyst in the rapid H/D exchange, and as will be further developed in the substituent and noncatalytic electrode studies, the carbonyl coordination mechanism with direct C_β–OAr bond cleavage is identified as the more plausible cleavage mechanism for the breakdown of the ketone β-O-4 linkage (Figure 9.62b).

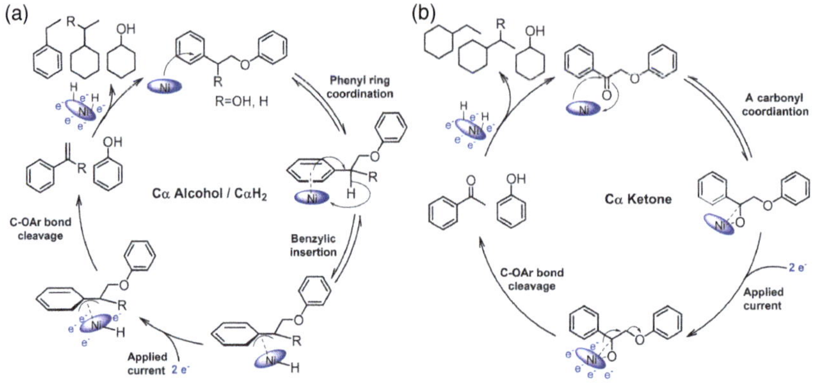

Figure 9.62 Multiple Mechanisms Mapped in Aryl Alkyl Ether Cleavage via Aqueous Electrocatalytic Hydrogenation over Skeletal Nickel. *Source:* Zhou et al. [148] with Permission, © 2020 American Chemical Society.

9.6.4 The Redox-neutral Process with First C_αO–H Activation by M=O Species

Before the CdS-catalyzed strategies for lignin depolymerization containing the C_α radical intermediate reported by the Wang group [120], in 2010, Toste and co-workers [149] reported a vanadium-catalyzed non-oxidative mechanism for the C_β–OAr cleavage (Figure 9.63a), in which the C_α radical was the key intermediate. Different from Co-Zn [140], Ir [141], and Ni-catalyzed [148] process shown above, Toste and co-workers [149, 150] demonstrated that changes in the ancillary ligands could divert the reactivity of vanadium–oxo complexes from the typical alcohol oxidation [151] to a C_β–OAr bond cleavage reaction. This novel reactivity was applied to a vanadium-catalyzed non-oxidative C_β–OAr bond cleavage reactions of dimeric lignin model compounds that produced aryl enones as degradation products. Based on experimental data, they proposed a one-electron process as the most plausible mechanism (Figure 9.63b). After ligand exchange (**a**→**c**) on the vanadium complex (V^V=O, **a**) with the benzylic hydroxy group of the β-O-4 alcohol (**b**), the benzylic hydrogen is abstracted to generate the ketyl radical (**d**), which eliminates the aryloxy radical (**f**). The elimination of the hydroxy group from the resulting enolate (**e**) produces enone (**g**) and a vanadium (IV) complex (**h**), which is re-oxidized to vanadium(V) by the aryloxy radical (**f**). This reaction is formally redox neutral. Nonetheless, the reaction can be accelerated in the air due to the prevention of catalyst precipitation (Figure 9.63a).

Then, in 2012, Hason and co-workers [66] found that under identical conditions with the air existence the oxidation products with C–C bond cleavage will change to the phenolic substrate with $C(sp^3)$–O cleavage if the catalyst is changed from **V-cat.2** to **V-cat.1** (Figure 9.64), which was similar to the aforementioned results reported by Toste [149]. Different from Toste's proposal, Hanson et al. suggested that vanadium-catalyzed aerobic oxidation of alcohols in the presence of bases might give the triplet V^{III} complex **I** [68]. According to their proposal, a competitive

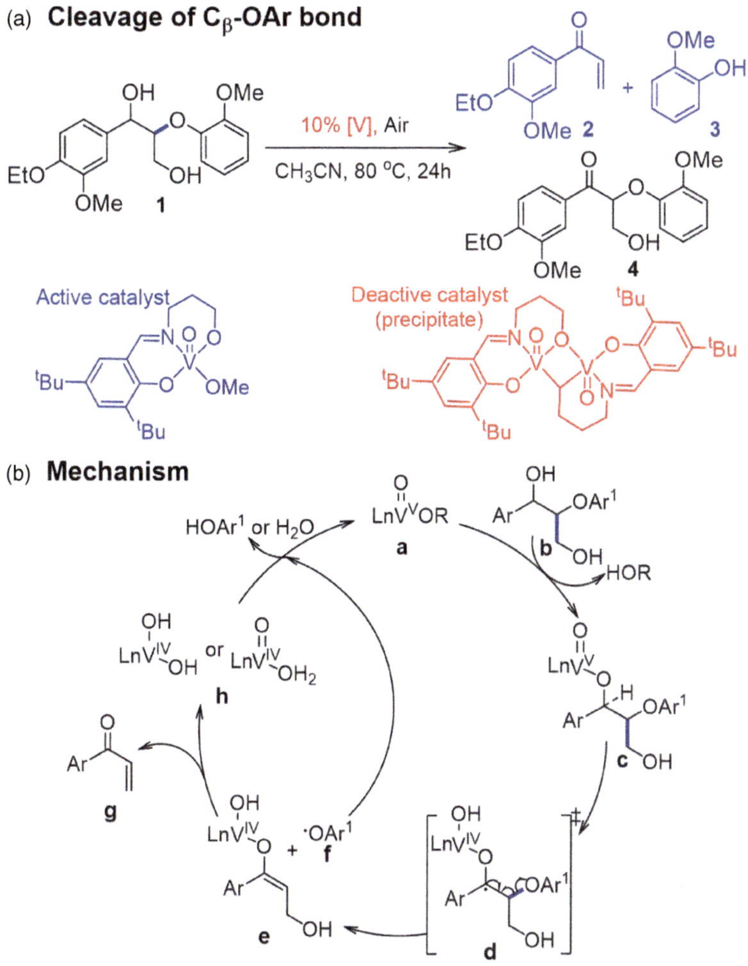

Figure 9.63 Vanadium-catalyzed Non-oxidative Cleavage of $_\beta$-O-4 Linkage. *Source:* Son and Toste [149] with Permission, © 2010 Wiley-VCH.

pathway leading to **H** and **I** from **C** via base-promoted benzylic deprotonation, spin crossover, and protonation of the oxo ligand is plausible in the presence of pyridine (Path B, Figure 9.64a). From **I**, paths B and A might converge in either the isomerization step between **I** and **D** or the C(sp^3)–O cleavage on **I** to form **E**. Besides the cleavage of the C_β–OAr bond, Hason and co-workers [66] also discovered the Ar–C(sp^3) cleavage of the phenolic β-O-4 lignin model compound in an air atmosphere using the bis(8-oxyquinolate) vanadium complex as the catalyst. Regarding the Ar–C(sp^3) cleavage process, a superoxo vanadium intermediate after the hydrogen atom abstraction from the lignin model compound by the vanadium catalyst was pointed out (Figure 9.64b). It is similar to the mechanism of Co-catalyzed Ar–C(sp^3) cleavage reactions [152]. Although this proposal seems plausible, the details remain obscure. More importantly, the mechanism of V-catalyzed C–O/C–C

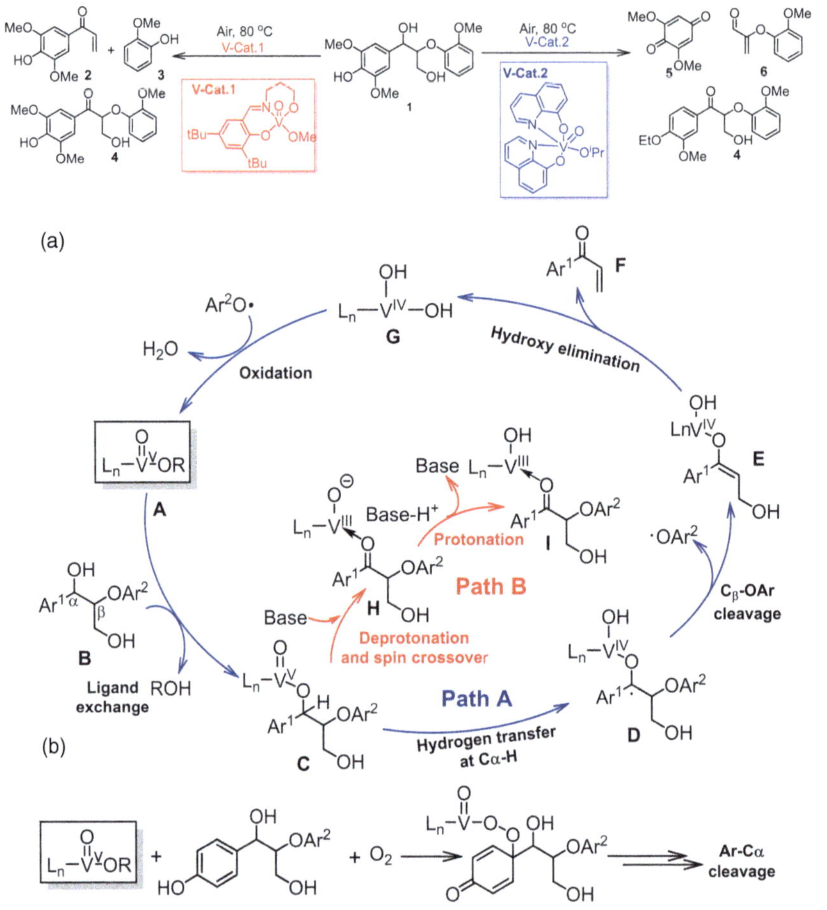

Figure 9.64 C–C or C–O Bond Cleavage in a Phenolic Lignin Model Compound: Selectivity Depends on Vanadium Catalyst. *Source:* Permission from Jiang et al. [153], © 2016 American Chemical Society.

cleavage of β-O-4 lignin model compound, and especially the origin of the ligand-controlled selectivity, is unclear.

Then, in 2016, Fu and co-workers [153] carried out a combined theoretical and experimental study that focused on the mechanism of the vanadium-catalyzed selective C(sp^3)–O and Ar–C(sp^3) bond cleavages of the phenolic β-O-4 lignin model compound. An updated mechanism involving V^{III}, V^{IV}, and V^{V} complexes was located for the C(sp^3)–O bond cleavage process (Figure 9.65, left), and a complete catalytic cycle involving a novel stepwise O–O/Ar–C(sp^3) cleavage mechanism was clarified for the Ar–C(sp^3) bond cleavage process (Figure 9.65, right). On the basis of results obtained, they claimed that the switchable selectivity at the initial stage of phenolic β-O-4 model oxidation results from the different T1 (ground triplet state)–HOMO separation/charge dispersion effects of ligands and the different formal oxidation states of the TOF-determining transition state (TDTS) in the C(sp^3)–O and Ar–C(sp^3) cleavage processes. One can check more detailed information for this study from the research article [153], which requires a strong research background on the DFT methods.

9.7 Lignin C_α–C_β Bond Cleavage after the First Activation of C_αO–H Bond to $C_\alpha O^\bullet$ Radical via PCET Strategies and LMCT Mechanisms

Focusing on the typical structure zone of R_1–CH(OH)–C–R_2 in β-O-4 and β-1 lignin linkages, in order to cleave the C_α–C_β bond, the adjacent alcohol group at C_α could work as the substrate site and be first transformed to an alkoxy radical, which would be an active intermediate that can further induce the β-scission of weakened C–C bond (Figure 9.66) [154, 155]. To achieve the generation of alkoxy radicals from alcohols with high dissociation free energies of O–H bonds (O–H BDFE≈105 kcal · mol^{-1}), compared with the direct hydrogen atom transfer from the hydroxyl groups of alcohols, the homolytic activation of strong bonds based on proton-coupled electron transfer (PCET) [156–159] with a lower reaction *Ea* will be an alternative approach and make the generation of $C_\alpha O^\bullet$ radical more favorable than the $C_\alpha^\bullet$ radical from the HAT of C_α–H. In addition, as another choice, when R_1 is an aromatic ring, the first generated cation radical can further transform into the C_αO• radical (Figure 9.66). Although the latter method begins with the first activation of the aromatic ring, we need to add some samples here to make a discussion about the $C_\alpha^\bullet$ radical more systemic. Or, the alcohol group can coordinate with the high valence metal center, then the single-electron-transfer/proton transfer (SET/PT) of the key intermediate M^{n+}–OR species via a ligand-to-metal charge-transfer (LMCT) mechanism can provide an $M^{(n-1)+}$ and a $C_\alpha O^\bullet$ radical before the β-scission with C_α–C_β bond cleavage (Figure 9.66).

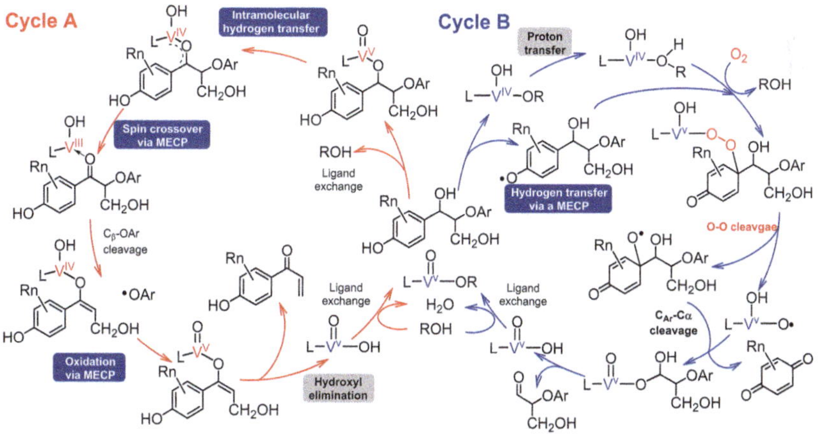

Figure 9.65 Overall Mechanistic Proposals for C(sp^3)–OAr and Ar–C(sp^3) Bond Cleavage of Lignin Model Compound by Different Vanadium Catalysts. Note: MECP means minimum energy crossing point. (Cycle A for the V-cat.1, Cycle B for the V-cat.2, what have been presented in Figure 9.64). *Source:* Permission from Jiang et al. [153], © 2016 American Chemical Society.

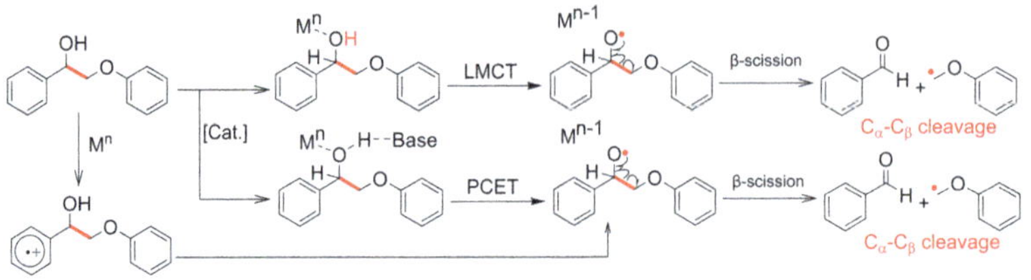

Figure 9.66 Catalytic Cleavage of the C–C Bond with an Adjacent Alcohol Group.

9.7.1 The HAT Process with the $C_\alpha O^\bullet$ Radical Intermediate

Although the direct HAT process at the lignin C_αO–H to provide a C_αO radical intermediate is not the favorable route, it can be confirmed in some certain processes. In 2014, Mitchell and co-workers [160] reported that alcohols could be converted into their corresponding carbonyl compounds using catalytic amounts of 1,4-hydroquinone (BQ) with a copper nanoparticle (Cu/AlO(OH)) electron transfer mediator under visible light irradiation. The fact that both alcohol oxidation and C–C bond cleavage in the lignin models are observed suggests that two competing reaction pathways are operating (Figure 9.67). Presumably, alcohol oxidation occurs by the accepted mechanism of initial hydrogen abstraction from the C_α–H bond by the photochemically excited triplet quinone (^{3}BQ) to give a C_α centered radical that can undergo a further hydrogen abstraction from the C_αO–H bond to afford the ketone product and hydroquinone (Figure 9.67, Route A). On the other hand, the C–C cleavage reaction is most probably initiated by hydrogen abstraction from the O–H bond to form an alkoxy radical that undergoes C–C homolytic fragmentation to the aldehyde and a new alkyl radical (Figure 9.67, Route B).

9.7.2 The PCET Process with the $C_\alpha O^\bullet$ Radical Intermediate

Taking the lignin β-O-4 model as an example, the oxidative potential for the benzyl C_α–H bond is about 0.4 V lower than that of the C_αO–H bond. Therefore, the preferential formation of alkoxy radical in lignin linkage necessitates the development of a novel catalytic strategy. In 2016, Knowle and co-worker [155] reported a photocatalytic protocol for the redox-neutral isomerization of cyclic alcohols to linear ketones via an alkoxy radical-induced C–C bond scission (Figure 9.68). The ternary catalyst system comprising a photoredox Ir catalyst ([Ir(dF(CF3)ppy)2(5,5'-d(CF3)bpy)] (PF_6), $E_{1/2}$=1.30 V vs Fc/Fc^+), a weak Brønsted base (collidine), and a thiol hydrogen atom donor (Ph-SH). Mechanistic studies demonstrate that key alkoxy radical intermediates in this reaction are generated via the direct homolytic activation of alcohol O–H bonds in an unusual intramolecular PCET process, wherein the electron travels to a proximal radical cation in concert with a proton transfer to a weak Brønsted base. During the catalytic cycle, an intramolecular PCET reaction mediated by Brønsted base and Ir* at excited state would follow that deprotonation of the hydroxyl group by the Brønsted base catalyst occurs in concert with one-electron reduction of the radical cation to provide the key alkoxy radical intermediate. The alkoxy radicals substantially enabled the ring-opening to generate an aryl ketone and a distal alkyl

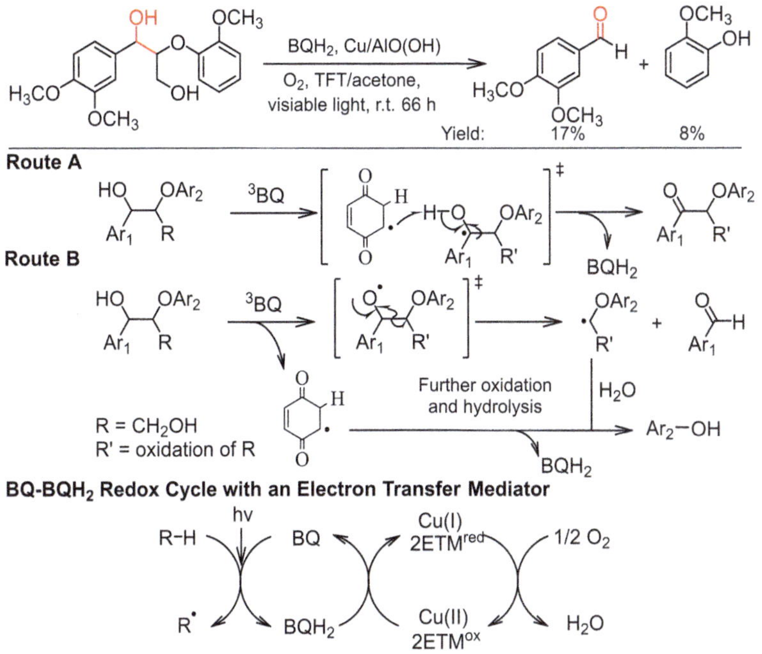

Figure 9.67 Proposed Mechanistic Cleavage Pathways Schematic BQ–BQH_2 Redox Cycle with an Electron Transfer Mediator (ETM). *Source:* Mitchell and Moody [160] with Permission, © 2014 American Chemical Society.

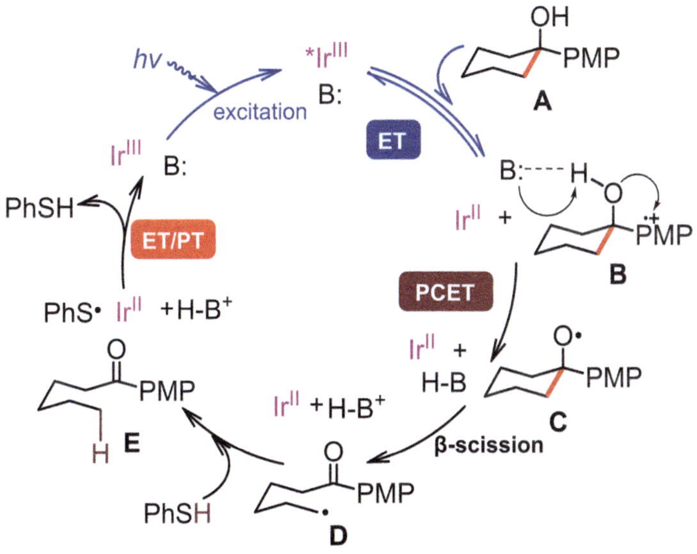

Figure 9.68 Catalytic Ring-opening of Cyclic Alcohols Enabled by PCET Activation. *Source:* Yayla et al. [155] with Permission, © 2016 American Chemical Society.

radical. The thiophenol, working as a catalytic H atom donor, then intercepts and reduces the nascent alkyl radical [161, 162], affording a linear ketone product. The generated PhS• radical be further reduced by the photocatalyst Ir^{II} and protonated by the conjugate acid of the Brønsted base to complete the catalytic cycle.

Then, in 2019, the Knowles group [163] presented detailed mechanism research with DFT calculation. Additionally, in 2020, Knowles et al. [164] applied the corresponding photocatalytic protocol discussed above in the cleavage of lignin C–C bonds. The method exhibits good efficiency and excellent selectivity for the activation and fragmentation of the β-O-4 and β-1 linkages (Figure 9.69a and b), even in the presence of numerous other PCET-active functional groups. They further demonstrated the feasibility of this depolymerization protocol on a preparative scale of 480 mg of ponderosa lignin with the optimized reaction conditions, which can provide 4.0% yields of corresponding monomers (Figure 9.69c).

Almost at the same period, in 2019, Zhang and co-workers [165] adapted the PCET method to activate the O–H bond followed by β-scission to cleave the C–C bond in both β-O-4 and β-1 linkages at room temperature under photoredox-neutral condition (Figure 9.70). Choosing a suitable base is vital for the PCET process to furnish the key alkoxy radical, and collidine is the optimal one in this case. The conversion initiates with electron transfer between excited photocatalyst $^{*}Ir^{III}$ and phenyl moiety of substrate, delivering a radical cation intermediate. Then the radical cation is transformed into the key alkoxy radical through a PCET process with the aid of collidine, which subsequently undergoes β-scission to produce benzaldehyde and another carbon free-radical. Finally, the H-atom donor thiophenol quenches the generated carbon radical and oxidizes the reduced Ir photocatalyst, thereby enclosing the catalytic cycle. Under visible light irradiation, both lignin β-O-4 and β-1 models can be converted into corresponding products with high yields (>90%). However, this method is not applicable to the direct conversion of birch-dioxane-lignin (BDL) due to the presence of numerous phenolic structures in lignin. The phenolic hydroxyl group is an excellent substrate for the PCET process, which significantly competes

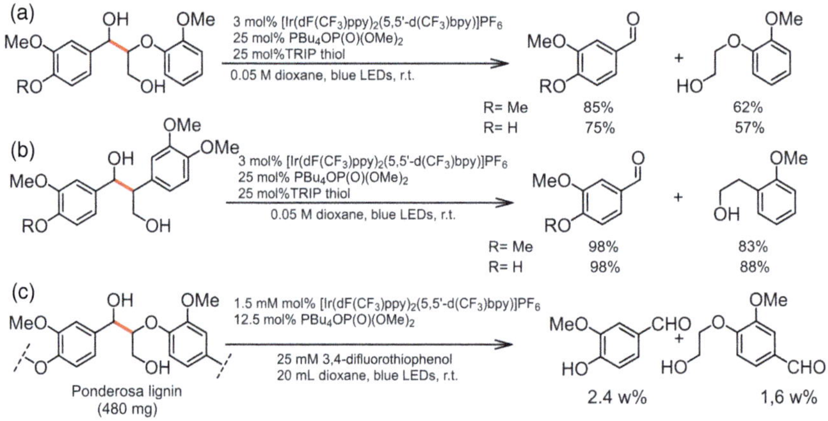

Figure 9.69 Light-driven Depolymerization of Lignin Enabled by PCET. *Source:* Reaction Data Abstracted from Nguyen et al. [164].

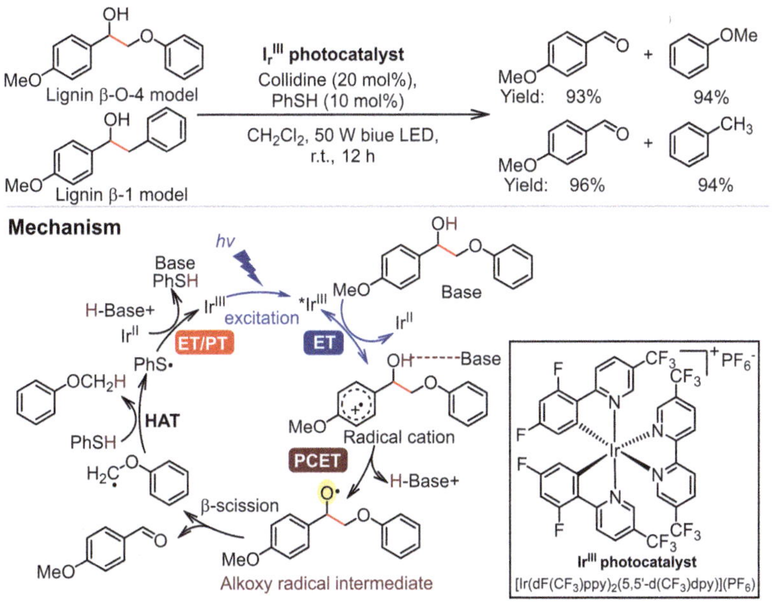

Figure 9.70 Photocatalytic C_α–C_β Bond Cleavage in Lignin $_\beta$-O-4 and $_\beta$-1 Linkages via PECT Process. *Source:* Wang et al. [165] with Permission, © 2019 Science China Press.

with the α–OH group to be activated. Only after a pre-treatment with iodomethane and K_2CO_3 to protect hydroxyl groups on the phenoxy ring, the BDL can be converted by this method to offer degraded products with a yield of 4.99 wt%. In addition to the Ir-based catalysts, an acridinium salt/copper catalytic system cc was developed to break down the C_α–C_β bond in lignin β-O-4 and β-1 models via a similar alkoxy radical-mediated mechanism. Under irradiation, the excited acridinium photocatalyst transforms the lignin model into an alkoxy radical with the aid of NaOAc that works as a base. The resulting alkoxy radical then undergoes β-scission to produce benzaldehyde and a carbon radical which was further converted into phenyl formate or another molecule of aldehyde under the aerobic condition in the presence of a copper catalyst. The detailed discussion is presented in Chapter 7.2.4.

9.7.3 The LMCT Process with the $C_\alpha O^\bullet$ Radical Intermediate

Besides the catalytic PCET process above with an iridium catalyst, in 2010, Baker and co-workers [63] reported a different (V^V=O)-catalyzed oxidative mechanism for lignin models conversion. A similar intermediate with a V^V–OC_α structure [63] which was involved in the (V^V=O)-catalyzed mechanism reported by Toste [149] could induce the direct cleavage of the C_α–C_β bond under O_2 condition. Based on products analysis, the single electron transfer between the V^V and oxygen in V^V–OC_α produced an active ArCH(O•)CH_2OAr, which further induced the selective C_α–C_β bond cleavage.

Then, photocatalysis was introduced in the (V^V=O)-catalyzed mechanism, in 2015, Gazi and co-workers [167] show that homogeneous vanadium photocatalysts can harvest visible light (>420 nm) irradiation to catalyze the chemoselective C_α–C_β bond cleavage of lignin model compounds under ambient conditions, leading to the generation of corresponding aromatic aldehydes and methoxylated phenyl formate (Figure 9.71). Isotope labeling experiments, product analyses, and intermediate radical trapping, together with density functional theory studies, suggest a unique pathway that involves a photogenerated T1 state during the C–C bond cleavage reactions with the LMCT mechanism (Figure 9.71). In addition, they further modified the ligand with different substituent groups, some of which with fluoro-substitution can obviously accelerate the photocatalysis with C_α–C_β bond cleavage [168].

Besides, in 2020, by using vanadium catalysts $VO(O^iPr)_3$ under visible light, Liu and co-workers [169] achieved the selective cleavage of C_α–C_β bonds in β-1 and β-O-4 interlinkages included in lignin models and extracts by an oxidative protocol (Figure 9.72). On the basis of the obtained results, they proposed a photo-driven single electron transfer mechanism for the oxidative cleavage of the β-1 lignin model. First, the benzylic hydroxyl group of the substrate coordinates with vanadium, leading to a complex ready to be excited by the visible light. Under light irradiation, the ligand to metal charge transfer induces reduction of vanadium, which simultaneously drives the direct C_α–C_β bond cleavage and produces benzaldehyde and benzyl radical. Then the benzyl radical can be further oxidized to benzaldehyde through a peroxide intermediate, finally generating two equivalents of monomeric aromatics from the substrate. Meanwhile, the reduced vanadium could be oxidized back to V^V to complete the catalytic cycle. When this V-catalyzed photocatalysis method was used in the lignin conversion, the aromatic monomers yield was below 2 w%.

For the fate of the distal alkyl radical from the C–C bond β-scission of the alkoxy radical intermediate, besides reduction with the hydrogen atom donor, it can be intercepted by C=C [170] double bonds to generate new C–C bonds after the reduction of the later carbon radical. In addition, the distal alkyl radical can be also captured by the C=N or N=N bonds [171–173], which can produce an N-centered radical that further transforms into an N-containing product after the

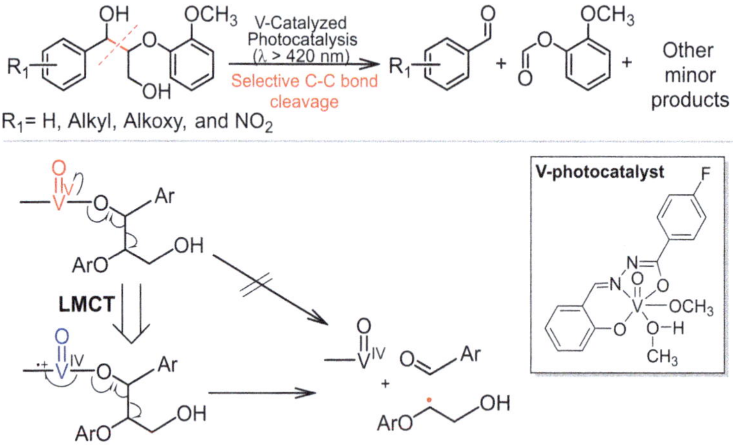

Figure 9.71 Selective Lignin β-O-4 C–C Bond Cleavage with a Vanadium Photocatalyst.

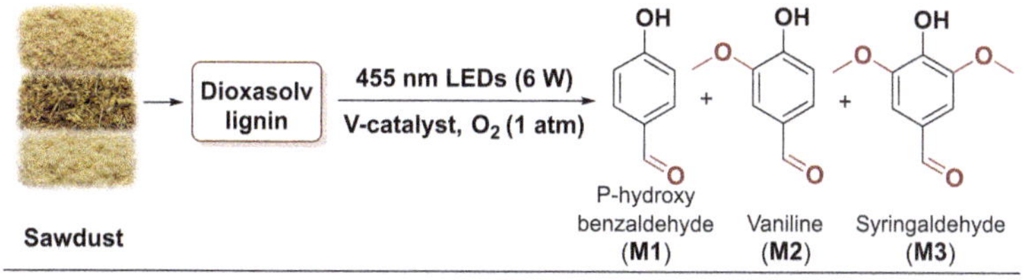

Entry	Substrate	Catalyst	Solvent / Time	Aromatic monomer (wt %)		
				M1	M2	M3
1	Birch lignin	$VO(O^iPr)_3$	CH_3CN / 24 h	-	0.34	0.30
2	Straw lignin	$VO(O^iPr)_3$	CH_3CN / 24 h	0.21	0.23	0.10
3	Straw lignin	$VO(acac)_2$	CH_3CN / 24 h	-	0.24	0.07
4	Birch lignin	$VO(acac)_2$	CH_3CN / 24 h	-	0.22	0.18
5	Birch lignin	$VO(O^iPr)_3$	Acetone: CH_3OH = 9:1 / 24 h	-	0.35	0.57
6	Birch lignin	$VO(acac)_2$	Acetone: CH_3OH = 9:1 / 24 h	-	**0.60**	**0.79**
7	Straw lignin	$VO(acac)_2$	Acetone: CH_3OH = 9:1 / 24 h	-	0.25	0.28
8	Poplar lignin	$VO(acac)_2$	Acetone: CH_3OH = 9:1 / 24 h	-	0.20	0.26

Conditions: Dioxasolv lignin (20 mg), solvent (1 mL), V-catalyst (10 wt%). "-" means not detected.

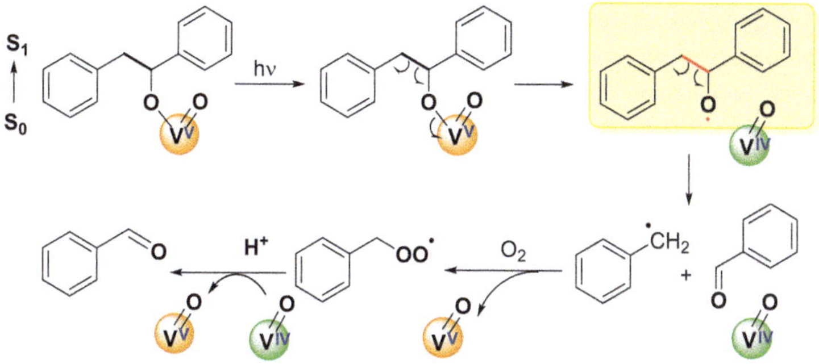

Figure 9.72 Visible-Light-Induced Oxidative Lignin C–C Bond Cleavage to Aldehydes with Homogeneous $VO(O^iPr)_3$. *Source:* Liu et al. [169] with Permission, © 2020 American Chemical Society.

single-electron-transfer/proton transfer (SET/PT) reduction. In 2016, Zuo and co-workers [171] developed a general strategy for the cleavage and amination of C–C bonds in cyclic alcohols employing simple cerium(III)-chloride-based photocatalyst with di-*tert*-butyl azodicarboxylate (DBAD). According to the proposed mechanism research (Figure 9.73), the reaction involves the β-scission of the key intermediate Ce^{IV}–OR species via a ligand-to-metal charge-transfer (LMCT) mechanism [173]. In detail, initial irradiation of the cerium chloride/alcohol complex **B** would generate the photoexcited state **C**. The photoexcited Ce^{III} complex **C** would subsequently engage in single-electron transfer with a nitrogen-centered radical intermediate **D** to generate a Ce^{IV} species **F**. Given the highly oxidizing nature of the Ce^{IV} species, it could promote the β-scission process of the adjacent C–C bond to generate the carbon-centered radical and the Ce^{III}-coordinated carbonyl **G**. The reactive radical species **G** could be captured by the N=N bond of DBAD to form the nitrogen-centered radical intermediate **D** and release the $CeCl_3$ for the next cycle. Radical **D** then undergoes single-electron-transfer/proton transfer (SET/PT) to furnish the product.

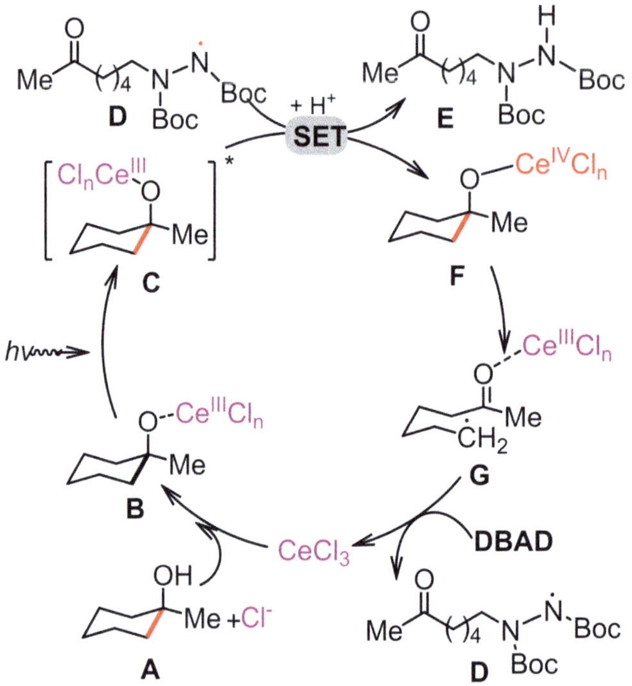

Figure 9.73 Photocatalytic C–C Bond Cleavage and Amination of Cycloalkanols by Cerium(III) Chloride Complex. *Source:* Guo et al. [171] with Permission, © 2016 Wiley-VCH.

In 2020, Zhang and co-worker [174] developed a one-step, $CeCl_3$-based photocatalytic strategy to not only achieve highly selective C_α–C_β bond cleavage but also produce functionalized aromatics, such as aldehyde (up to 97%) and N-containing products (up to 95%) in high to excellent yield from both β-1 and a wide range of β-O-4 lignin model compounds at room temperature (Figure 9.74). This method is also effective for the depolymerization of real lignin, which can provide a 12% yield of monomers from the corresponding cleavage and amination of C_α–C_β bonds in pine lignin. Additionally, this $CeCl_3$-based photocatalyst system could maintain excellent catalytic performance even after 10 sequential cycles with new starting materials. On the basis of the characterization of lignin conversion with this system and previous research about the photoinduced ligand-to-metal charge-transfer mechanism (LMCT) with Ce ion [171–173], the proposed mechanism is as follows (Figure 9.74): (I-II) Due to the interaction of the chloride anion with the 5d orbitals of cerium, the addition of n-Bu_4NCl to the reaction mixture induced an intense light absorption. $CeCl_3$ then coordinated to the lignin model **A** to form a $Ce^{III}Cl_n$/lignin complex **B** through the deprotonation of the hydroxyl group, which subsequently went through the single-electron oxidation with DBAD* to generate a $Ce^{IV}Cl_n$/ lignin species **C** along with the release of DBAD−·; (III) The photoinduced LMCT homolysis of $Ce^{IV}Cl_n$/lignin species afforded the key alkoxy radical intermediate along with the release of $Ce^{III}Cl_3$ for the next catalytic cycle; (IV) Such alkoxy radical intermediate **D** could substantially weaken the adjacent C–C bond, thus enabling

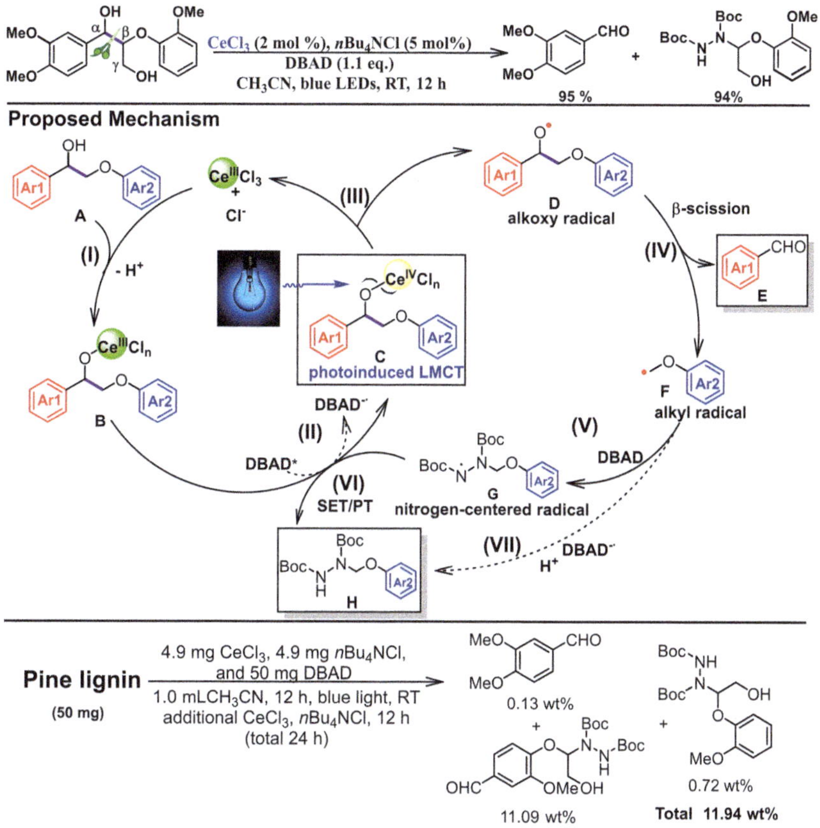

Figure 9.74 $CeCl_3$-promoted Photocatalytic Cleavage of C_α-C_β Bond in Lignin Model Compounds and Lignin. *Source:* Wang et al. [174] with Permission, © 2020 CCS Chemistry.

the β-scission of C_α–C_β bond [155] to generate the aldehyde product **E** and alkyl radical **F**; (V-VI) Subsequently, the reactive alkyl radical species could be coupled with DBAD to form the nitrogen-centered radical intermediate **G** [175, 176], which went through the single-electron-transfer/proton transfer (SET/PT) process with species **B** to generate the hydrazinium product **H** and species **C** for the next cycle; and (VII) In another competitive pathway, alkyl radical **F** could also react with $DBAD^{-\cdot}$ to produce the hydrazinium product.

9.8 Chapter Summary

In Chapter 9, strategies for lignin catalytic depolymerization that begin with the activation of C_α–H, C_α–OH, or C_αO–H bonds via a non-ionized route are summarized. Given the fact that the first cleavage of C_αO–H and C_α–H bonds can generate a β-O-4 ketone intermediate that can weaken the C_β–OAr bond and slightly strengthen the C_α–C_β bond but with an active C_β–H, the lignin depolymerization strategies of catalytic transfer hydrogenation (Chapter 9.1), dehydrogenation/oxidation-hydrogenation (reduction) (Chapter 9.2), and the multiple oxidation process (Chapter 9.3) are first summarized. Then, given the Baeyer-Villiger oxidation and Beckmann rearrangement can be used in the cleavage of C–C bonds in lignin linkages that contain a ketone substrate site, the corresponding strategies, and their derivative methods are summarized in Chapter 9.4. In addition, the first oxidation of C_α–OH can promote the lignin hydrolysis, an embellishing β-O-4 linkages hydrolysis method involving the C_α–OH first oxidation and C_γ–OH transformation is discussed in Chapter 9.5. Besides the first cleavage of C_αO–H and C_α–H bonds to C_α=O at the beginning of the β-O-4 model transformation, some catalytic strategies begin with the C_α–H or C_αO–H activation to the corresponding active radical intermediates can also lead to the cleavage of the lignin fragments. The first activation of C_α–OH(–H) to a $C_\alpha^\bullet$ radical intermediate can obviously reduce the BDE of the C_β–OAr bond for the following self-cleavage of β-O-4 linkage (Chapter 9.6). Besides, the first activation of C_αO–H bond to the $C_\alpha O^\bullet$ via a HAT, PCET oxidation or redox-neutral LMCT process can lead to the selective cleavage of C_α–C_β bonds (Chapter 9.7). The keywords of this chapter could be Transfer hydrogenation, Multiple oxidation, Oxidation-reduction, Baeyer-Villiger oxidation, Beckmann rearrangement, Radical intermediate, C–C Bond cleavage, Ether bond Cleavage.

References

1 Nichols, J.M., Bishop, L.M., Bergman, R.G., and Ellman, J.A. (2010). Catalytic C–O bond cleavage of 2-aryloxy-1-arylethanols and its application to the depolymerization of lignin-related polymers. *J. Am. Chem. Soc.* 132: 12554–12555.

2 Wu, A., Patrick, B.O., Chung, E., and James, B.R. (2012). Hydrogenolysis of β-O-4 lignin model dimers by a ruthenium-xantphos catalyst. *Dalton Trans.* 41: 11093–11106.

3 Chmely, S.C., Kim, S., Ciesielski, P.N. et al. (2013). Mechanistic study of a Ru-xantphos catalyst for tandem alcohol dehydrogenation and reductive aryl-ether cleavage. *ACS Catal.* 3: 963–974.

4 Huo, W., Li, W., Zhang, M. et al. (2014). Effective C–O bond cleavage of lignin β-O-4 model compounds: A new $RuHCl(CO)(PPh_3)_3$/KOH catalytic system. *Catal. Lett.* 144: 1159–1163.

5 Shaw, L., Somisara, D.M.U.K., How, R.C. et al. (2017). Electronic and bite angle effects in catalytic C–O bond cleavage of a lignin model compound using ruthenium Xantphos complexes. *Catal. Sci. Technol.* 7: 619–626.

6 vom Stein, T., Weigand, T., Merkens, C. et al. (2013). Trimethylenemethane-ruthenium(II)-triphos complexes as highly active catalysts for catalytic C–O bond cleavage reactions of lignin model compounds. *ChemCatChem* 5: 439–441.

7 Liu, C. and Wilson, A.K. (2015). Cleavage of the β-O-4 linkage of lignin using group 8 pincer complexes: A DFT study. *J. Mol. Catal. A: Chem.* 399: 33–41.

8 Liu, Y.X., Li, C.Z., Miao, W. et al. (2019). Mild redox-neutral depolymerization of lignin with a binuclear Rh complex in water. *ACS Catal.* 9: 4441–4447.

9 Liu, Y., Li, C., Miao, W. et al. (2020). Rhodium-terpyridine catalyzed redox-neutral depolymerization of lignin in water. *Green Chem.* 22: 33–38.

10 Gao, F., Webb, J.D., Sorek, H. et al. (2016). Fragmentation of lignin samples with commercial Pd/C under ambient pressure of hydrogen. *ACS Catal.* 6: 7385–7392.

11 Zhou, X., Mitra, J., and Rauchfuss, T.B. (2014). Lignol cleavage by Pd/C under mild conditions and without hydrogen: a role for benzylic C–H activation? *ChemSusChem* 7: 1623–1626.

12 Galkin, M.V., Sawadjoon, S., Rohde, V. et al. (2014). Mild heterogeneous palladium-catalyzed cleavage of β-O-4 '-ether linkages of lignin model compounds and native lignin in air. *ChemCatChem* 6: 179–184.

13 Galkin, M.V. and Samec, J.S.M. (2014). Selective route to 2-propenyl aryls directly from wood by a tandem organosolv and palladium-catalysed transfer hydrogenolysis. *ChemSusChem* 7: 2154–2158.

14 Galkin, M.V., Dahlstrand, C., and Samec, J.S.M. (2015). Mild and robust redox-neutral Pd/C-catalyzed lignol β-O-4 bond cleavage through a low-energy-barrier pathway. *ChemSusChem* 8: 2187–2192.

15 Horiuti, I. and Polanyi, M. (1934). Exchange reactions of hydrogen on metallic catalysts. *T. Faraday Soc.* 30: 1164–1172.

16 Zaera, F. (2013). Key unanswered questions about the mechanism of olefin hydrogenation catalysis by transition-metal surfaces: a surface-science perspective. *Phys. Chem. Chem. Phys.* 15: 11988–12003.

17 Srifa, P., Galkin, M.V., Samec, J.S.M. et al. (2016). Detecting important intermediates in Pd catalyzed depolymerization of a lignin model compound by a combination of DFT calculations and constrained minima hopping. *J. Phys. Chem. C* 120: 23469–23479.

18 Lu, J., Wang, M., Zhang, X. et al. (2016). β-O-4 bond cleavage mechanism for lignin model compounds over Pd catalysts identified by combination of first-principles calculations and experiments. *ACS Catal.* 6: 5589–5598.

19 de Andrade, A.M., Srifa, P., Broqvist, P., and Hermansson, K. (2020). Lignin intermediates on palladium: Insights into keto-enol tautomerization from theoretical modelling. *ChemSusChem* 13: 6574–6581.

20 Xu, W.Y., Miller, S.J., Agrawal, P.K., and Jones, C.W. (2012). Depolymerization and hydrodeoxygenation of switchgrass lignin with formic acid. *ChemSusChem* 5: 667–675.

21 Ould Hamou, C.A., Réocreux, R., Sautet, P. et al. (2017). Adsorption and decomposition of a lignin β-O-4 linkage model, 2-phenoxyethanol, on Pt(111): Combination of experiments and first-principles calculations. *J. Phys. Chem. C* 121: 9889–9900.

22 Zhang, J.W., Lu, G.P., and Cai, C. (2017). Self-hydrogen transfer hydrogenolysis of β-O-4 linkages in lignin catalyzed by MIL-100(Fe) supported Pd-Ni BMNPs. *Green Chem.* 19: 4538–4543.

23 Li, Q. and Lopez, N. (2018). Chirality, rigidity, and conjugation: A first-principles study of the key molecular aspects of lignin depolymerization on Ni-based catalysts. *ACS Catal.* 8: 4230–4240.

24 Luo, N., Wang, M., Li, H. et al. (2017). Visible-light-driven self-hydrogen transfer hydrogenolysis of lignin models and extracts into phenolic products. *ACS Catal.* 7: 4571–4580.

25 Zhang, B., Guo, T., Liu, Y. et al. (2021). Sustainable production of benzylamines from lignin. *Angew. Chem. Int. Ed.* 60: 20666–20671.

26 Lancefield, C.S., Ojo, O.S., Tran, F., and Westwood, N.J. (2015). Isolation of functionalized phenolic monomers through selective oxidation and C–O bond cleavage of the β-O-4 Linkages in lignin. *Angew. Chem. Int. Ed.* 54: 258–262.

27 Yao, Z., Ye, D., Liu, H., and Chen, K. (2007). Transformation of aryl acyloin O-alkyl and O-phenyl derivatives to ketones. *Synthetic Commun.* 37: 149–156.

28 Zhu, R., Wang, B., Cui, M.S. et al. (2016). Chemoselective oxidant-free dehydrogenation of alcohols in lignin using Cp*Ir catalysts. *Green Chem.* 18: 2029–2036.

29 Gall, D.L., Kim, H., Lu, F. et al. (2014). Stereochemical features of glutathione-dependent enzymes in the *Sphingobium* sp. strain SYK-6 β-aryl etherase pathway. *J. Biol. Chem.* 289: 8656–8667.

30 Pereira, J.H., Heins, R.A., Gall, D.L. et al. (2016). Structural and biochemical characterization of the early and late enzymes in the lignin β-aryl ether cleavage pathway from *Sphingobium* sp. SYK-6. *J. Biol. Chem.* 291: 10228–10238.

31 Helmich, K.E., Pereira, J.H., Gall, D.L. et al. (2016). Structural basis of stereospecificity in the bacterial enzymatic cleavage of β-aryl ether bonds in lignin. *J. Biol. Chem.* 291: 5234–5246.

32 Kontur, W.S., Olmsted, C.N., Yusko, L.M. et al. (2019). A heterodimeric glutathione S-transferase that stereospecifically breaks lignin's β (R)-aryl ether bond reveals the diversity of bacterial β-etherases. *J. Biol. Chem.* 294: 1877–1890.

33 Picart, P., Mueller, C., Mottweiler, J. et al. (2014). From gene towards selective biomass valorization: Bacterial β-etherases with catalytic activity on lignin-like polymers. *ChemSusChem* 7: 3164–3171.

34 Klinger, G.E., Zhou, Y., Foote, J.A. et al. (2020). Nucleophilic thiols reductively cleave ether linkages in lignin Mmdel polymers and lignin. *ChemSusChem* 13: 4394–4399.

35 Klinger, G.E., Zhou, Y., Hao, P. et al. (2019). Biomimetic Reductive Cleavage of Keto Aryl Ether Bonds by Small-Molecule Thiols. *ChemSusChem* 12: 4775–4779.

36 Reiter, J., Strittmatter, H., Wiemann, L.O. et al. (2013). Enzymatic cleavage of lignin β-O-4 aryl ether bonds via net internal hydrogen transfer. *Green Chem.* 15: 1373–1381.

37 Lu, F.C. and Ralph, J. (1997). DFRC method for lignin analysys. 1. New method for beta aryl ether cleavage: Lignin model studies. *J. Agr. Food. Chem.* 45: 4655–4660.

38 Zhang, C., Li, H., Lu, J. et al. (2017). Promoting lignin depolymerization and restraining the condensation via an oxidation–hydrogenation strategy. *ACS Catal.* 7: 3419–3429.

39 Zhou, L.P., Zhang, C.F., Fang, T. et al. (2011). Selective oxidation of alcohols catalyzed by a transition metal-free system of NHPI/DDQ/$NaNO_2$. *Chin. J. Catal.* 32: 118–122.

40 Zhang, C.F. and Wang, F. (2020). Catalytic lignin depolymerization to aromatic chemicals. *Acc. Chem. Res.* 53: 470–484.

41 Wang, M., Zhang, X., Li, H. et al. (2018). Carbon modification of nickel catalyst for depolymerization of oxidized lignin to aromatics. *ACS Catal.* 8: 1614–1620.

42 Rashidi, M., Konarova, M., Aslam, W., and Beltramini, J.N. (2018). TiN-Cu heterogeneous nanocatalysts for effective depolymerisation of oxidised lignin. *ChemistrySelect* 3: 3379–3385.

43 Nguyen, J.D., Matsuura, B.S., and Stephenson, C.R. (2014). A photochemical strategy for lignin degradation at room temperature. *J. Am. Chem. Soc.* 136: 1218–1221.

44 Karkas, M.D., Bosque, I., Matsuura, B.S., and Stephenson, C.R.J. (2016). Photocatalytic oxidation of lignin model systems by merging visible-light photoredox and palladium catalysis. *Org. Lett.* 18: 5166–5169.

45 Magallanes, G., Kärkäs, M.D., Bosque, I. et al. (2019). Selective C–O bond cleavage of lignin systems and polymers enabled by sequential palladium-catalyzed aerobic oxidation and visible-light photoredox catalysis. *ACS Catal.* 9: 2252–2260.

46 Stromskaya, G.I. and Chupka, E.I. (1983). Electrochemical reduction of lignin in liquid ammonia. 1. Change in the chromophore and molecularweight composition during electrochemical reduction in liquid ammonia. *Koksnes Kimija* 2: 60–63.

47 Bosque, I., Magallanes, G., Rigoulet, M. et al. (2017). Redox catalysis facilitates lignin depolymerization. *ACS Cent. Sci.* 3: 621–628.

48 Shiraishi, T., Takano, T., Kamitakahara, H., and Nakatsubo, F. (2012). Studies on electro-oxidation of lignin and lignin model compounds. Part 2: N-Hydroxyphthalimide (NHPI)-mediated indirect electro-oxidation of non-phenolic lignin model compounds. *Holzforschung* 66: 311–315.

49 Enright, M.J., Gilbert-Bass, K., Sarsito, H., and Cossairt, B.M. (2019). Photolytic C–O bond cleavage with quantum dots. *Chem. Mater.* 31: 2677–2682.

50 Tanaka, K., Calanag, R.C.R., and Hisanaga, T. (1999). Photocatalyzed degradation of lignin on TiO_2. *J. Mol. Catal. A: Chem.* 138: 287–294.

51 Luo, N., Wang, M., Li, H. et al. (2016). Photocatalytic oxidation–hydrogenolysis of lignin β-O-4 models via a dual light wavelength switching strategy. *ACS Catal.* 6: 7716–7721.

52 Li, H., Gao, Z., Lei, L. et al. (2020). Photocatalytic transfer hydrogenolysis of aromatic ketones using alcohols. *Green Chem.* 22: 3802–3808.

53 Fang, J., Ye, P., Wang, M. et al. (2018). Hydrogenolysis and hydrogenation of β-O-4 ketones by a simple photocatalytic hydrogen transfer reaction. *Catal. Commun.* 107: 18–23.

54 Chen, C., Liu, P., Xia, H. et al. (2020). Photocatalytic cleavage of β-O-4 ether bonds in lignin over Ni/TiO_2. *Molecules* 25: 2109.

55 Imamura, K., Kato, H., Wada, Y. et al. (2018). Photocatalytic chemoselective cleavage of C-O bonds under hydrogen gas- and acid-free conditions. *Chem. Commun.* 54: 7298–7301.

56 Nair, V., Dhar, P., and Vinu, R. (2016). Production of phenolics via photocatalysis of ball milled lignin-TiO_2 mixtures in aqueous suspension. *RSC Adv.* 6: 18204–18216.

57 Li, S.Y., Li, Z.J., Yu, H. et al. (2020). Solar-driven lignin oxidation via hydrogen atom transfer with a dye-sensitized TiO_2 photoanode. *ACS Energy Lett.* 5: 777–784.

58 Luo, J., Zhang, X., Lu, J., and Zhang, J. (2017). Fine tuning the redox potentials of carbazolic porous organic frameworks for visible-light photoredox catalytic degradation of lignin β-O-4 models. *ACS Catal.* 7: 5062–5070.

59 Zeman, C.J., Kim, S., Zhang, F., and Schanze, K.S. (2020). Direct observation of the reduction of aryl halides by a photoexcited perylene diimide radical anion. *J. Am. Chem. Soc.* 142: 2204–2207.

60 Li, S., Hao, Z., Wang, K. et al. (2020). Visible light-enabled selective depolymerization of oxidized lignin by an organic photocatalyst. *Chem. Commun.* 56: 11243–11246.

61 Hanson, S.K., Baker, R.T., Gordon, J.C. et al. (2009). Aerobic Oxidation of pinacol by vanadium(V) dipicolinate complexes: Evidence for reduction to vanadium(III). *J. Am. Chem. Soc.* 131: 428–429.

62 Hanson, S.K., Baker, R.T., Gordon, J.C. et al. (2010). Mechanism of alcohol oxidation by dipicolinate vanadium(V): Unexpected role of pyridine. *J. Am. Chem. Soc.* 132: 17804–17816.

63 Hanson, S.K., Baker, R.T., Gordon, J.C. et al. (2010). Aerobic oxidation of lignin models using a base metal vanadium catalyst. *Inorg. Chem.* 49: 5611–5618.

64 Hanson, S.K., Wu, R., and Silks, L.A.P. (2011). Mild and selective vanadium-catalyzed oxidation of benzylic, allylic, and propargylic alcohols using air. *Org. Lett.* 13: 1908–1911.

65 Sedai, B., Diaz-Urrutia, C., Baker, R.T. et al. (2011). Comparison of copper and vanadium homogeneous catalysts for aerobic oxidation of lignin mdels. *ACS Catal.* 1: 794–804.

66 Hanson, S.K., Wu, R.L., and Silks, L.A. (2012). C-C or C-O bond cleavage in a phenolic lignin model compound: Selectivity depends on vanadium catalyst. *Angew. Chem. Int. Ed.* 51: 3410–3413.

67 Sedai, B., Diaz-Urrutia, C., Baker, R.T. et al. (2013). Aerobic oxidation of β-1 lignin model compounds with copper and oxovanadium catalysts. *ACS Catal.* 3: 3111–3122.

68 Wigington, B.N., Drummond, M.L., Cundari, T.R. et al. (2012). A biomimetic pathway for vanadium-catalyzed aerobic oxidation of alcohols: Evidence for a base-assisted dehydrogenation mechanism. *Chem. Eur. J.* 18: 14981–14988.

69 Zhang, G., Scott, B.L., Wu, R. et al. (2012). Aerobic oxidation reactions catalyzed by vanadium complexes of bis(phenolate) ligands. *Inorg. Chem.* 51: 7354–7361.

70 Hanson, S.K. and Baker, R.T. (2015). Knocking on wood: Base metal complexes as catalysts for selective oxidation of lignin models and extracts. *Acc. Chem. Res.* 48: 2037–2048.

71 Sedai, B. and Baker, R.T. (2014). Copper catalysts for selective C-C bond cleavage of β-O-4 lignin model compounds. *Adv. Synth. Catal.* 356: 3563–3574.

72 Wang, M., Lu, J.M., Zhang, X.C. et al. (2016). Two-step, catalytic C-C bond oxidative cleavage process converts lignin models and extracts to aromatic acids. *ACS Catal.* 6: 6086–6090.

73 Wang, M., Li, L.H., Lu, J.M. et al. (2017). Acid promoted C–C bond oxidative cleavage of β-O-4 and β-1 lignin models to esters over a copper catalyst. *Green Chem.* 19: 702–706.

74 Wang, M., Lu, J.M., Li, L.H. et al. (2017). Oxidative C(OH)—C bond cleavage of secondary alcohols to acids over a copper catalyst with molecular oxygen as the oxidant. *J. Catal.* 348: 160–167.

75 Liu, H., Wang, M., Li, H. et al. (2017). New protocol of copper-catalyzed oxidative C(CO)—C bond cleavage of aryl and aliphatic ketones to organic acids using O_2 as the terminal oxidant. *J. Catal.* 346: 170–179.

76 Jagadeesh, R.V., Surkus, A.-E., Junge, H. et al. (2013). Nanoscale Fe_2O_3-based catalysts for selective hydrogenation of nitroarenes to anilines. *Science* 342: 1073–1076.

77 Westerhaus, F.A., Jagadeesh, R.V., Wienhofer, G. et al. (2013). Heterogenized cobalt oxide catalysts for nitroarene reduction by pyrolysis of molecularly defined complexes. *Nat. Chem.* 5: 537–543.

78 Jagadeesh, R.V., Junge, H., and Beller, M. (2014). Green synthesis of nitriles using non-noble metal oxides-based nanocatalysts. *Nat. Commun.* 5: 4123.

79 Jagadeesh, R.V., Stemmler, T., Surkus, A.-E. et al. (2015). Cobalt-based nanocatalysts for green oxidation and hydrogenation processes. *Nat. Protoc.* 10: 916–926.

80 Liu, S.J., Bai, L.C., van Muyden, A.P. et al. (2019). Oxidative cleavage of β-O-4 bonds in lignin model compounds with a single-atom Co catalyst. *Green Chem.* 21: 1974–1981.

81 Luo, H., Wang, L., Shang, S. et al. (2020). Cobalt Nanoparticles-catalyzed widely applicable successive C—C bond cleavage in alcohols to access esters. *Angew. Chem. Int. Ed.* 59: 19268–19274.

82 Luo, H., Weeda, E.P., Alherech, M. et al. (2021). Oxidative catalytic fractionation of lignocellulosic biomass under non-alkaline conditions. *J. Am. Chem. Soc.* 143: 15462–15470.

83 Yang, Y.Y., Fan, H.L., Song, J.L. et al. (2015). Free radical reaction promoted by ionic liquid: a route for metal-free oxidation depolymerization of lignin model compound and lignin. *Chem. Commun.* 51: 4028–4031.

84 Evans, D.F. (1953). Molecular association of oxygen and aromatic substances. *J. Chem. Soc.* 345–347.

85 Zhang, J., Liu, Y., Chiba, S., and Loh, T.-P. (2013). Chemical conversion of β-O-4 lignin linkage models through Cu-catalyzed aerobic amide bond formation. *Chem. Commun.* 49: 11439–11441.

86 Li, H., Liu, M., Liu, H. et al. (2020). Amine-mediated bond cleavage in oxidized lignin models. *ChemSusChem* 13: 4660–4665.

87 Liu, X., Zhang, H., Wu, C. et al. (2018). Copper-catalyzed synthesis of benzanilides from lignin model substrates 2-phenoxyacetophenones under an air atmosphere. *New J. Chem.* 42: 1223–1227.

88 Kang, Y., Lu, X.M., Zhang, G.J. et al. (2019). Metal-free photochemical degradation of lignin-derived aryl ethers and lignin by autologous radicals through ionic liquid induction. *ChemSusChem* 12: 4005–4013.

89 Gao, W.-J., Lam, C.M., Sun, B.-G. et al. (2017). Selective electrochemical C–O bond cleavage of β-O-4 lignin model compounds mediated by iodide ion. *Tetrahedron* 73: 2447–2454.

90 Patil, N.D. and Meier, M.S. (2013). Depolymerization of lignin by selective cleavage of the Cα-Cβ linkage via Baeyer-Villiger oxidation. *Abstr. Pap. Am. Chem. S.* 245.

91 Zhu, T., Kadla, J.F., Chang, H.M., and Jameel, H. (2003). Reactions of lignin with peroxymonophosphoric acid: The degradation of lignin model compounds. *Holzforschung* 57: 44–51.

92 Li, X.-H., Meng, X.-G., Liu, Y., and Peng, X. (2013). Direct oxidation of secondary alcohol to ester by performic acid. *Green Chem.* 15: 3332–3336.

93 Argyropoulos, D.S., Sun, Y., Berry, R.M., and Bouchard, J. (1996). Reactions of dimethyldioxirane with lignin model compounds. *J. Pulp. Pap. Sci.* 22: J84–J90.

94 Bouchard, J., Maine, C., Berry, R.M., and Argyropoulos, D.S. (1996). Kraft pulp bleaching using dimethyldioxirane: Stability of the oxidants. *Can. J. Chem.* 74: 232–237.

95 Wang, Y., Du, Y., He, J., and Zhang, Y. (2018). Transformation of lignin model compounds to N-substituted aromatics via Beckmann rearrangement. *Green Chem.* 20: 3318–3326.

96 Li, H., Wang, M., Liu, H. et al. (2018). NH_2OH–mediated lignin conversion to isoxazole and nitrile. *ACS Sustainable Chem. Eng.* 6: 3748–3753.

97 Crestini, C., Pro, P., Neri, V., and Saladino, R. (2005). Methyltrioxorhenium: a new catalyst for the activation of hydrogen peroxide to the oxidation of lignin and lignin model compounds. *Bioorg. Med. Chem.* 13: 2569–2578.

98 Bianchini, G., Crucianelli, M., Crestini, C., and Saladino, R. (2006). Catalytic MTO-based C–H insertion reactions of hydrogen peroxide: an investigation on the polymeric support role in heterogeneous conditions. *Top. Catal.* 40: 221–227.

99 Crestini, C., Caponi, M.C., Argyropoulos, D.S., and Saladino, R. (2006). Immobilized methyltrioxo rhenium (MTO)/H_2O_2 systems for the oxidation of lignin and lignin model compounds. *Bioorg. Med. Chem.* 14: 5292–5302.

100 Rahimi, A., Azarpira, A., Kim, H. et al. (2013). Chemoselective metal-free aerobic alcohol oxidation in lignin. *J. Am. Chem. Soc.* 135: 6415–6418.

101 Patil, N.D., Yao, S.G., Meier, M.S. et al. (2015). Selective cleavage of the Cα–Cβ linkage in lignin model compounds via Baeyer-Villiger oxidation. *Org. Biomol. Chem.* 13: 3243–3254.

102 Jennings, J.A., Parkin, S., Munson, E. et al. (2017). Regioselective Baeyer-Villiger oxidation of lignin model compounds with tin beta zeolite catalyst and hydrogen peroxide. *RSC Adv.* 7: 25987–25997.

103 Wang, Y.L., Wang, Q.Y., He, J.H., and Zhang, Y.T. (2017). Highly effective C–C bond cleavage of lignin model compounds. *Green Chem.* 19: 3135–3141.

104 Santos, W.C.C., Dias, K.A., Santos, L.P. et al. (2018). Evaluating gold and selenium Chemistry for selective transformations of lignin model compounds. *Adv. Synth. Catal.* 360: 1376–1383.

105 Hao, K., Zhang, L.-L., Song, L. et al. (2020). Highly active Mo-V-based bifunctional catalysts for catalytic conversion of lignin dimer model compounds at room temperature. *Inorg. Chem. Commun.* 116: 107910.

106 Li, H.J., Bunrit, A., Lu, J.M. et al. (2019). Photocatalytic cleavage of aryl ether in modified lignin to non-phenolic aromatics. *ACS Catal.* 9: 8843–8851.

107 Tay, N.E.S. and Nicewicz, D.A. (2017). Cation radical accelerated nucleophilic aromatic substitution via organic photoredox catalysis. *J. Am. Chem. Soc.* 139: 16100–16104.

108 Liu, J.Z., Qiu, X., Huang, X.Q. et al. (2019). From alkylarenes to anilines via site-directed carbon-carbon amination. *Nat. Chem.* 11: 71–77.

109 Mei, Q., Liu, H., Shen, X. et al. (2017). Selective utilization of the methoxy group in lignin to produce acetic acid. *Angew. Chem. Int. Ed.* 56: 14868–14872.

110 Mei, Q.Q., Shen, X.J., Liu, H.Z. et al. (2019). Selective utilization of methoxy groups in lignin for N-methylation reaction of anilines. *Chem. Sci.* 10: 1082–1088.

111 Rahimi, A., Ulbrich, A., Coon, J.J., and Stahl, S.S. (2014). Formic-acid-induced depolymerization of oxidized lignin to aromatics. *Nature* 515: 249–252.

112 Fouad, F.M. and Farrell, P.G. (1978). Primary and secondary kinetic isotope effects in E2-elimination reactions. *Tetrahedron Lett.* 19: 4735–4738.

113 Das, A., Rahimi, A., Ulbrich, A. et al. (2018). Lignin conversion to low-molecular-weight aromatics via an aerobic oxidation-hydrolysis sequence: Comparison of different lignin sources. *ACS Sustainable Chem. Eng.* 6: 3367–3374.

114 Qu, S.L., Dang, Y.F., Song, C.Y. et al. (2015). Depolymerization of oxidized lignin catalyzed by formic acid exploits an unconventional elimination mechanism involving 3c-4e bonding: A DFT mechanistic study. *ACS Catal.* 5: 6386–6396.

115 Chan, M.K., Ye, Q., Png, Z.M. et al. (2017). Valorization of lignin: Effective conversion of depolymerized lignin to oil by simple chemical modifications. *Waste Biomass Valor.* 8: 2029–2036.

116 Gasser, C.A., Cvancarova, M., Ammann, E.M. et al. (2017). Sequential lignin depolymerization by combination of biocatalytic and formic acid/formate treatment steps. *Appl. Microbiol. Biotechnol.* 101: 2575–2588.

117 Dai, J., Patti, A.F., Longé, L. et al. (2017). Oxidized lignin depolymerization using rormate ionic liquid as catalyst and solvent. *ChemCatChem* 9: 2684–2690.

118 Oregui-Bengoechea, M., Gandarias, I., Arias, P.L., and Barth, T. (2017). Unraveling the role of formic acid and the type of solvent in the catalytic conversion of lignin: A holistic approach. *ChemSusChem* 10: 754–766.

119 Zhang, C., Lu, J., Zhang, X. et al. (2016). Cleavage of the lignin β-O-4 ether bond via a dehydroxylation-hydrogenation strategy over a NiMo sulfide catalyst. *Green Chem.* 18: 6545–6555.

120 Wu, X.J., Fan, X.T., Xie, S.J. et al. (2018). Solar energy-driven lignin-first approach to full utilization of lignocellulosic biomass under mild conditions. *Nat. Catal.* 1: 772–780.

121 Autrey, S.T., Alnajjar, M.S., Nelson, D.A., and Franz, J.A. (1991). Absolute rate constants for the beta-scission reaction of the 1-phenyl-2-phenoxypropyl radical - a model for radical reactions of lignin. *J. Org. Chem.* 56: 2197–2202.

122 Kandanarachchi, P.H., Autrey, T., and Franz, J.A. (2002). Model compound studies of the β-O-4 linkage in lignin: Absolute rate expressions for beta-scission of phenoxyl radical from 1-phenyl-2-phenoxyethanol-1-yl radical. *J. Org. Chem.* 67: 7937–7945.

123 Bourbonnais, R. and Paice, M.G. (1990). Oxidation of nonphenolic substrates - an expanded role for laccase in lignin biodegradation. *FEBS Lett.* 267: 99–102.

124 Munk, L., Sitarz, A.K., Kalyani, D.C. et al. (2015). Can laccases catalyze bond cleavage in lignin? *Biotechnol. Adv.* 33: 13–24.

125 Kawai, S., Nakagawa, M., and Ohashi, H. (1999). Aromatic ring cleavage of a non-phenolic beta-O-4 lignin model dimer by laccase of Trametes versicolor in the presence of 1-hydroxybenzotriazole. *FEBS Lett.* 446: 355–358.

126 Kawai, S., Asukai, M., Ohya, N. et al. (1999). Degradation of a non-phenolic β-O-4 substructure and of polymeric lignin model compounds by laccase of Coriolus versicolor in the presence of 1-hydroxybenzotriazole. *Fems. Microbiol. Lett.* 170: 51–57.

127 Kawai, S., Nakagawa, M., and Ohashi, H. (2002). Degradation mechanisms of a nonphenolic β-O-4 lignin model dimer by Trametes versicolor laccase in the presence of 1-hydroxybenzotriazole. *Enzyme. Microb. Tech.* 30: 482–489.

128 Kawai, S., Umezawa, T., and Higuchi, T. (1988). Degradation mechanisms of phenolic β-1 lignin substructure model compounds by laccase of Coriolus versicolor. *Arch. Biochem. Biophys.* 262: 99–110.

129 Srebotnik, E. and Hammel, K.E. (2000). Degradation of nonphenolic lignin by the laccase/1-hydroxybenzotriazole system. *J Biotechnol* 81: 179–188.

130 Heap, L., Green, A., Brown, D. et al. (2014). Role of laccase as an enzymatic pretreatment method to improve lignocellulosic saccharification. *Catal. Sci. Technol.* 4: 2251–2259.

131 Hilgers, R., Twentyman-Jones, M., van Dam, A. et al. (2019). The impact of lignin sulfonation on its reactivity with laccase and laccase/HBT. *Catal. Sci. Technol.* 9: 1535–1542.

132 Li, K.C., Xu, F., and Eriksson, K.E.L. (1999). Comparison of fungal laccases and redox mediators in oxidation of a nonphenolic lignin model compound. *Appl. Environ. Microb.* 65: 2654–2660.

133 Fabbrini, M., Galli, C., and Gentili, P. (2002). Comparing the catalytic efficiency of some mediators of laccase. *J Mol Catal B-Enzym* 16: 231–240.

134 Ibarra, D., Romero, J., Martinez, M.J. et al. (2006). Exploring the enzymatic parameters for optimal delignification of eucalypt pulp by laccase-mediator. *Enzyme. Microb. Tech.* 39: 1319–1327.

135 Barreca, A.M., Fabbrini, M., Galli, C. et al. (2003). Laccase/mediated oxidation of a lignin model for improved delignification procedures. *J Mol Catal B-Enzym* 26: 105–110.

136 Moldes, D., Diaz, M., Tzanov, T., and Vidal, T. (2008). Comparative study of the efficiency of synthetic and natural mediators in laccase-assisted bleaching of eucalyptus kraft pulp. *Bioresour. Technol.* 99: 7959–7965.

137 Hilgers, R., van Dam, A., Zuilhof, H. et al. (2020). Controlling the competition: Boosting laccase/HBT-catalyzed cleavage of a β-O-4′ linked lignin model. *ACS Catal.* 10: 8650–8659.

138 Parsell, T.H., Owen, B.C., Klein, I. et al. (2013). Cleavage and hydrodeoxygenation (HDO) of C–O bonds relevant to lignin conversion using Pd/Zn synergistic catalysis. *Chem. Sci.* 4: 806–813.

139 Klein, I., Marcum, C., Kenttamaaa, H., and Abu-Omar, M.M. (2016). Mechanistic investigation of the Zn/Pd/C catalyzed cleavage and hydrodeoxygenation of lignin. *Green Chem.* 18: 2399–2405.

140 Dou, X., Li, W., Zhu, C. et al. (2020). Cleavage of aryl–ether bonds in lignin model compounds using a Co-Zn-beta catalyst. *RSC Adv.* 10: 43599–43606.

141 Haibach, M.C., Lease, N., and Goldman, A.S. (2014). Catalytic cleavage of ether C–O bonds by pincer iridium complexes. *Angew. Chem. Int. Ed.* 53: 10160–10163.

142 Dou, X., Jiang, X., Li, W. et al. (2020). Highly efficient conversion of kraft lignin into liquid fuels with a Co-Zn-beta zeolite catalyst. *Appl. Catal., B* 268: 118429.

143 Müller, U., Rätzsch, M., Schwanninger, M. et al. (2003). Yellowing and IR-changes of spruce wood as result of UV-irradiation. *J. Photochem. Photobiol. B, Biol.* 69: 97–105.

144 Fabbri, C., Bietti, M., and Lanzalunga, O. (2005). Generation and reactivity of ketyl radicals with lignin related structures. On the importance of the ketyl pathway in the photoyellowing of lignin containing pulps and papers. *J. Org. Chem.* 70: 2720–2728.

145 Wu, X.J., Xie, S.J., Liu, C.X. et al. (2019). Ligand-controlled photocatalysis of CdS quantum dots for lignin valorization under visible light. *ACS Catal.* 9: 8443–8451.

146 Yoo, H., Lee, M.-W., Lee, S. et al. (2020). Enhancing photocatalytic β-O-4 bond cleavage in lignin model compounds by silver-exchanged cadmium sulfide. *ACS Catal.* 10: 8465–8475.

147 Lin, J., Wu, X., Xie, S. et al. (2019). Visible-light-driven cleavage of C—O Linkage for lignin valorization to functionalized aromatics. *ChemSusChem* 12: 5023–5031.

148 Zhou, Y.T., Klinger, G.E., Hegg, E.L. et al. (2020). Multiple mechanisms mapped in aryl alkyl ether cleavage via aqueous electrocatalytic hydrogenation over skeletal nickel. *J. Am. Chem. Soc.* 142: 4037–4050.

149 Son, S. and Toste, F.D. (2010). Non-oxidative vanadium-catalyzed C–O bond cleavage: Application to degradation of lignin model compounds. *Angew. Chem. Int. Ed.* 49: 3791–3794.

150 Chan, J.M.W., Bauer, S., Sorek, H. et al. (2013). Studies on the vanadium-catalyzed nonoxidative depolymerization of *Miscanthus giganteus*-derived lignin. *ACS Catal.* 3: 1369–1377.

151 Ma, Y.Y., Du, Z.T., Liu, J.X. et al. (2015). Selective oxidative C–C bond cleavage of a lignin model compound in the presence of acetic acid with a vanadium catalyst. *Green Chem.* 17: 4968–4973.

152 Zombeck, A., Drago, R.S., Corden, B.B., and Gaul, J.H. (1981). Activation of molecular-oxygen - Mechanistic studies of the oxidation of hindered phenols with cobalt-dioxygen complexes. *J. Am. Chem. Soc.* 103: 7580–7585.

153 Jiang, Y.Y., Yan, L., Yu, H.Z. et al. (2016). Mechanism of vanadium-catalyzed selective C–O and C–C cleavage of lignin model compound. *ACS Catal.* 6: 4399–4410.

154 Gu, X., Zhang, W., and Salomon, R.G. (2012). Fragmentation of beta-hydroxy hydroperoxides. *J. Org. Chem.* 77: 1554–1559.

155 Yayla, H.G., Wang, H., Tarantino, K.T. et al. (2016). Catalytic ring-opening of cyclic alcohols enabled by PCET activation of strong O–H bonds. *J. Am. Chem. Soc.* 138: 10794–10797.

156 Weinberg, D.R., Gagliardi, C.J., Hull, J.F. et al. (2012). Proton-coupled electron transfer. *Chem. Rev.* 112: 4016–4093.

157 Warren, J.J., Tronic, T.A., and Mayer, J.M. (2010). Thermochemistry of proton-coupled electron transfer reagents and its implications. *Chem. Rev.* 110: 6961–7001.

158 Gentry, E.C. and Knowles, R.R. (2016). Synthetic applications of proton-coupled electron transfer. *Acc. Chem. Res.* 49: 1546–1556.

159 Reece, S.Y. and Nocera, D.G. (2009). Proton-coupled electron transfer in biology: results from synergistic studies in natural and model systems. *Annu. Rev. Biochem.* 78: 673–699.

160 Mitchell, L.J. and Moody, C.J. (2014). Solar photochemical oxidation of alcohols using catalytic hydroquinone and copper nanoparticles under oxygen: Oxidative cleavage of lignin models. *J. Org. Chem.* 79: 11091–11100.

161 Nguyen, T.M. and Nicewicz, D.A. (2013). Anti-Markovnikov hydroamination of alkenes catalyzed by an organic photoredox system. *J. Am. Chem. Soc.* 135: 9588–9591.

162 Romero, N.A. and Nicewicz, D.A. (2014). Mechanistic insight into the photoredox catalysis of anti-markovnikov alkene hydrofunctionalization reactions. *J. Am. Chem. Soc.* 136: 17024–17035.

163 Ota, E., Wang, H., Frye, N.L., and Knowles, R.R. (2019). A redox strategy for light-driven, out-of-equilibrium isomerizations and application to catalytic C–C bond cleavage reactions. *J. Am. Chem. Soc.* 141: 1457–1462.

164 Nguyen, S.T., Murray, P.R.D., and Knowles, R.R. (2020). Light-driven depolymerization of native lignin enabled by proton coupled electron transfer. *ACS Catal.* 10: 800–805.

165 Wang, Y., Liu, Y., He, J., and Zhang, Y. (2019). Redox-neutral photocatalytic strategy for selective C–C bond cleavage of lignin and lignin models via PCET process. *Sci. Bull.* 64: 1658–1666.

166 Zhou, W., Nakahashi, J., Miura, T., and Murakami, M. (2018). Light/copper relay for aerobic fragmentation of lignin model compounds. *Asian. J. Org. Chem.* 7: 2431–2434.

167 Gazi, S., Ng, W.K.H., Ganguly, R. et al. (2015). Selective photocatalytic C–C bond cleavage under ambient conditions with earth abundant vanadium complexes. *Chem. Sci.* 6: 7130–7142.

168 Gazi, S., Đokić, M., Moeljadi, A.M.P. et al. (2017). Kinetics and DFT studies of photoredox carbon–carbon bond cleavage reactions by molecular vanadium catalysts under ambient conditions. *ACS Catal.* 4682–4691.

169 Liu, H.F., Li, H.J., Luo, N.C., and Wang, F. (2020). Visible-light-induced oxidative lignin C-C bond cleavage to aldehydes using vanadium catalysts. *ACS Catal.* 10: 632–643.

170 Zhao, K., Yamashita, K., Carpenter, J.E. et al. (2019). Catalytic ring expansions of cyclic alcohols enabled by proton-coupled electron transfer. *J. Am. Chem. Soc.* 141: 8752–8757.

171 Guo, J.J., Hu, A., Chen, Y. et al. (2016). Photocatalytic C–C bond cleavage and amination of cycloalkanols by cerium (III) chloride complex. *Angew. Chem. Int. Ed.* 55: 15319–15322.

172 Hu, A., Guo, J.-J., Pan, H. et al. (2018). δ-Selective functionalization of alkanols enabled by visible-light-induced ligand-to-metal charge transfer. *J. Am. Chem. Soc.* 140: 1612–1616.

173 Hu, A., Guo, J.-J., Pan, H., and Zuo, Z. (2018). Selective functionalization of methane, ethane, and higher alkanes by cerium photocatalysis. *Science* 361: 668–672.

174 Wang, Y., He, J., and Zhang, Y. (2020). $CeCl_3$-Promoted simultaneous photocatalytic cleavage and amination of Cα–Cβ bond in lignin model compounds and native lignin. *CCS Chem.* 2: 107–117.

175 Amaoka, Y., Kamijo, S., Hoshikawa, T., and Inoue, M. (2012). Radical amination of C (sp^3)–H bonds using N-hydroxyphthalimide and dialkyl azodicarboxylate. *J. Org. Chem.* 77: 9959–9969.

176 Ryu, I., Tani, A., Fukuyama, T. et al. (2013). Efficient C–H/C–N and C–H/C–CO–N conversion via decatungstate-photoinduced alkylation of diisopropyl azodicarboxylate. *Org. Lett.* 15: 2554–2557.

10

Lignin Linkages Cleavage Beginning with C_β–H, C_γ–H, or C_γO–H Direct Activation

10.1 Lignin C–C/C_β–OAr Bond Cleavage Beginning with C_β–H Bond Direct Activation

To achieve the catalytic cleavage of the C–C bond in lignin linkages, besides the first SET oxidation at the aromatic ring (Chapter 7.2), based-promoted depolymerization-oxidation (Chapter 8.1) and multiple oxidation (Chapter 9.3) with the first activation of C_α–OH, and the PCET process beginning with the activation of C_αO–H (Chapter 9.7), the direct C_β–H activation could also provide access for the introduction of the cleavage reagent (Figure 10.1).

However, the direct and selective activation of the C_β–H bonds in lignin β-O-4 and β-1 linkages is an interesting but challenging work, which needs to compete with the thermodynamically superior processes of C_α–OH oxidation/dehydrogenation to C_α=O [2–5]. Besides the thermocatalytic process with C_β–H activation after the activation of Cα–OH to C_α=O (Chapter 9.3) [4], and some enzymatic process with the direct HAT of the C_β–H [2], the main route of some photocatalytic systems was found to begin with the first activation of C_β–H bond, which can induce the cleavage of C–C bonds in β–1 and β–O-4 linkages. In 2017, Hou and co-workers [1] found that a hybrid CuO_x/ceria/anatase ($CuO_x/CeO_2/TiO_2$) nanotube catalyst can efficiently catalyze the photocleavage of the C_α–C_β bond in lignin β-1 models under visible light irradiation, leading to high yields of substituted benzaldehydes (Figure 10.2a). The results from the reaction with trapping agents suggest that photoexcited electrons, holes, and superoxide anion radicals are essential species for the reaction. Mechanism investigation by the deuterium labeling experiments indicates hydrogen abstraction from β-carbon by photogenerated holes is a vital step (Figure 10.2b). They then proposed a mechanism for the C–C bond cleavage (Figure 10.2c). A photogenerated electron is transferred to O_2, producing a superoxide anion radical. The hole abstracts a hydrogen from C_β to generate a C_β-centered radical. The superoxide anion radical then adds to a C_β-centered radical, forming an unstable peroxide intermediate. The latter will undergo C_α–C_β cleavage with the elimination of H_2O to form benzaldehyde through a six-membered ring transition

Lignin Conversion Catalysis: Transformation to Aromatic Chemicals, First Edition.
Chaofeng Zhang and Feng Wang.

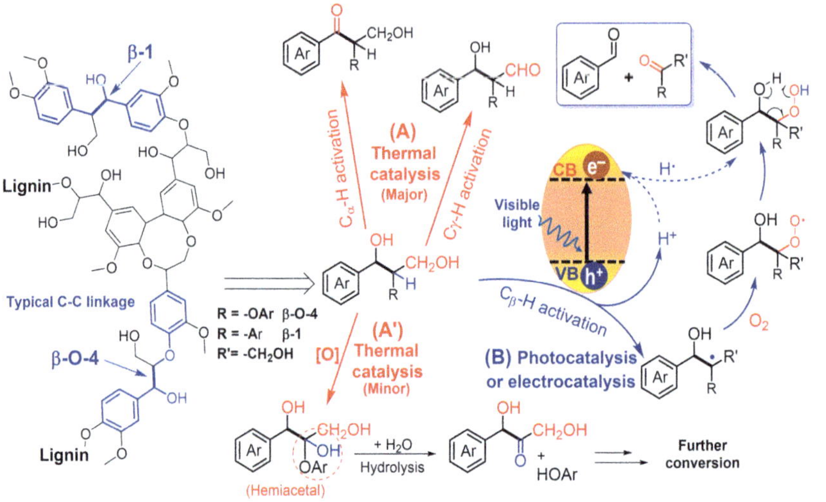

Figure 10.1 Lignin Linkages Cleavage Beginning with C_αO–H, or C_β–H Activation.

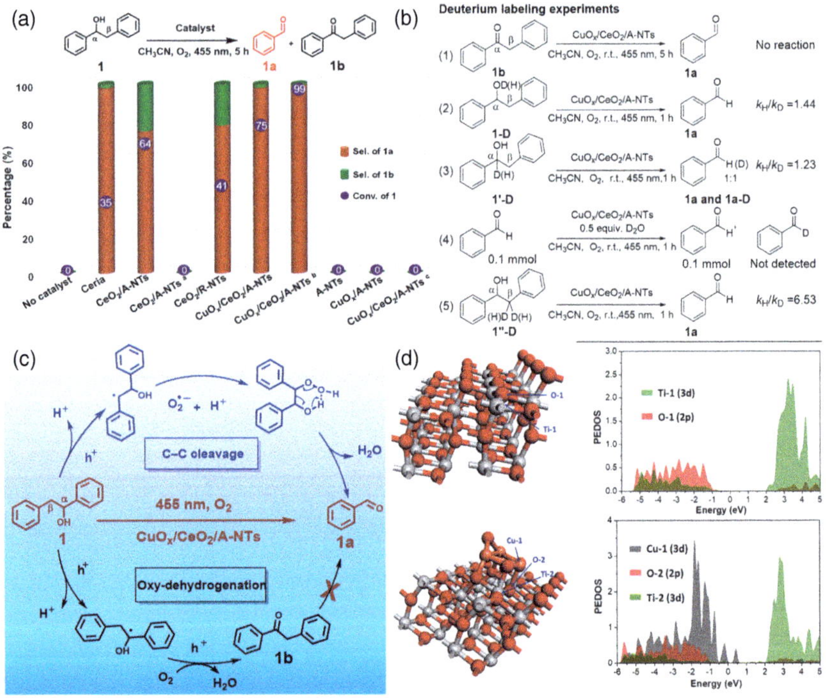

Figure 10.2 Photocatalytic C–C Bond Cleavage of β-1 Lignin Models to Aromatic Aldehydes over the CuO_x/Ceria/A-NTs under Blue LEDs. *Source*: Hou et al. [1] with Permission, © 2017 American Chemical Society.

state. Detailed characterizations were carried out to understand the catalysis. Characterization results by high angle annular dark-field scanning transmission electron microscopy (HAADF-STEM) and energy-dispersive X-ray spectroscopy element (EDX) mapping reveal that CuO_x clusters are highly dispersed on the exposed anatase surface as well as on the nanosized ceria domains. Investigations by Raman and ultraviolet-visible diffuse reflectance spectra (UV–vis DRS), together with density functional theory (DFT) calculations, further verify that the CuO_x clusters present on the ceria domains increase the concentration of surface defects (Ce^{3+} ions and oxygen vacancies) and accordingly improve the photocatalytic activity of O_2 activation for the C_α–C_β bond cleavage. Besides, UV-vis characterizations and DFT calculation about the partial electronic density of states (PEDOS) of CuO_x/anatase shows that the CuO_x clusters decorating will lead to an upward shift in the valence band (VB) edge of anatase and a decrease in the activity of holes. The above change on CuO_x/anatase can suppress the side reaction (Figure 10.2c, oxy-dehydrogenation without C–C bond cleavage).

Then, in 2018, Liu and co-workers [5] reported a mild photocatalytic oxidative strategy for C–C bond cleavage of lignin β-O-4 and β-1 linkages using a mesoporous graphitic carbon nitride catalyst (mpg-C_3N_4) (Figure 10.3). Identifications by solid-state NMR techniques and DFT calculations indicate that π-π stacking interactions are most likely present between the flexible carbon nitride surface and the lignin model molecule (Figure 10.3c). Besides, low charge recombination efficiency and high specific surface area (206.5 $m^2\ g^{-1}$) of the catalyst also contribute to its high catalytic activity. Mechanistic investigations reveal that photogenerated holes, as the main active species, trigger the oxidation and C–C bond cleavage of lignin models. In detail, mpg-C_3N_4 is first excited by visible light irradiation, generating holes and electrons separately in the valence and conduction band. For the reactant molecule interacting with the mpg-C_3N_4 surface, a hydrogen from Cβ is abstracted under the role of photogenerated holes and surface basic sites on mpg-C_3N_4, generating a C_β-centered radical **A** (Figure 10.3b, Mechanism), which further combines with O_2 and hydrogen to give a peroxide intermediate **C**. The following electron transfer in **C** through a six-membered ring transition state would induce the C_α–C_β and O–O bond cleavage, forming aromatic aldehyde and phenyl formate as the major products. Part of the aldehyde product can be further oxidized to benzoic acid.

Although the direct C_β–H activation is not a thermodynamically superior process in most lignin catalytic systems, few thermal catalytic researches showed the existence of this catalytic mechanism. In 2016, Ma and co-worker [3] reported a different mechanism with direct C_β–H activation, which is a little different from the above photocatalysis. Before this work, Baker and co-workers 6, 7] have studied vanadium-catalyzed aerobic oxidative cleavage of 2-phenoxy-1-phenylethanol. However, in their publications, there was no description of the presence of benzaldehyde as a product. In the following Ma's study of $VO(acac)_2$-catalyzed oxidation of 2-phenoxy-1-phenylethanol [3], the formation of benzaldehyde was remarkable in that it occurred despite its low selectivity, in contrast to previous reports. Different from the vanadium-mediated PECT process with a $C_\alpha O^\bullet$ radical intermediate that induced the cleavage of C–C bonds [6, 8, 9], Ma and co-worker studied the reaction pathways with benzaldehyde generation and further provide a hemiacetal mechanism for the cleavage of lignin C–OAr bond and C–C bond (Figure 10.4). In Route B, the V-based catalytic system could mediate the direct C_β–H transformation to C_β–OH. The unstable hemiacetal intermediate could further induce the cleavage of the C–OAr bond via

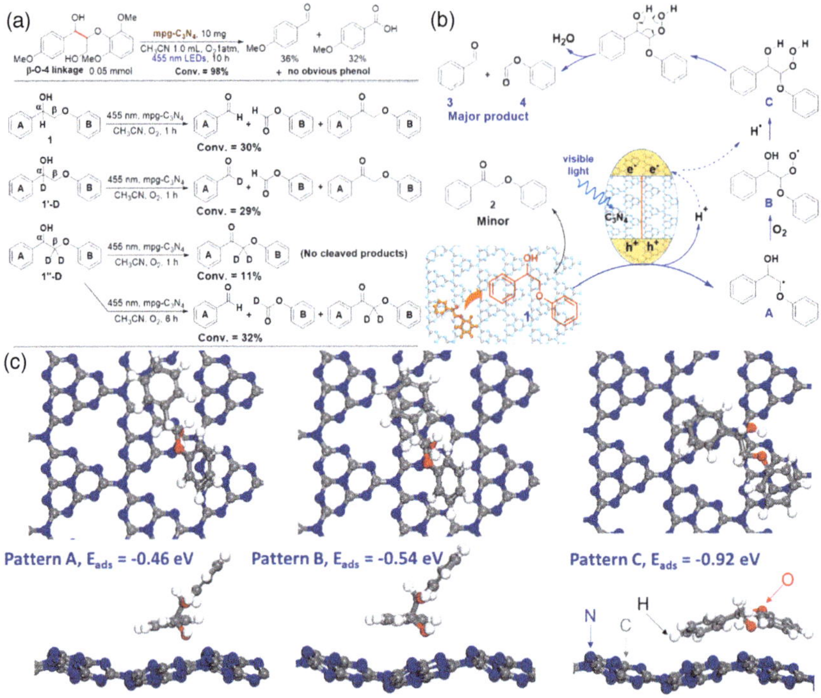

Figure 10.3 Photocatalytic Cleavage of C–C Bond in Lignin Models under Visible Light on Mesoporous Graphitic Carbon Nitride through π–π Stacking Interaction. *Source:* Liu et al. [5] with Permission, © 2018 American Chemical Society.

hydrolysis or induce the cleavage of the C–C bond in the β-O-4 linkage via selective oxidation. The generated fragment could further transform into benzaldehyde and benzoic acid. A similar mechanism could be confirmed by the selective oxidation of the α-O-4 lignin model, which could also provide benzaldehyde. Compared with direct C–O bond cleavage, the prior formation of hemiacetal-like structures via oxidation would be expected to facilitate the cleavage of the aryl ether bond in aralkyl aryl ethers under mild reaction conditions.

In addition, in 2021, Cui and co-workers [10] studied the electro-catalytic oxidation of β-O-4 linkage with the atomically dispersed Pt–N_3C_1 sites planted on nitrogen-doped carbon nanotubes (Pt_1/N-CNTs) as the anode, which provided a better selectivity in lignin C–C bond cleavage than traditional electrocatalyst of bulk metal and metal oxides. Furthermore, Pt_1/N-CNTs exhibits 99% substrate conversion with an 81% yield of benzaldehyde. For the superior performance compared with previously reported electrocatalysts, systematic experimental investigation together with DFT calculation suggests that it arises from the atomically dispersed Pt–N_3C_1 sites facilitating the formation of a key C_β radical intermediate, further inducing a radical/radical cross-coupling path to break the C_α–C_β bond (Figure 10.5). In detail, first, the tert-butoxyl radical (^{t}BuO$^\bullet$) **R1** and tertbutyl peroxyl radical (^{t}BuOO$^\bullet$) **R2** are generated from TBHP electrooxidation (**I**). Subsequently, C_β radical **(Rβ)** is produced via HAT of the C_β–H bond of lignin model compound **A** on the Pt_1/N CNTs anode

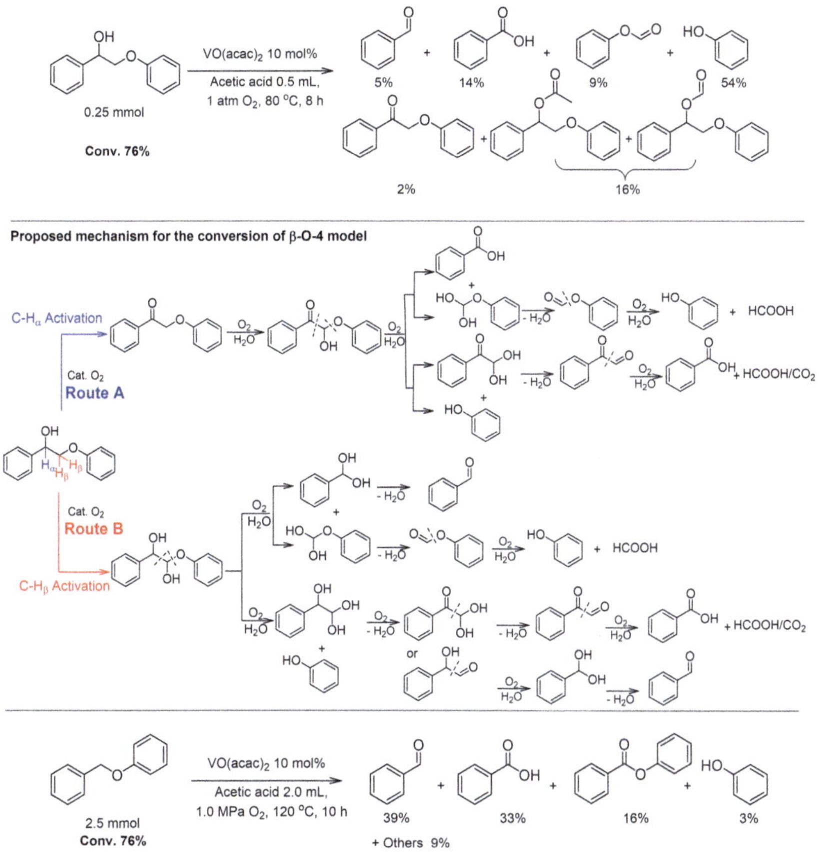

Figure 10.4 VO(acac)$_2$-catalyzed Oxidative Cleavage of 2-Phenoxy-1-phenylethanol in Acetic Acid. *Source:* Ma et al. [3] with Permission, © 2016 The Royal Society of Chemistry.

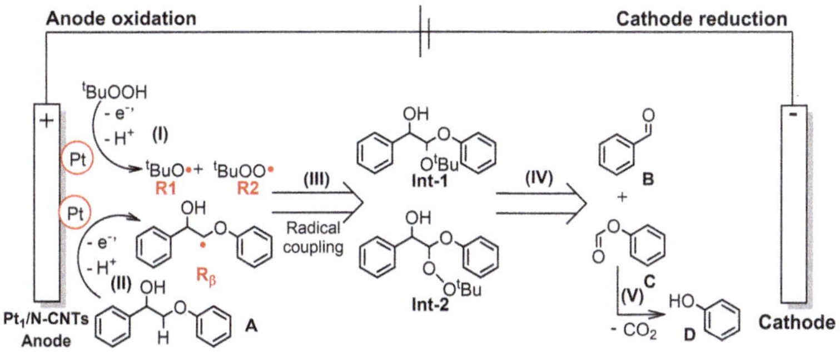

Figure 10.5 Pt$_1$/N-CNTs Catalyze the Electrooxidation of β-O-4 Linkage with C_α–C_β Bond Cleavage. *Source:* Cui et al. [10] with Permission, © 2021 American Chemical Society.

surface (**II**). Lastly, **R1**/**R2** reacts with the unstable **Rβ** to generate intermediate **Int-1** and **Int-2** via a radical/radical cross-coupling reaction (**III**). The reaction is followed by electron transferring in **Int-1** and **Int-2**, inducing the cleavage of the C_α–C_β bond and the generation of aromatic aldehyde, phenol, CO_2.

10.2 First $-C_\gamma H_2OH$ Activation to $-C_\gamma HO/-C_\gamma OOR$ Inducing Lignin C–C Bonds Selective Cleavage and Its Derivative Methods

As we discussed above, the first oxidation of secondary alcohol (C_αHOH) to ketones, $C_\alpha HO^\bullet$ or $C_\alpha^\bullet$ radicals has inspired many methods and strategies to cleave lignin β-O-4 and other linkages. Then, the methods with first C_β–H activation is discussed above, the next question is about whether the C_γ first activation in the branch chain of phenylpropane units can induce the lignin linkages cleavage. This subsection will focus on the methods for this question.

According to previous reports about the transformation of polyol and C_5–C_6 sugar compounds, the Retro-aldol condensation is an effective strategy to cleave the various biomass C–C bonds (Figure 10.6) [11]. In 2012, Li and co-workers [12] reported the direct catalytic conversion of raw woody biomass into two groups of chemicals over a carbon-supported Ni-W_2C/C catalyst. The carbohydrate fraction in the woody biomass, i.e., cellulose and hemicellulose, were converted to ethylene glycol and other diols with a total yield of up to 75.6% (based on the amount of cellulose & hemicellulose), while the lignin component was converted selectively into monophenols with a yield of 46.5% (based on lignin). For the mechanism of the C_5–C_6 sugar compounds from the hydrolysis of cellulose and hemicellulose, they highlight the Retro-aldol process in the C–C bond cleavage. The discussion about the lignin conversion is not discussed in Li's article [12]. Based on their further study [13], the lignin conversion is mainly via an acid hydrolysis+hydrogenation process. Besides the acid hydrolysis process, given the fact that the lignin β-O-4 and β-1 linkages containing 1,3-diol structure, first transforming the $C_\gamma H_2OH$ to carbonyl structure like C_γHO, C_γOOH or C_γOOR be also a potential method to induce the C–C bonds cleavage in β-O-4 and β-1linkage (Figure 10.6).

While, the direct selective activation of C_γH–OH with the existence of C_αH–OH is a challenging work, which needs a delicate manoeuvre to control the main reaction route. In 2015, Jürgen Klankermayer [14] and co-workers describe a novel Ru-catalyzed redox-neutral C–C bond cleavage of 1, 3-diol lignin model compounds. Different from the previous Ru-catalyzed transfer hydrogenation mechanism [15–18], aiming at C–OAr bond cleavage, the new Ru-catalyzed mechanism, involving a dehydrogenation-initiated retro-aldol reaction [11] for the C_α–C_β bond cleavage, was proposed in line with control experimental data and DFT calculations. As shown in Figure 10.7, the specific stereochemical structure of triphos ligand with Ru center and $C_\gamma H_2OH$ structure of the β-O-4 lignin model compounds were both necessary to generate the intermediate **A** in Figure 10.7, which further went through a dehydrogenation-retro-aldol conversion and caused the C_α–C_β bond cleavage via the C_γ aldehyde of intermediate **B**. Then, the 2-phenoxyethen-1-ol, not the generated aromatic aldehyde from C_α–C_β bond cleavage, is further selectively hydrogenated to 2-phenoxyethan-1-ol. After this work, in 2019, Zhang and co-workers [19] further investigated the unique catalytic activity and selectivity of these ruthenium complexes towards the redox-neutral depolymerization of β-O-4 lignin model compound with the DFT calculation methods.

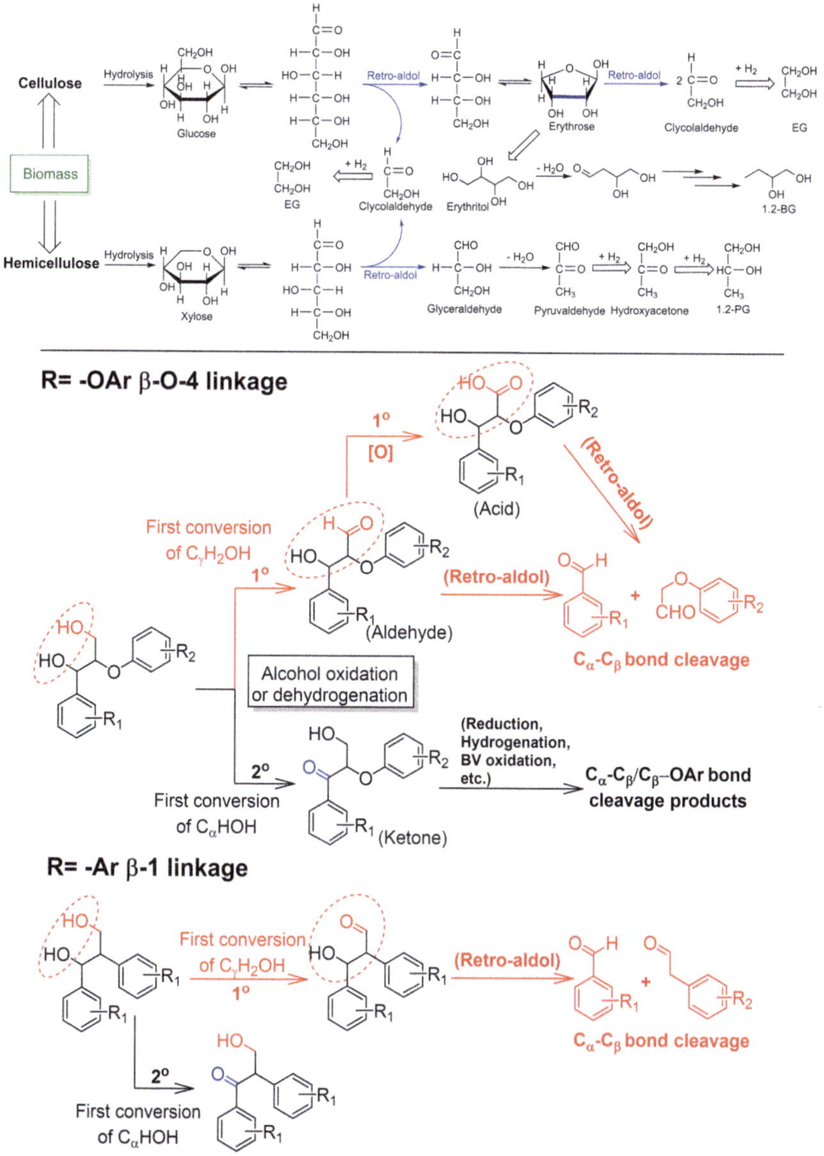

Figure 10.6 Retro-aldol Strategies for Biomass C–C Bond Cleavage. *Source:* Li et al. [12] with Permission, © 2012 The Royal Society of Chemistry.

To achieve the first $C_\gamma H_2OH$ conversion to $C_\gamma HO$, besides the selective dehydrogenation methods discussed above, in 2017, Dabral and co-workers [20] develop a one-pot two-step degradation method of lignin β-O-4 model compounds, which was initiated by preferred oxidation of the primary over the secondary hydroxyl groups with a TEMPO/DAIB system. The C_α–C_β bond of oxidized products with a $C_\gamma HO$ and a $C_\alpha HOH$ group was then cleaved by proline-catalyzed retro-aldol reactions (Figure 10.8). This degradation methodology produced simple aromatic aldehydes,

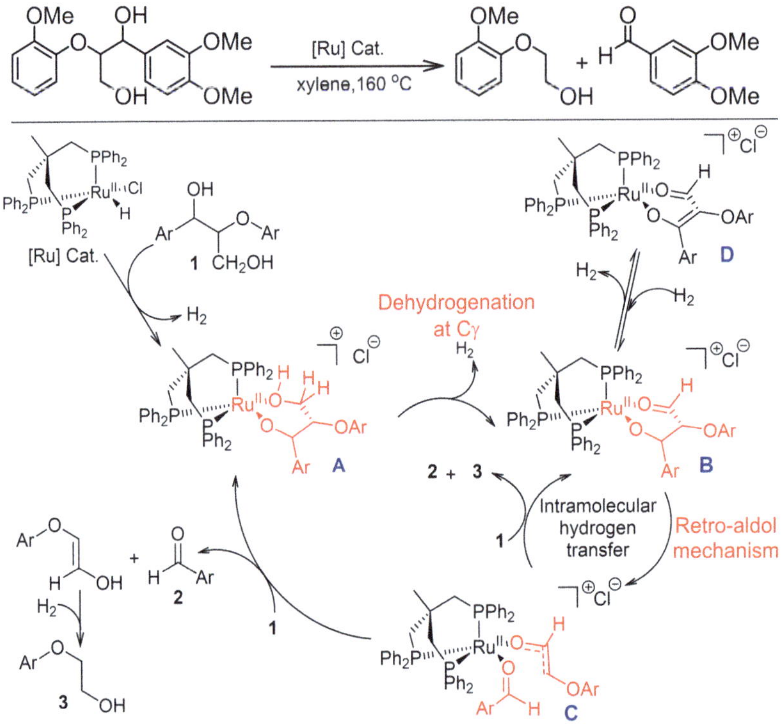

Figure 10.7 The Transfer Hydrogen-based Retro-aldol Mechanism for the C–C Bond Cleavage. *Source:* vom Stein et al. [14] with Permission, © 2015 Wiley-VCH.

substituted 2-phenoxyacetaldehyde, and phenols in good yields from lignin model compounds at room temperature.

Besides the homogeneous Ru-centred dehydrogenation system and TEMPO/DAIB oxidation system shown above, in 2018, Bruijnincx and co-workers [21] reported a Cp*Ir-bipyridonate system to achieve the C–C bond cleavage in β-O-4 alcohol under basic conditions. This Ir-centered system was capable of promoting the depolymerization of β-O-4 model compounds and isolated lignins via a sequence of selective primary C_γ alcohol dehydrogenation to C_γHO, retro-aldol (C_α–C_β) bond cleavage, and in situ stabilization of the aldehyde products by transfer hydrogenation to alcohols or dehydrogenation to carboxylic acids via a gem-diol intermediate (Figure 10.9).

Then, in 2019, the Stahl group [22] developed an electrochemical process for chemoselective oxidation of primary alcohols (-C_γHOH) in lignin to the corresponding carboxylic acids. The electrochemical oxidation reactions proceeded under mildly basic conditions and employed 2,2,6,6-tetramethyl-1-piperidine N-oxyl (TEMPO) and 4-acetamido-TEMPO (ACT) as catalytic mediators to selectively oxidize the primary alcohols (-C_γHOH) to carboxylic acid at 0.80 V vs Ag/AgCl, the generated carboxylic acid at C_γ can further cause the cleavage of the C_α–C_β in β-O-4 linkage via the retro-aldol mechanism (Figure 10.10). Besides the lignin models, the electrocatalytic method was then applied to the oxidation of lignin. Further base titration

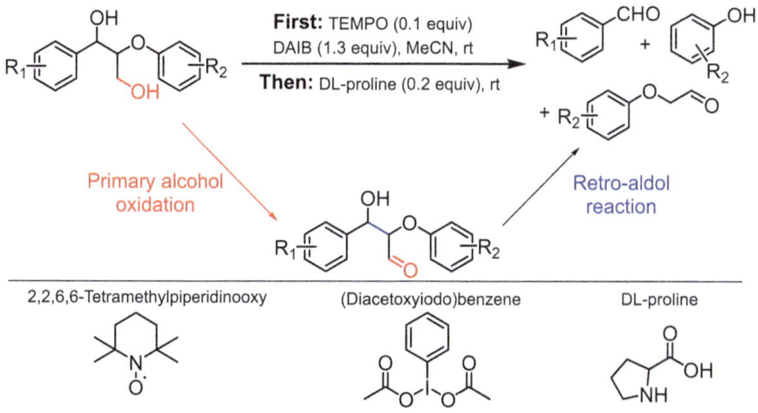

Figure 10,8. Organocatalytic Chemoselective Primary Alcohol Oxidation and Subsequent Cleavage of C_α–C_β Bond of Lignin β-O-4 Linkage.

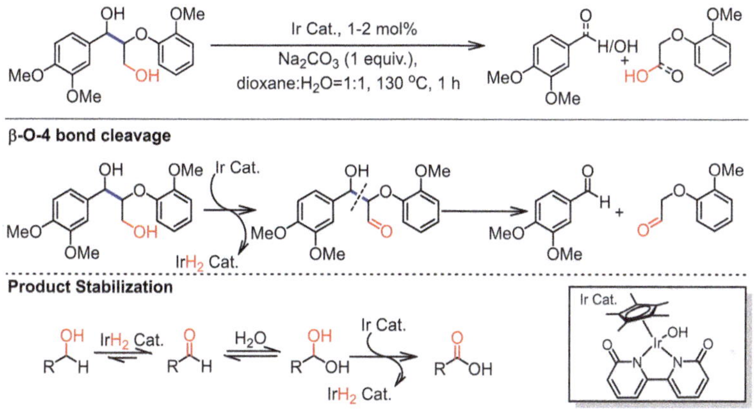

Figure 10.9 Iridium-catalyzed Primary Alcohol Oxidation and Hydrogen Shuttling for the Lignin Depolymerization. *Source:* Lancefield et al. [21] with Permission, © 2018 The Royal Society of Chemistry.

revealed a significant increase in the acid content, and the oxidized lignin had much higher water solubility relative to the extracted lignin. Treating the electro-oxidized lignin fragments under an optimized acidic condition could result in the depolymerization of the lignin fragments into characterized aromatic monomers with a nearly 30 wt% yield.

Besides cleaving the C–C bonds in β-O-4 alcohol linkage via the retro-aldol strategy, other linkages with the same 1,3-diol structure could also be cleaved via the same strategy. In 2017, Li and co-workers [23] reported a selective copper-N-heterocyclic carbene (copper-NHC)-catalyzed method for aerobic cleavage of β-1 lignin model compounds. For the (SIMes)CuCl-catalyzed oxidation system, the C_α–C_β bond cleavage reaction proceeds via a retro-aldol reaction after the selective oxidation of the primary alcohol (-C_γHOH) to aldehyde (-C_γHO) with the catalysis

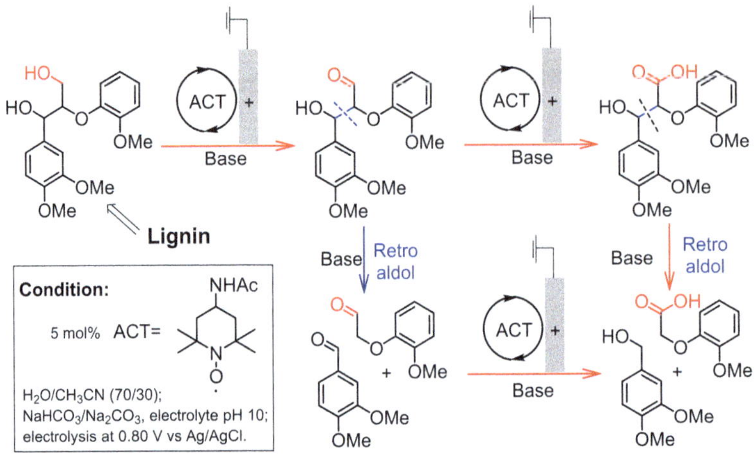

Figure 10.10 Electrochemical Aminoxyl-mediated Oxidation of Primary Alcohols in Lignin to Carboxylic Acids and Further Depolymerization of Oxidized Lignin. *Source:* Rafiee et al. [22] with Permission, © 2019 American Chemical Society.

of Cu catalyst. Using this catalyst system, the cleavage of β-1 model compounds gave >99% conversions and satisfactory yields (70~90%) of the corresponding aldehydes (Figure 10.11).

Besides the methods shown above with first -$C_\gamma H_2OH$ activation, the first-etherification of the $C_\alpha H_2OH$ can enhance the selectivities of the $C_\gamma H_2OH$ oxidation to carbonyl structure [24–29]. In 2019, Westwood and co-workers [27] developed a butanosolv process for the pretreatment of agricultural waste biomass (Figure 10.12). After refluxing for 6 h, pure butoxylated xylose (anomeric mixture) was isolated directly from the pretreatment in a high yield with 95 v% butanol solution at final

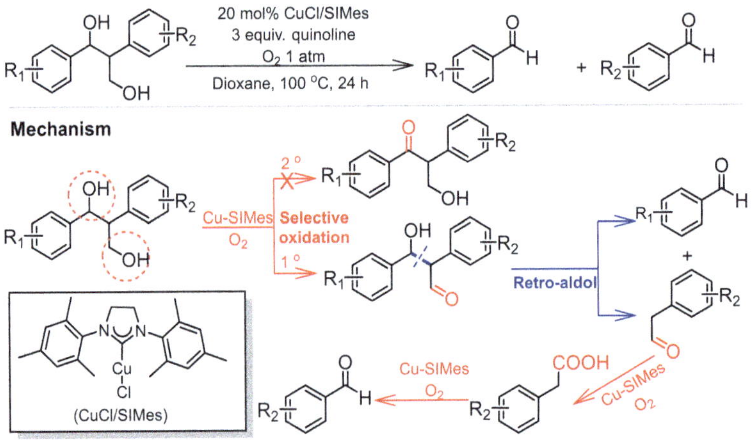

Figure 10.11 Selective Copper-N-heterocyclic Carbene-catalyzed Aerobic Cleavage of β-1 Lignin Models to Aldehydes. *Source:* Zhou et al. [23] with Permission, © 2017 American Chemical Society.

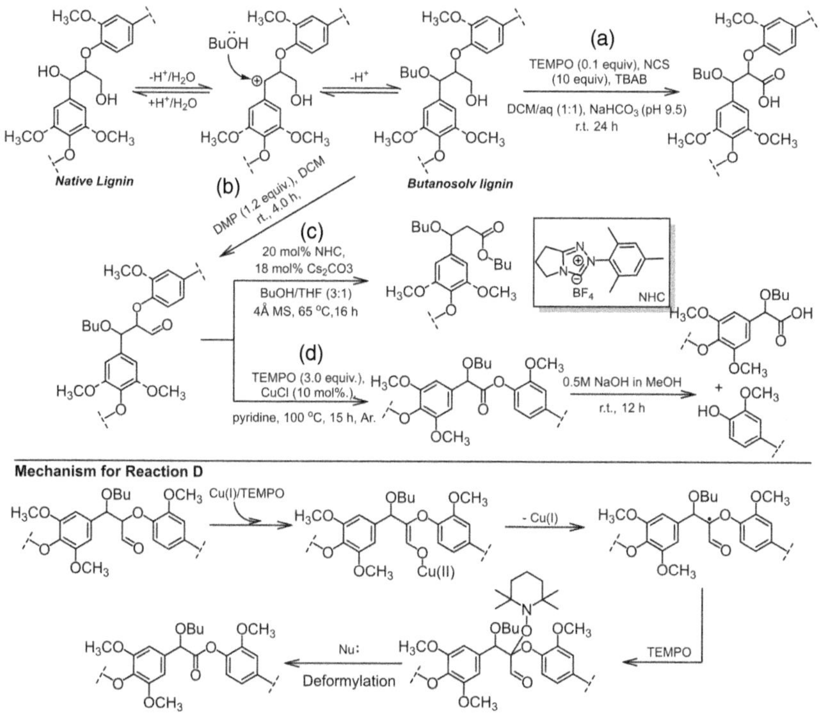

Figure 10.12 The Transformation of Alcohol(butanol)-pretreated Lignin Linkages.

acid molarity of 0.2 M. Besides, selective γ-oxidation of the butoxylated β-O-4 units over the TEMPO/N-chlorosuccinimide (NCS) system gave γ-carboxylic acids despite the presence of phenolic end groups in the lignin (Figure 10-12a) [27]. Then the same group [25] reported a new two-step depolymerization method for lignin involving primary alcohol oxidation followed by an NHC mediated redox esterification-depolymerization step (Figure 10.12c). According to the following research [26, 30], the first butanol-etherification can restrain the generation of monomers from β-O-4 cleavage over the Cu-catalyzed oxidation. While the optimized CuCl/TEMPO system can provide esters with $C_\beta(=O)$–OAr structure as the main products in the absence of oxygen (Figure 10.12d). A CuCl/TEMPO-promoted deformylation mechanism was proposed (Figure 10.12). Besides, the ester ($C_\beta(=O)$–OAr) intermediates can occur hydrolysis with lignin C_β–OAr bond cleavage.

Similar to the strategy containing first-etherification of $C_\alpha H_2OH$ shown above, in 2019 and 2020, Deuss and co-workers [28, 31] reported that the modification of lignin with α-ethoxylated β-O-4 motif could facilitate C–O bond cleavage via a stepwise conversion protocol (Figure 10.13). The process involved the first copper-catalyzed selective aerobic oxidation of $-C_\gamma HOH$ inα-ethoxylated β-O-4 motif and subsequent rhodium-catalyzed decarbonylation, and the obtained partially defunctionalized β-O-4 lignin motif after de-etherification to release the $-C_\alpha HOH$ allowed the effective homogeneous ruthenium-catalyzed hydrogen neutral C–O bond cleavage to ketones and phenols.

10.3 Lignin C_β–OAr Bond Cleavage Beginning with C_γ First Sulphonation

Besides the direct -$C_\gamma H_2OH$ Activation to -C_γHO/-C_γOOR that induce the cleavage of lignin C–C bonds, in some uncommon methods, the first pre-modification of -$C_\gamma H_2OH$ can also induce the cleavage of lignin C_β–OAr bonds. According to previous reports [32, 33], a sulfonyl group at the α-position of lignin facilitated the cleavage of β-O-4 linkages; in other words, that the sulfonyl group activated the β-elimination (E1cB type reaction) from the α-position. Furthermore, Lindberg and Lundström [34] found that the ether bond (C_1-O-C_5) in methyl 6-deoxy-6-ptolylsulfonyl-β-D-glucopyranoside is cleaved under weakly alkaline conditions. The substructure of the glucopyranoside is similar to that of β-O-4/γ-sulfone lignin because a sulfonyl group is introduced in the neighboring primary carbon of the ether bond. This report suggests that the β-elimination could be also assisted by a sulfonyl group in the γ-position.

Then, in 1980 and 1982, Matsumoto and co-workers [35, 36] developed a four-step β-O-4 linkage degradation: (1) γ-Tosylation; (2) Substitution reaction with iodine; (3) Substitution reaction with sulfonate (sulfinylation); and (4) Mild alkali degradation. The key reaction (cleavage of β-O-4 linkage) was effected by mild alkali (NaOH) splitting the ether bond in the presence of an electron-withdrawing sulfinyl group at the γ-position. The key reaction proceeded easily and afforded the cleaved product in high yield. In this case, sulfinyl compound was obtained by the substitution of the γ-iodine model compound in the presence of sodium *p*-toluenesulfinate, another paper showed that sulfone compound could be obtained in a high yield with substitution of a bromide compound with the sodium *p*-toluenesulfinate [37]. Then, in 2003, Katahira and co-worker [38, 39] have proposed the so-called TIZ method (Figure 10.14) for the splitting of β-O-4 linkages, consisting of the reaction steps: (1) γ-Tosylation; (2) Substitution reaction with iodine; (3) Reduction cleavage by zinc powder. The reactions are carried out under almost neutral conditions and the

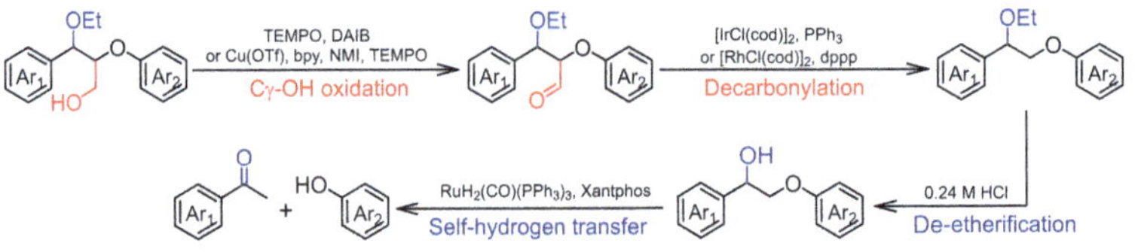

Figure 10.13 Sequential Catalytic Modification of the Lignin α-Ethoxylated β-O-4 Motif to Facilitate C–O Bond Cleavage.

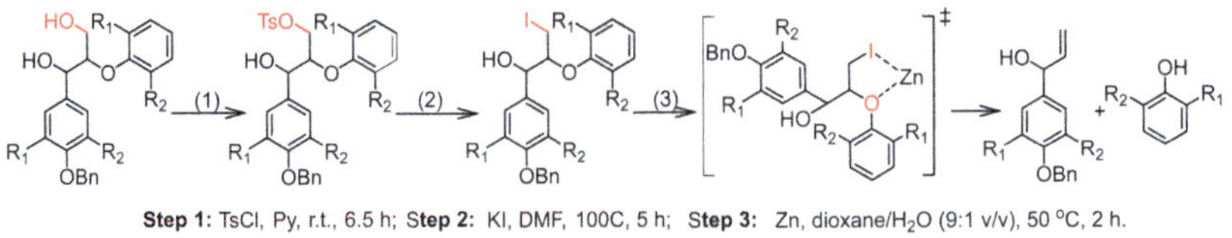

Figure 10.14 The Degradation of β-O-4 Model Compounds by a TIZ Method.

β-O-4 cleavage initiated from the γ-position retains some information about the α-linkage.

Although the above γ-tosylation induced methods can be simplified to three steps with Katahira's TIZ method, the use of zinc powder as the reductant stoichiometric is problematic if the splitting strategies are applied to native lignin or wood utilization. In 2012, Ando and co-workers [40] developed a multiple-step method for the cleavage of β-O-4 linkages in lignin, which was named the γ-TTSA strategy. The γ-TTSA strategy consisted of four steps: (1) γ-Tosylation; (2) Thioetherification (substitution reaction with 1-dodecanethiol (Dod-SH) to γ-thioether); (3) Sulfonylation (oxidation from γ-thioether to γ-sulfone), and (4) mild alkali degradation of γ-sulfone (Figure 10.15a). In detail, the first selective tosylation of $C_\gamma H_2OH$ not $C_\alpha HOH$ in β-O-4 linkage was achieved by adding tosyl chloride into the dry pyridine solution of β-O-4 alcohol at room temperature. Then, the obtained C_γ methyl benzene sulfonate ester was treated with Dod-SH and K_2CO_3 in refluxed acetone to achieve the thioetherification at C_γ. The thioether was further selective oxidized to γ-sulfone by *m*-chloroperbenzoic acid (mCPBA) in dioxane/H_2O (9/1, v/v) at room temperature. The γ-sulfone occurred selective C_β–OAr bond cleavage in the optimum 0.03 M NaOH solution (dioxane/H_2O= 9/1) at 25 °C, which was further neutralized with 0.03 M AcOH solution. Based on their study, the mild alkali degradation of the γ-sulfone group proceeds efficiently, i.e., the electron-withdrawing sulfone group at γ-position contributed to the cleavage of C_β–OAr bond in β-O-4 linkages. Ando and co-workers [41] believed that the γ-TTSA method might be especially useful for the elucidation of lignin-carbohydrate complex structures. FT-IR and HSQC NMR analyses of a lignin polymer model (DHP), submitted to the four-step γ-TTSA method consisting of γ-tosylation, thioetherification, sulfonylation, and mild alkali degradation, confirmed the results of similar experiments with a dimeric lignin model compound. After the first three steps, the β-O-4 and β-5 substructures in DHP were transformed into γ-sulfone derivatives without any side reaction. Finally, the β-O-4 linkage was cleaved selectively (Figure 10.15b).

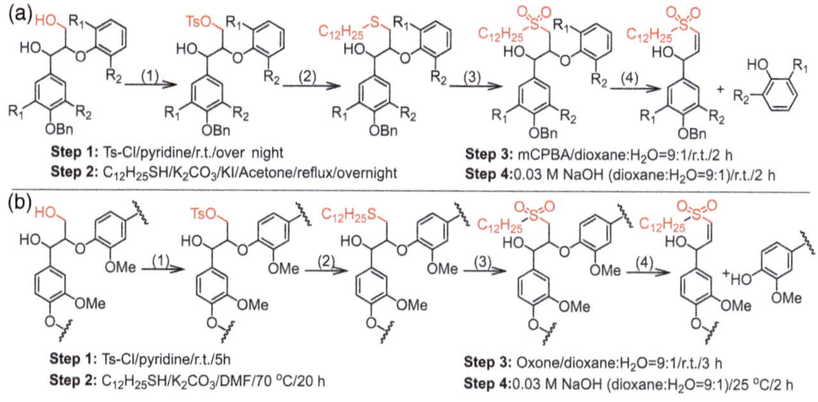

Figure 10.15 Multi-steps γ-TTSA Method for β-O-4 Linkage Cleavage.

10.4 Chapter Summary

This chapter mainly summarized the lignin linkages cleavage strategies beginning with C_β–H, C_γ–H, or C_γO–H direct activation. Photocatalytic strategies were summarized first, including the direct activation of C_β–H bonds and the introduction of cleavage agent O* leading to the cleavage of the C_α–C_β bonds in β-1 and β-O-4 linkages. Then, there is a hemiacetal process with the Cβ–H direct activation to C_β–OH that can easily undergo hydrolysis to cleave the Cβ(OH)–OAr bond. Next, some retro-aldol strategies with first -$C_\gamma H_2OH$ activation to -C_γHO/-C_γOOR (R=H or alkyl) were introduced. Then, as the modification of the direct -$C_\gamma H_2OH$ activation, the first-etherification of the -$C_\alpha H_2OH$ with following -C_βH(OAr)–$C_\gamma H_2OH$ oxidation to -C_β(=O)–OAr and another oxidation-decarbonylation-de-etherification-self-hydrogenation transfer strategy to achieve the cleavage of C_β–OAr bond were also discussed. In addition, some strategies for the lignin C–OAr bond cleavage with first -C_γHOH sulphonation were introduced at the end of this chapter. The keywords of this chapter could be Lignin linkages cleavage, Lignin C_β–H bond activation, Alcohol selective (oxy)dehydrogenation, Lignin C_γHOH activation, Retro-aldol cleavage, Lignin sulphonation.

References

1 Hou, T., Luo, N., Li, H. et al. (2017). Yin and Yang dual characters of CuO_x clusters for C–C bond oxidation driven by visible light. *ACS Catal.* 3850–3859.

2 Reddy, G.V.B., Sridhar, M., and Gold, M.H. (2003). Cleavage of nonphenolic β-1 diarylpropane lignin model dimers by manganese peroxidase from Phanerochaete chrysosporium-Evidence for a hydrogen abstraction mechanism. *Eur. J. Biochem.* 270: 284–292.

3 Ma, Y.Y., Du, Z.T., Xia, F. et al. (2016). Mechanistic studies on the $VO(acac)_2$-catalyzed oxidative cleavage of lignin model compounds in acetic acid. *RSC Adv.* 6: 110229–110234.

4 Wang, M., Li, L.H., Lu, J.M. et al. (2017). Acid promoted C–C bond oxidative cleavage of β-O-4 and β-1 lignin models to esters over a copper catalyst. *Green Chem.* 19: 702–706.

5 Liu, H., Li, H., Lu, J. et al. (2018). Photocatalytic cleavage of C–C bond in lignin models under visible light on mesoporous graphitic carbon nitride through π–π stacking interaction. *ACS Catal.* 8: 4761–4771.

6 Hanson, S.K., Baker, R.T., Gordon, J.C. et al. (2010). Aerobic oxidation of lignin models using a base metal vanadium catalyst. *Inorg. Chem.* 49: 5611–5618.

7 Hanson, S.K. and Baker, R.T. (2015). Knocking on wood: Base metal complexes as catalysts for selective oxidation of lignin models and extracts. *Acc. Chem. Res.* 48: 2037–2048.

8 Gazi, S., Ng, W.K.H., Ganguly, R. et al. (2015). Selective photocatalytic C–C bond cleavage under ambient conditions with earth abundant vanadium complexes. *Chem. Sci.* 6: 7130–7142.

9 Liu, H.F., Li, H.J., Luo, N.C., and Wang, F. (2020). Visible-light-induced oxidative lignin C–C bond cleavage to aldehydes using vanadium catalysts. *ACS Catal.* 10: 632–643.

10 Cui, T., Ma, L., Wang, S. et al. (2021). Atomically dispersed Pt–N_3C_1 sites enabling efficient and selective electrocatalytic C–C bond cleavage in lignin models under ambient conditions. *J. Am. Chem. Soc.* 143: 9429–9439.

11 Andrews, M.A. and Klaeren, S.A. (1989). Selective hydrocracking of monosaccharide carbon-carbon single bonds under mild conditions. Ruthenium hydride catalyzed formation of glycols. *J. Am. Chem. Soc.* 111: 4131–4133.

12 Li, C.Z., Zheng, M.Y., Wang, A.Q., and Zhang, T. (2012). One-pot catalytic hydrocracking of raw woody biomass into chemicals over supported carbide catalysts: simultaneous conversion of cellulose, hemicellulose and lignin. *Energy Environ. Sci.* 5: 6383–6390.

13 Guo, H., Zhang, B., Li, C. et al. (2016). Tungsten carbide: A remarkably efficient catalyst for the selective cleavage of lignin C–O bonds. *ChemSusChem* 9: 3220–3229.

14 vom Stein, T., den Hartog, T., Buendia, J. et al. (2015). Ruthenium-catalyzed C–C bond cleavage in lignin model substrates. *Angew. Chem. Int. Ed.* 54: 5859–5863.

15 Nichols, J.M., Bishop, L.M., Bergman, R.G., and Ellman, J.A. (2010). Catalytic C–O bond cleavage of 2-aryloxy-1-arylethanols and its application to the depolymerization of lignin-related polymers. *J. Am. Chem. Soc.* 132: 12554–12555.

16 Chmely, S.C., Kim, S., Ciesielski, P.N. et al. (2013). Mechanistic study of a Ru-xantphos catalyst for tandem alcohol dehydrogenation and reductive aryl-ether cleavage. *ACS Catal.* 3: 963–974.

17 vom Stein, T., Weigand, T., Merkens, C. et al. (2013). Trimethylenemethane-ruthenium(II)-triphos complexes as highly active catalysts for catalytic C–O bond cleavage reactions of lignin model compounds. *ChemCatChem* 5: 439–441.

18 Wu, A., Patrick, B.O., Chung, E., and James, B.R. (2012). Hydrogenolysis of β-O-4 lignin model dimers by a ruthenium-xantphos catalyst. *Dalton Trans.* 41: 11093–11106.

19 Zhao, X., Yang, Y., Zhu, R. et al. (2019). Mechanistic picture of the redox-neutral C-C bond cleavage in 1,3-dilignol lignin model compound catalyzed by [Ru(Cl)(H)(PPh3)3]/triphos. *Mol Catal* 471: 77–84.

20 Dabral, S., Hernandez, J.G., Kamer, P.C.J., and Bolm, C. (2017). Organocatalytic chemoselective primary alcohol oxidation and subsequent cleavage of lignin model compounds and lignin. *ChemSusChem* 10: 2707–2713.

21 Lancefield, C.S., Teunissen, L.W., Weckhuysen, B.M., and Bruijnincx, P.C.A. (2018). Iridium-catalysed primary alcohol oxidation and hydrogen shuttling for the depolymerisation of lignin. *Green Chem.* 20: 3214–3221.

22 Rafiee, M., Alherech, M., Karlen, S.D., and Stahl, S.S. (2019). Electrochemical aminoxyl-mediated oxidation of primary alcohols in lignin to carboxylic acids: Polymer modification and depolymerization. *J. Am. Chem. Soc.* 141: 15266–15276.

23 Zhou, Z.-z., Liu, M., and Li, C.-J. (2017). Selective copper–N-heterocyclic carbene (Copper-NHC)-catalyzed aerobic cleavage of β-1 lignin models to aldehydes. *ACS Catal.* 7: 3344–3348.

24 Lancefield, C.S., Panovic, I., Deuss, P.J. et al. (2017). Pre-treatment of lignocellulosic feedstocks using biorenewable alcohols: towards complete biomass valorisation. *Green Chem.* 19: 202–214.

25 Xiao, G., Lancefield, C.S., and Westwood, N.J. (2019). Selective depolymerisation of γ-oxidised lignin via NHC catalysed redox esterification. *ChemCatChem* 11: 3182–3186.

26 Xiao, G., Montgomery, J.R.D., Lancefield, C.S. et al. (2020). Copper-mediated conversion of complex ethers to esters: Enabling biopolymer depolymerisation under mild conditions. *Chem. Eur. J.* 26: 12397–12402.

27 Panovic, I., Lancefield, C.S., Phillips, D. et al. (2019). Selective primary alcohol oxidation of lignin streams from butanol-pretreated agricultural waste biomass. *ChemSusChem* 12: 542–548.

28 Zhang, Z., Zijlstra, D.S., Lahive, C.W., and Deuss, P.J. (2020). Combined lignin defunctionalisation and synthesis gas formation by acceptorless dehydrogenative decarbonylation. *Green Chem.* 22: 3791–3801.

29 Panovic, I., Miles-Barrett, D.M., Lancefield, C.S., and Westwood, N.J. (2019). Preparation and reaction of β-O-4 γ-aldehyde-containing butanosolv lignins. *ACS Sustainable Chem. Eng.* 7: 12098–12104.

30 Salonen, H.E.P., Mecke, C.P.A., Karjomaa, M.I. et al. (2018). Copper catalyzed alcohol oxidation and cleavage of β-O-4 lignin model systems: From development to mechanistic examination. *ChemistrySelect* 3: 12446–12454.

31 Zhang, Z., Lahive, C.W., Zijlstra, D.S. et al. (2019). Sequential catalytic modification of the lignin α-ethoxylated β-O-4 motif to facilitate C–O bond cleavage by ruthenium-xantphos catalyzed hydrogen transfer. *ACS Sustainable Chem. Eng.* 7: 12105–12116.

32 Jerkeman, P. and Lindberg, B. (1964). Sulphones of lignin models, synthesis and reactions in alkali. *Acta Chem. Scand* 18: 1477–1482.

33 Ljunggren, S., Ljungquist, P.O., and Wenger, U. (1983). The significance of α-sulfone and α-sulfonate groups for the cleavage of β-aryl ether structures in lignin. *Acta Chem. Scand. B* 37.

34 Lindberg, B. and Lundström, H. (1966). 6-Deoxy-6-p-tolylsulphonyl-D-glueopyranosides. *Acta Chem. Scorni* 20.

35 Matsumoto, Y., Ishizu, A., Iiyama, K., and Nakano, J. (1982). Selective cleavage of arylglycerol-β-aryl ether type structure in lignin II. Determination of the arylglycerol-β-aryl ether type structure in milled wood lignin of spruce. *Mokuzai Gakkaishi* 249–254.

36 Matsumoto, Y., Ishizu, A., and Nakano, J. (1980). Selective cleavage of arylglycerol-β-aryl ether type structure in lignin I. *Model experiment. Mokuzai Gakkaishi* 26: 806–810.

37 Lenihan, B.D. and Shechter, H. (1998). Synthesis and conversions of substituted o- [(trimethylsilyl) methyl] benzyl *p*-tolyl sulfones to *o*-quinodimethanes and products thereof. *J. Org. Chem.* 63: 2072–2085.

38 Katahira, R., Ujihara, M., and Nakatsubo, F. (2003). A novel selective cleavage method for β-O-4 substructure in lignins named TIZ method. I. Degradation of guaiacyl and syringyl models. *J. Wood. Chem. Technol.* 23: 71–87.

39 Katahira, R., Kamitakahara, H., Takano, T., and Nakatsubo, F. (2006). Stability of α-ether substructures in lignins under TIZ reaction conditions. *Cellul. Chem. Technol.* 40: 19–34.

40 Ando, D., Takano, T., and Nakatsubo, F. (2012). Multi-steps degradation method for β-O-4 linkage in lignins: γ-TTSA method. Part 1: Reaction of non-phenolic dimeric β-O-4 model compounds. *Holzforschung* 66: 331–339.

41 Ando, D., Nakatsubo, F., Takano, T. et al. (2013). Multistep degradation method for β-O-4 linkage in lignins: γ-TTSA method. Part 2: reaction of lignin model polymer (DHP). *Holzforschung* 67: 249–256.

11

Lignin Linkages Cleavage Considering Fragments Condensation

11.1 Different Mechanisms of Lignin Fragments Condensation

As we discussed in previous chapters, the key issues for lignin depolymerization to aromatic monomers lie in the development of highly selective and active catalysts to effectively cleave the ubiquitous C–C or C–O linkages, while leaving the aromatic benzene rings unconverted. The next step is to develop effective depolymerization strategies. However, before the catalysis test, elements of lignin structure change should be taken into account [1] in which lignin organic functional groups change, linkage structure transforms,and the corresponding molecular weight changes during the lignin pretreatment and in-situ reaction [2, 3]. Therefore, one should know how to take advantage of related changes and avoid concomitant problems. For lignin depolymerization to aromatic monomers, two factors in particular should be considered: (1) Promoting the selective cleavage of the ubiquitous lignin C–O/C–C bonds, increasing the conversion; and (2) Restraining the lignin fragments from recondensation into more stable multimers to increase the selectivity and monomer yields [4]. This chapter will mainly focus on the lignin linkages cleavage beginning with C_α–OH protection and other lignin depolymerization strategies, considering fragments condensation. Meanwhile, a summary of the methods for restraining the lignin fragments' condensation will be also provided. Beforehand, a brief discussion about different mechanisms of lignin fragments condensation is necessary.

11.1.1 Acid-catalyzed Condensation

For the different condensation mechanisms of lignin fragments, the acid-catalyzed condensation is one of the most studied mechanisms [2–6], which usually occurs at the existence of H^+ species or acid sites of the solid catalysts, especially at a high reaction temperature.

As shown in Figure 11.1, with β-O-4 linkages as the model, four main acid-catalyzed recondensation processes, beginning with the generation of C_α carbocation

Lignin Conversion Catalysis: Transformation to Aromatic Chemicals, First Edition.
Chaofeng Zhang and Feng Wang.

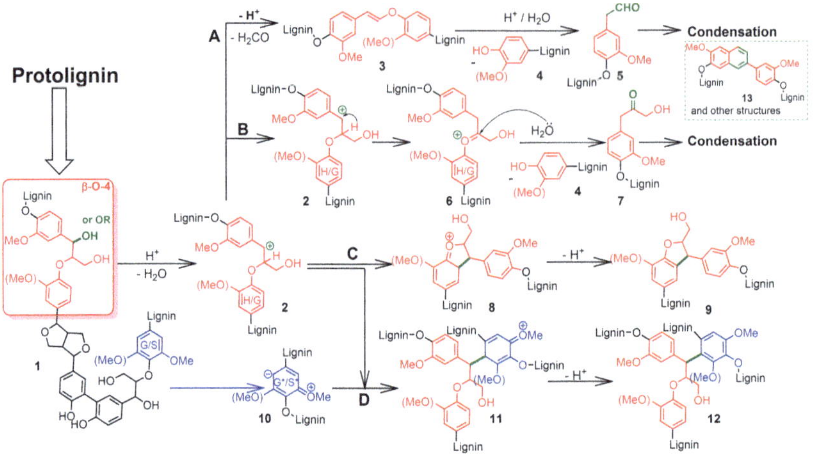

Figure 11.1 Lignin Fragments Condensations or New Stable C–C Bond Generation Caused by C_α Carbocation. Reprinted from Ref [4] with Permission, © 2017 American Chemical Society.

(C_α^+), exist during the lignin β-O-4 linkages transformation as follows: (1) Lignin condensation caused by the generation of aromatic aldehydes: The aromatic aldehyde fragments generated from the acidolysis of β-O-4 structure [7–9] via the first transformation of $C_\alpha H(-OH)CHR-OAr$ to $C_\alpha=C_\beta HR-OAr$ (R=H or CH_2OH) can occur condensation with other active C–H bond, such as bimolecular condensation to the biphenyl structure with new stable C–C bond generation (Figure 11.1A); (2) Lignin condensation caused by the generation of aromatic Hibbert ketones: the Hibbert ketones or the lignin-bound Hibbert ketones generated from the acidolysis of oxonium ion (**6** in Figure 11.1B) [10] after the H_β-shift of C_α carbocation can also occur condensation with other active C–H bond; (3) Lignin condensation caused by the intramolecular attack: The C_α^+ of β-O-4 carbocation can attack the adjacent aromatic rings with G or H type and generate a benzofuran structure after the formation of new C–C bond (Figure 11.1C) [7]. (4) Lignin condensation caused by the intermolecular attack: The C_α^+ of generated β-O-4 carbocation intermediate can condense with another G or S type of aromatic ring in another lignin molecule or the farside of the same fragment to achieve the aromatic substitution of C_α (Figure 11.1D) [5]. Based on the analysis of these processes, the generation of β-O-4 carbocation (C_α^+) is the precondition of the acid-catalyzed recondensations, and the following new C–C bonds generation will suppress further lignin depolymerization.

11.1.2 Base-catalyzed Condensation

Besides acid-catalyzed lignin depolymerization and their fragments condensation shown above, the base not only can catalyze the cleavage of lignin but also catalyze the condensation of the corresponding fragments. During the base-catalyzed lignin degradation, the condensation reaction takes place simultaneously and both reaction pathways share the same intermediates [11]. The first critical fragment/intermediate during base-catalyzed lignin fragment condensation is the quinone methide [11], which is generated from the alkaline cleavage of α-aryl ether bonds in phenolic arylpropane units (Figure 11.2a).

In detail, the hydrogen sulfide and sodium hydroxide anions can promote the transformation of phenolic units to the lignin carbanions, which can further transform to quinone methide after releasing a HO^- or RO^- at C_α. The generated quinone methide can further release a formaldehyde at C_γ to vinyl ether [12]. In Figure 11.2a, the generated quinone methide intermediate can degrade to guaiacol and coniferyl alcohol with the nucleophile attack at C_α of quinone methide, which process needs to compete with the condensation reaction between quinones methide and lignin carbanions via the Michael addition (Figure 11.2b). In addition, quinone methides

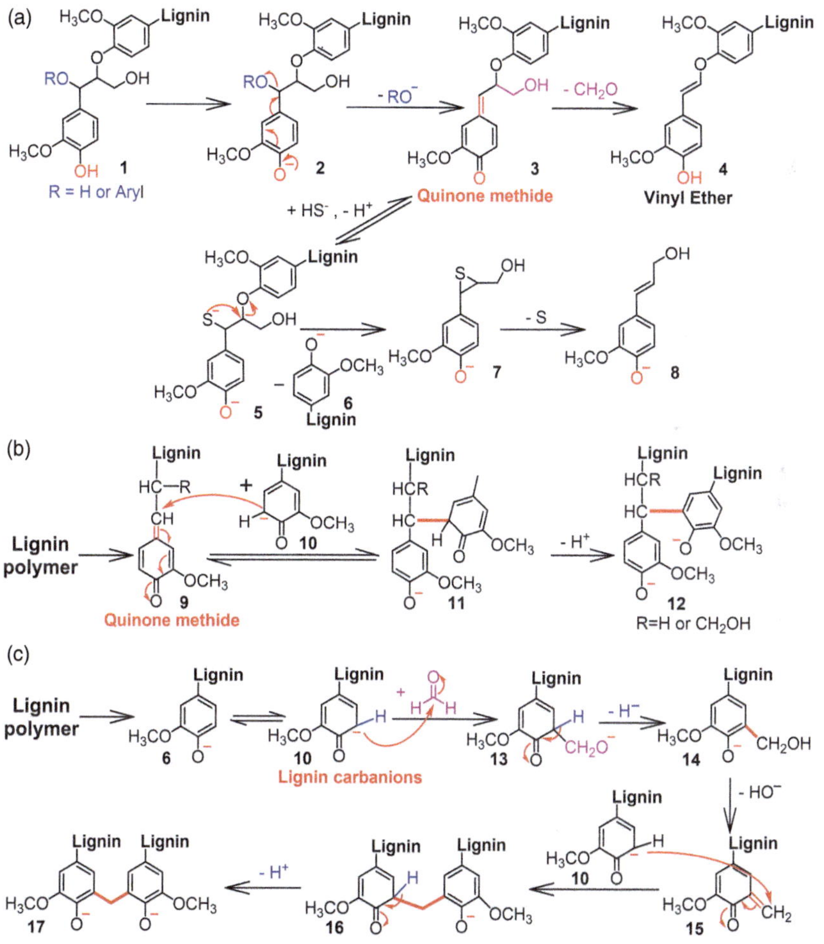

Figure 11.2 Alkaline Cleavage of C_α–OR Bonds in Phenolic Arylpropane Units to Quinone Methides and the Generation of Formaldehyde (a), Base-catalyzed Condensation of Phenolic Units with Quinone Methides (b) or Formaldehyde (c).

are not the only acceptors. Formaldehyde constitutes another type of acceptor. Figure 11.2c illustrates the reaction between one formaldehyde and two phenolate units. The primary addition product, a hydroxybenzyl alcohol, generated from the addition of the lignin carbanion from phenolic ion and formaldehyde, further converts to an *o*-quinone methide. Another lignin carbanion reacts with it to generate a phenolic diarylmethane structure, which can yield further condensation.

11.1.3 Radical Condensation and Other Lignin Linkages Transformation

Besides the generation of lignin carbocations/carbanions and phenolic anions that can induce the condensation of lignin fragments, the lignin conversion usually involves various radical fragments, The uncontrolled condensation of these radicals is another field that needs more attention, especially with respect to getting a higher yield of monomers from lignin depolymerization by restraining the condensation.

Given the sensitivity of the functional group in lignin fragments and their monomers with active H atoms, such as the phenolic hydroxyl group and the aldehyde group, they can easily occur HAT process and provide an O-centered or C-centered radical. These radicals can occur random coupling like the generation of lignin polymer (Chapter 2, referring to the generation of 4-O-5, 5-5, etc.). These radicals can also attract the group like C=C bonds or aromatic rings to generate stable C–C/C–O bonds [13], which can hinder the lignin depolymerization or provide the insoluble polymers via the generation of over-oxidized quinone (Figure 11.3a) [14, 15]. Furthermore, the sufficiently strong thermal condition and light irradiation can cause the self-decomposition of the lignin structure to various radicals without catalysts [16], which can further combine together to generate new C–C/C–O bonds (Figure 11.3b). For example, treating spun lignin fiber in the air at 100-250 °C can promote the transformation of β-O-4 linkage to oxidized β-1 linkages with the shift of the Lignin-C_β–O bond to Lignin-C_β–C_{Ar} (Figure 11.3b) [17, 18]. In detail, the high temperature can cause the decomposition of β-O-4 lignin to one C_β carbon radical and one phenolic radical, then the C_β carbon radical attack the aromatic ring of the phenolic radical and the generated intermediate return to the aromatic β-1 structure [17]. In addition, the thermal treatment can induce the C–C/C–O bond shift in the lignin radical structure, such as the C_β carbon radical mentioned above, which can occur the β-scission to various radicals and active chemicals like aldehydes and alkenes and make the condensation more complex (Figure 11.3b). Comparatively speaking, some semiconductor-catalyzed reduction processes under mild conditions can make the reaction process controllable, but there is still a small number of byproducts with C–C condensation. For example, the in-situ generated surface Ti^{3+} on TiO_2 surface can induce the reduction cleavage of C_β–OAr bond in the oxidized β-O-4 linkage with a C_α=O to a C_β radical and one phenol, the excess C_β radicals can undergo coupling to a new C–C bond (Figure 11.3c) [19]. Besides the photochemical process, radical condensations can also happen in some hydrogenation processes. For example, during the catalytic hydrogenolysis of kraft lignin, the free H radical can attack the C=C bond of dehydrated β-O-4 linkage to a C_α radical, which

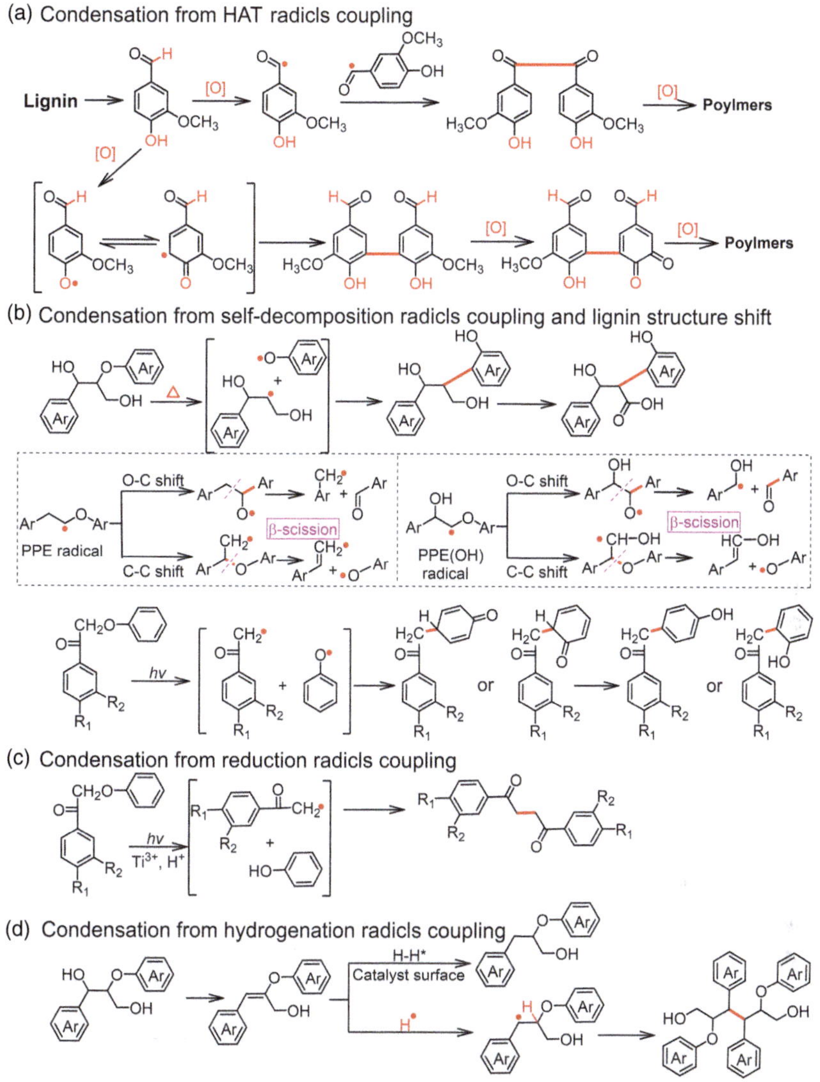

Figure 11.3 Some Examples of β-O-4 Lignin Fragments Recondensation Involving the Participation of Various Radicals and Lignin Linkages Transformations with C–O/C–C Shift.

can occur coupling with another C_α radical to generate a new stable C–C bond generation (Figure 11.3d) [20]. Given the complexity of lignin transformation via various radicals, Figure 11.3 just lists four typical condensation approaches, viewing from the generation of radical from β-O-4 linkage transformation. If one expands the reaction systems from thermal catalysis and photocatalysis to electro-catalysis, enzymatic catalysis of other linkages [21], and more condensation reactions with new C–C/C–O bonds generation could be found.

In addition, besides the condensation transformation of β-O-4 linkage, the other linkages also can undergo transformation. For example, the C_α–OH of β-1 can also

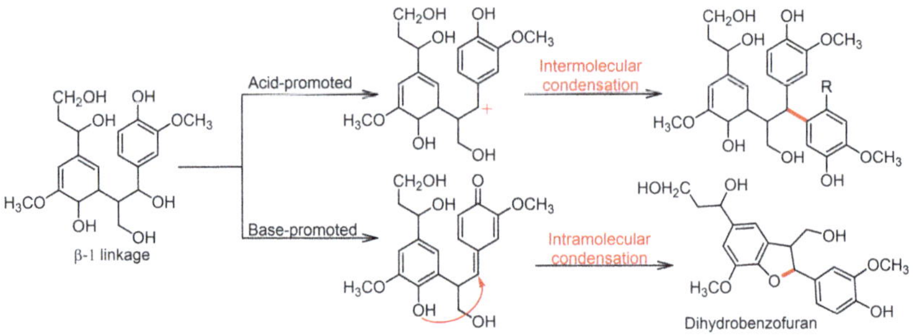

Figure 11.4 The Transformation of β-1 Linkages to Dihydrobenzofuran Structure.

transform to a carbocation ion, which will attack another aromatic ring to provide the diphenylmethane structure (Figure 11.4). In addition, the β-1 linkages with terminal phenolic hydroxyl group can also transform to the quinone methides, which can further occur intramolecular condensation to the dihydrobenzofuran structure (Figure 11.4). While, compared with the β-O-4 linkage, the other linkages just account for a smaller number of the total lignin linkages. Therefore, the condensation discussion about these linkages is less. Nonetheless, these condensations need to be taken into account during lignin depolymerization for getting a higher monomers yield.

11.2 Methods for Restraining Lignin Fragments Condensation

Given the existence of the lignin fragments condensation, we herein summarize and provide six research approaches for the efficient lignin depolymerization with restraining the potential fragments condensation: (1) Strategies with pre-protection of the active groups to inhibit the condensation; (2) Strategies with the first transformation the active groups to promote the depolymerization; (3) Strategies with the in-situ conversion of active intermediates and scavenging the unwanted species; (4) Modifying and designing the catalysts to restrain the condensation by space stereoscopic effect or the control of the intermediate; (5) Intensifying reaction systems with photo, electro, photo-electro, microwave assistance or process reinforcement with a flow reactor or a reactor with in-situ separation/abstraction ability; and (6) Employing an “ideal lignin” as the substrates which can initially prevent/avoid the most condensation reactions.

11.2.1 Pre-protection of the Active Groups

As shown in Figure 11.1, the unwanted acid-promoted generation of C_α carbonation can lead to the further generation of active aldehydes, Hibbert ketones as well as electrophilic aromatic substitution reactions, during which the stable C–C bond generation will suppress lignin degradation. To restrain condensation reactions, the pre-protection of the active C_α–OH to an acid-proof group is a promising method [22–24].

In 2016, Ouyang group [24] reported that both $Fe_2(SO_4)_3$ and $Al_2(SO_4)_3$ exhibited excellent catalytic methylation reactivity of lignin benzylic alcohols with methanol under microwave irradiation. Furthermore, compared with the direct hydrogenation over the Pd/C with HCOOH as the hydrogen donor, the first C_α–OH methylation increased the β-O-4 bond cleavage rate by 55.9% (Figure 11.5). Their preliminary study of lignin depolymerization at 280 °C illustrated that methylation pretreatment of benzylic alcohols facilitated lignin depolymerization to produce aromatic monomers and reduced the oxygen content of aromatic monomers. The yield of monophenolic products from the direct hydrogenation of wheat alkali lignin is 10.9 w%. While the value is increased to 16.8 w% by employing the two-step methylation-hydrogenation method.

Besides the etherization with the methanol, Westwood and co-workers [25] developed different methods for the pre-treatment of lignocellulosic feedstock with biorenewable alcohols, such as ethanol and *n*-butanol. Although the above two-step methylation-hydrogenation method is a promoting method for the lignin depolymerization to provide high yields of monomers, the corresponding methods need to use strong acid to first release the C_α–OH of lignin. If the C_α–OH releasing process can be achieved under an acid-free condition and the C_α–OH could be in-situ protected with the alcohol solvent, the β-O-4 linkage could be efficiently preserved during the lignocellulose pretreatment, which can further promote the following lignin depolymerization [22].

In 2016, Xu group [23] found that the Ni/C catalyst could efficiently catalyze the depolymerization of woody biomass to produce natural phenolic alcohols in a methanol-water co-solvent. Further study suggests that the Ni-based catalyst can efficiently in-situ etherify the C_α–OH group in lignin β-O-4 motifs under a hydrogen atmosphere. The NMR results and density functional theory (DFT) calculations suggested that replacement of the C_α–OH proton in GGGE (one β-O-4 model, Figure 11.6) with a methyl group breaks intramolecular hydrogen bonds ($C_\alpha OH\cdots O_\beta$) between the C_β–OAr oxygen and the C_α–OH, which can facilitate the C_β–OAr cleavage. 2D HSQC NMR analysis showed that water could also accelerate the etherification of raw lignin with methanol through in situ formation of an acid catalyst. The above results suggested that breaking the intramolecular hydrogen bonds could accelerate the C_β–OAr cleavage, keeping the original structure of lignin unchanged. In addition, high yield production of natural phenolic alcohols (38.7 wt%) and a high total yield of phenolic monomers (51.4 w%) from woody biomass has been

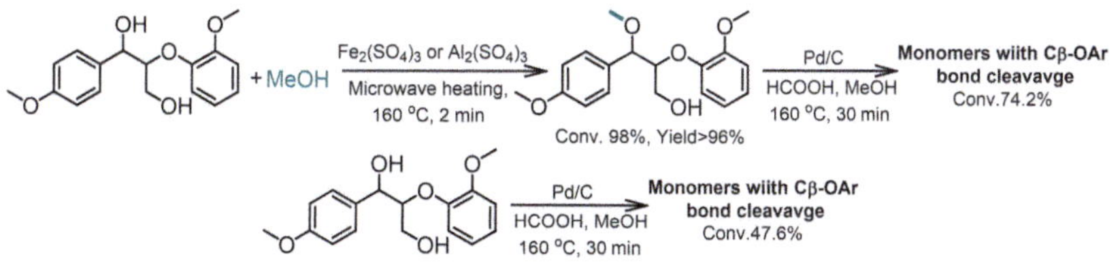

Figure 11.5 Depolymerization of Lignin by Microwave-assisted Methylation of Benzylic Alcohols. Reaction Data Abstracted from Ref [24].

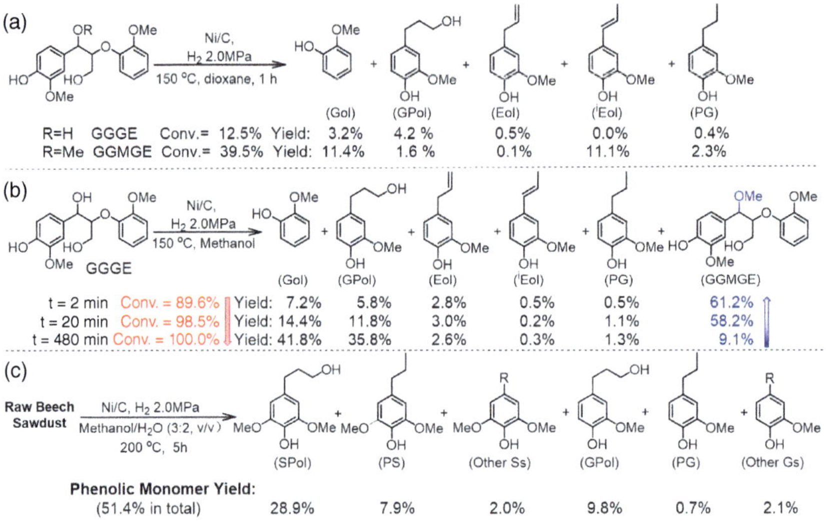

Figure 11.6 Production of Phenolic Alcohols from Woody Biomass via an In-situ Etherization-hydrogenation Process over the Nickel-Based Catalyst. Reaction Data Abstracted from Ref [23].

achieved using the Ni/C catalyst in a methanol-water co-solvent at 200 °C and 2 MPa H_2 in 5 h (Figure 11.6c).

As shown in Figure 11.1, to avoid the condensation process with new C–C bonds generation, besides preventing the formation of the C_α carbocation via the C_α–OH etherization, another reinforcement method is blocking the potential sites on the aromatic rings that can combine with the C_α carbocation (Figure 11.1 Route C and D). In 2016, Luterbacher and co-workers [5] reported that adding formaldehyde during biomass pretreatment produces a soluble lignin fraction that can be converted to guaiacyl and syringyl monomers at near theoretical yields during subsequent hydrogenolysis over Ru/C at 250 °C (47 mole % of Klason lignin for beech and 78 mole %for a high-syringyl transgenic poplar). These yields were three to seven times those obtained without formaldehyde. For the potential reasons, as shown in Figure 11.7, the pretreatment of lignin with the formaldehyde can provide a 1,3-dioxane structure and block the C_α–OH and C_γ–OH. Meanwhile, the reaction between the aromatic ring and formaldehyde can generate an $ArCH_2OH$, which can block the sites for the generation of diphenylmethane and dihydrobenzofuran structure with the C_α carbocation. Therefore, the formaldehyde treatment can efficiently prevent the lignin condensation reactions caused by the C_α carbocation. Coordinating this formaldehyde protection treatment with the following selective hydrogenation over the Ru/C, the authors achieved the efficient whole utilization of the native lignocellulose.

Besides the formaldehyde stabilization method mentioned above to pre-transform the lignin α,γ-diol to the 1,3-dioxane structure [5, 26], Luterbacher and co-workers further extended the application of this α,γ-diol lignin stabilization strategies [6] into the lignin depolymerization to monomers with different

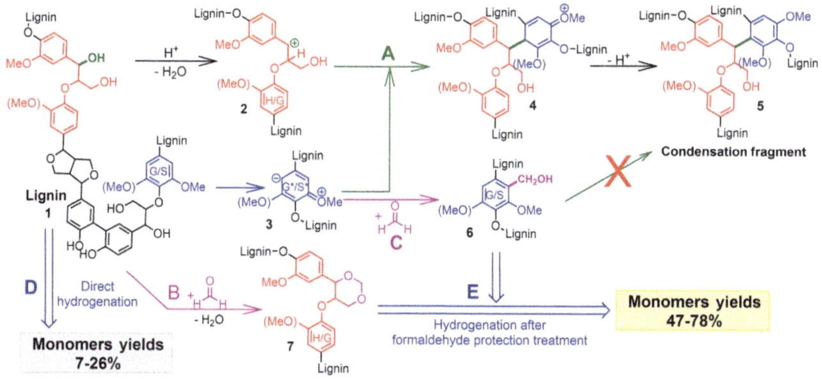

Figure 11.7 Formaldehyde Stabilization Facilitates Lignin Monomers Production from Lignin Hydrogenation Depolymerization and Its Derivative Methods. Reaction Data Were Abstracted from Ref [5].

aldehydes, ketones, carbonate esters, or phenylboronic acids to prevent condensation, which involves the critical six-atom ring structure (Figure 11.8).

For the different performance of the protection reagents, in 2019, Zhang and coworkers [27] present a density functional theory calculation study that reveals the intrinsic mechanism why formaldehyde (FA) can efficiently hinder lignin condensation (Figure 11.9). Using veratrylglycerol-β-guaiacyl ether (VG) as a lignin model compound, they confirmed that the formation of C–C linkage between two VG units is highly exothermic (-43.7 kcal·mol^{-1}) and involves a medium-heightened barrier (20.1 kcal·mol^{-1}), explaining the extensive and irreversible lignin condensation observed during lignin depolymerization. In the presence of FA, acetaldehyde, or propionaldehyde, VG prefers to react with these aliphatic aldehydes with a much lower barrier (13.1~15.9 kcal·mol^{-1}) to form A and B with high thermodynamic and kinetic stability, which protect the reactive sites of VG (1,3-diol groups and the ortho/para positions to alkoxy groups on lignin aromatic rings) and thereby hinder lignin condensation. In contrast, for the benzaldehyde-containing system, the barrier is calculated to be 22.4 kcal·mol^{-1}, which is higher than the barrier involved in the condensation of VG. In their proposed mechanism, FA as an efficient protecting

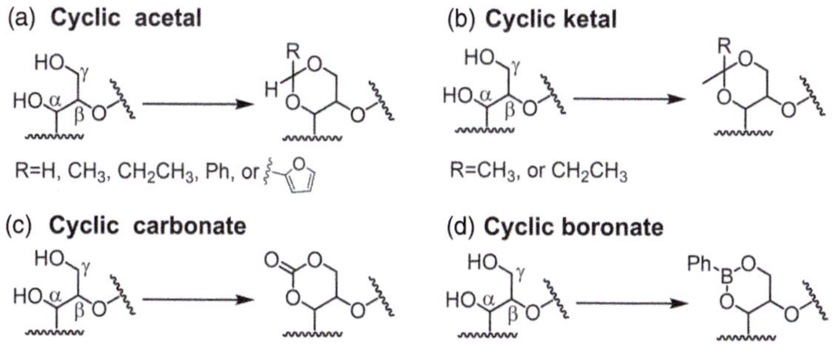

Figure 11.8 α,γ-Diol Lignin Stabilization Strategies besides the Formaldehyde Method.

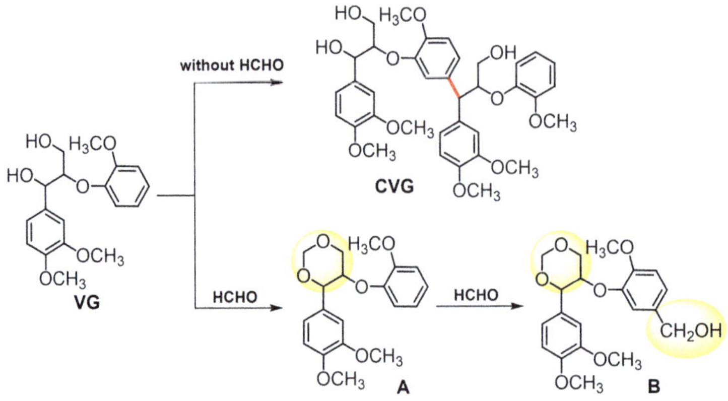

Figure 11.9 Mechanistic Insights into Formaldehyde-Blocked Lignin Condensation.

agent blocks the reactive sites of VG, which may be attributed to the small steric hindrance and strong π–π and d–π interactions. In addition, the net energy change forming A and B is the main overall thermodynamic driving force for lignin depolymerization.

Furthermore, to achieve the pre-protection of the active groups and inhibit the condensation of fragments condensation caused by the C_α carbocation, besides the use of monohydric alcohols shown above, the dihydric alcohols could be also a potential choice. One possibility is the use of polyethylene glycol (PEG)-modified glycol lignin [22, 28]. In addition, besides the etherization method, the pre-esterification of secondary and primary alcohol hydroxyl in lignin linkages combining with the following selective hydrogenation is also a promising method [29]. Furthermore, similar strategies can be extended to the silylation of lignin diols (Figure 11.10) [30, 31].

Moreover, besides the modification at the side chain, blocking the reaction sites on the aromatic ring may be another choice to restrain the condensation. While the

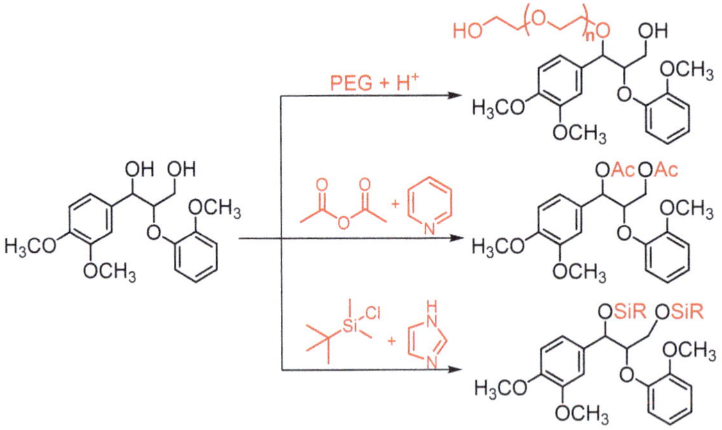

Figure 11.10 Pre-protection of Lignin: Diol Etherization, Acetylation, and Silylation.

corresponding pretreatment is rarely reported. One relevant method is using the "Ideal Lignin" which will be discussed in Chapter 11.2.6, and another method is the in-situ alkylation with an alcohol solvent. In 2016, Hensen and co-workers [32] studied the Lewis acid-catalyzed depolymerization of soda lignin in supercritical ethanol/water mixtures. The ethanol was found to be involved in the alkylation of the lignin depolymerization products which was related to the strong catalytic activity of the triflate metal salts, increasing the product yield by inhibiting repolymerization of the products.

11.2.2 First Transformation of the Active Groups to Promote the Depolymerization

As shown above, the generation of C_α carbocation (C_α^+) can cause the recondensation of the lignin fragments, besides the methods shown above to pre-protect the active C_α–OH via etherification with formaldehyde or alcohol, or via esterization with anhydride, carbonate esters, or phenylboronic acids, another strategy is the first transformation the active groups that can restrain its back to C_α–OH, such as the oxidation-(reduction)hydrogenation strategy containing first C_αH–OH transformation to C_α=O, C_α defunctionalization-hydrogenation strategy containing first C_αH–OH transformation to $C_\alpha H_2$, some DFRC methods used in the lignin analysis, and so on.

For the oxidation-(reduction)hydrogenation strategies (Figure 11.11a-1), it was difficult to make the hydrogenation system prefer to catalyze the cleavage of C_β–OAr rather than hydrogenate the C_α=O bond to alcohol or hydrogenate aromatic rings. The stoichiometric reduction with Zn/NH_4Cl, catalytic hydrogenation system with NiMo, Ni/C, etc., and photoreduction system mediated by Ir, TiO_2, $ZnIn_2S_3$, CdS, etc. were reported, which have been discussed in Chapter 9.2. In addition, besides the reduction and hydrogenation methods to cleave the C_β–OAr bond in oxidized lignin structure, Stahl group [33] reported that the HCOOH/HCOONa can promote the hydrolysis of pre-oxidized lignin (Figure 11.11a-2), which have been discussed in Chapter 9.5.

For the C_α defunctionalization-hydrogenation strategy, in 2015, Marks group [29, 34] applied the tandem M(OTf)n + Pd/C system to cleaving the β-O-4 arylether bonds of lignin models. They first modified the hydroxyl groups via

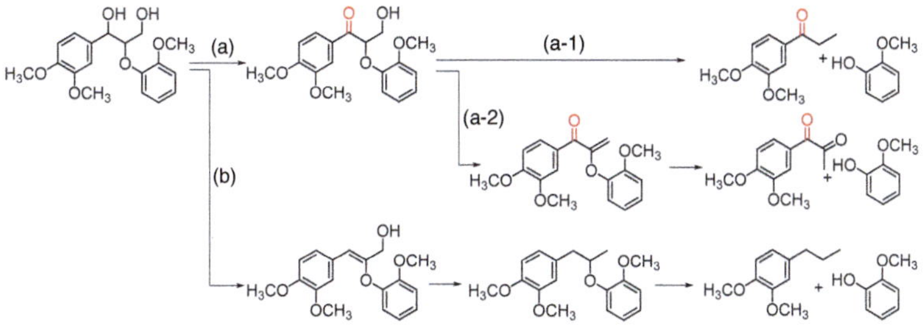

Figure 11.11 First Transformation of the Active Groups to Promote the Depolymerization.

acetylation. The $M(OTf)_n$ first promotes the elimination of a CH_3COOH between C_α and C_β, generating an olefin intermediate. Then the C=C bond is hydrogenated to saturated bond ($C_\alpha H_2$), and the C–OAr ether bonds in the generated intermediates further been cleaved via the acidolysis-hydrogenation with $M(OTf)_n$ + Pd/C system (Figure 11.11b), which has discussed in Chapter 6.1.

Besides the first C_αH–OH transformation to C_α=O and $C_\alpha H_2$, some DFRC (derivatization followed by reductive cleavage) method [35–41] first use the reaction between C_αH–OR/C_γH–OR and acetyl bromide (AcBr) to achieve the C_αH–OH cleavage to C_αH–Br and the esterization protection of the C_γH–OH, the C_β–OAr bond in obtained bromide product can occur reductive cleavage with Zn powders in acetic acid (Figure 11.12).

Compared with the pre-protection of the active groups to inhibit the condensation, the first transformation of the active groups to promote depolymerization sometimes can have advantages. As the important etherization-protection example, Luterbacher and co-workers [5] achieve the goal by etherifying the C_αH–OH and blocking the electron-rich positions at the positions ortho or para to methoxyl in G or S type of aromatic ring with formaldehyde. This is an efficient method to restrain the lignin recondensation, while the etherification of C_αH–OH structure will make the hydrogenolysis of C_β–OPh bond harder, and high reaction temperature or more active hydrogenation catalysts with noble metal are needed. On the contrary, oxidation-(reduction)hydrogenation strategies can carry out below 200 °C with non-noble metal or even at room temperature during the photocatalysis process for a weaker C_β–OAr bond.

If the concept of the first transformation of the active groups expands to radicals, focusing on the first transformation of the C_αH–OH structure, there will be more examples (Figure 11.13). In 2016, Zhang and co-workers [42] proposed an efficient dehydroxylation-hydrogenation strategy for the C_β–OAr bond cleavage (Figure 11.13a). Further mechanism study suggested that the β-O-4-A model $PhCH(OH)CH_2OPh$ first lost a hydroxyl group (C_α–OH) at medium strong acid sites of NiMo catalyst. Then the generated $PhCH\delta^+CH_2OPh$ transformed to $PhCH{\bullet}CH_2OPh$ (β-O-4-R) by obtaining an electron from the NiMo catalyst redox cycle. DFT calculation suggested the dehydroxylation of C_α–OH to a β-O-4-R radical could greatly decrease the C_β–OAr BDE from 274.0 $kJ{\cdot}mol^{-1}$ to 66.9 $kJ{\cdot}mol^{-1}$ and thereby facilitate C_β–OAr cleavage. Then, Wang and co-workers [43–45] reported some photocatalytic lignin transformation methods, during which the photocatalyst CdS or $Zn_4In_2S_7$ could selectively abstract the H from the carbon of C_αH–OH and generate a C_α carbon radical ($PhC{\bullet}(OH)CH_2OPh$) (Figure 11.13b), which could decrease the BDE of C_β–OAr bond and induce the self-cleavage of C_β–OAr bond

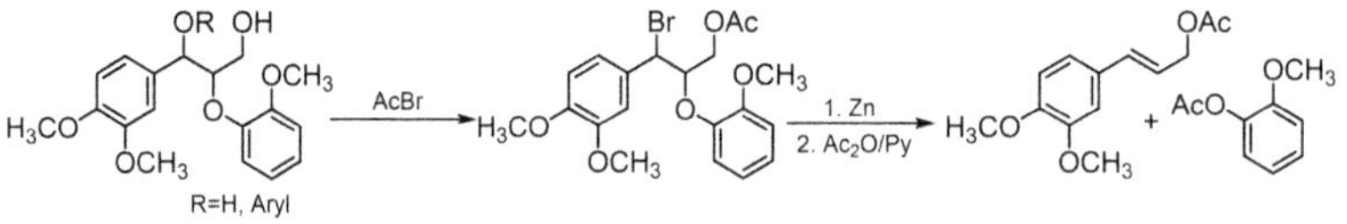

Figure 11.12 Selective Ether Bonds Cleavage in Lignins by the DFRC Method.

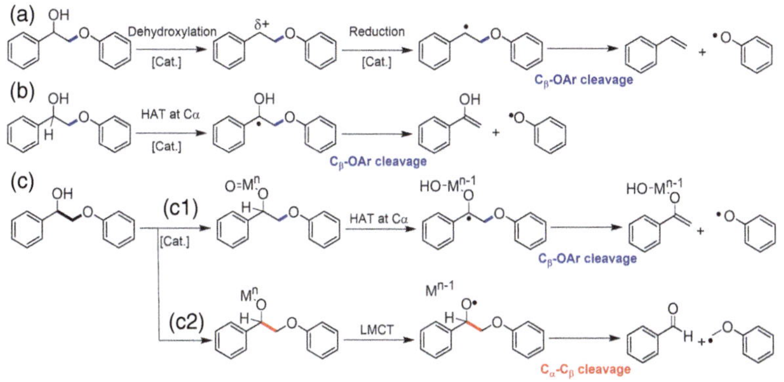

Figure 11.13 The Lignin Depolymerization Strategies with β-O-4 Fragments Generated from C_αH–OH Firstly Activation as the Critical Intermediate for C_β–OAr and C_α–C_β Bonds Cleavage.

like above intermediate PhCH•CH_2OPh. Furthermore, besides the C_α carbon radical intermediate, Toste and co-workers [46] reported a (V^V=O)-catalyzed non-oxidative mechanism for β-O-4 C_β–OAr bond cleavage under N_2 condition, during which the V^V=O could abstract the H from C_α carbon of the coordinate β-O-4 alkoxide via a V^V–OC_α structure (Figure 11.13c1). Then, the generated C_α carbon radical induced the cleavage of the C_β–OAr bond. In 2010, the same year, Baker and co-workers [47] reported a different (V^V=O)-catalyzed oxidative mechanism for lignin models conversion. A similar intermediate with a V^V–OC_α structure which was involved in the (V^V=O)-catalyzed mechanism reported by Toste [46] could induce the direct cleavage of the C_α–C_β bond under O_2 condition. Based on products analysis, the single electron transfer between the V^V and oxygen in V^V–OC_α produced an active ArCH(O•)CH_2OAr, which further induced the selective C_α–C_β bond cleavage (Figure 11.13c2). Then, (V^V=O)-catalyzed [48, 49] and $Ce^{(IV)}$-catalyzed [50] photocatalysis was developed for the cleavage of the C_α–C_β bonds in β-1 and β-O-4 interlinkages via the ArCH(O•)CH_2OAr intermediate.

11.2.3 In-situ Converting the Active Intermediates and Scavenging the Unwanted Species

As we discussed in Chapter 11.1, some lignin fragment condensation mechanisms are caused by active intermediates or phenolic/aldehyde products. To increase the final yield of monomers, in-situ converting the active intermediates and scavenging the unwanted species is another choice.

In 2004, Adschiri and co-workers [51] reported that depolymerization of lignin was examined in a water-phenol mixture at 400 °C. No char formation was found and the 99 w% of the products were soluble in tetrahydrofuran after a reaction time of 1 h. Lercher [52], Labidi [53], and their co-workers have attempted this approach and added boric acid, phenol, and *p*-cresol in water to entrap reactive fragments, such as phenolic compounds and formaldehyde, and mask active sites, such as C_α in the lignin structure. In these approaches, complete conversion of lignin without char formation was realized. To restrain the repolymerization caused by the β-O-4 carbocation at thc lignin extraction, in 2007, Li et al. [54] reported that adding of a reactive phenol, 2-naphthol, in the steam explosion of biomass components was shown to inhibit the repolymerization reaction strongly, resulting in a highly improved delignification by subsequent solvent extraction and an extracted lignin of uniform structure. Besides the above carbocation capping reagents, in 2017, Li and co-workers [55] achieved the efficient depolymerization of organosolv lignin at 250 °C for 30 min with a 66.7% lignin conversion, and a 14.5% yield of phenolic monomer. The transformation with no obvious char formation used the remarkable catalyst of cooperative IL pairs, consisting of [bSmim][HSO4] (IL catalyst) and [bmim][CF3SO3] (IL additive, an excellent IL solvent for lignin dissolution). It was noteworthy that the cooperative IL pairs both played important roles in lignin depolymerization, in which the IL catalyst presented excellent ether linkage cleavage capability, whereas the IL additive exhibited efficient carbocation capping and scavenging.

In addition, to restrain the recondensation, guiding the transformation of the β-O-4 carbocation by in-situ converting the active intermediates is as efficient as the methods by avoiding the generation of the β-O-4 carbocation. Besides the products generated from lignin C–OAr bond cleavage, acidolysis of lignocellulose can provide numerous byproducts owing to the recondensation of enol ether intermediates formed after the first dehydration of the β-O-4 linkages. In 2015, Watanabe and co-workers [56] studied the in-situ trapping of enol intermediates with alcohol during acid-catalyzed depolymerization of lignin. In 2015 and 2016, Barta and co-workers [8] reported a series of triflic acid-catalyzed methods to achieve the depolymerization of lignin β-O-4 linkages with low recondensation, which were based on the rapid transformation of in-situ generated substituted phenylacetaldehyde. Capturing the unstable phenylacetaldehyde by reaction with diols and by in situ catalytic hydrogenation or decarbonylation lead to three distinct groups of aromatic compounds in high yields acetals, ethyl aromatics, and methyl aromatics (Figure 11.14). Notably, the same product groups were obtained when these approaches were successfully extended to lignin, and the formation of higher molecular weight side products was markedly suppressed, indicating that the in-situ aldehyde

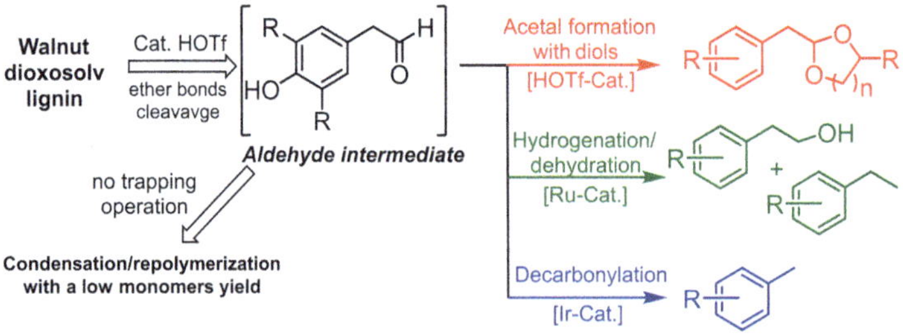

Figure 11.14 In Situ Conversion of the Reactive Aldehyde Intermediate in the Lignin Acidolysis. Adapted from Ref [8] with Permission, © 2015 American Chemical Society.

intermediates conversion is an efficient method to increase the yield of monomers from lignin acidolysis.

Besides the triflic acid, Barta and co-workers further studied the promotion effect of metal triflates for the production of aromatics from lignin [57]. The methodology relies on the use of catalytic amounts of easy-to-handle metal triflates ($M(OTf)_x$). They evaluated the reactivity of a broad range of metal triflates using simple lignin model compounds and more advanced lignin model compounds. The product aromatic monomers were either phenolic C_2-acetals obtained by stabilization of the aldehyde cleavage products by reaction with ethylene glycol or methyl aromatics obtained by catalytic decarbonylation with an Ir-catalyst. When the method was ultimately tested on lignin, especially $Fe(OTf)_3$ proved very effective and the phenolic C_2-acetal products were obtained in an excellent, 19 wt% yield [57, 58]. In the same year, Bruijnincx and co-workers [59] further studied a tandem catalysis strategy for ether linkage cleavage within lignin, involving ether hydrolysis by water-tolerant Lewis acids followed by aldehyde decarbonylation by an Rh complex. When the rate of C_α–OH hydrolysis and decarbonylation were matched, the method can be successfully applied to lignin models, dioxasolv lignins, as well as poplar sawdust, to give the anticipated decarbonylation products. Promisingly, product selectivity can be tuned by variation of the Lewis-acid strength and lignin source. The tandem catalytic approach involves lignin hydrolysis/decarbonylation, using water-tolerant Lewis acids [M(OTf)3, M=Sc, Y, In, Ga, and Yb]. Rh-catalyzed decarbonylation then effectively scavenges the aldehyde hydrolysis products to afford the desired decarbonylated 4-methylphenols in reactions with model compounds, various lignins. Remarkably, a second alternative trapping pathway is observed with the actual lignins, leading instead to the formation of 4-(1-propenyl)phenols (Figure 11.15), which are attractive high-value/low-volume lignin products. As shown in the proposed mechanism, initial Lewis acid catalyzed dehydration of the α-hydroxyl functionality in **1** forms an isomeric mixture of enol ethers **3**, followed by a slower Lewis acid-catalyzed hydrolysis step to the corresponding acetaldehyde **6**. A sufficiently active decarbonylation catalyst then gives the corresponding methyl-substituted product **7**. In addition, an alternative pathway is also observed, tentatively originating from the enol ether **3**, leading to the formation of a styrene derivative **4** by insertion of a formal hydrogen equivalent. For the detailed mechanism, the authors do not provide in the article, but one can refer to Zhang's research about the dehydroxylation–hydrogenation strategy for the cleavage of lignin β-O-4 ether bond [42]. According to DFT calculation, the addition of a H radical at the C_β of enol ether to generate a C_α radical can reduce the BDE of C_β–OAr from 336.6 kJ mol^{-1} to 66.9 kJ mol^{-1}, which can induce the self-cleavage of the target ether bond.

Furthermore, besides the active C_α carbocation (C_α^+) and aldehyde intermediates that can induce the fragments' condensation, the phenolic hydroxy group is another potential species that can cause the condensation via the base-catalyzed condensation and radical condensation and other lignin linkages transformations (Chapter 11.1). Compared with the researches about controlling the transformation of C_α carbocation and aldehyde intermediates, the reports about the in-situ converting phenolic group are fewer.

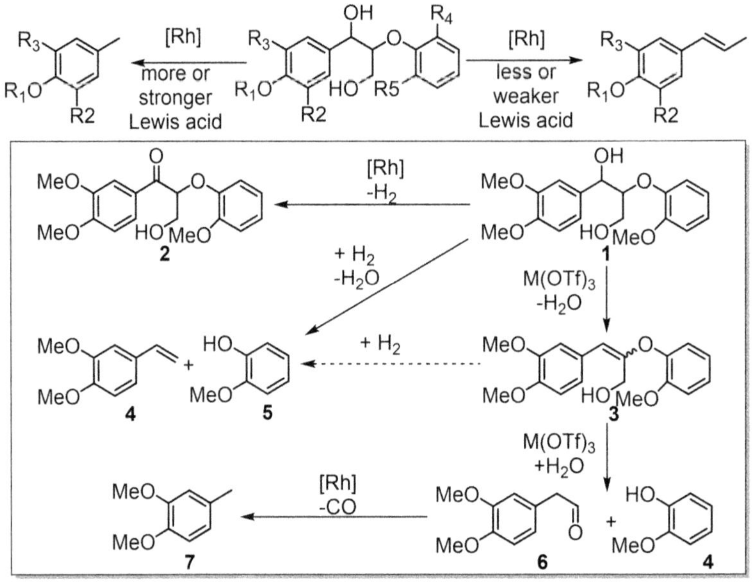

Figure 11.15 Lewis Acid-catalyzed β-O-4 Linkage Cleavage with Rh-catalyzed Decarbonylation. Adapted from Ref [59] with Permission, © 2016 Wiley-VCH.

At the beginning of this research, the target of lignin phenolic group modification is to obtain a functional lignin polymer [60–62]. In 2016, the Ford group [63] evaluate the addition of dimethyl carbonate (DMC) to a biomass conversion system that uses a Cu-doped porous metal oxide ($Cu_{20}PMO$) catalyst in supercritical methanol (scMeOH) to disassemble lignin with little to no char formation. While $Cu_{20}PMO$ catalyzes C–O hydrogenolysis of aryl–ether bonds linking lignin monomers, it also catalyzes arene methylation and hydrogenation. The MeOH/DMC cosolvent system can significantly suppress the arene hydrogenation of the phenolic intermediates responsible for much of the undesirable product diversity via O-methylation of phenolic –OH groups to form more stable aryl-OCH_3 species. Consequently, product proliferation was greatly reduced and aromatic yields greatly enhanced with lignin models, 2-methoxy-4-propylphenol, benzyl phenyl ether, and 2-phenoxy-1-phenylethan-1-ol.

11.2.4 Catalysts Modification and Design

Lignin depolymerization releases reactive monomers/fragments by the cleavage of weak C–O/C–C bonds. These monomers/fragments can further recombine to form undesired polymers through the formation of recalcitrant C–C bonds, which will decrease the efficiency of lignin depolymerization. As shown above, different strategies have been developed to pre-protect or pre-transform the active groups, or in-situ stabilize the reactive monomers/fragments to inhibit the condensation. Besides the conversion strategies design, modifying, and designing the efficient catalysts can provide more opportunities for getting a higher monomers yield from lignin depolymerization.

Zeolites are porous systems with high surface areas and shape/size selectivity along with the ability to accommodate different metals in their frameworks, resulting in acid-base or redox properties. Furthermore, because of the internal space limitation in the zeolite porous structure, zeolites might be able to lower the rate of bimolecular condensations by pore size constraints and diffusion control [64], which may eliminate the need for trapping agents to control the lignin fragments condensation. In 2020, Samec and co-workers [65] reported a zeolite-assisted method for the lignin-first fractionation from lignocellulose, which can overcome the lignin recondensation via shape-selective catalysis (Figure 11.16) The lignocellulosic feedstock underwent solvolytic depolymerization and released reactive allylic alcohols, which were transformed inside the pores of zeolites at Brønsted acid active sites into stable monophenolic products. Experiments with model lignin compounds showed that the recondensation of aldehydes and allylic alcohols produced by the cleavage of β-O-4 bonds was efficiently inhibited by the use of protonic β zeolite. The developed methodology was further extended to birch wood to yield 20 w% monophenolic compounds and value-added products from carbohydrates at 220 °C in 2 h. Before Samec's report, in 2019, Jiang and co-workers [66] found that when the acid-treated mesoporous MCM-41 molecular sieves were adopted as catalysts for degradation of dealkalized lignin dissolved in acidic ionic liquid 1-methyl-3-benzylimidazolium chloride (BnMIMCl), the efficiency of lignin degradation was as high as 74.02% at 200 °C for 4 h; meanwhile, over 80% of the liquid products were tested as phenols. For the efficiency, as discussed in the article, on one hand, the acid sites doped the nanochannels activate lignin macromolecules to conduct degradation reaction, on the other hand, the confined spaces are benefit for restricting the repolymerization of the degraded lignin fragments.

11.2.5 Intensifying the Reaction Systems

In addition, to control the lignin fragments' recondensation, besides the strategies methods and catalysts design which is shown above, engineering and technology

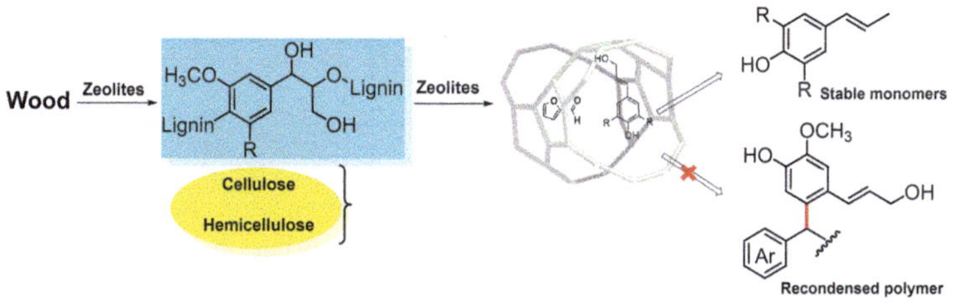

Figure 11.16 Preventing Lignin Fragments Condensation by Diffusion Control Based on the Size and Topology of the Zeolite Pores. Adapted from Ref [65] with Permission, © 2020 Wiley-VCH.

methods are employed for intensifying the reaction systems. The methods can be divided into the following typical methods: (1) Increasing the reaction rate and reducing the reaction time, such as introducing the microwave assistance; (2) Accurately controlling the extent of reaction and reaction time, such as using the flow-through reactors; (3) Isolating the reaction substrate and products, such as using the microemulsion method; and (4) In-situ separating the depolymerization products from the substrates, such as the membrane filtration assistance method.

The microwave irradiation method, an alternative method for biomass conversion, is increasingly used to accelerate the decomposition rate of lignin model compounds [67] and improve the selectivity of bio-oils [68]. And the conversion of lignin into low molecular weight aromatics under microwave irradiation is gathering more attention [69–73]. Because the microwave can increase the reaction rate, so the total reaction time can be reduced, which in other words means the time left for the byproducts generation can be shortened and fewer lignin fragments condensation reactions occur. Therefore, as another speculation, the combined photo-electro method can be more efficient than the photo- or electro method only, and the photothermal process may increase the efficiency of the photocatalytic process.

In addition, to control the extent of reaction and reaction time, the flow-through reactor is another choice. In 2017, Samec et al. [74] achieved the lignin depolymerization to monophenolic compounds in a flow-through system, in which pulping and transfer hydrogenolysis steps were separated in time and space. Without the hydrogenolysis step or addition of trapping agents to the pulping, it is possible to obtain partially depolymerized lignin (21 wt% monophenolic compounds) that is prone to further processing. By applying a transfer hydrogenolysis step with a Pd/C catalyst, a 37 wt% yield of lignin-derived monophenolic compounds was obtained. In 2017-2018, Beckham and Román-Leshkov [75, 76] investigated the kinetics of the lignin RCF (reductive catalytic fractionation) process by using flow-through reactors to decouple the two limiting mechanistic steps, namely lignin solvolysis and reduction. Their study demonstrates RCF occurs by two limiting processes that can be independently controlled. By connecting the solvolysis reactor that can selective in-situ extract the lignin fragments from lignocellulose with the further reduction reactor, two independent steps in flow-through reactors can be physically separated and isolated. Few sugar fragments can be introduced in the following reduction reactor, and the total amount of lignin fragments flowing into the reduction reactor can be controlled, which means the potential condensation can be efficiently controlled by manipulating the rates of the two reactors.

To control the condensation of lignin fragments, isolating the substrates and products in a tiny zone is another strategy. One can use the microemulsion method, which means the catalyst, limited lignin fragments, reductant/oxidant, and generated products can be capsuled in a small zone to achieve sufficient depolymerization of lignin fragments and reduce the condensation. In this aspect, Li and co-workers [77–79] applied the emulsion microreactor in the lignin oxidation and hydrogenation to provide phenolic monomers. As a similar method with a mixed solvent, one can use a two-phase reaction system which may separate the products from the substrate phase immediately and reduce the overreaction of the products and restrain the condensation. In 2020, Chen and co-workers [80] reported a process of selectively oxidative degradation of lignin model compounds operated in a two-phase reaction system. Lignin model compounds of 4-benzyloxyphenol (PBP) or guaiacylglycerol-β-guaiacyl ether (GGE) in a bottom phase of 1-butyl-3-methylimidazole chloride ([BMIM]Cl) ionic liquid was selectively oxidized by H_2O_2 in the presence of a solid acid (SO_4^{2-}/Fe_2O_3-ZrO_2), and the degradation products immediately diffused into the upper organic solvent phase (butyl acetate). In this reaction system, the yield of the products was improved due to the prolonged life of $HO^\bullet$ in ionic liquid, and the product selectivity was maintained due to the timely product separation, and the ionic liquid and the catalyst were easily recycled.

To separate the products from the reaction system in a timely manner, one can also add a filtration membrane in the reaction system (Figure 11.17). In 2020, Villain-gambier and co-workers [81] reported a membrane bioreactor for lignin catalytic oxidation. In their work, the oxidative depolymerization of lignin was catalyzed by a commercial laccase both in a batch experiment and in a membrane bioreactor using the same catalytic conditions. The membrane bioreactor was previously optimized to reach high permeation flux (25 $L \cdot h^{-1} \cdot m^{-2}$) during lignin diafiltration. The lignin was exclusively condensed in the batch experiment leading to high molecular weight macromolecules (from 9 to 16 kDa), but the lignin depolymerization in the bioreactor was effective and produce fragments of less than 1 kDa thanks to the in-situ extraction/separation of the reaction products. Their results demonstrate that the reactor configuration is playing an essential role in triggering or preventing lignin condensation. The in-situ membrane extraction of the reactive fragments of lignin from the bulk medium can be useful against detrimental repolymerization.

Given the challenges that the native lignin depolymerization research meets now, the development of reaction-intensifying systems is necessary to achieve the final transition of lignin utilization from laboratory research to real industrial applications. Therefore, reaction-intensifying research is as important as the lignin depolymerization strategies research.

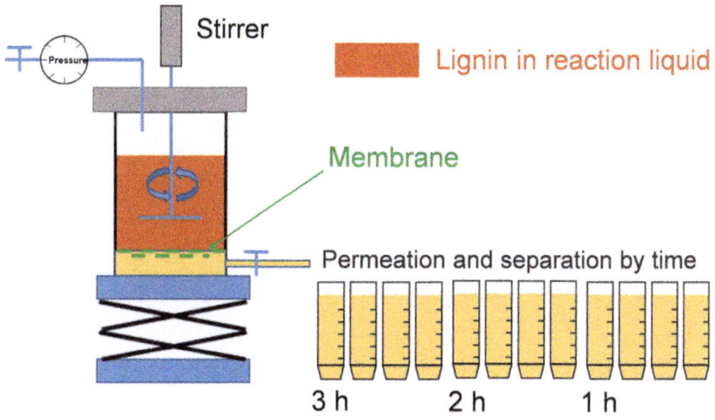

Figure 11.17 The Reactor Installation with Membrane Separation.

11.2.6 Employing an "Ideal Lignin" as the Substrates

To restrain the potential fragment condensation, the methods shown above focus primarily on how to control the generation of the unwanted intermediates and restrain their undesirable reaction routes; beyond strategy, catalyst, and reactor design, the consideration of native lignin structure design is usually ignored. To achieve efficient lignin depolymerization, besides the ideal catalyst, reactor, and strategy, we believe that one "ideal lignin" substrate can yield twice the result with half the effort. Let's return to the condensation mechanism discussion in Chapter 11.1, with β-O-4 linkage as the model, the lignin fragments condensation position is mainly located at the C_α (acid-condensation), C_2 (acid-condensation), or C_3/C_5 (base-condensation, radical-condensation) of the phenolic ring. Therefore, if the native lignin substrate can block the above carbon positions, especially for the carbon positions of the phenolic rings, the condensation can be efficiently restrained. It is now a realistic juncture to posit the characteristics of an ideal lignin archetype for biomass processing. For the depolymerization of the lignin polymer to monomers, lignin should have at least the following three characteristics. First, if acidic pretreatment is used, then it should be stable under acidic conditions to prevent condensation and the generation of undesired new C–C bonds. Second, it should contain only ether (C–O) interunit linkages in its backbone so that it can be fully depolymerized. Finally, it should be generated in planta from a single phenylpropanoid monomer to allow the production of the simplest array of compounds. According to the current lignin research process, two ideal lignin is the alternative of the "ideal lignin".

The S-dominant lignin with a high β-O-4 linkage content and a high ratio of S-unit. On the basis of the information stemming from the lignin biosynthetic research community over the last decade, and with the revelations regarding lignins' structural malleability from studies on lignin pathway mutants and transgenics as well as on various "natural" plants discovered to have unusual lignins, researchers have been able to contemplate designing lignins to achieve higher monomers yields and simpler product mixtures [82]. For example, the recent use of formaldehyde protection during lignin extraction from high-S poplar lignin that has up to 98% syringyl S units and ~90% β–O–4 linkages prevented condensation reactions and allowed an unprecedentedly high monomer yield (78%) under hydrogenolytic conditions [5, 83].

Catechyl lignin with a phenolic group protected C_α. The second "ideal lignin" is usually been called C-lignin [84], which can be regarded as a unique β-O-4 linkage with a benzodioxane structure (Figure 11.18). The unusual catechyl lignin usually presents in the seed coats of vanilla (*Vanilla planifolia*) [85] and various members of the Cactaceae of the genus Melocactus [86]. For its biosynthesis, in the special case, the lack of O-methyltransferase (OMT) activity for conversion from catechyl C to guaiacyl G and, subsequently, on to syringyl S, aromatic-level precursors, results in 100% C units in the cell wall. This C-lignin was, somewhat surprisingly, found to be essentially a homopolymer synthesized almost purely by β–O–4 coupling of caffeyl

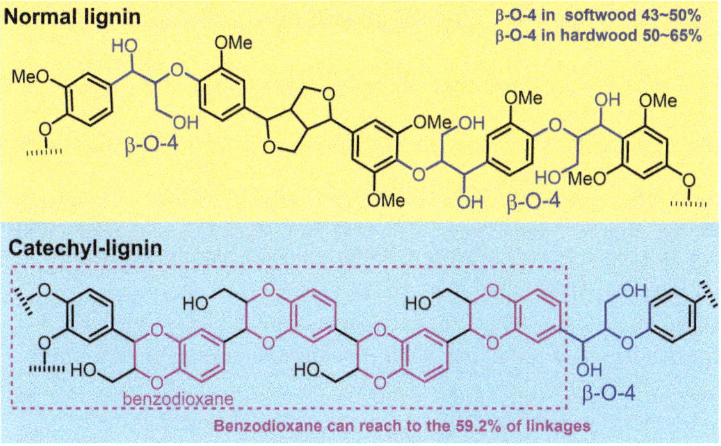

Figure 11.18 The Normal Lignin Structure and Benzodioxane Structure of C-lignin.

alcohol with the growing polymer chain, producing benzodioxanes as the dominant unit in the polymer. Ralph et. al. [84] discovered that C-lignin is highly acid-resistant. Furthermore, hydrogenolysis of C-lignin resulted in the cleavage of all benzodioxane structures to produce catechol-type monomers in near-quantitative yield with a selectivity of 90% to a single monomer [84]. In Song's recent research [87], an atomically dispersed Ru catalyst, which can serve as an efficient catalyst for the hydrogenolysis of C-lignin via the cleavage of C—O bonds in benzodioxane linkages, giving catechols in high yields with TONs up to 345. A unique selectivity to propenylcatechol (77%) is obtained.

11.3 Chapter Summary

The main topic of this chapter 11 is promoting the lignin linkages cleavage with considering the potential lignin fragments condensation. Chapter 11.1 provides a brief summary of potential lignin fragments condensation mechanism, including acid-catalyzed condensation, base-catalyzed condensation, radical fragments condensation, and other lignin linkages transformations. When condensation mechanisms are made clear, Chapter 11.2 introduces the developed methods for restraining the potential condensation of lignin fragments. The methods are divided into six categories as follows: (1) Pre-protection of the active groups to inhibit the condensation; (2) First transformation of the active groups to promote the depolymerization; (3) In-situ converting the active intermediates and scavenging the unwanted species; (4) Modifying and designing the catalysts; (5) Intensifying reaction systems; and (6) Employing an "ideal lignin" as the substrates. By the end, The keywords of this chapter could be Lignin depolymerization, Lignin fragments condensation, Lignin modification, Catalyst design, Reaction intensification, Ideal lignin.

References

1 Yang, H., Yoo, C.G., Meng, X. et al. (2020). Structural changes of lignins in natural Populus variants during different pretreatments. *Bioresour. Technol.* 295: 122240.

2 Questell-Santiago, Y.M., Galkin, M.V., Barta, K., and Luterbacher, J.S. (2020). Stabilization strategies in biomass depolymerization using chemical functionalization. *Nat. Rev. Chem.* 4: 311–330.

3 Luo, X.L., Li, Y.D., Gupta, N.K. et al. (2020). Protection Strategies Enable Selective Conversion of Biomass. *Angew. Chem. Int. Ed.* 59: 11704–11716.

4 Zhang, C., Li, H., Lu, J. et al. (2017). Promoting lignin depolymerization and restraining the condensation via an oxidation–hydrogenation strategy. *ACS Catal.* 7: 3419–3429.

5 Shuai, L., Amiri, M.T., Questell-Santiago, Y.M. et al. (2016). Formaldehyde stabilization facilitates lignin monomer production during biomass depolymerization. *Science* 354: 329–333.

6 Lan, W., Amiri, M.T., Hunston, C.M., and Luterbacher, J.S. (2018). Protection group effects during α,γ-diol lignin stabilization promote high-selectivity monomer production. *Angew. Chem. Int. Ed.* 57: 1356–1360.

7 Sturgeon, M.R., Kim, S., Lawrence, K. et al. (2014). A mechanistic investigation of acid-catalyzed cleavage of aryl-ether linkages: Implications for lignin depolymerization in acidic environments. *ACS Sustainable Chem. Eng.* 2: 472–485.

8 Deuss, P.J., Scott, M., Tran, F. et al. (2015). Aromatic monomers by in situ conversion of reactive intermediates in the acid-catalyzed depolymerization of lignin. *J. Am. Chem. Soc.* 137: 7456–7467.

9 Lahive, C.W., Deuss, P.J., Lancefield, C.S. et al. (2016). Advanced model compounds for understanding acid-catalyzed lignin depolymerization: Identification of renewable aromatics and a lignin-derived solvent. *J. Am. Chem. Soc.* 138: 8900–8911.

10 Miles-Barrett, D.M., Neal, A.R., Hand, C. et al. (2016). The synthesis and analysis of lignin-bound Hibbert ketone structures in technical lignins. *Org. Biomol. Chem.* 14: 10023–10030.

11 Chakar, F.S. and Ragauskas, A.J. (2004). Review of current and future softwood kraft lignin process chemistry. *Ind. Crops Prod.* 20: 131–141.

12 Renders, T., Schutyser, W., Van den Bosch, S. et al. (2016). Influence of acidic (H_3PO4) and alkaline (NaOH) additives on the catalytic reductive fractionation of lignocellulose. *ACS Catal.* 6: 2055–2066.

13 Bjørsvik, H.R. and Minisci, F. (1999). Fine chemicals from lignosulfonates. 1. Synthesis of vanillin by oxidation of lignosulfonates. *Org. Process. Res. Dev.* 3: 330–340.

14 Lundquist, K. and Kristersson, P. (1985). Exhaustive laccase-catalysed oxidation of a lignin model compound (vanillyl glycol) produces methanol and polymeric quinoid products. *Biochem. J.* 229: 277–279.

15 Bozell, J.J., Hames, B.R., and Dimmel, D.R. (1995). Cobalt-Schiff base complex catalyzed oxidation of para-substituted phenolics. *Preparation of benzoquinones. J. Org. Chem.* 60: 2398–2404.

16 Fabbri, C., Bietti, M., and Lanzalunga, O. (2005). Generation and reactivity of ketyl radicals with lignin related structures. On the importance of the ketyl pathway in the photoyellowing of lignin containing pulps and papers. *J. Org. Chem.* 70: 2720–2728.

17 Norberg, I., Nordström, Y., Drougge, R. et al. (2013). A new method for stabilizing softwood kraft lignin fibers for carbon fiber production. *J. Appl. Polym. Sci.* 128: 3824–3830.

18 Ragauskas, A.J., Beckham, G.T., Biddy, M.J. et al. (2014). Lignin valorization: Improving lignin processing in the biorefinery. *Science* 344: 1246843.

19 Luo, N., Wang, M., Li, H. et al. (2016). Photocatalytic oxidation–hydrogenolysis of lignin β-O-4 models via a dual light wavelength switching strategy. *ACS Catal.* 6: 7716–7721.

20 Qi, S.C., Hayashi, J., Kudo, S., and Zhang, L. (2017). Catalytic hydrogenolysis of kraft lignin to monomers at high yield in alkaline water. *Green Chem.* 19: 2636–2645.

21 Fan, H.L., Yang, Y.Y., Song, J.L. et al. (2015). Free-radical conversion of a lignin model compound catalyzed by Pd/C. *Green Chem.* 17: 4452–4458.

22 Kubo, S., Yamada, T., Hashida, K., and Ono, H. (2007). Grafting of ethylene glycol chains in lignin during the solvolysis for biomass conversion using ethylene carbonate/ethylene glycol system. *Chem. Lett.* 36: 502–503.

23 Chen, J., Lu, F., Si, X. et al. (2016). High yield production of natural phenolic alcohols from woody biomass using a nickel-based catalyst. *ChemSusChem* 9: 3353–3360.

24 Zhu, G., Qiu, X., Zhao, Y. et al. (2016). Depolymerization of lignin by microwave-assisted methylation of benzylic alcohols. *Bioresour. Technol.* 218: 718–722.

25 Lancefield, C.S., Panovic, I., Deuss, P.J. et al. (2017). Pre-treatment of lignocellulosic feedstocks using biorenewable alcohols: towards complete biomass valorisation. *Green Chem.* 19: 202–214.

26 Karkas, M.D. (2017). Lignin hydrogenolysis: Improving lignin disassembly through formaldehyde stabilization. *ChemSusChem* 10: 2111–2115.

27 Mu, X.L., Han, Z., Liu, C.B., and Zhang, D.J. (2019). Mechanistic insights into formaldehyde-blocked lignin condensation: A DFT study. *J. Phys. Chem. C* 123: 8640–8648.

28 Thi Thi, N., Tobimatsu, Y., Takahashi, S. et al. (2018). Isolation and characterization of polyethylene glycol (PEG)-modified glycol lignin via PEG solvolysis of softwood biomass in a large-scale batch reactor. *ACS Sustainable Chem. Eng.* 6: 7841–7848.

29 Lohr, T.L., Li, Z., and Marks, T.J. (2015). Selective ether/ester C-O cleavage of an acetylated lignin model via tandem catalysis. *ACS Catal.* 5: 7004–7007.

30 Buono, P., Duval, A., Verge, P. et al. (2016). New insights on the chemical modification of lignin: Acetylation versus silylation. *ACS Sustainable Chem. Eng.* 4: 5212–5222.

31 Monsigny, L., Feghali, E., Berthet, J.-C., and Cantat, T. (2018). Efficient reductive depolymerization of hardwood and softwood lignins with Brookhart's iridium((III)) catalyst and hydrosilanes. *Green Chem.* 20: 1981–1986.

32 Guvenatam, B., Heeres, E.H.J., Pidko, E.A., and Hensen, E.J.M. (2016). Lewis acid-catalyzed depolymerization of soda lignin in supercritical ethanol/water mixtures. *Catal. Today* 269: 9–20.

33 Rahimi, A., Ulbrich, A., Coon, J.J., and Stahl, S.S. (2014). Formic-acid-induced depolymerization of oxidized lignin to aromatics. *Nature* 515: 249–252.

34 Li, Z., Assary, R.S., Atesin, A.C. et al. (2013). Rapid ether and alcohol C–O bond hydrogenolysis catalyzed by tandem high-valent metal triflate + supported Pd catalysts. *J. Am. Chem. Soc.* 136: 104–107.

35 Lu, F.C. and Ralph, J. (1997). DFRC method for lignin analysys. 1. New method for β-aryl ether cleavage: Lignin model studies. *J. Agr. Food. Chem.* 45: 4655–4660.

36 Lu, F.C. and Ralph, J. (1998). The DFRC method for lignin analysis. 2. Monomers from isolated lignins. *J. Agr. Food. Chem.* 46: 547–552.

37 Lu, F.C. and Ralph, J. (1998). The DFRC method for lignin analysis. 3. NMR studies. *J. Wood. Chem. Technol.* 18: 219–233.

38 Peng, J.P., Lu, F.C., and Ralph, J. (1998). The DFRC method for lignin analysis. 4. Lignin dimers isolated from DFRC-degraded loblolly pine wood. *J. Agr. Food. Chem.* 46: 553–560.

39 Peng, J.P., Lu, F.C., and Ralph, J. (1999). The DFRC method for lignin analysis. 5. Isochroman lignin trimers from DFRC-degraded Pinus taeda. *Phytochemistry* 50: 659–666.

40 Ralph, J. and Lu, F.C. (1998). The DFRC method for lignin analysis. 6. A simple modification for identifying natural acetates on lignins. *J. Agr. Food. Chem.* 46: 4616–4619.

41 Lu, F.C. and Ralph, J. (1999). The DFRC method for lignin analysis. 7. Behavior of cinnamyl end groups. *J. Agr. Food. Chem.* 47: 1981–1987.

42 Zhang, C., Lu, J., Zhang, X. et al. (2016). Cleavage of the lignin β-O-4 ether bond via a dehydroxylation-hydrogenation strategy over a NiMo sulfide catalyst. *Green Chem.* 18: 6545–6555.

43 Wu, X.J., Fan, X.T., Xie, S.J. et al. (2018). Solar energy-driven lignin-first approach to full utilization of lignocellulosic biomass under mild conditions. *Nat. Catal.* 1: 772–780.

44 Lin, J., Wu, X., Xie, S. et al. (2019). Visible-light-driven cleavage of C—O linkage for lignin valorization to functionalized aromatics. *ChemSusChem* 12: 5023–5031.

45 Wu, X.J., Xie, S.J., Liu, C.X. et al. (2019). Ligand-controlled photocatalysis of CdS quantum dots for lignin valorization under visible light. *ACS Catal.* 9: 8443–8451.

46 Son, S. and Toste, F.D. (2010). Non-oxidative vanadium-catalyzed C-O bond cleavage: Application to degradation of lignin model compounds. *Angew. Chem. Int. Ed.* 49: 3791–3794.

47 Hanson, S.K., Baker, R.T., Gordon, J.C. et al. (2010). Aerobic oxidation of lignin models using a base metal vanadium catalyst. *Inorg. Chem.* 49: 5611–5618.

48 Gazi, S., Ng, W.K.H., Ganguly, R. et al. (2015). Selective photocatalytic C—C bond cleavage under ambient conditions with earth abundant vanadium complexes. *Chem. Sci.* 6: 7130–7142.

49 Liu, H.F., Li, H.J., Luo, N.C., and Wang, F. (2020). Visible-light-induced oxidative lignin C—C bond cleavage to aldehydes using vanadium catalysts. *ACS Catal.* 10: 632–643.

50 Wang, Y., He, J., and Zhang, Y. (2020). $CeCl_3$-promoted simultaneous photocatalytic cleavage and amination of C_α–C_β bond in lignin model compounds and native lignin. *CCS Chem.* 2: 107–117.

51 Okuda, K., Umetsu, M., Takami, S., and Adschiri, T. (2004). Disassembly of lignin and chemical recovery—rapid depolymerization of lignin without char formation in water–phenol mixtures. *Fuel Process. Technol.* 85: 803–813.

52 Roberts, V.M., Stein, V., Reiner, T. et al. (2011). Towards 1uantitative catalytic lignin depolymerization. *Chem. Eur. J.* 17: 5939–5948.

53 Toledano, A., Serrano, L., and Labidi, J. (2014). Improving base catalyzed lignin depolymerization by avoiding lignin repolymerization. *Fuel* 116: 617–624.

54 Li, J.B., Henriksson, G., and Gellerstedt, G. (2007). Lignin depolymerization/repolymerization and its critical role for delignification of aspen wood by steam explosion. *Bioresour. Technol.* 98: 3061–3068.

55 Li, Y.W., Cai, Z.P., Liao, M.C. et al. (2017). Catalytic depolymerization of organosolv sugarcane bagasse lignin in cooperative ionic liquid pairs. *Catal. Today* 298: 168–174.

56 Kaiho, A., Kogo, M., Sakai, R. et al. (2015). In situ trapping of enol intermediates with alcohol during acid-catalysed depolymerisation of lignin in a nonpolar solvent. *Green Chem.* 17: 2780–2783.

57 Deuss, P.J., Lahive, C.W., Lancefield, C.S. et al. (2016). Metal triflates for the production of aromatics from lignin. *ChemSusChem* 9: 2974–2981.

58 Deuss, P.J., Lancefield, C.S., Narani, A. et al. (2017). Phenolic acetals from lignins of varying compositions via iron(III) triflate catalysed depolymerisation. *Green Chem.* 19: 2774–2782.

59 Jastrzebski, R., Constant, S., Lancefield, C.S. et al. (2016). Tandem catalytic depolymerization of lignin by water-tolerant Lewis acids and rhodium complexes. *ChemSusChem* 9: 2074–2079.

60 Sadeghifar, H., Cui, C.Z., and Argyropoulos, D.S. (2012). Toward Thermoplastic Lignin Polymers. Part 1. Selective masking of phenolic hydroxyl groups in kraft lignins via methylation and oxypropylation chemistries. *Ind. Eng. Chem. Res.* 51: 16713–16720.

61 Sen, S., Patil, S., and Argyropoulos, D.S. (2015). Methylation of softwood kraft lignin with dimethyl carbonate. *Green Chem.* 17: 1077–1087.

62 Lui, M.Y., Lokare, K.S., Hemming, E. et al. (2016). Microwave-assisted methylation of dihydroxybenzene derivatives with dimethyl carbonate. *RSC Adv.* 6: 58443–58451.

63 Barrett, J.A., Gao, Y., Bernt, C.M. et al. (2016). Enhancing aromatic production from reductive lignin disassembly: in situ O-methylation of phenolic intermediates. *ACS Sustainable Chem. Eng.* 4: 6877–6886.

64 Dusselier, M., Van Wouwe, P., Dewaele, A. et al. (2015). Shape-selective zeolite catalysis for bioplastics production. *Science* 349: 78–80.

65 Subbotina, E., Velty, A., Samec, J.S., and Corma, A. (2020). Zeolite-Assisted lignin-First fractionation of lignocellulose: Overcoming lignin recondensation through shape-Selective catalysis. *ChemSusChem* 13: 4528–4536.
66 Kang, S., Bu, D., Xue, L. et al. (2019). Degradation of lignin catalyzed by nanochannels solid acids in ionic liquid. *Polym. Degrad. Stab.* 166: 1–7.
67 Zhu, G., Ouyang, X., Yang, Y. et al. (2016). Selective cleavage of aryl ether bonds in dimeric lignin model compounds. *RSC Adv.* 6: 17880–17887.
68 Wan, Y., Chen, P., Zhang, B. et al. (2009). Microwave-assisted pyrolysis of biomass: Catalysts to improve product selectivity. *J. Anal. Appl. Pyrolysis* 86: 161–167.
69 Toledano, A., Serrano, L., Labidi, J. et al. (2013). Heterogeneously catalysed mild hydrogenolytic depolymerisation of lignin under microwave irradiation with hydrogen-donating solvents. *ChemCatChem* 5: 977–985.
70 Dong, C., Feng, C., Liu, Q. et al. (2014). Mechanism on microwave-assisted acidic solvolysis of black-liquor lignin. *Bioresour. Technol.* 162: 136–141.
71 Shen, D., Liu, N., Dong, C. et al. (2015). Catalytic solvolysis of lignin with the modified HUSYs in formic acid assisted by microwave heating. *Chem. Eng. J.* 270: 641–647.
72 Qu, C., Ito, K., Katsuyama, I. et al. (2020). Directly microwave-accelerated cleavage of C—C and C—O bonds of lignin by copper oxide and H_2O_2. *ChemSusChem* 13: 4510–4518.
73 Liu, X., Bouxin, F.P., Fan, J. et al. (2020). Microwave-assisted catalytic depolymerization of lignin from birch sawdust to produce phenolic monomers utilizing a hydrogen-free strategy. *J. Hazard. Mater.* 123490.
74 Kumaniaev, I., Subbotina, E., Savmarker, J. et al. (2017). Lignin depolymerization to monophenolic compounds in a flow-through system. *Green Chem.* 19: 5767–5771.
75 Anderson, E.M., Stone, M.L., Katahira, R. et al. (2017). Flowthrough reductive catalytic fractionation of biomass. *Joule* 1: 613–622.
76 Anderson, E.M., Stone, M.L., Hulsey, M.J. et al. (2018). Kinetic studies of lignin solvolysis and reduction by reductive catalytic fractionation decoupled in flow-through reactors. *ACS Sustainable Chem. Eng.* 6: 7951–7959.
77 Liu, S., Lin, Z., Cai, Z. et al. (2018). Selective depolymerization of lignosulfonate via hydrogen transfer enhanced in an emulsion microreactor. *Bioresour. Technol.* 264: 382–386.
78 Cai, Z.P., Li, Y.W., He, H.Y. et al. (2015). Catalytic depolymerization of organosolv lignin in a novel water/oil emulsion reactor: Lignin as the self-surfactant. *Ind. Eng. Chem. Res.* 54: 11501–11510.
79 Kong, J., Li, L., Zeng, Q. et al. (2021). Oxidation of organosolv lignin in a novel surfactant-free microemulsion reactor. *Bioresour. Technol.* 321: 124466.
80 Chen, Y., Xue, A., Jiang, H. et al. (2020). A two-phase reaction system for selective oxidative degradation of lignin model compounds. *BioResources* 15: 6526–6538.
81 Steinmetz, V., Villain-gambier, M., Klem, A. et al. (2020). In-situ extraction of depolymerization products by membrane filtration against lignin condensation. *Bioresour. Technol.* 123530.
82 Mottiar, Y., Vanholme, R., Boerjan, W. et al. (2016). Designer lignins: harnessing the plasticity of lignification. *Curr. Opin. Biotech.* 37: 190–200.

83 Franke, R., McMichael, C.M., Meyer, K. et al. (2000). Modified lignin in tobacco and poplar plants over-expressing the Arabidopsis gene encoding ferulate 5-hydroxylase. *Plant J.* 22: 223–234.

84 Li, Y., Shuai, L., Kim, H. et al. (2018). An "ideal lignin" facilitates full biomass utilization. *Sci. Adv.* 4: eaau2968.

85 Chen, F., Tobimatsu, Y., Havkin-Frenkel, D. et al. (2012). A polymer of caffeyl alcohol in plant seeds. *P. Natl. Acad. Sci.* 109: 1772–1777.

86 Chen, F., Tobimatsu, Y., Jackson, L. et al. (2013). Novel seed coat lignins in the Cactaceae: Structure, distribution and implications for the evolution of lignin diversity. *Plant J.* 73: 201–211.

87 Wang, S., Zhang, K., Li, H. et al. (2021). Selective hydrogenolysis of catechyl lignin into propenylcatechol over an atomically dispersed ruthenium catalyst. *Nat. Commun.* 12: 416.

Part V

Outcome and Outlook

12

Summary on Lignin Utilization and Perspectives on Preparation of Aromatic Chemicals

12.1 Brief Summary on Lignin Utilization as Materials

As has been mentioned in Parts I and II, lignin is the second most abundant natural biopolymer after cellulose and between 40 and 50 million cubic meters/year can be produced worldwide as a mostly noncommercialized waste product [1]. Such a remarkable quantity of renewable carbon resources has drawn an enormous amount of attention from academia to industry worldwide every year for use in many applications, either directly or chemically modified. Lignin can be directly used as a solid fuel in the paper industry for its high calorific value. The modified lignosulfonates can be used as the binder in the pelletizing of coal/solid fuel briquettes, adhesive for preparation of ceramic, and the manufacture of packing material. Because of the water-absorbing capacity, lignosulfonates can be also used as the water reducing admixture in the building industry, water reservoirs for dry soils, and some biocompatible cellulose–lignin-based hydrogels used in health care products and cosmetics. Given its reversible chelation ability, the lignosulfonates can be an excellent candidate to sequester or liberate metal ions in aquatic media or soils, which means lignosulfonate can be also used as a carrier to provide essential ionic nutrients to plants and as a material to remove dangerous pollutants, such as heavy metals, in soil or water. The ammoxidized kraft lignin can be used as a slow-release fertilizer. The modified lignin can be also used as a dispersant, emulsifier, and surfactant. Given the antioxidant, antiviral, antitumor, and anticarcinogenic activities, native lignin and some of its preparations and derivatives have been shown to have promising results in fields such as biosciences and medicine because of their interactions with living organisms. Lignin is a known free-radical scavenger and uvioresistant material that can be used as sunscreen cream. Given the polymer structure of the lignin, lignin can be modified using chemical treatment such as phenolation, demethylation, and hydroxymethylation to expose more active (phenolic) hydroxyl groups working as nucleophilic attack groups for the preparation of phenol-formaldehyde binders, polyolefin–lignin polymers, polyester-lignin polymers, acrylamide-lignin polymers, polyurethane-lignin polymers, epoxy resin, polyurethanes, and

Lignin Conversion Catalysis: Transformation to Aromatic Chemicals, First Edition.
Chaofeng Zhang and Feng Wang.

polyesters. In addition, sulfonated lignin can be used as the frame in conducting polymer synthesis with the polymerization of N-containing aromatic chemicals that can be used as binders, coatings material for anticorrosion, electrostatic dissipative, and other uses. Lignosulfonate and polypyrrole form a renewable interpenetrating polymeric network that can be used as a cathode in batteries for charge and energy storage. In summary, lignin is versatile in the field of materials science and engineering [2–5].

12.2 Outlets of Lignin Resources Beyond Aromatic Chemicals

In addition to the lignin utilization discussed above that emphasized the materials, lignin has value from the perspective of chemical preparation. Lignin is composed of three different phenylpropyl monomers that can be regarded as a bank of aromatic chemicals: coumarin (H-unit), coniferol (G-unit), and syringol (S-unit). For the preparation of chemicals for lignin conversion, the critical issue is to develop efficient catalysts and conversion strategies to selectively break stable ether-oxygen bonds or carbon-carbon bonds in the lignin structure to prepare raw lignin monomers that can be conveniently used. Compared with the complete depolymerization and hydrodeoxygenation of lignin structural units to prepare C_6~C_{15} alkanes, considering the cost of hydrogenolysis and the value of the product as a fuel, lignin transformation to aromatic monomers with maintaining the aromaticity seems to be more economical. Although great progress for lignin depolymerization to aromatic chemicals has been made in the past decade, the preparation of aromatic chemicals from lignin also faces challenges that mainly come from the cost comparison with the mature petrochemical resource routes. The main problem lies in the choice of products and the cost of isolation from the lignin depolymerization mixture. The above reasons lead to an understanding of lignin transformation and utilization that illustrates that lignin can make anything but money, raising a critical issue about how to make lignin depolymerization utilization profitable.

To change this view, it is necessary to put forward new ideas to integrate and optimize the lignin depolymerization and export applications for the systematic utilization of lignin. The basic idea is to customize the process of lignin depolymerization into aromatic monomers, starting from the analysis of downstream products that are economically feasible raw products. Specifically, the following product-oriented formula needs to be considered:

$$\textbf{\textit{Profit}}_{(lignin\ route)} = \textbf{\textit{Product value}}_{(demand*unit\ price)} - \textbf{\textit{Production cost}}_{(Pretreatment+Conversion+Purification)} - \textbf{\textit{Others}}$$

Focusing on this formula, there are two promising choices for profitable lignin utilization: (1) lignin can be used to prepare some chemicals in a low demand but of high-value; or (2) the target chemicals can be in huge demand and valuable enough to make the process profitable. With these two possibilities, the preparation of high-value medicine and synthetic blocks from depolymerization monomers and artificial polymer monomer preparation from natural lignin polymer should be the two typical examples [6–9]. In terms of the transformation of stable lignin fragments to high-value medicine and synthetic block, In 2002, Bjørsvik and co-workers [8] provided a general overview of some first, second, and third-generation fine chemicals with origins from the lignin oxidation process. The synthetic organic processes of the substances are presented as well as several applications for the synthesis of a variety of pharmaceutical chemicals. In 2018, Barta and co-workers [7] presented a catalytic strategy that can transform lignocellulose into a range of attractive products. From the lignin, pharmaceutical building blocks are obtained from the delicate transformation of aromatic monomers. In addition, besides the utilization of ether linkages and some C—C linkages which can be cleaved under relatively mild conditions, usually leaving some stable lignin linkages. One example is the biaryl 5-5 linkage. Taking into account the difficulty and low benefit from the cleavage of the biaryl C—C bond to monomers, the idea of using the lignin biaryl structure derivatives to generate high-value synthetic building blocks is challenging but attractive. In 2019, Li and co-workers [9] developed a photoinduced biaryl compounds dearomatization method for spirolactones preparation without the C_{Ar}–C_{Ar} bond cleavage. Besides the pharmaceutical building blocks, the utilization of lignocellulose resources to replace petrochemical resources for polymer monomers and polymer materials preparation has attracted widespread attention [10–16]. In addition to directly using modified lignin for polymerization synthesis, and given the potential market demand and cost of fine chemical synthesis from protolignin or traditional fossil resources, achieving the transformation of protolignin polymer to meet the large unmet need artificial polymers via monomers like aromatic phenol [17–19], bisphenol A [20, 21], styrene [22], aromatic acids PTA [12, 13], or and adipic acid [23, 24], cyclohexanone-based monomers [25, 26] with the broken of the aromatic ring could be a potentially valuable economical process. For more information about the functionalization and defunctionalization of lignin-derived monomers to prepare compounds with pharmacological activity and polymer monomers, one can check the recent reviews from Barta Group [6] and Yan Group [27].

12.3 Standardized Lignin Substrate and Standardized Products

The monomeric products from lignin conversion usually cannot be used as the chemicals stocks directly and the further chemical transformation or functionalization of these crude products is necessary to prepare platform chemicals and precursors for the preparation of high-value medicine and synthetic blocks, and artificial polymerization [27]. Currently, one of the most critical challenges of lignin utilization comes from the separation process of various products even when these products can be obtained in certain kinds of compounds with high yields. Subtle differences in side-chain structure with different carbon numbers or the number of methoxy groups on the aromatic ring will lead to the result that the mixture cannot be regarded as a single component in the further application, especially for a polymerization process that requires a high purity of the monomer. Therefore, the separation of products is a critical but complex process for lignin depolymerization utilization. Today, the difficult separation process is also a critical limiting factor for the lignin utilization in the field of chemical preparation to compete with the petrochemical industry that is under advantageous conditions. In addition, another critical challenge comes from the lignin structure change from one plant to another, which may mean that certain lignin depolymerization/utilization technologies cannot work efficiently for all of the lignin resources. Facing the above two key issues, here are two effective key thoughts that can be summarized as standardized lignin substrate and standardized products.

12.3.1 Standardized Lignin Substrate

According to the above discussion, the ideal lignin for aromatic chemicals preparation is one from which a single monolignol or the percentage of one kind of monolignol can be >85%. In addition, considering that the subsequent formation of lignin polymers by monolignols is a free radical polymerization process (in other words, that even the single monolignol can provide various linkage types), maintaining a higher proportion of easily transformable chemical bond connections in the structure, such as β-O-4 and β-1, and the lignin structure itself can avoid side reaction problems such as polymerization and make the lignin conversion more efficient. An understanding can be obtained that controlling the complexity of the lignin substrate from the source (that is, the predetermined plant structure) can effectively simplify the subsequent lignin catalytic depolymerization and increase the corresponding product yield.

As we have discussed in Chapter 2, the generation of monolignols and their processors is conducted by various enzymes that are further controlled by the expression of different genes. For the past decades, to produce lignins that are easily separable cellulose and other components from original biomass for paper industry or biorefineries, and biofuels/biofermentation factories (ethanol, butanol), a research line has been opened to develop genetically modified plant species. Transgenic species show some differences in lignin structure with respect to natural plants. These differences consist of the ratio of basic monolignols, and, in some cases, of bond types between monolignols and the linkages between lignin and carbohydrates.

As one example, in 2009, Stewart and co-workers [28] reported that the genetically modified poplar is extremely rich in S units in the lignin produced by cinnamate-4-hydroxylase/5-hydroxylase (C4H:F5H) overexpression. The plants analyzed appear to have the highest S content on record, with some 97.5% of the lignin deriving from sinapyl alcohol. The approximate 40 of the S/(G+H) ratio suggests that the decrease of radical resonance intermediates for the polymerization which is blocked by the high –OMe groups, and the lignin chains are predominantly linear, which is supported by comparative NMR determination of isolated lignin fractions (Figure 12.1). As might be anticipated from the substantial compositional shifts documented between the control and the transgenic trees, the structure of the

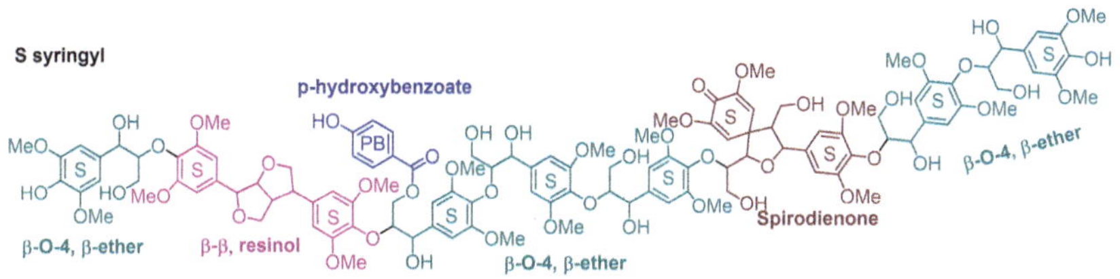

Figure 12.1 The Lignin Polymers Model from the C4H:F5H-up-regular Transgenic Trees. *Source:* Stewart et al. [28] with Permission, © 2009 American Society of Plant Biologists.

polymer can be dramatically altered with this methodology, with substantial differences from natural native lignins of nontransgenic species. Whereas the β-ether link remains predominant, phenylcoumarans units become rare in this lignin type (at high S levels), and spirodienones S increase. The contribution from β-β′ of resinols units was compensatorily higher. These results remain consistent with the accepted theory of lignification based on combinatorial radical coupling reactions under simple chemical control. It can be predicted that the monomers products from high-S poplar lignin depolymerization are mainly syringol-based compounds. At the same time, the plants with low-S unit lignin and some species with the high-H unit are also cultivated with genetic technology. Besides changing the production ratio of lignin structural monomers, thereby affecting the connection types of lignin structural units to a certain extent, genetic engineering can also play an important role in lignin extraction [29].

Besides the acquired genetically modified plants that can provide unique lignin resources, some natural plants also contain unusual lignins due to their genetic defects with the most famous being Catechyl lignin (C-lignin) [30, 31] that is present in the seed coats of vanilla (*Vanilla planifolia*) [32], various members of the Cactaceae of the genus *Melocactus* [33], and *Ricinus communis* (castor) [34]. In some special cases, the lack of O-methyltransferase (OMT) activity for conversion from catechyl C to guaiacyl G and, subsequently, on to syringyl S, aromatic-level precursors, results in 100% C units in the cell wall (CW). This C-lignin was found to be essentially a homopolymer synthesized almost purely by β-O-4 coupling of caffeyl alcohol with the growing polymer chain, producing benzodioxanes as the dominant unit in the polymer (Figure 12.2). For the chemical properties of C-lignin, because of the lack of an accessible and eliminable benzylic hydroxyl group in C-lignin units, condensation reactions due to the formation of benzyl cations might

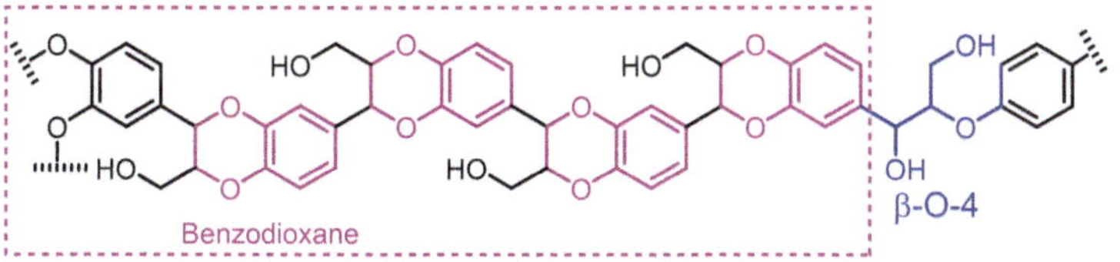

Figure 12.2 The Typical Structure Model of the Catechyl lignin.

be mitigated under acidic conditions, which indicated the stability of C-lignin. As checked by Ralph [30], C-lignin is highly acid-resistant, and the C-lignin survives even the harshest of acidic pretreatment methods-the KL isolation procedure includes a 1-hour treatment in 72% (w/w) sulfuric acid, followed by dilution to 4% (w/w) sulfuric acid and autoclaving at 121°C for 1 hour-while retaining its original lignin structure. In addition, given the fact that Castor (*R. communis*) seeds have been widely used for the production of non-edible oils, the C-lignin from the waste of castor seed coats is a plentiful and inexpensive C-lignin feedstock (0.5–1 $/kg) instead of other resources like vanilla seeds (350–500 $/kg), and the C-lignin extraction from castor seeds through the combination of enzymatic and mild acidolysis treatment is efficient, making full use of the C-lignin will be a meaningful promising process.

Although the lignins from genetically modified and genetically mutated plants can achieve lignin transformation substrate customization to a certain extent, considering the cost of gene breeding species, some concerns about gene diffusion, and the fact that the lignin structure of most wild plants does not have the characteristics of standardized structural units, which puts forward the new requirements for the lignin extraction to prepare standardized lignin. Aiming at the comprehensive utilization of lignocellulose, the current popular scheme is to extract and separate the three components one by one, such as the lignin first concept. According to the requirements of the new biorefinery technology presented by Professor Sels [35, 36], the extraction process of lignin should be efficient and the β-O-4 structure in the structure needs to be kept as unchanged as possible, which is the premise of the following selective and efficient lignin depolymerization to monomers. Most of the high-efficiency and quality lignin extraction under mild conditions requires the use of acid and alkali as a catalyst. Although it can avoid the structural changes of lignin caused by the high-temperature extraction process, excessive acid and alkali may also cause condensation of the lignin structure. This technical specification puts forward new requirements for the exploration of this extraction scheme. All in all, for the systematic utilization of lignocellulose, the ideal situation is to eat dry and squeeze clean without leaving waste. In addition, if the extraction solution and acid-base catalyst can be recycled [37, 38], it will be a more meaningful extraction process.

As one example, in 2017, Zhu and co-workers [37] reported that the *p*-toluenesulfonic acid (*p*-TsOH) could be in the low-cost and efficient fractionation of wood through the rapid and near-complete dissolution of lignin. Approximately 90% of poplar wood lignin can be dissolved at 80 °C in 20 min. The lignin dissolved in the acid liquor stream can be easily precipitated as lignin nanoparticles by diluting the spent acid liquor to below the minimal hydrotrope concentration. NMR analyses of the dissolved lignin revealed that *p*-TsOH can depolymerize lignin via ether bond cleavage and can separate carbohydrate-free lignin from the wood. In addition, the water solubility of *p*-TsOH is relatively low, which can facilitate efficient recovery using commercially proven crystallization technology by cooling the concentrated spent acid solution to ambient temperatures. In 2020, Zhu and co-workers [38], further evaluated maleic acid (MA) as a green hydrotrope with a minimal hydrotropic concentration of approximately 25 wt% for efficient wood fractionation at atmospheric pressure and ≤100 °C. The MA hydrotropic fractionation (MAHF) resulted in esterified lignin at C_γ and expose the C_α–OH with a low degree of condensation and a very light color, confirmed by the 2D ^1H–^{13}C HSQC (heterogeneous single quantum correlation) and HMBC (heteronuclear multiple bond correlation) NMR analyses of reaction products. As shown in Figure 12.3, the surface charge of lignocellulosic MAHF water-insoluble solids (WIS), induced by lignin esterification (carboxylation), enhanced enzymatic sugar yield by reducing nonproductive cellulase binding to lignin through pH-mediated electrostatic repulsion and also enhanced the lubrication effect of lignin in mechanical nanofibrillation for producing cellulose nanofibrils from WIS. The fractionated liquor after the dilution can provide the light color lignin with a low degree of condensation. Preliminary studies indicated that dissolved xylan can be dehydrated into furfural by MA in the fractionated liquor at a good yield of 70% and MA can be reused for repeated fractionation with minimal loss of less than 5%.

12.3.2 Standardized Products

When the standardized lignin substrate is obtained, the next work is developing efficient strategies for the transformation of the products, which need to provide standardization of products with the same structure.

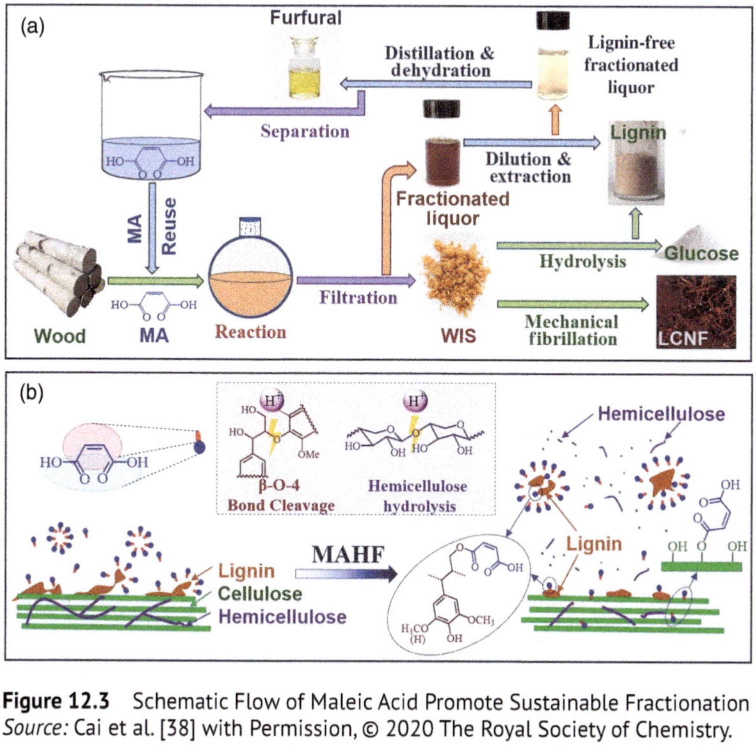

Figure 12.3 Schematic Flow of Maleic Acid Promote Sustainable Fractionation of Wood. *Source:* Cai et al. [38] with Permission, © 2020 The Royal Society of Chemistry.

As we have introduced above, for the lignin formed from only one kind of phenolic precursor, the distribution of the depolymerization product will be relatively simple because of the structural unity of the aromatic ring. For the standardized lignin substrate with a high-S unit, Shuai and co-workers [39] reported that adding formaldehyde during biomass pretreatment produces a soluble lignin fraction that can be converted to syringyl monomers at near theoretical yields during subsequent hydrogenolysis (78 mole% for a high-syringyl transgenic poplar). To compare the advantages of genetically modified plants than the normal lignins in the lignin extraction and utilization, in 2018, Abu-Omar and co-workers [40, 41] evaluated three different solvent systems (methanol, acetone, and acetic acid) for the extraction of native lignin from wild-type and genetically modified poplar species (wild type, high-S, and low-S). As shown in Figure 12.4, 63% of the original lignin extracted by the methanol from high S poplar was depolymerized to aromatic monomers over a heterogeneous Ni/C catalyst (10 wt% of catalyst in methanol solvent, under 35 bar H_2 pressure at 225 °C). Products distribution analysis indicated that the main product is the 4-propenyl syringol. The 85% selectivity to one product showed the advantage of genetically modified plant species with a high ratio of one unit in further products isolation.

As another "ideal lignin" substrate for aromatic preparation, Ralph and co-workers [30] discovered that C-lignin, a benzodioxane homopolymer without condensed units, is highly acid-resistant. Hydrogenolysis at 200 °C of C-lignin resulted in the cleavage of all benzodioxane structures to produce catechol-type monomers in near-quantitative yield with a selectivity of 90% to a single catechylpropanol monomer (M1, 4-(3-hydroxypropyl)benzene-1,2-diol) over the Pd/C in methanol (Figure 12.5). In the same year, Beckham and Román-Leshkov [42] used Ni/C to treat vanilla seed coats directly, and the C-lignin component was depolymerized into a series of catechol monomers with 7.9% yields (based on the weight of vanilla seed coats). In 2021, Song and co-workers [31] reported that anatomically dispersed Ru catalyst over the ZnO/C support prepared from the N_2 calcination of the Zn-based MOF precursor, which can serve as an efficient catalyst for the hydrogenolysis of C-lignin via the cleavage of C—O bonds in benzodioxane linkages, giving catechols in high yields and a unique selectivity to propenylcatechol (77%) with an unsaturated

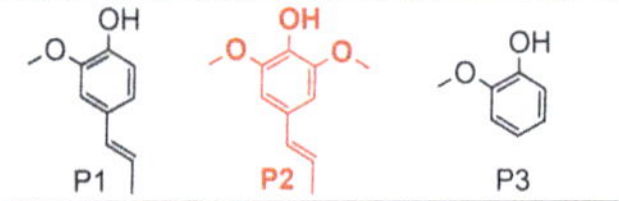

Poplar variant	Mw (kg·mol $^{-1}$)	Mn (kg·mol $^{-1}$)	Total Yield (%)	% Yield of P1	% Yield of P2	% Yield of P3
Wild type	2.2	1.5	60	8	43	9
High-S	2.0	1.4	63	2	54	7
Low-S	2.3	1.7	56	14	34	8

Figure 12.4 Catalytic Hydrogenation Depolymerization of Organosolv (Methanol) Lignin from Different Polar Variant over the 10 wt% Ni/C. *Source:* Reaction Data Abstracted from Luo and Abu-Omar [41]. *Note:* Reaction condition: Organosolv lignin (50 mg) with 10 wt% Ni/C in 20 mL of MeOH, at 225 °C and 35 bar H_2 for 12 h.

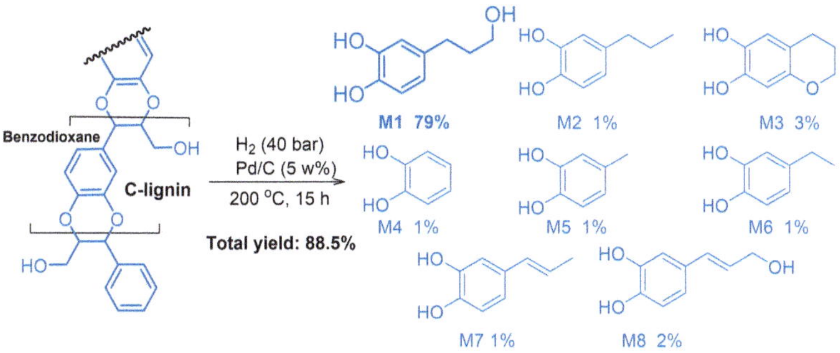

Figure 12.5 Hydrogenolysis Monomer Yields of C-lignin over the Pd/C in Methanol. *Source:* Reaction Data Abstracted from Li et al. [30].

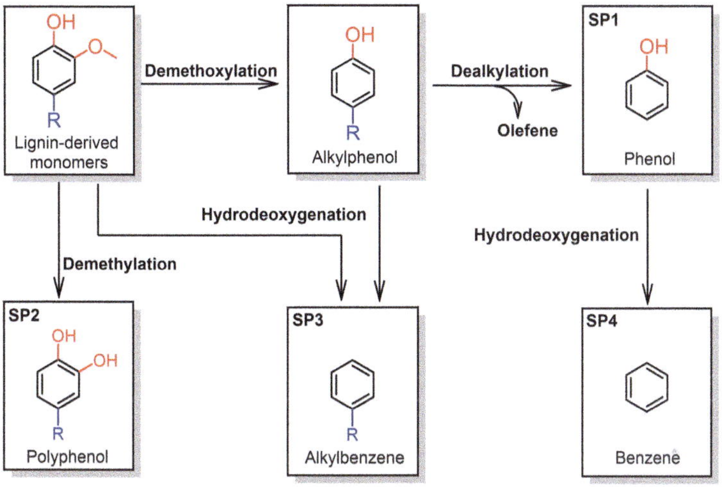

Figure 12.6 Defunctionalization of Lignin Depolymerization Monomers to Standardized Products.

propenyl group (-CH=$CHCH_3$) in the side chain. For more information about the mechanism with various C–O bonds cleavage, one can check the reaction routes study in Song's article [31].

For most native lignins depolymerization, especially for the hydrolysis/reduction lignin with a relatively higher monomers yield than other methods, H-, G-, and S-unit products with different structural groups exist in the same mixture, to realize the lignin depolymerization monomers to the standardized product, one of the potential solutions is to cut the side carbon chain, or remove the oxygen-containing groups via the demethoxylation (or demethylation) reactions, which in turn provide methoxy phenol, and phenol (polyphenol) as the standardized product (Figure 12.6). Otherwise, the removal of all oxygen-containing structures on the aromatic ring will provide alkyl aromatics as the hydrodeoxygenation products, which can be incorporated into petrochemicals to prepare the corresponding toluene, benzene, and *p*-xylene.

Although some recent catalytic systems can efficiently produce aromatic monomers via lignin C—O/C—C bonds hydrogenolysis cleavage, these methods generally produce phenol mixtures with C_1—C_3 alkyl tails on the *para* sites. These alkyl tails will affect the further application of lignin extraction, which raises a new claim for lignin utilization strategy. For the "tail cutting" research, in 2016, Verboekend and Sels [19] explored the selective conversion of alkylphenols derived from coal and lignin to phenol and olefins (Figure 12.7). It was demonstrated here that (mixtures of) the most abundant alkylphenols derived from lignocellulose, that is, ethyl and propylphenol can be highly selectively converted to the corresponding phenol and olefin with C_{Ar}–C bond cleavage. The highly efficient and robust dealkylation is achieved using unmodified metal-free ZSM-5 zeolite catalysts and the co-feeding of water, in which co-feeding of water is crucial to maintain catalytic activity. The catalytic results pave the way for further study on the fundamental origin of the shape-selective behavior and the competitive adsorptive of water and phenols. During the catalytic dealkylation process, the role of water is attributed to the competitive adsorption of water and phenol.

Then, in 2018, Verboekend and Sels [17] presented a further detailed study about the steam-assisted dealkylation of 4-*n*-propylphenol (4-*n*-PP) into phenol and propylene over several micro- and mesoporous acidic aluminosilicates in the gas phase. A series of acidic zeolites with different topologies (e.g., FER, TON, MFI, BEA, and FAU) were studied, of which ZSM-5 outperforms the others. The catalytic results, including zeolite topology and water stability effects, are rationalized in terms of thermodynamics and kinetics. According to the proposed mechanism, the dealkylation reaction accords to the classic carbenium chemistry including isomerization, disproportionation, transalkylation, and dealkylation, as the most important reactions (Figure 12.8). In addition, the exceptional selectivity of ZSM-5 is attributed to the pore confinement, avoiding disproportionation/transalkylation as a result of a transition state shape selectivity. The presence of water maintains a stable catalysis, especially for ZSM-5 with low acid density. The working hypothesis of this stabilization is that water precludes diphenyl ether(s) formation in the pores by reducing the lifetime of the phenolics at the active site due to the high heat of adsorption of water on H-ZSM-5, besides counteracting the equilibrium of the phenolics condensation

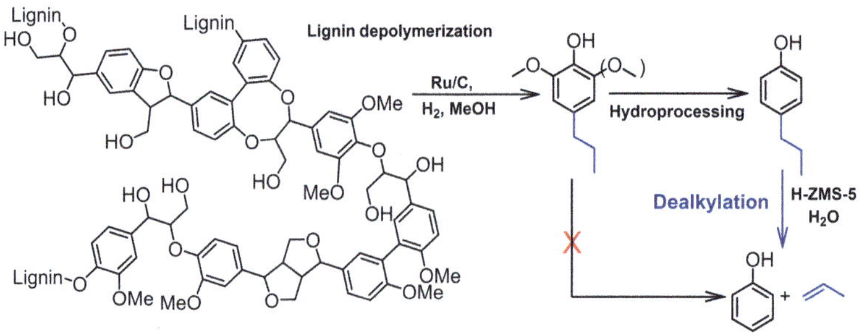

Figure 12.7 Catalytic Upgrading of Alkylphenols Derived to Phenol and Olefin. *Source:* Verboekend et al. [19] with Permission, © 2016 The Royal Society of Chemistry.

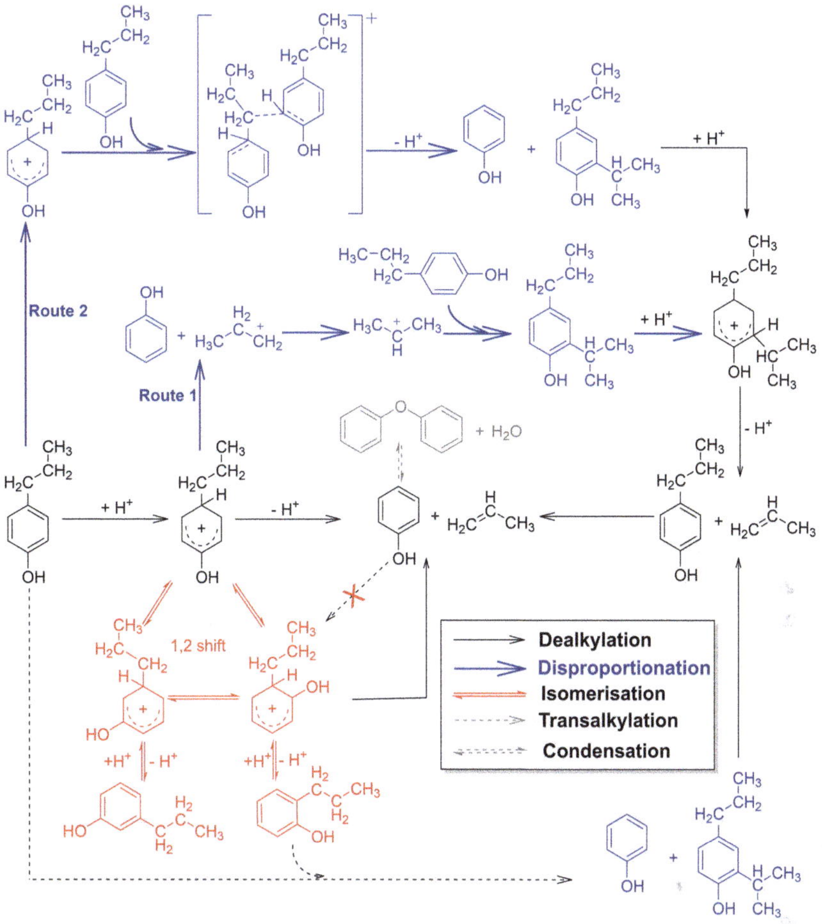

Figure 12.8 Proposed Reaction Pathway of Dealkylation, Isomerization, Disproportionation, Transalkylation, and Condensation of 4-n-Propylphenol. *Source:* Liao et al. [17] with Permission, © 2018 American Chemical Society.

reaction. The water effect is unique for the combination of (alkyl)phenols and ZSM-5. Focusing on the dealkylation process of the phenols, Verboekend and Sels [43] further studied the shape selectivity vapor-phase conversion of lignin-derived 4-ethylphenol (EP) to phenol and ethylene over the acidic aluminosilicates, which paid more attention to the impact of catalysts acid properties and pore constraint. According to the detailed thermodynamic analysis, the endothermal dealkylation of alkylphenol to phenol and olefin favored at high reaction temperature (e.g., $T > 350$ °C), while isomerization, disproportionation, and transalkylation were thermodynamically preferred at low temperature (e.g., 200 °C< T < 350 °C K). To improve the reaction efficiency and make the catalyst more stable, in 2020, Verboekend and Sels [44] prepared a series of hierarchical ZSM-5 zeolites by the alkaline treatment, which showed improved catalytic activity compared to the microporous parent material for dealkylation of 4-*n*-propylphenol to phenol and propylene without compromising the product selectivity. According to the characterization, the hierarchization reduces the amount and strength of the Brønsted acid sites and induces a change in Al sites, leading to the formation of Lewis acids. The strong Lewis acid sites have higher activity than the Brønsted acid sites at sufficiently high temperatures, which is the result of the larger entropic contribution of activation. Therefore, a catalyst named ZSM-5-H-0.2M, containing a high content of Lewis and Brønsted acid sites along with preservation of the high strength of the acid sites is the highest active catalyst. The smaller amount of strong acid sites and higher external surface area of the hierarchical structures minimizes fast deactivation, and therefore less water is required to stabilize a long time on stream dealkylation catalysis due to the better diffusion characteristics of the hierarchical material with a microporous structure.

For the dealkylation of phenols from lignin depolymerization, besides the one beginning with the classic carbenium intermediate, in 2018, Maes and co-workers [45] provide a new O- and C-dealkylation mechanism for the transformation of ferulic acid into bio-catechol, catalyzed HCl (or H_2SO_4) in pressurized hot water (250 °C, 50 bar N_2). As shown in Figure 12.9, treatment of the substrate ferulic acid **A** with acid in water leads to the formation of benzylic alcohol **B** through the addition of water to the alkene. The phenolic hydroxy group directs protonation on the carbon atom ipso to the alkyl chain. The obtained intermediate **C** then allows C_{Ar}–C_{α} bond cleavage through a retro vinylogical aldol reaction, delivering guaiacol **D** and aldehyde **E**. The **E** then transforms to acetaldehyde (**G**) after the decarboxylation. The first principle static and dynamic molecular simulations support the proposed C-dealkylation and the calculations results also revealed that this cleavage reaction competes with proton transfer of unstable intermediate **C** to the solvent. The O-demethylation reaction of **D** is initiated by protonation of the arene into **H**. Subsequent addition of water gives hemiacetal **I**, which upon elimination of methanol provides 6-hydroxycylohexa-2,4-dien-1-one (**J**), in which aromaticity is restored through tautomerization to deliver product catechol (**K**).

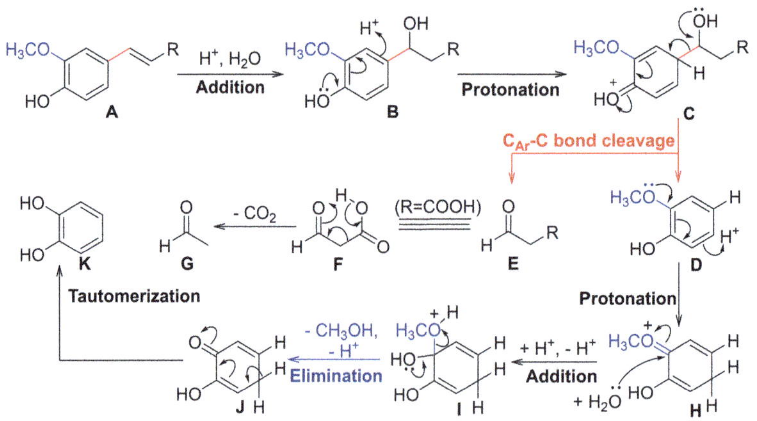

Figure 12.9 Proposed Mechanism for the O- and C-dealkylation of Ferulic Acid into Catechol. *Source:* Bomon et al. [45] with Permission, © 2019 Wiley-VCH.

In addition, in 2020, Meng and Han [46] developed a protocol to produce phenol from lignin by direct transformation of C_{sp2}–C_{sp3} and C–O bonds under mild conditions, which was achieved by direct cleavage of C_{sp2}–C_{sp3} bond to remove the propyl structure and hydrolysis of the aliphatic C_β–O bond to form an OH group on the aromatic ring, and HY_{30} zeolite ((Si/Al = 30) showed an outstanding catalytic performance for transformation of both bonds. The yield of phenol could reach 10.9 weight % with a selectivity of 91.8%. In this unique route, water was the only reactant besides lignin. A scaled-up experiment showed that 4.1 g of pure phenol could be obtained from 50.0 g of lignin. As discussed above, the reaction involves the hydrolysis of the aliphatic C_β–O group to the OH group and the direct cleavage of the C_{sp2}–C_{sp3} bond. The pathway for the hydrolysis of the aliphatic C_β–O bond over the acidic catalyst is well known. Then, according to the results of the control experiments and DFT calculations, the chemical environment at the C_α position of the aliphatic chain was critical for the success of the strategy, and the C_{sp2}–C_{sp3} bond of the phenols monomers from the first C_β–O bond hydrolysis was cleaved via a dehydroxylation, γ-methyl shift, and C–C β scission pathway (Figure 12.10). The C_α hydroxyl group is initially protonated on a Brønsted acid site of HY_{30}, leading to the formation of oxonium ion (structure II), which then transforms into a carbonium ion (structure III) with the elimination of H_2O. Thermodynamically, γ-methyl on the side chain is then shifted to the aliphatic C_α position by the function of zeolite, evolving to the chemically adsorbed structure III, and subsequent proton abstraction by zeolite delivers structure V. The structure V then undergoes a second

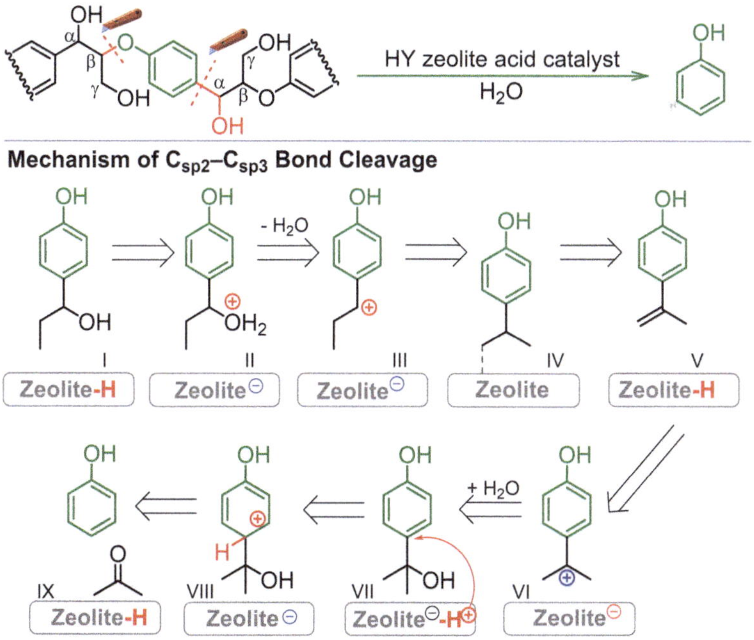

Figure 12.10 Lignin Valorization to Phenol by Direct Transformation of C_{sp2}–C_{sp3} and C–O Bonds. *Source:* Yan et al. [46] with Permission, © 2020 The Authors.

protonation at the Cα position to form carbonium ion (structure VI), and this would be pursued by the trap of H_2O, which gives tertiary alcohol geometry (structure VII) and concurrently accomplishes another proton rejuvenation of the Brønsted acid site. Afterward, endothermically turning the benzene ring of the tertiary alcohol toward the Brønsted acid site gives the appropriate geometry to enable the protonation of the C_{sp2} position (structure VIII). The resulting carbonium ion VIII undergoes desired C_{sp2}–C_{sp3} bond β scission to exothermically generate phenol and acetone [47].

Different from the zeolite systems shown above which prefer to catalyze the dealkylation/transformation of the H-unit monomers, not the G- or S-unit monomers to the corresponding phenol, in 2020, Meng and Han [48] discovered that homogeneous $La(OTf)_3$ could catalyze the transformation of lignin with guaiacol as the only liquid product (Figure 12.11). In the reaction, $La(OTf)_3$ catalyzed the hydrolysis of lignin ether linkages to form alkyl-syringol and alkyl-guaiacol, which further underwent decarbonization and demethoxylation to produce guaiacol with a yield of up to 25.5 wt%, and the remaining residue was solid contained only G units. In the scale-up experiment, 0.32 g isolated guaiacol can be obtained from 1.5 g organosolv lignin (a 21.2 wt% yield) at 270 °C for 24 h. According to the time-course of the product distributions for the C–C bond cleavage of the ethyl group in 4-ethyl guaiacol with the $La(OTf)_3$ catalyst in MeOH, the 4-methylguaiacol was initially the major liquid product due to the cleavage of the C_α–C_β bond of 4-ethylguaiacol, which further transform to the guaiacol with the cleavage of the left C_{Ar}–C bond. Correspondingly, methane was generated as the main gaseous product for the decarbonization process.

Returning to the utilization of lignin resources from lignocellulose, a strategy different from the above strategies involving the C_{Ar}–C bond cleavage with acid catalysts was developed in 2018 when Wang and co-workers [18] achieved the one-pot

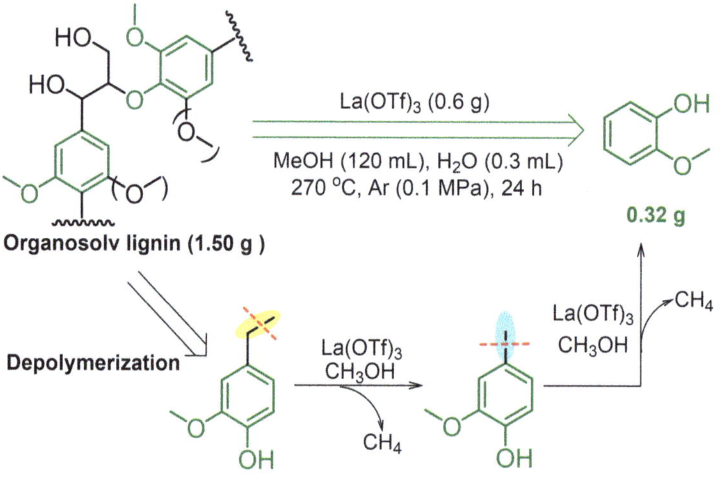

Figure 12.11 $La(OTf)_3$ Catalyzed the Selective Transformation of Lignin to Guaiacol. *Source:* Reaction Data Abstract from Shen et al. [48].

preparation of alkyl-free phenols from lignin, which combined the first $C_\alpha-C_\beta$ bond oxidative cleavage to acids and subsequent decarboxylation with Cu(II) (Figure 12.12). During the model compounds study, they found $CuCl_2$ could be used as both an oxidant for the $C_\alpha-C_\beta$ oxidative cleavage in β-O-4 linkage and catalyst for the decarboxylation in water at 200 °C for 6 h. Nonphenolic β-O-4 dimers with H units underwent $C_\alpha-C_\beta$ bond cleavage to acid and phenol. In addition, the present method is selective for the cleavage of H-based β-O-4 monomer and shows low efficiency in the cleavage of S- and G-based linkages. By the combination of C–OAr bond hydrogenolysis to release phenolic groups over Ru/CeO_2 and $C_{aryl}-C_\alpha$ bond oxidative cleavage and following decarboxylation with $CuCl_2$, various raw biomass sources, including poplar, pine, birch, peanut, bamboo willow, and straw, were tested, and 1-33 mg of phenol from lignin, 70-224 mg of LA (levulinic acid), and 2-48 mg of GVL (γ-valerolactone) from cellulose per gram of biomass were obtained, respectively. The poplar showed the highest yield of phenol with 33 $mg\cdot g^{-1}$, with up to a 13 wt % yield. It should be noted that the *p*-hydroxybenzoate (PB) units in poplar lignin can be also effectively converted into phenol with around 80% yields.

As one typical work for the lignin utilization to standardized products, in 2020, Liao and Sels [49] provided a utilization plan for the sustainable wood biorefinery for low-carbon footprint chemicals production. The integrated biorefinery can convert 78 w% of birch into xylochemicals (small molecules which can be gained

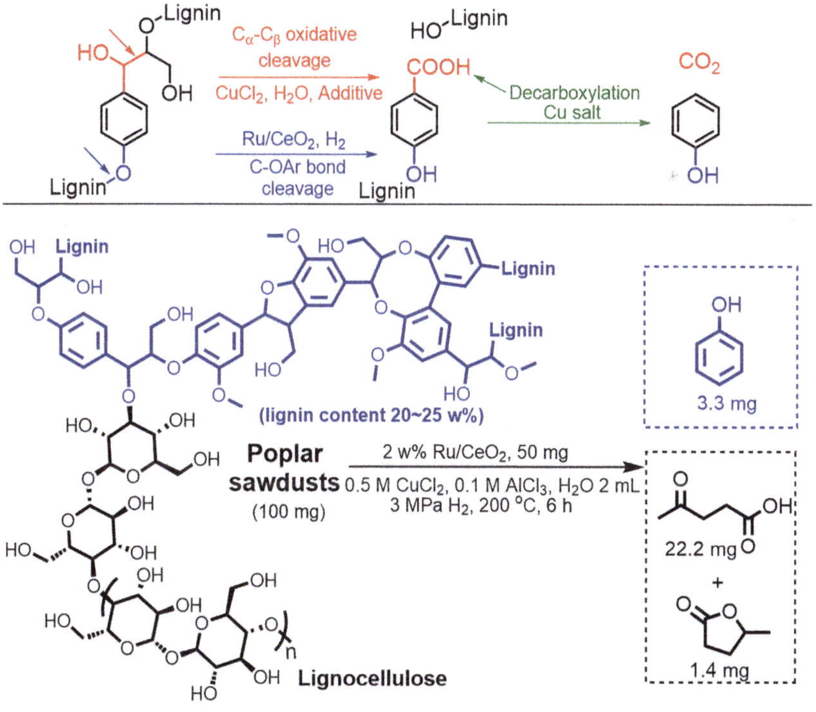

Figure 12.12 Lignin Dealkylation to Phenol via Hydrogenation-oxidation-decarboxylation Strategy. *Source:* Wang et al. [18] with Permission, © 2018 American Chemical Society.

from wood). Reductive catalytic fractionation of the wood produces a carbohydrate pulp amenable to bioethanol production and a lignin oil. After extraction of the lignin oil and solvent recycle, the crude, unseparated mixture of phenolic monomers is catalytically funneled into 20 wt % of phenol and 9 wt % of propylene (on the basis of lignin weight) by gas-phase hydroprocessing and dealkylation; the residual phenolic oligomers (30 wt %) are used in printing ink as replacements for controversial para-nonylphenol. The techno-economic analysis predicts an economically competitive production process, and a life-cycle assessment estimates a lower carbon dioxide footprint relative to that of fossil-based production. Beyond the content of this pioneering work, two different thoughts of the lignin oil utilization inspire us that the standardized products can be not only the pure chemical such as the phenol+propylene but also can be a useful mixture of phenolic oligomers as the printing ink.

For the preparation of the alkylphenol without methoxy group, besides the strategies containing multiple steps of first lignin depolymerization to the oil mixture, further demethoxylation, and separation, in 2020, Guo and Wang [50] further reported a one-step and hydrogen-free production of 4-alkylphenols directly from native lignin via self-reforming-driven depolymerization and hydrogenolysis over $Pt/NiAl_2O_4$. Using this strategy, high yields of 4-alkylphenols, 17.3 wt %, were obtained from birch lignin. Reaction pathway and mechanism studies revealed that this strategy initiates from the reforming of aliphatic hydroxyls, followed by the cleavage of C–O linkages, and ends via demethoxylation over $Pt/NiAl_2O_4$. Moreover, the subsequent aqueous phase reforming of the as-formed methanol accelerates the whole process. This strategy realizes the one-pot production of 4-alkylphenols from lignin by fully utilizing the structural hydrogen in lignin and H_2O, providing a low-cost and safe method of lignin valorization (Figure 12.13). In 2021, Guo and Wang [51] expand this strategy for lignin self-decomposition over the $M/NiAl_2O_4$ (M = Pt, Pd, Ru) catalysts and to investigate the rate-determined step, structure–activity relationships, the adsorbed species, and the essence of different metals, and finally screen a good catalyst with excellent performance. A temporal profile of reaction using a model compound and kinetic measurements disclosed that demethoxylation is the rate-determining step, in which $Ru/NiAl_2O_4$ has superior ability probably due to the faster removal of products that enables the rapid recovery of active sites for demethoxylation when compared to $Pt/NiAl_2O_4$. In contrast, side reactions and

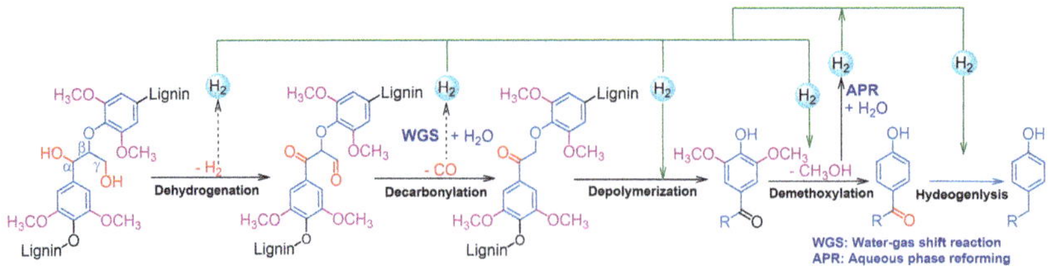

Figure 12.13 Reaction Pathways of the Self-Reforming-Driven Depolymerization and Hydrogenolysis of Lignin to 4-Alkylphenols. *Source:* Li et al. [50] with Permission, © 2020 American Chemical Society.

a low rate of demethoxylation were observed over $Pd/NiAl_2O_4$. In the self-reforming-driven hydrogenolysis of birch lignin and lignin oil, Ru/ $NiAl_2O_4$ exhibited again its excellence in gaining 4-alkylphenols with yields of 40.7 and 92.8 mol % (5.7 and 44.5 wt %), from lignin and lignin oil, respectively. This comprehensive study would help to design catalysts with higher activity for the self-reforming process of biomass resources.

As the further transformation, the phenols can be selectively converted to the alkylbenzenes after the full thermolysis over the zeolite catalysts or hydrodeoxygenation over the metal catalyst. In 2003, Li and co-workers [52] reported a two-step method for the production of ethylbenzene from lignin through the directional catalytic pyrolytic depolymerization of lignin into the aromatic monomers followed by the selective alkylation of the aromatic monomers with ethanol. For the first step, the aromatics selectivity of benzene derived from the catalytic depolymerization of lignin reached about 90.2 C-mol% over the composite catalyst of Re-Y/HZSM-5. For the alkylation of the aromatic monomers in the second step, the highest selectivity of ethylbenzene was about 72.3 C-mol% over the HZSM-5 catalyst.

In addition, in 2017, Wang and co-workers [53] reported that the porous Ru/ Nb_2O_5 catalyst enabled the complete hydrodeoxygenation of the oxygen content from organosolv lignin to selectively produce arenes. The conversion of birch lignin to monomer C_7–C_9 hydrocarbons is nearly quantitative based on its monomer content, with a total mass yield of 35.5 wt% and an exceptional arene selectivity of 71 wt%. Inelastic neutron scattering and DFT calculations confirm that the Nb_2O_5 support is catalytically unique compared with other traditional oxide supports, and the disassociation energy of Caromatic–OH bonds in phenolics is significantly reduced upon adsorption on Nb_2O_5, resulting in its distinct selectivity to arenes.

Besides the preparation of simple aromatic compounds of phenol, guaiacol, and alkylphenol without methoxy group, and arenes with carbon chain from lignin, based on their previous research about the self-supported hydrogenolysis of aromatic ethers to produce arenes over the RuW alloy [54] and the direct transformation of lignin C_{sp2}–C_{sp3} and C–O bonds [46] over the HY catalyst, in 2021, Meng and Han [55] reported an efficient strategy for the in situ refining of lignin that exclusively produces benzene by using a high-silica HY zeolite (HY_{30}, Si/Al=30) supported RuW alloy catalyst (Figure 12.14), which involves the total transformation of C_{sp2}–C_{sp3}/C_{sp2}–O into C–H bonds without side hydrogenation of the aromatic rings. The catalyst Brønsted acid-catalyzed the transformation of C_{sp2}–C_{sp3} bonds on the local structure of lignin molecule with H_2O, which is similar to their previous process that has been discussed in Figure 12.10. In addition, RuW catalyzed hydrogenolysis of the C_{sp2}–O bonds using the locally abstracted hydrogen from lignin molecule [54], affording a benzene yield of 18.8% on a lignin weight basis in the water system. The high-performance protocol can be readily scaled up to produce 8.5 g of benzene product from 50.0 g lignin without any saturation byproducts.

Of course, as representative research on the conversion of lignin and coal-based chemicals, the selective and efficient conversion and purification of aromatic phenols, including hydrodeoxygenation and water-phase reforming [56], have been

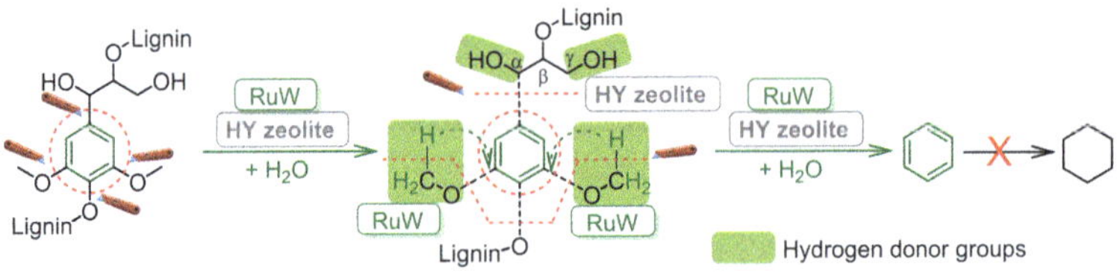

Figure 12.14 Catalytic Lignin Decomposition to Benzene over the RuW/Zeolite System. *Source:* Meng et al. [55] with Permission. © 2021 The Authors.

widely studied. We will not make a specific summary here and one can check the corresponding reviews.

As discussed above, the defunctionalization process can reduce the complexity of subsequent separation to a certain extent, and the obtained products can be used in tandem with the existing utilization process based on the petrochemical resources. However, it should be pointed that these defunctionalization operations will reduce the value of the resulting products to a certain extent. At the same time, considering that the corresponding biomass resource utilization route is not mature enough, the corresponding chemical price is high. Compared with the fossil resource route, in the current context in which CO_2 release is not totally considered, the cost is still high. Facing this question, besides simplifying the monomers products from lignin depolymerization, transfunctionalizing of the organic groups at the same time is another promising method to prepare the standardized product. One of the important examples is the above-mentioned terephthalic acid (TPA) generation process from lignin. In 2016, Zhu and co-workers [12] studied the transformation of vanillic acid and syringic acid to terephthalic acid via a two-step process using a fixed-bed reactor (Figure 12.15). The cascade route includes hydrogenation demethoxylation and carboxylation reactions. Activated carbon (AC)-supported $MoWBO_x$ and $PdNiO_x$ were determined to be suitable catalyst precursors for the process. The selectivity of 71.6% of *p*-hydroxybenzoic acid was achieved with a >99% acid conversion by using the catalyst $MoWBO_x$/AC under a 40 bar (H_2+CO_2, H_2/CO_2=0.5) mixture gas at 400 °C. Furthermore, the high yield of 58.7% of terephthalic acid with 88.4% selectivity was achieved by using catalyst precursor of $PdNiO_x$/AC during the second carboxylation reaction at 400 °C with 40 bar mixture gas (CO_2/N_2 = 1/1).

Beginning with the lignocellulose, in 2019, Yan and co-workers [13] reported a three-step strategy for the production of terephthalic acid (TPA) from lignin-derived

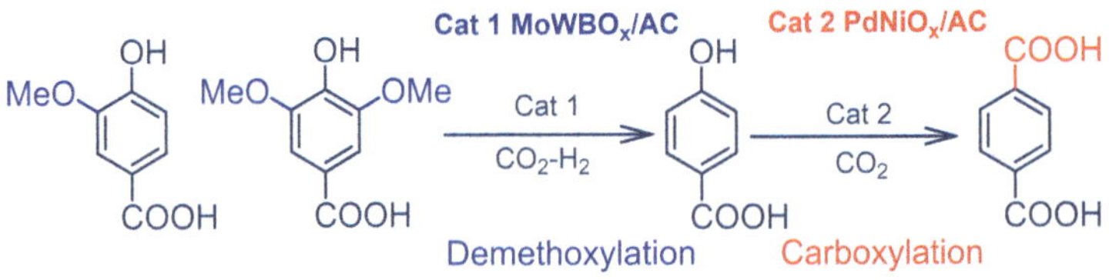

Figure 12.15 Production of Terephthalic Acid (TPA) from Lignin-Based Phenolic Acids by a Cascade Fixed-Bed Process. *Source:* Bai et al. [12] with Permission, © 2016 American Chemical Society.

monomer mixtures (Figure 12.16), in which redundant, non-uniform substitutes such as methoxy groups are removed and the desired carboxy groups are introduced. This strategy begins with the hydro-treatment of corn-stover-derived lignin oil over a supported molybdenum catalyst to selectively remove methoxy groups. On a lignin basis, 16.1 wt % of 4-alkylphenols were obtained after the MoO_x/AC-catalyzed demethoxylation step. The generated 4-alkylphenols are converted into 4-alkylbenzoic acids by carbonylation with carbon monoxide at 60 °C. The homogeneous Pd catalyzed carbonylation reaction provides a 14.8 wt% yield of 4-alkylbenzoic acids. The Co-Mn-Br catalyst then oxidizes various alkyl chains into carboxy groups, transforming the 4-alkylbenzoic acid mixture into a single product TPA. For this route, the overall yields of TPA based on the lignin content of corn stover could reach 15.5 wt%, and importantly, TPA with greater than 99% purity was obtained simply by first decanting the reaction mixture and then washing the solid product with water.

The isolation and separation of phenolic compounds are challenging and will gain in importance as we become more dependent on phenolics from lignocellulose-derived bio-oil to meet our needs for aromatic compounds, as well as for some high-valued functionalized monomers and dimers. Research on the products separation systems of biomass transformation has attracted much attention and corresponding new materials and new technologies have been developed. For the isolation of various phenols from the lignin oil mixture of lignin depolymerization, besides the normal organic solvent, the application of ionic liquid and deep eutectic solvent [57–59] are developed. Chromatography technology is also used in the phenolic compounds separation and identification. These approaches include HPLC high-performance liquid chromatography (HPLC) [60, 61], centrifugal partition chromatography [62, 63], high-speed counter-current chromatography [64], micellar electrokinetic capillary chromatography [65], and others. While given the amount of the lignin depolymerization oil mixture, the continuous liquid-liquid extraction methods [66] and the corresponding scheme after the improvement for the separation of phenolic compounds will be suitable for scaled-up applications. In addition, besides utilizing the liquid extraction and chemical bonding or hydrogen bonds effects between phenolic substrate molecules and reagents to achieve the separation of lignin depolymerization monomers, separation by molecular size is also a commonly used method. In 2015, Dirk De Vos and co-workers [67] showed

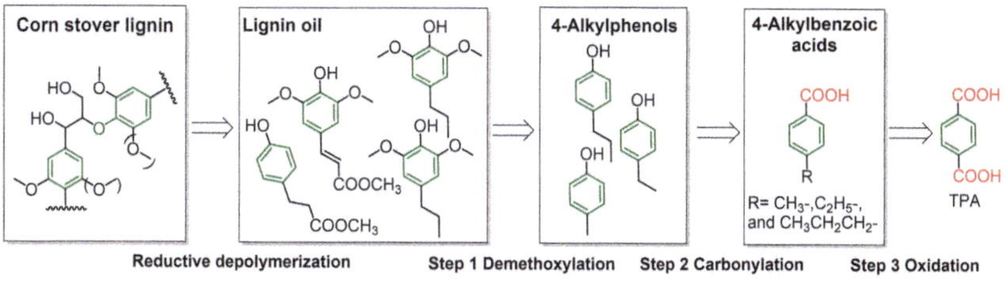

Figure 12.16 A Three-step Strategy for the TPA Production from Corn Stover Lignin. *Source:* Song et al. [13] with Permission, © 2019 Wiley-VCH.

that highly stable and hydrophobic Zr metal-organic frameworks of the MIL-140 type are effective adsorbent materials for the separation of different phenolics and far outperform other classes of porous solids (silica, zeolites, carbons). The mechanism of the hydroquinone–catechol separation on MIL-140C was studied in detail by combining experimental results with computational techniques. Unlike many other metal-organic frameworks, MIL-140 is remarkably stable and maintained structure, porosity, and performance after five adsorption-desorption cycles. Therefore, regarding the overall conversion of lignin, the process of lignin conversion to prepare different monomers has been developed, and the improvement of substrate separation technology is another key step to further promote the corresponding process to practical application in the preparation of standardized products.

12.4 Concluding Remarks

In this chapter, we first briefly introduce the research process of lignin utilization to prepare versatile materials. Then, viewing from the lignin depolymerization to aromatic chemicals and the point how to make the lignin depolymerization profitable, we provide our viewpoint for the outlets of lignin resource, which highlights the importance of high-value medicine and synthetic block preparation or the lignin utilization from natural polymer to artificial polymer. To accelerate the lignin utilization from lab to industry application, we discussed the critical roles of the standardized lignin substrate and standardized products, during which we hold the idea that funneling and functionalization of a mixture of lignin-derived monomers into a single high-value chemical is fascinating and promising, while a mixed product that can be used as a commercial goods is also acceptable.

By the end of 2021, although many exciting results have been achieved, the sufficient utilization of lignin to prepare chemicals still faces enormous challenges. To improve the technical feasibility, intensive investigations are crucially imperative. In the end, for further lignin catalytic utilization to aromatic chemicals, we herein provide the following perspectives:

1) Scaled lignin transformation needs a standard substrate, which depends on the elaborate pre-treatment to protect/transform the organic groups and lignin abstracting methods from the lignocellulose that can keep more β-O-4 linkages and pre-cut lignin into stable fragments for further catalytic depolymerization [38, 68]. The standardized lignin substrate from genetically modified or common plants can be in the solid state, but, in some situations, the lignin substrate need to be organic solvent-soluble (such as for the photocatalytic and electrocatalytic process to proceed under mild conditions during which the high-M_w lignin polymer is hard to be dissolved;
2) Catalytic strategies and methods should make lignin depolymerization profitable, which depends on how to increase the lignin conversion and monomers yields with robust and low-cost catalysts after strategies optimization to meet

request lines. In this respect, the target lignin depolymerization products can focus on the high demand monomers for artificial polymerization [69–71] or low demand but high-valued pharmaceutical chemicals [8], functional materials [72–74], highly functionalized motifs with or without heteroatom [75–77]; In other words, the lignin depolymerization to chemicals turns from the lignin substrate-oriented transformation to the product-oriented select;

3) Different from the traditional thermal cracking technology, the catalytic self-degradation of lignin promoted by the addition of extra renewable energy like photo-thermal solar energy will be the next focus. One example is the catalytic lignin self-transformation with the assistance of the water/solvent, during which the lignin can be firstly broken into fragments with the cooperation of the solvent and catalyst, then the hydrogen donor groups can occur further transformation to active H* species to participate the further C–O linkages cleavage to monomers;
4) Regarding the overall conversion of lignin, the process of lignin conversion to prepare different monomers has been developed. The high-efficiency strategy for single product preparation [78] combined with the low-cost but efficient product separation technology is a key step to further promote the corresponding lignin depolymerization process to the practical application;
5) The innovations about the reactors and new technology-process intensification methods, such as the flow system [79–85], microwave-assistance system [86–88], in situ separation technology, etc. can combine and enhance the advantages of thermal catalysis, photocatalysis [89–93], electrocatalysis [26, 94–97], photo-electro catalysis [98, 99], photo-thermal catalysis, thermal-electro catalysis [100], and even enzymatic catalysis [101–103], biotechnology, genetic engineering [104, 105], which may make lignin conversion efficient at a mild condition;

At the end of this book, we intend to abstract the lignin catalytic conversion into a common life sport (playing billiards) by highlighting the correlation between the two things. At the beginning of the match, the billiard balls are placed in different states, corresponding to the initial polymer state of the lignin monomers. The table is equivalent to the reaction vessel and solvent for the conversion of lignin. The white ball is equivalent to the catalyst (Cat). Corresponding to the characteristics of the catalyst, after the game, the white Cat ball is still the white ball and must be left on the table. During the game, billiard balls in various configurations are punched into different holes, representing the various linkage are converted into chemicals, materials, fuels, and polymers of different uses. In these processes, the power of the white ball hitting is equivalent to the energy introduced into the catalytic reaction, which is Ea and $h\nu$. How to hit the billiard balls, the angle of the white ball and the choice of the target ball to accept the first collision corresponds to the strategies designing of lignin depolymerization. Predicting the trajectory of the ball and discussing the corresponding possibilities like the mechanism discussion of the lignin depolymerization. It is the same, people do not like fragments recondensation during lignin depolymerization, as the balls aggregation in the billiards match. In addition, the word challenge in "Lignin Challenge" can be regarded as a difficult thing to and it can be also regarded as a match or a tournament. The two words like a battle cry, lignin utilization research not only need great ambitions to face the challenge but also should have know-how and inspiration to overcome the challenge.

In the end, lignin is a sophisticated and complex material that nature bestows upon us. It would be a colossal oversight not to use this highly accessible renewable substance and an expensive one in these times as raw materials become scarce. Although the lignin conversion can trace back to the 1920s and it is still a challenging topic 100 years later, we believe the process in lignin utilization is like that of a forest fire: faster when fanned by the winds of new ideas and brighter when passing through other areas of science, which will finally provide a successful solution for human society in the demand of sustainable chemicals and energy.

References

1 Calvo-Flores, F.G., Dobado, J.A., Isac-García, J., and Martín-MartíNez, F.J. (2015). *Lignin and Lignans as Renewable Raw Materials: Chemistry, Technology and Applications*, P249. Wiley and Sons, Ltd, Hoboken.

2 Hou, D.X., Li, T., Chen, X. et al. (2019). Hydrophobic nanostructured wood membrane for thermally efficient distillation. *Sci. Adv.* 5: eaaw3203.

3 Song, J., Chen, C., Zhu, S. et al. (2018). Processing bulk natural wood into a high-performance structural material. *Nature* 554: 224–228.

4 Xia, Q., Chen, C., Yao, Y. et al. (2021). In situ lignin modification toward photonic wood. *Adv. Mater.* 33: 2001588.

5 Li, T., Song, J.W., Zhao, X.P. et al. (2018). Anisotropic, lightweight, strong, and super thermally insulating nanowood with naturally aligned nanocellulose. *Sci. Adv.* 4: eaar3724.

6 Sun, Z., Fridrich, B., de Santi, A. et al. (2018). Bright side of lignin depolymerization: Toward new platform chemicals. *Chem. Rev.* 118: 614–678.

7 Sun, Z.H., Bottari, G., Afanasenko, A. et al. (2018). Complete lignocellulose conversion with integrated catalyst recycling yielding valuable aromatics and fuels. *Nat. Catal.* 1: 82–92.

8 Bjorsvik, H.R. and Liguori, L. (2002). Organic processes to pharmaceutical chemicals based on fine chemicals from lignosulfonates. *Org. Process. Res. Dev.* 6: 279–290.

9 Li, H.J., Subbotina, E., Bunrit, A. et al. (2019). Functionalized spirolactones by photoinduced dearomatization of biaryl compounds. *Chem. Sci.* 10: 3681–3686.

10 Delidovich, I., Hausoul, P.J.C., Deng, L. et al. (2016). Alternative monomers based on lignocellulose and their use for polymer production. *Chem. Rev.* 116: 1540–1599.

11 Dai, T., Li, C.Z., Li, L. et al. (2018). Selective production of renewable para-xylene by tungsten carbide catalyzed atom-economic cascade reactions. *Angew. Chem. Int. Ed.* 57: 1808–1812.

12 Bai, Z.Y., Phuan, W.C., Ding, J. et al. (2016). Production of terephthalic acid from lignin-based phenolic acids by a cascade fixed-bed process. *ACS Catal.* 6: 6141–6145.

13 Song, S., Zhang, J.G., Gozaydin, G., and Yan, N. (2019). Production of terephthalic acid from corn stover lignin. *Angew. Chem. Int. Ed.* 58: 4934–4937.

14 Satoh, K., Saitoh, S., and Kamigaito, M. (2007). A linear lignin analogue: Phenolic alternating copolymers from naturally occurring β-methylstyrene via aqueous-controlled cationic copolymerization. *J. Am. Chem. Soc.* 129: 9586–9587.

15 Lu, R., Lu, F., Chen, J.Z. et al. (2016). Production of diethyl terephthalate from biomass-derived muconic acid. *Angew. Chem. Int. Ed.* 55: 249–253.

16 Abu-Omar, M., Zhao, S., Whelton, A., and Huang, X. (2018). Renewable polymers and ligands derived from lignin. *Abstr. Pap. Am. Chem. S.* 255.

17 Liao, Y.H., Zhong, R.Y., Makshina, E. et al. (2018). Propylphenol to phenol and propylene over acidic zeolites: Role of shape selectivity and presence of steam. *ACS Catal.* 8: 7861–7878.

18 Wang, M., Liu, M., Li, H. et al. (2018). Dealkylation of lignin to phenol via oxidation-hydrogenation strategy. *ACS Catal.* 8: 6837–6843.

19 Verboekend, D., Liao, Y.H., Schutyser, W., and Sels, B.F. (2016). Alkylphenols to phenol and olefins by zeolite catalysis: a pathway to valorize raw and fossilized lignocellulose. *Green Chem.* 18: 297–306.

20 Ferrini, P., Koelewijn, S.F., Van Aelst, J. et al. (2017). Zeolites as sustainable catalysts for the selective synthesis of renewable bisphenols from lignin-derived monomers. *ChemSusChem* 10: 2249–2257.

21 Koelewijn, S.F., Cooreman, C., Renders, T. et al. (2018). Promising bulk production of a potentially benign bisphenol A replacement from a hardwood lignin platform. *Green Chem.* 20: 1050–1058.

22 Liu, D., Sun, J., Simmons, B.A., and Singh, S. (2018). N-Heterocyclic carbene promoted decarboxylation of lignin-derived aromatic acids. *ACS Sustainable Chem. Eng.* 6: 7232–7238.

23 Van de Vyver, S. and Roman-Leshkov, Y. (2013). Emerging catalytic processes for the production of adipic acid. *Catal. Sci. Technol.* 3: 1465–1479.

24 Corona, A., Biddy, M.J., Vardon, D.R. et al. (2018). Life cycle assessment of adipic acid production from lignin. *Green Chem.* 20: 3857–3866.

25 Schutyser, W., Van den Bosch, S., Dijkmans, J. et al. (2015). Selective Nickel-Catalyzed Conversion of Model and Lignin- Derived phenolic compounds to cyclohexanone-based polymer building blocks. *ChemSusChem* 8: 1805–1818.

26 Zhou, Y.L., Gao, Y.J., Zhong, X. et al. (2019). Electrocatalytic upgrading of lignin-derived nio-oil based on surface-engineered PtNiB nanostructure. *Adv. Funct. Mater.* 29: 1807651.

27 Wong, S.S., Shu, R.Y., Zhang, J.G. et al. (2020). Downstream processing of lignin derived feedstock into end products. *Chem. Soc. Rev.* 49: 5510–5560.

28 Stewart, J.J., Akiyama, T., Chapple, C. et al. (2009). The effects on lignin structure of overexpression of ferulate 5-hydroxylase in hybrid poplar. *Plant. Physiol.* 150: 621–635.

29 Cai, Y.H., Zhang, K.W., Kim, H. et al. (2016). Enhancing digestibility and ethanol yield of Populus wood via expression of an engineered monolignol 4-O-methyltransferase. *Nat. Commun.* 7: 11989.

30 Li, Y., Shuai, L., Kim, H. et al. (2018). An "ideal lignin" facilitates full biomass utilization. *Sci. Adv.* 4: eaau2968.

31 Wang, S., Zhang, K., Li, H. et al. (2021). Selective hydrogenolysis of catechyl lignin into propenylcatechol over an atomically dispersed ruthenium catalyst. *Nat. Commun.* 12: 416.

32 Chen, F., Tobimatsu, Y., Havkin-Frenkel, D. et al. (2012). A polymer of caffeyl alcohol in plant seeds. *Proc. Natl. Acad. Sci. U.S.A.* 109: 1772–1777.

33 Chen, F., Tobimatsu, Y., Jackson, L. et al. (2013). Novel seed coat lignins in the Cactaceae: structure, distribution and implications for the evolution of lignin diversity. *Plant J.* 73: 201–211.

34 Tobimatsu, Y., Chen, F., Nakashima, J. et al. (2013). Coexistence but independent biosynthesis of catechyl and guaiacyl/syringyl lignin polymers in seed coats. *Plant Cell* 25: 2587–2600.

35 Renders, T., Van den Bosch, S., Koelewijn, S.F. et al. (2017). Lignin-first biomass fractionation: the advent of active stabilisation strategies. *Energy Environ. Sci.* 10: 1551–1557.

36 Abu-Omar, M.M., Barta, K., Beckham, G.T. et al. (2021). Guidelines for performing lignin-first biorefining. *Energy Environ. Sci.* 14: 262–292.

37 Chen, L., Dou, J., Ma, Q. et al. (2017). Rapid and near-complete dissolution of wood lignin at ≤80 °C by a recyclable acid hydrotrope. *Sci. Adv.* 3: e1701735.

38 Cai, C., Hirth, K., Gleisner, R. et al. (2020). Maleic acid as a dicarboxylic acid hydrotrope for sustainable fractionation of wood at atmospheric pressure and ≤100 °C: mode and utility of lignin esterification. *Green Chem.* 22: 1605–1617.

39 Shuai, L., Amiri, M.T., Questell-Santiago, Y.M. et al. (2016). Formaldehyde stabilization facilitates lignin monomer production during biomass depolymerization. *Science* 354: 329–333.

40 Abu-Omar, M. and Luo, H. (2018). Lignin extraction and catalytic upgrading from genetically modified poplar. *Abstr. Pap. Am. Chem. S.* 255.

41 Luo, H. and Abu-Omar, M.M. (2018). Lignin extraction and catalytic upgrading from genetically modified poplar. *Green Chem.* 20: 745–753.

42 Stone, M.L., Anderson, E.M., Meek, K.M. et al. (2018). Reductive catalytic fractionation of C-lignin. *ACS Sustainable Chem. Eng.* 6: 11211–11218.

43 Liao, Y., d'Halluin, M., Makshina, E. et al. (2018). Shape selectivity vapor-phase conversion of lignin-derived 4-ethylphenol to phenol and ethylene over acidic aluminosilicates: Impact of acid properties and pore constraint. *Appl. Catal., B* 234: 117–129.

44 Liao, Y.H., Zhong, R.Y., d'Halluin, M. et al. (2020). Aromatics production from lignocellulosic biomass: Shape selective dealkylation of lignin-derived phenolics over hierarchical ZSM-5. *ACS Sustainable Chem. Eng.* 8: 8713–8722.

45 Bomon, J., Van Den Broeck, E., Bal, M. et al. (2020). Brønsted acid catalyzed tandem defunctionalization of biorenewable ferulic acid and derivates into bio-catechol. *Angew. Chem. Int. Ed.* 59: 3063–3068.

46 Yan, J., Meng, Q.L., Shen, X.J. et al. (2020). Selective valorization of lignin to phenol by direct transformation of Csp_2–Csp_3 and C–O bonds. *Sci. Adv.* 6: eabd1951.

47 Toppi, S., Thomas, C., Sayag, C. et al. (2005). On the radical cracking of n-propylbenzene to ethylbenzene or toluene over Sn/Al_2O_3–Cl catalysts under reforming conditions. *J. Catal.* 230: 255–268.

48 Shen, X., Meng, Q., Mei, Q. et al. (2020). Selective catalytic transformation of lignin with guaiacol as the only liquid product. *Chem. Sci.* 11: 1347–1352.

49 Liao, Y., Koelewijn, S.-F., Van den Bossche, G. et al. (2020). A sustainable wood biorefinery for low-carbon footprint chemicals production. *Science* 367: 1385–1390.

50 Li, L., Dong, L., Li, D. et al. (2020). Hydrogen-free production of 4-alkylphenols from lignin via self-reforming-driven depolymerization and hydrogenolysis. *ACS Catal.* 10: 15197–15206.

51 Li, L., Zhang, T., Guo, Z. et al. (2021). Unraveling the role of metal in $M/NiAl_2O_4$ (M= Pt, Pd, Ru) catalyst for the self-reforming-driven hydrogenolysis of lignin. *Ind. Eng. Chem. Res.* 60: 11699–11706.

52 Fan, M.H., Jiang, P.W., Bi, P.Y. et al. (2013). Directional synthesis of ethylbenzene through catalytic transformation of lignin. *Bioresour. Technol.* 143: 59–67.

53 Shao, Y., Xia, Q., Dong, L. et al. (2017). Selective production of arenes via direct lignin upgrading over a niobium-based catalyst. *Nat. Commun.* 8: 16104.

54 Meng, Q.L., Yan, J., Liu, H.Z. et al. (2019). Self-supported hydrogenolysis of aromatic ethers to arenes. *Sci. Adv.* 5: eaax6839.

55 Meng, Q., Yan, J., Wu, R. et al. (2021). Sustainable production of benzene from lignin. *Nat. Commun.* 12: 4534.

56 Otromke, M., Theiss, L., Wunsch, A. et al. (2015). Selective and controllable purification of monomeric lignin model compounds via aqueous phase reforming. *Green Chem.* 17: 3621–3631.

57 Ji, Y., Hou, Y., Ren, S. et al. (2017). Highly efficient separation of phenolic compounds from oil mixtures by imidazolium-based dicationic ionic liquids via forming deep eutectic solvents. *Energy Fuel* 31: 10274–10282.

58 Zhang, Y., Li, Z., Wang, H. et al. (2016). Efficient separation of phenolic compounds from model oil by the formation of choline derivative-based deep eutectic solvents. *Sep. Purif. Technol.* 163: 310–318.

59 Jiao, T.T., Zhuang, X.L., He, H.Y. et al. (2015). Separation of phenolic compounds from coal tar via liquid-liquid extraction using amide compounds. *Ind. Eng. Chem. Res.* 54: 2573–2579.

60 Rodriguez-Delgado, M.A., Malovana, S., Perez, J.P. et al. (2001). Separation of phenolic compounds by high-performance liquid chromatography with absorbance and fluorimetric detection. *J. Chromatogr. A* 912: 249–257.

61 Kalili, K.M. and de Villiers, A. (2011). Recent developments in the HPLC separation of phenolic compounds. *J. Sep. Sci.* 34: 854–876.

62 Santos, J.H.P.M., Almeida, M.R., Martins, C.I.R. et al. (2018). Separation of phenolic compounds by centrifugal partition chromatography. *Green Chem.* 20: 1906–1916.

63 Alherech, M., Omolabake, S., Holland, C.M. et al. (2021). From Lignin to Valuable aromatic chemicals: Lignin depolymerization and monomer separation via centrifugal partition chromatography. *ACS Cent. Sci.* 7: 1831–1837.

64 Wang, J., Gao, H., Zhao, J. et al. (2010). Preparative separation of phenolic compounds from Halimodendron halodendron by high-speed counter-current chromatography. *Molecules* 15: 5998–6007.

65 Rodrıguez-Delgado, M., Pérez, M., Corbella, R. et al. (2000). Optimization of the separation of phenolic compounds by micellar electrokinetic capillary chromatography. *J. Chromatogr. A* 871: 427–438.

66 Jiao, T.T., Li, C.S., Zhuang, X.L. et al. (2015). The new liquid-liquid extraction method for separation of phenolic compounds from coal tar. *Chem. Eng. J.* 266: 148–155.

67 Van de Voorde, B., Borges, D.D., Vermoortele, F. et al. (2015). Isolation of renewable phenolics by adsorption on ultrastable hydrophobic MIL-140 metal-organic frameworks. *ChemSusChem* 8: 3159–3166.

68 Cronin, D.J., Chen, X., Moghaddam, L., and Zhang, X. (2020). Deep eutectic solvent extraction of high-purity lignin from a corn stover hydrolysate. *ChemSusChem* 13: 4678–4690.

69 Upton, B.M. and Kasko, A.M. (2016). Strategies for the conversion of lignin to high-value polymeric materials: Review and perspective. *Chem. Rev.* 116: 2275–2306.

70 Isikgor, F.H. and Becer, C.R. (2015). Lignocellulosic biomass: a sustainable platform for the production of bio-based chemicals and polymers. *Polym. Chem.* 6: 4497–4559.

71 Esposito, D. and Antonietti, M. (2015). Redefining biorefinery: the search for unconventional building blocks for materials. *Chem. Soc. Rev.* 44: 5821–5835.

72 Wang, Z., Yuan, L., and Tang, C. (2017). Sustainable elastomers from renewable biomass. *Acc. Chem. Res.* 50: 1762–1773.

73 Shen, X. and Sun, R. (2021). Recent advances in lignocellulose prior-fractionation for biomaterials, biochemicals, and bioenergy. *Carbohydr. Polym.* 261: 117884.

74 Shi, Z.J. and Ma, M.G. (2019). Synthesis, structure, and applications of lignin-based carbon materials: A review. *Sci. Adv. Mater.* 11: 18–32.

75 Pelckmans, M., Renders, T., Van de Vyver, S., and Sels, B.F. (2017). Bio-based amines through sustainable heterogeneous catalysis. *Green Chem.* 19: 5303–5331.

76 Li, H., Bunrit, A., Li, N., and Wang, F. (2020). Heteroatom-participated lignin cleavage to functionalized aromatics. *Chem. Soc. Rev.* 49: 3748–3763.

77 Rong, Y., Ji, N., Yu, Z. et al. (2021). Lignin amination valorization: heterogeneous catalytic synthesis of aniline and benzylamine from lignin-derived chemicals. *Green Chem.* 23: 6761–6788.

78 Sudarsanam, P., Ruijten, D., Liao, Y. et al. (2020). Towards lignin-derived chemicals using atom-efficient catalytic routes. *Trends in Chemistry* 2: 898–913.

79 Zhou, H., Xu, J.Y., Fu, Y.J. et al. (2019). Rapid flow-through fractionation of biomass to preserve labile aryl ether bonds in native lignin. *Green Chem.* 21: 4625–4632.

80 Anderson, E.M., Stone, M.L., Hulsey, M.J. et al. (2018). Kinetic studies of lignin solvolysis and reduction by reductive catalytic fractionation decoupled in flow-through reactors. *ACS Sustainable Chem. Eng.* 6: 7951–7959.

81 Li, S., Hao, Z., Wang, K. et al. (2020). Visible light-enabled selective depolymerization of oxidized lignin by an organic photocatalyst. *Chem. Commun.* 56: 11243–11246.

82 Smith, C.Z., Utley, J.H.P., and Hammond, J.K. (2011). Electro-organic reactions. Part 60[1]. The electro-oxidative conversion at laboratory scale of a lignosulfonate into vanillin in an FM01 filter press flow reactor: preparative and mechanistic aspects. *J. Appl. Electrochem.* 41: 363–375.

83 Kumaniaev, I., Subbotina, E., Savmarker, J. et al. (2017). Lignin depolymerization to monophenolic compounds in a flow-through system. *Green Chem.* 19: 5767–5771.

84 Anderson, E.M., Stone, M.L., Katahira, R. et al. (2017). Flowthrough reductive catalytic fractionation of biomass. *Joule* 1: 613–622.

85 Abdelaziz, O.Y. and Hulteberg, C.P. (2020). Lignin depolymerization under continuous-flow conditions: Highlights of recent developments. *ChemSusChem* 13: 4382–4384.

86 Qu, C., Ito, K., Katsuyama, I. et al. (2020). Directly microwave-accelerated cleavage of C–C and C–O bonds of lignin by copper oxide and H_2O_2. *ChemSusChem* 13: 4510–4518.

87 Liu, X., Bouxin, F.P., Fan, J. et al. (2020). Microwave-assisted catalytic depolymerization of lignin from birch sawdust to produce phenolic monomers utilizing a hydrogen-free strategy. *J. Hazard. Mater.* 2: 123490.

88 Toledano, A., Serrano, L., Labidi, J. et al. (2013). Heterogeneously catalysed mild hydrogenolytic depolymerisation of lignin under microwave irradiation with hydrogen-donating Solvents. *ChemCatChem* 5: 977–985.

89 Wu, X., Luo, N., Xie, S. et al. (2020). Photocatalytic transformations of lignocellulosic biomass into chemicals. *Chem. Soc. Rev.* 49: 6198–6223.

90 Colmenares, J.C. and Luque, R. (2014). Heterogeneous photocatalytic nanomaterials: Prospects and challenges in selective transformations of biomass-derived compounds. *Chem. Soc. Rev.* 43: 765–778.

91 Li, S.H., Liu, S.Q., Colmenares, J.C., and Xu, Y.J. (2016). A sustainable approach for lignin valorization by heterogeneous photocatalysis. *Green Chem.* 18: 594–607.

92 Bosque, I., Magallanes, G., Rigoulet, M. et al. (2017). Redox catalysis facilitates lignin depolymerization. *ACS Cent. Sci.* 3: 621–628.

93 Zhang, J. (2018). Conversion of lignin models by photoredox catalysis. *ChemSusChem* 11: 3071–3080.

94 Bawareth, B., Di Marino, D., Nijhuis, T.A., and Wessling, M. (2018). Unravelling electrochemical lignin depolymerization. *ACS Sustainable Chem. Eng.* 6: 7565–7573.

95 Garedew, M., Lin, F., Song, B. et al. (2020). Greener routes to biomass waste valorization: Lignin transformation through electrocatalysis for renewable chemicals and fuels production. *ChemSusChem* 13: 4214–4237.

96 Garedew, M., Lam, C.H., Petitjean, L. et al. (2021). Electrochemical upgrading of depolymerized lignin: a review of model compound studies. *Green Chem.* 23: 2868–2899.

97 Wijaya, Y.P., Smith, K.J., Kim, C.S., and Gyenge, E.L. (2020). Electrocatalytic hydrogenation and depolymerization pathways for lignin valorization: toward mild synthesis of chemicals and fuels from biomass. *Green Chem.* 22: 7233–7264.

98 Li, T., Mo, J.Y., Weekes, D.M. et al. (2020). Photoelectrochemical decomposition of lignin model compound on a $BiVO_4$ photoanode. *ChemSusChem* 3622–3626.

99 Li, S., Kim, S., Davis, A.H. et al. (2021). Photocatalytic chemoselective C–C bond cleavage at room temperature in dye-sensitized photoelectrochemical cells. *ACS Catal.* 11: 3771–3781.

100 Zirbes, M., Quadri, L.L., Breiner, M. et al. (2020). High-temperature electrolysis of kraft lignin for selective vanillin formation. *ACS Sustainable Chem. Eng.* 8: 7300–7307.

101 ten Have, R. and Teunissen, P.J.M. (2001). Oxidative mechanisms involved in lignin degradation by white-rot fungi. *Chem. Rev.* 101: 3397–3413.

102 Hofrichter, M. (2002). Review: lignin conversion by manganese peroxidase (MnP). *Enzyme Microb. Technol.* 30: 454–466.

103 Franssen, M.C.R., Steunenberg, P., Scott, E.L. et al. (2013). Immobilised enzymes in biorenewables production. *Chem. Soc. Rev.* 42: 6491–6533.

104 Wilkerson, C.G., Mansfield, S.D., Lu, F. et al. (2014). Monolignol ferulate transferase introduces chemically labile linkages into the lignin backbone. *Science* 344: 90–93.

105 Rinaldi, R., Jastrzebski, R., Clough, M.T. et al. (2016). Paving the way for lignin valorisation: Recent advances in bioengineering, biorefining and catalysis. *Angew. Chem. Int. Ed.* 55: 8164–8215.

Index

Lignin Conversion Catalysis: Transformation to Aromatic Chemicals, First Edition.
Chaofeng Zhang and Feng Wang.

o

p

r

s